Lecture Notes in Physics

The Editorial Policy for Edited Volumes

The series *Lecture Notes in Physics* (LNP), founded in 1969, reports new developments in physics research and teaching - quickly, informally but with a high degree of quality. Manuscripts to be considered for publication are topical volumes consisting of a limited number of contributions, carefully edited and closely related to each other. Each contribution should contain at least partly original and previously unpublished material, be written in a clear, pedagogical style and aimed at a broader readership, especially graduate students and nonspecialist researchers wishing to familiarize themselves with the topic concerned. For this reason, traditional proceedings cannot be considered for this series though volumes to appear in this series are often based on material presented at conferences, workshops and schools.

Acceptance

A project can only be accepted tentatively for publication, by both the editorial board and the publisher, following thorough examination of the material submitted. The book proposal sent to the publisher should consist at least of a preliminary table of contents outlining the structure of the book together with abstracts of all contributions to be included. Final acceptance is issued by the series editor in charge, in consultation with the publisher, only after receiving the complete manuscript. Final acceptance, possibly requiring minor corrections, usually follows the tentative acceptance unless the final manuscript differs significantly from expectations (project outline). In particular, the series editors are entitled to reject individual contributions if they do not meet the high quality standards of this series. The final manuscript must be ready to print, and should include both an informative introduction and a sufficiently detailed subject index.

Contractual Aspects

Publication in LNP is free of charge. There is no formal contract, no royalties are paid, and no bulk orders are required, although special discounts are offered in this case. The volume editors receive jointly 30 free copies for their personal use and are entitled, as are the contributing authors, to purchase Springer books at a reduced rate. The publisher secures the copyright for each volume. As a rule, no reprints of individual contributions can be supplied.

Manuscript Submission

The manuscript in its final and approved version must be submitted in ready to print form. The corresponding electronic source files are also required for the production process, in particular the online version. Technical assistance in compiling the final manuscript can be provided by the publisher's production editor(s), especially with regard to the publisher's own LaTeX macro package which has been specially designed for this series.

LNP Homepage (springerlink.com)

On the LNP homepage you will find:
—The LNP online archive. It contains the full texts (PDF) of all volumes published since 2000. Abstracts, table of contents and prefaces are accessible free of charge to everyone. Information about the availability of printed volumes can be obtained.
—The subscription information. The online archive is free of charge to all subscribers of the printed volumes.
—The editorial contacts, with respect to both scientific and technical matters.
—The author's / editor's instructions.

N. Akhmediev A. Ankiewicz (Eds.)

Dissipative Solitons

Editors

Nail Akhmediev
Optical Sciences Group
Research School
of Physical Sciences and Engineering
Institute of Advanced Studies
Australian National University
Canberra, A.C.T., 0200
Australia

Adrian Ankiewicz
Optical Sciences Group
Research School
of Physical Sciences and Engineering
Institute of Advanced Studies
Australian National University
Canberra, A.C.T., 0200
Australia

N. Akhmediev Adrian Ankiewicz (Eds.), *Dissipative Solitons*,
Lect. Notes Phys. **661** (Springer, Berlin Heidelberg 2005), DOI 10.1007/b11728

Library of Congress Control Number: 2004114848

ISSN 0075-8450
ISBN-10 3-540-23867-0 Springer Berlin Heidelberg New York
ISBN-13 978-3-540-23782-2 Springer Berlin Heidelberg New York

Springer is a part of Springer Science+Business Media

springeronline.com

Printed in Germany

Typesetting:: by the authors and TechBooks using a Springer LaTeX macro package
Cover design: *design & production*, Heidelberg

Printed on acid-free paper
57/3141/jl - 5 4 3 2 1 0

Preface

The term "soliton" was invented to describe nonlinear solitary wave (localized) solutions of integrable equations such as the Korteveg de Vries and nonlinear Schrödinger equations. However, as localized nonlinear solutions also exist for a wide range of physical situations, this notion was soon extended to cover solitary wave solutions in conservative and non-conservative systems. Non-conservative or dissipative solitons have characteristics that are markedly different from those of conservative systems. The study of their properties has dramatically expanded in the last two decades, making the subject of dissipative solitons an almost independent area of research.

Most previous studies of optical solitons in conservative systems have focussed on the fact that the nonlinearity of the material counteracts the diffraction (spatial systems) or dispersion (temporal systems). This balance results in a localized structure that has been named a "soliton". A localized structure in a dissipative system requires a continuous energy flow into the system, and, in particular, into the localized structure, in order to keep it "alive". Hence, a separate balance, namely that between the energy input and energy output, is more important than than that between the nonlinearity and dispersion (or diffraction). For dissipative systems, this idea of viewing structure formation as a consequence of energy flow introduces a new paradigm and allows us to understand pulses and forms appearing in many areas of physics, chemistry and biology.

Nicols and Prigogine showed, in the book "Self-organization in nonequilibrium systems" [John Wiley and sons, New York, 1977], that systems far from equilibrium are usually governed by nonlinear equations, and that the input of energy or matter allows for stable structures to form. They suggested that pre-biotic evolution could have involved numerous instabilities leading to the development of complexity. Thus the flow of energy and matter produce order of structure and function. This form of self-organization thus links the animate and the inanimate.

There have been assorted forerunners of this idea, some dating from ancient times. For example, in the first Chinese medical text (500 B.C.), it is explained that the flow of *ch'i* (energy) produces a balance between the *yin* and *yang* organs of the body. So these organs can be viewed as dissipative structures. If the flow stops, then these organs cannot continue with their

correct shapes, and illness occurs. The *I ching* (book of changes) emphasizes that changes and transformations occur because of movement and flow.

The increase in entropy was interpreted by Boltzmann as increasing disorganization. So an isolated system can only evolve to have entropy equal to or higher than the value it previously had. It is then clear that we need non-equilibrium open systems to allow structures to form spontaneously. They need to be able to exchange energy and matter with their environment. It is this imposition of outside energy that allows Bénard cells to form when a liquid is heated from below. Hexagons can form when the naturally-occurring fluctuations are "amplified", thus leading to convection. Thus, while Newtonian mechanics deals with trajectories of point particles, and the second law of thermodynamics (mid 19th century) introduced irreversibility or directivity of time, the formation of dissipative structures due to energy flow in nonlinear systems can be regarded as a fully new paradigm of dynamics.

The notion of a dissipative soliton is a natural extension of these structures. It can be localized in space and time and have many features known from soliton theory. The cross-fertilization of ideas from both fields must be useful and productive.

This multi-author book is written by experts in this area. The chapters cover the progress in the field of dissipative solitons that has been made recently. It includes a study of dissipative solitons in a variety of media, in optics and reaction-diffusion systems, as well as some mathematical apects of the theory of dissipative solitons.

There are many properties of dissipative solitons that are common to all these fields, and, unavoidably, there are differences. In one book, all these issues cannot be covered. The choice of material must necessarily be restricted. However, we have tried to make the choice of topics as wide as possible.

We start the book with a chapter discussing basic features of dissipative solitons of the one-dimensional Ginzburg-Landau and Swift-Hohenberg equations. We concentrate on those aspects of the problem that can be studied on the basis of a qualitative analysis of nonlinear dynamical systems.

Spatial dissipative solitons are discussed in two chapters. Firstly, the chapter by Boardman, Velasco and Egan deals with the subject of dissipative solitons in magneto-optics. Then, spatial dissipative solitons in semiconductor materials are discussed in the chapter by Ultanir, Stegeman, Michaelis, Lange and Lederer.

The three following chapters are devoted to two-dimensional dissipative solitons in optics. Ackemann and Firth give a review of phenomena related to dissipative solitons in driven optical cavities containing a nonlinear medium (cavity solitons) and similar phenomena (feedback solitons), where a driven nonlinear optical medium is in front of a single feedback mirror. Rozanov presents a review of the features of dissipative solitons in optical systems with nonlinear amplification and absorption, with no driving (holding) radiation, including cases with and without feedback. Taranenko, Slekys and

Weiss review pattern formation and experiments on semiconductor resonator solitons which are aimed at applications. Some close analogies of resonator solitons with the biological structures are also discussed.

Dissipative solitons in the time domain is the topic of Chaps. 7–10. The general properties of dissipative solitons in the time domain, and their particular application in all-optical transmission links, are explained by Peschel, Michaelis, Bakonyi, Onishchukov and Lederer. Laser systems with passive mode-locking are the ideal devices for studying optical dissipative solitons. Short – pulse generation by solid-state passively mode-locked lasers is reviewed by Cundiff. Soto-Crespo and Grelu present theoretical and experimental aspects of multi-soliton generation phenomena in fiber lasers. Various types of mode-locking in fiber lasers via nonlinear mode-locking are discussed in the chapter by Kutz.

Reaction-diffusion systems is a topic that includes a large class of systems from the areas of chemistry, physics, geology and biology. Dissipative solitons in reaction-diffusion systems are covered by Purwins, Bödeker and Liehr.

Dissipative solitons in discrete lattices with gain and loss provide one more facet of the subject, and these solitons have their own distinctive features and properties. Efremidis and Christodoulides present a review of recent results concerning dissipative lattices of Ginzburg-Landau type. The existence of dissipative solitons in various models of nonlinear lattices is examined by Abdullaev.

Bose-Einstein condensates form another type of system that admits dissipative solitons. The chapter by Konotop presents recents results obtained in this "hot" area of research.

The last three chapters are devoted to mathematical aspects of dissipative solitons. The search for closed-form analytic expressions for the solitary waves of nonlinear non-integrable partial differential equations, using non-perturbative techniques, is the topic of the chapter written by Conte and Musette. Stability issues are very important for dissipative solitons. The chapter by Kapitula presents ways of solving this problem for dissipative systems by using perturbation analysis and the Evans function. The chapter by S. R. Choudhury, considers bifurcations and strongly amplitude-modulated pulses of the complex Ginzburg-Landau equation.

Clearly, one book cannot cover all aspects of this rapidly-developing area. We hope that the publication of this book will not only assemble the issues related to dissipative solitons in one place, but will also raise new questions and facilitate further developments in this fascinating area of research.

The editors acknowledge support from the Australian Research Council.

Canberra, *Nail Akhmediev*
January 2005 *Adrian Ankiewicz*

Contents

List of Contributors

Fatkhulla Kh. Abdullaev
Physical-Technical Institute of the Uzbek Academy of Sciences
700084, Tashkent-84
G.Mavlyanov str.2-b, Uzbekistan
fatkh@physic.uzsci.net

Thorsten Ackemann
Institut für Angewandte Physik
Westfälische
Wilhelms-Universität Münster
Corrensstraße 2/4
48149 Münster, Germany
t.ackemann@uni-muenster.de

Nail Akhmediev
Optical Sciences Group
Research School of Physical Sciences and Engineering
Institute of Advanced Studies
Australian National University
Canberra, ACT 0200, Australia
nna124@rsphysse.anu.edu.au

Adrian Ankiewicz
Optical Sciences Group
Research School of Physical Sciences and Engineering
Institute of Advanced Studies
Australian National University
Canberra, ACT 0200, Australia
ana124@rsphysse.anu.edu.au

Zoltan Bakonyi
Friedrich-Schiller-Universität Jena
Institute of Applied Physics
Max-Wien-Platz 1
07743 Jena, Germany
zbakonyi@iap.uni-jena.de

Allan D. Boardman
Joule Physics Laboratory
Institute for Materials Research
University of Salford
Salford, Manchester
M5 4WT, United Kingdom
a.d.bordman@salford.ac.uk

H. U. Bödeker
Institut für Angewandte Physik
Westfälische
Wilhelms-Universität Münster
Corrensstraße 2/4
48149 Münster, Germany
boedeker@nwz.uni-muenster.de

S. Roy Choudhury
Department of Mathematics
MAP207
University of Central Florida
Orlando, FL 32816-1364, USA
choudhur@longwood.cs.ucf.edu

Demetrios N. Christodoulides
School of Optics
Center for Research and Education for Laser and Optics
University of Central Florida
4000 Central Florida Blvd., Orlando
FL., U.S.A.
demetri@creol.ucf.edu

Robert Conte
Service de physique de l'état condensé (URA no. 2464)
CEA–Saclay
F–91191 Gif-sur-Yvette
Cedex, France
Conte@drecam.saclay.cea.fr

Steven T. Cundiff
JILA, National Institute of Standards and Technology
and University of Colorado
Boulder
CO 80309-0440, USA
cundiffs@jila.colorado.edu

Nikolaos K. Efremidis
School of Optics
Center for Research and Education for Laser and Optics
University of Central Florida
4000 Central Florida Blvd., Orlando
FL., U.S.A.
nefrem@mail.ucf.edu

Peter Egan
Joule Physics Laboratory
Institute for Materials Research
University of Salford
Salford, Manchester, M5 4WT
United Kingdom
p.egan@salford.ac.uk

William J. Firth
Department of Physics
and Applied Physics
University of Strathclyde
Glasgow G4 0NG, United Kingdom
willie@phys.strath.ac.uk

Philippe Grelu
Laboratoire de Physique
de l'Université de Bourgogne
UMR 5027, B.P. 47870
21078 Dijon, France.
Philippe.Grelu@u-bourgogne.fr

Todd Kapitula
Department of Mathematics
and Statistics
University of New Mexico
Albuquerque, NM 87131, USA
kapitula@math.unm.edu

Vladimir V. Konotop
Centro de Física Teórica e Computacional
and
Departamento de Física
Universidade de Lisboa
Complexo Interdisciplinar
Av. Prof. Gama Pinto 2
Lisbon 1649-003, Portugal
konotop@cii.fc.ul.pt

J. Nathan Kutz
Department of Applied Mathematics
University of Washington, Seattle
WA 98195-2420, USA
kutz@amath.washington.edu

Christoph H. Lange
Friedrich-Schiller-Universität Jena
Max-Wien-Platz 1
07743 Jena, Germany
c.h.lange@uni-jena.de

Falk Lederer
Friedrich-Schiller-Universität Jena
Max-Wien-Platz 1
07743 Jena, Germany
pfl@uni-jena.de

A. W. Liehr
Institut für Angewandte Physik
Correnssstraße 2/4
48149 Münster, Germany
obi@uni-muenster.de

Dirk Michaelis
Fraunhofer Institute of Applied Optics and Precision Mechanics
Albert-Einstein-Strasse 7
07745 Jena, Germany
dirk@physse.nlwl.uni-jena.de

Micheline Musette
Dienst Theoretische Natuurkunde
Vrije Universiteit Brussel
Pleinlaan 2, B–1050 Brussels
Belgium
MMusette@vub.ac.be

Georgi Onishchukov
Friedrich-Schiller-Universität Jena
Institute of Applied Physics
Max-Wien-Platz 1
07743 Jena, Germany
George.Onishchukov@uni-jena.de

Ulf Peschel
Friedrich-Schiller-Universität Jena
Institute of Condensed Matter Theory and Optics
Max-Wien-Platz 1
07743 Jena, Germany
p6peul@uni-jena.de

H.-G. Purwins
Institut für Angewandte Physik
Corrensstraße 2/4
48149 Münster, Germany
purwins@nwz.uni-muenster.de

Nikolay N. Rosanov
Research Institute for Laser Physics
Birzhevaya Liniya 12
Saint Petersburg, Russia
nrosanov@yahoo.com

G. Slekys
Physikalisch-Technische Bundesanstalt
38116 Braunschweig, Germany
g.slekys@ptb.de

J. M. Soto-Crespo
Instituto de Óptica, C.S.I.C.
Serrano 121
28006 Madrid, Spain
iodsc09@io.cfmac.csic.es

George I. Stegeman
School of Optics and CREOL
University of Central Florida
4000 Central Florida Blvd.
P.O. Box 162700, Orlando
FL 32816-2700, USA
george@creol.ucf.edu

Victor B. Taranenko
Physikalisch-Technische Bundesanstalt
38116 Braunschweig, Germany
Victor.Taranenko@ptb.de

Erdem Ultanir
School of Optics and CREOL
University of Central Florida
4000 Central Florida Blvd.
P.O. Box 162700, Orlando
FL 32816-2700, USA
eultanir@mail.ucf.edu

Larry Velasco
Joule Physics Laboratory
Institute for Materials Research
University of Salford, Salford
Manchester
M5 4WT, United Kingdom
l.n.velascohernandez@pgt.salford.ac.uk

Carl O. Weiss
Physikalisch-Technische Bundesanstalt
38116 Braunschweig, Germany
carl.weiss@ptb.de

Dissipative Solitons in the Complex Ginzburg-Landau and Swift-Hohenberg Equations

N. Akhmediev and A. Ankiewicz

Optical Sciences Group, Research School of Physical Sciences and Engineering, Institute of Advanced Studies, Australian National University, Canberra, ACT 0200, Australia
nna124@rsphysse.anu.edu.au
ana124@rsphysse.anu.edu.au

Abstract. We explain the meaning of dissipative solitons and place them in a framework which shows their use in various scientific fields. Indeed, dissipative solitons form a new paradigm for the investigation of phenomena involving stable structures in nonlinear systems far from equilibrium. We consider those aspects of the problem that can be studied on the basis of a qualitative analysis of nonlinear systems.

1 What are Dissipative Solitons?

A dissipative soliton is a localized structure which exists for an extended period of time, even though parts of the structure experience gain and loss of energy and/or mass. This "structure" could be a profile of light intensity, temperature, magnetic field, etc. These solitons exist in "open" systems which are far from equilibrium. Thus energy and matter can flow into the system through its boundaries. The structure exists indefinitely in time, as long as the parameters in the system stay constant. It may evolve (i.e. change its shape periodically or otherwise) but it disappears when the source of energy or matter is switched off, or if the parameters of the system move outside the possible range of existence of the soliton.

In contrast to solitons in conservative systems, solitons in systems far from equilibrium are dynamical objects that have non-trivial internal energy flows. Since they are produced by dissipative systems, they depend strongly on an energy supply from an external source. Even if it is a stationary object, a dissipative soliton continuously re-distributes energy between its parts. A pump of energy is essential, and this means that the structures are defined by the rules of the system (gain, loss, dispersion, nonlinearity, etc.), rather than by the initial conditions [1]. Stationary solitons (pulses, fronts, etc.) can form where the overall gain and loss are balanced. A wide range of initial conditions can thus evolve into a dissipative soliton. These structures will typically appear in biochemical, optical and thermal systems, as they are "generic" and do not require particular formations to create them.

N. Akhmediev and A. Ankiewicz: *Dissipative Solitons in the Complex Ginzburg-Landau and Swift-Hohenberg Equations*, Lect. Notes Phys. **661**, 1–17 (2005)
www.springerlink.com

At the lowest levels, when a system can be described by a mathematical model employing a single equation, the rules may be simple and may reduce to a balance between gain and loss, as well as a balance between dispersion and nonlinearity. On the other hand, we can extend the analogies to more complicated systems and can consider animal species in nature as elaborate forms of solitons. These "structures" are localized, their internal processes must be balanced, they exist for a certain range of parameters (temperature, pressure, humidity, etc.), and, most importantly, they cease to exist if the supply of energy is switched off. Of course, an entire animal is a persistent structure which maintains its form with a constant flow of energy and matter. The same can be said of individual organs within an animal, since each maintains its shape and function over time. This relates to the ancient Chinese view discussed in the Preface. Thus, objects at each biological level – cell, organ and animal – can be considered as dissipative solitons. In each case, the unit would die if the dynamic flow stopped.

Examples in optics include ultra-short pulses in passively mode-locked lasers, spatial structures in wide-aperture laser systems and soliton propagation in long-haul all-optical transmission systems. All these systems require continuous energy supply for their operation and careful parameter adjustments for the localized structures to be stable. Investigations in these areas have allowed us to enrich the notion of the dissipative soliton.

In this chapter, partial differential equations which support solitons form the main underlying core. The use of this approach has the benefit that each component effect is clearly evident and identifiable. Furthermore, analytic solutions have recently been found in many relevant cases, and these allow for convenient comparison with experiment. In other cases, modern computation allows for simulations to be carried out quite quickly. This has lead to the discovery of stable and quasi-stable structures that were not imagined before computers appeared. With regard to this aspect, we note that even solitons in conservative integrable systems (e.g. the KdV equation) were first discovered through computer simulations. This discovery allowed for the later development of the complete theory, i.e. the inverse scattering transform.

The dissipative structure can be a profile of light intensity or temperature or a distribution of particles like sand or cells. It is clear that a particular sand distribution will form if we add sand to some central area and remove sand from a ring some distance away from that centre. Here gravity and nonlinear interactions between particles will determine the final shape of the sand dune. In the cellular case [2], communication between the cells leads to spatial patterns and controls localization.

A dissipative structure does not need to have a stationary profile. For example, "exploding" solitons feature a phase which is "laminar" or almost constant for a certain time, and then the structure is apparently destroyed as it breaks up [3]. The remarkable thing is that it then recovers its shape, so that the original form is reconstituted. Thus this dissipative structure continually renews itself.

2 Mathematical Model

The cubic-quintic complex Ginzburg-Landau equation (CGLE) can be written [4]:

$$\mathrm{i}\psi_z + \frac{D}{2}\,\psi_{tt} + |\psi|^2\psi + \nu|\psi|^4\psi = \mathrm{i}\delta\psi + \mathrm{i}\epsilon|\psi|^2\psi + \mathrm{i}\beta\psi_{tt} + \mathrm{i}\mu|\psi|^4\psi\,. \quad (1)$$

When used to describe passively mode-locked lasers, z is the cavity round-trip number, t is the retarded time, ψ is the normalized envelope of the field, D is the group velocity dispersion coefficient, with $D = \pm 1$, depending on whether the group velocity dispersion (GVD) is anomalous or normal, respectively, δ is the linear gain-loss coefficient, $\mathrm{i}\beta\psi_{tt}$ accounts for spectral filtering ($\beta > 0$), $\epsilon|\psi|^2\psi$ represents the nonlinear gain (which arises, e.g., from saturable absorption), the term with μ represents, if negative, the saturation of the nonlinear gain, while the one with ν corresponds, also if negative, to the saturation of the nonlinear refractive index. In the rest of the chapter we shall always assume $D = 1$. The addition of the fourth-order term $+\mathrm{i}s\psi_{tttt}$ transforms the CGLE to the complex Swift-Hohenberg equation (CSHE)

$$\mathrm{i}\psi_z + \frac{D}{2}\,\psi_{tt} + |\psi|^2\psi + \nu|\psi|^4\psi = \mathrm{i}\delta\psi + \mathrm{i}\epsilon|\psi|^2\psi + \mathrm{i}\beta\psi_{tt} + \mathrm{i}\mu|\psi|^4\psi + \mathrm{i}s\psi_{tttt}\,,$$

and this also has soliton solutions. In the case of optical systems, the parameter s is related to higher-order spectral filtering [5]. Interestingly, solitons with simple shapes can still exist even with equations which are apparently quite complicated, like the Swift-Hohenberg (CSHE) equation [6].

The CGLE and CSHE are generic equations describing systems near subcritical bifurcations [7, 8]. They relate to a wide range of dissipative phenomena in physics, such as binary fluid convection [9], electro-convection in nematic liquid crystals [10], patterns near electrodes in gas discharges [11] and oscillatory chemical reactions [12].

For an optical pulse circulating in a fibre laser, the centre of the pulse can experience gain, while the tails are subject to loss. In this case, the complex optical field envelope can also be viewed in the frequency domain, so that the gain may depend on frequency. This would commonly occur when a piece of fibre forming an erbium-doped fibre amplifier (EDFA) is spliced into the circuit. This has (continuous) pumping at a wavelength lower than that of the signal, and produces amplification only for wavelengths close to 1550 nm. Hence it selects against the other wavelengths. In the Ginzburg-Landau and S-H equations, the term $i\,\psi_{tt}$ in the time domain clearly corresponds to $-(\omega-\omega_0)^2\,\psi$ in the frequency domain, where ω is the frequency and

ω_0 is the centre frequency of the EDFA gain band. This means that the gain spectrum would be parabolic in shape. The addition of the higher-order term $i\,s\,\psi_{tttt}$, with s being a constant, means that we are including a "correction", proportional to $(\omega-\omega_0)^4\,\psi$, to the gain spectrum, so that it is modelled more accurately. Similarly, the fourth-order correction $(\nu|\psi|^4\,\psi)$ to the (2nd-order) Kerr nonlinearity $(|\psi|^2\,\psi)$ means that features like the saturation of nonlinearity can be modelled. Regular fibre attenuation tends to be independent of wavelength, on the other hand. In addition to frequency-dependent gain, other nonlinear effects also occur – thus a delay in response corresponds to a Raman effect term.

3 Are there Integrable Dissipative Systems?

The CGLE and CSHE admit a selected range of exact soliton solutions [6, 13]. These exist when certain relations between the parameters of the equation are satisfied. Moreover, some techniques for obtaining solutions in a regular way can be developed [14]. However, this certainly does not imply that the equations are integrable. The question arises, can dissipative systems be integrable at all? Some examples allow us to claim that the answer is "yes".

Burgers' equation provides an illustrative example of such a system. It can be written:

$$u_t = u_{xx} - u\,u_x = \frac{d}{dx}\left(u_x - \frac{u^2}{2}\right) . \tag{2}$$

This is a nonlinear dissipative equation with a complicated form of diffusion on the right-hand-side, and it has (soliton) front-type solutions.

The Cole-Hopf transformation $(u = -2\phi_x/\phi)$ converts it to the heat or diffusion equation $(\phi_t = \phi_{xx})$, which is linear. This allows us to obtain one of the simplest dissipative solitons, just by observing the exponential solution of the heat/diffusion equation:

$$\phi(x,t) = j + \exp[b\,(bt - x)] , \tag{3}$$

where j and b are arbitrary constants. Using the transformation above directly leads to the front (kink) solution of Burgers" equation:

$$u = \frac{2b}{1 + j\,\exp[-b\,(bt - x)]} . \tag{4}$$

The constant asymptotes, zero and $2b$, are joined in the middle by a smooth curve which is a front (or kink). If $j = 1$, it reduces to

$$u = b\,\exp[b\,(bt - x)/2]\,\mathrm{sech}[b\,(bt - x)/2] .$$

In principle, the Cole-Hopf transformation allows us to solve Koshi's problem with more complicated initial conditions.

In reality, most relevant nonlinear partial differential equations cannot be reduced to linear equations in any known way, and the paths to their solutions are mostly based on numerical simulations. Nevertheless, computers can also be used with some general concepts in mind. In some cases, the knowledge accumulated in the qualitative analysis of nonlinear dynamical finite-dimensional systems turns out to be useful. In the following sections, we give some analogies and ideas along these lines that might be very helpful.

4 Stationary Solitons as Fixed Points

As with other nonlinear dynamical systems having a finite number of degrees of freedom, the first step to take is finding z-independent solutions with $\partial\psi/\partial z = 0$. This gives us stationary solutions or "fixed points". As we mentioned above, solitons belong to the class of localized solutions. Equation (1) (as well as the CSHE) has a variety of them. These are stationary solitons, sources, sinks, moving solitons and fronts with fixed velocity [13, 15]. A multiplicity of solutions can exist for the same parameters. Solitons can exist in several forms, and many of them can be stable for a certain range of values of the equation parameters [5, 16]. Two examples of stationary solutions of the CSHE are shown in Fig. 1.

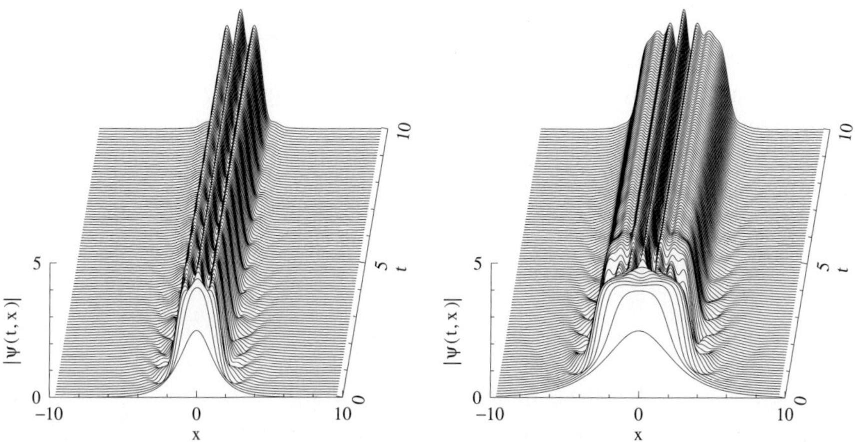

Fig. 1. Excitation of stationary soliton solutions of the CSHE. Parameters of the simulation are: $\beta = -0.3$, $\epsilon = 1.6$, $\nu = 0$, $\mu = -0.1$, $\delta = -0.5$ and $\gamma_2 = 0.05$ [5]

These are excited from Gaussian initial conditions. They can equally well be excited with other initial conditions. Each of the solitons is a stable fixed point (stable focus) in an infinite-dimensional phase space. As such, it attracts trajectories which start near that fixed point. The basin of attraction depends on the parameters of the equation. At some values of the parameters, the fixed points can be unstable (e.g. an unstable focus or node) or neutrally stable. In that case, solitons do exist, but localized initial conditions do not tend to converge to them.

5 Stability Analysis

As a second step, methods of qualitative analysis of nonlinear dynamical systems require investigation of the stability for each fixed point. To do this, we start with the stationary solution, and analyse its linear stability. Let us suppose that the stationary soliton solution of the CGLE is: $\psi(z,t) = \psi_0(t)e^{iqz}$, where $\psi_0(t)$ is a complex function of t with exponentially decaying tails, and that q, its propagation constant, is real. A technique for finding this function has been described – for example, see [16]. The stationary solution is a singular point of this dynamical system in an infinite-dimensional phase space. Then, the evolution of the solution in the vicinity of this singular point can be described by

$$\psi(z,t) = [\psi_0(t) + f(t)\mathrm{e}^{\lambda z} + g(t)\mathrm{e}^{\lambda^* z}]\mathrm{e}^{\mathrm{i}qz} \tag{5}$$

where $f(t)$ and $g(t)$ are small perturbation functions (we assume $|f,g| \ll |\psi_0|$ at any t), and λ is the associated perturbation growth rate. In general, each λ is a complex number and f and g are complex functions. For the soliton solutions of the CGLE, when the dissipative and higher-order terms are small, the stability analysis can be done analytically [17]. However, when the dissipative terms are not small, the analytic approach becomes problematic. Hence, at this stage, we can only rely on numerical calculations of the eigenvalues and eigenfunctions of the linearized problem. Substituting (5) into the CGLE (1), we obtain:

$$\begin{aligned}
&(\mathrm{i}\lambda - \mathrm{i}\delta - q)f\mathrm{e}^{\lambda z} + (\mathrm{i}\lambda^* - \mathrm{i}\delta - q)g\mathrm{e}^{\lambda^* z} \\
&\quad + \left(\frac{D}{2} - \mathrm{i}\beta\right) f_{tt}\mathrm{e}^{\lambda z} + \left(\frac{D}{2} - \mathrm{i}\beta\right) g_{tt}\mathrm{e}^{\lambda^* z} \\
&\quad +3(\nu - \mathrm{i}\mu)|\psi_0|^4(f\mathrm{e}^{\lambda z} + g\mathrm{e}^{\lambda^* z}) \\
&\quad +2(\nu - \mathrm{i}\mu)|\psi_0|^2\psi_0^2(f^*\mathrm{e}^{\lambda^* z} + g^*\mathrm{e}^{\lambda z}) \\
&\quad +2(1 - \mathrm{i}\epsilon)|\psi_0|^2(f\mathrm{e}^{\lambda z} + g\mathrm{e}^{\lambda^* z}) \\
&\quad +(1 - \mathrm{i}\epsilon)\psi_0^2(f^*\mathrm{e}^{\lambda^* z} + g^*\mathrm{e}^{\lambda z}) = 0
\end{aligned} \tag{6}$$

Separating terms with different functional dependencies on t, we obtain the following two coupled ordinary differential equations:

$$\begin{aligned}
Af + Bf_{tt} + Cg^* &= \lambda f \\
A^*g^* + B^*g^*_{tt} + C^*f &= \lambda g^*
\end{aligned} \tag{7}$$

where $A = \delta - \mathrm{i}q + 2(\epsilon + \mathrm{i})|\psi_0|^2 + 3(\mu + \mathrm{i}\nu)|\psi_0|^4$,

$$B = \beta + \mathrm{i}\frac{D}{2} \quad \text{and} \quad C = \left[\epsilon + \mathrm{i} + 2(\mu + \mathrm{i}\nu)|\psi_0|^2\right] \psi_0^2 .$$

To solve equations (7) numerically, we have to discretize them, by evaluating the functions of t at $N/2$ equi-distant points. In this way, we obtain N algebraic equations, and therefore N complex eigenvalues. By changing the value of N, we can observe the behaviour of the eigenvalues and approximate the solutions of the original continuous problem. In particular, we can identify the points corresponding to the discrete and continuous parts of the spectrum. An example of such a spectrum is presented in Sect. 11.

6 Energy Flow Across a Soliton

One of the main features that distinguishes dissipative solitons from those in conservative systems is their dependence on an external energy supply. As a result, solitons in dissipative systems have parts which generate energy and parts which dissipate it. The simplest way to show this is based on the continuity relation for the CGLE equation, in the form

$$\frac{\partial \rho}{\partial z} + \frac{\partial j}{\partial t} = P , \tag{8}$$

where ρ is the energy density, $\rho = |\psi|^2$. The corresponding flux, j, is

$$j = \frac{\mathrm{i}}{2} \left(\psi \psi_t^* - \psi_t \psi^*\right) ,$$

and the density of energy generation P is

$$P = 2\delta|\psi|^2 + 2\epsilon|\psi|^4 + 2\mu|\psi|^6 - 2\beta|\psi_t|^2 + \beta\left(|\psi|^2\right)_{tt} .$$

This last term distinguishes (8) from the continuity relation for conservative systems, where P is zero. For a stationary solution, the energy density $|\psi|^2$ does not depend on z. Therefore, the first term in (8) is zero. Hence, the energy flux j is defined by the regions of energy generation and loss. The energy flows from the parts where the energy is generated to the parts where it is dissipated.

Figure 2 shows the curve for the energy flux j across the soliton (dotted line) for a particular soliton solution with equation parameters given with the plot. The energy flux is zero at the center of the soliton, negative on the left, and positive on the right, reflecting the fact that the energy flows from the center to the tails in this particular case. In the figure, we additionally show $P(t)$, as a dashed line. The soliton itself for this set of parameters is indicated by the solid line.

The non-zero energy flux across a soliton also results in the strong phase chirp that solitons experience in dissipative systems. In contrast, stationary solitons in conservative systems have zero chirp.

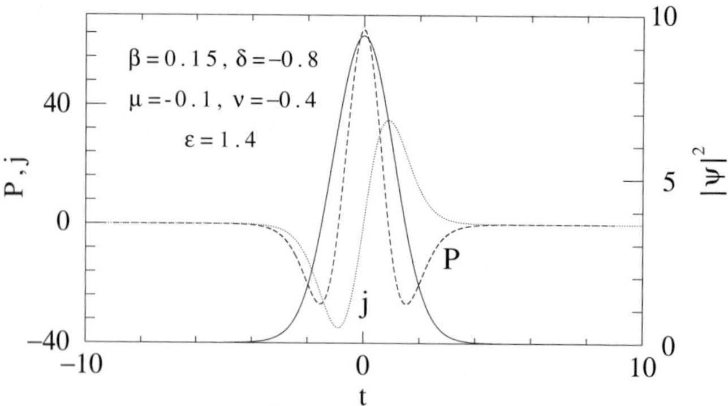

Fig. 2. Energy generation $P(t)$ (*dashed line*) and energy flux $j(t)$ (*dotted line*) across a soliton. The soliton profile itself, $|\psi|$, for the same set of parameters, is shown by the *solid* line (r.h.s. scale)

7 Pulsating Soliton as a Limit Cycle

While a stationary soliton is a fixed point of a phase space, a soliton that periodically changes its shape in time (i.e. a pulsating soliton [18]) can be considered as a limit cycle in an infinite-dimensional phase space. It can also be stable or unstable. Stable pulsating solitons exist for an indefinite time, in the same way as stationary solutions.

An example of a pulsating soliton is shown in Fig. 3a. It shows perfectly periodic behaviour, with the period in z being around 14. It has a different shape at each z, since it evolves, but it recovers its exact initial shape after a period. If we take *any* two parameters of the soliton (e.g. squared amplitude and the energy $Q = \int_{-\infty}^{\infty} |\psi(z,t)|^2 dt$) as functions of z, and present them

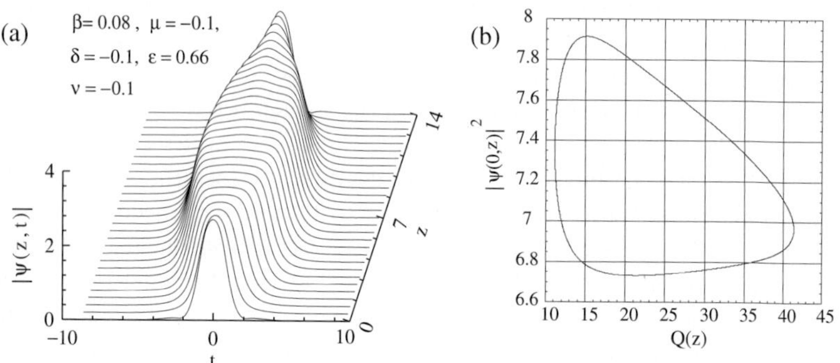

Fig. 3. (**a**) Plain pulsating soliton solution of the CGLE. Only one period is shown. (**b**) Square of soliton amplitude $|\psi|^2$ versus soliton energy. The parameters are: $D = +1$, $\epsilon = 0.66$, $\delta = -0.1$, $\beta = 0.08$, $\mu = -0.1$, and $\nu = -0.1$

on a two-dimensional plane as a parametric plot, we obtain a trajectory that will comprise a cycle repeating itself indefinitely (see Fig. 3b). Similar behaviour will be observed in an N-dimensional phase space if we choose any N parameters of the soliton (e.g. N higher-order momenta). The number of parameters can be chosen to be infinite. Clearly, this pulsating soliton is a limit cycle in an infinite-dimensional phase space. When starting from an initial condition which is not a soliton profile at any particular z but is reasonably close to one, the trajectory converges to the limit cycle, provided it is stable.

Thus far, there is no technique that would allow us to map the parameter space and analytically find regions that correspond to stationary and pulsating solutions. Up till now, this has mostly been done using numerical simulations [13].

8 Period Doubling of Pulsating Solitons

Pulsating solitons can exhibit more complicated behaviour when the parameters of the equation are changed. In particular, simple pulsations can be transformed into period-doubled pulsations. This occurs due to bifurcations at certain boundaries in the space of the equation parameters. An example of a pulsating solution which has suffered such a transformation is given in Fig. 4a. This solution has double periodicity, since the original shape recurs after two pulsations, rather than after one, as in Fig. 3a. The transition from one to the other occurs as a period-doubling bifurcation when one or two of the parameters of the equation are changed. At the point of bifurcation, the single loop shown in Fig. 3b splits into a double loop, as shown in Fig. 4b. Such splitting will take place if we parametrically plot any two other soliton parameters on a plane. We can also imagine such plot in an infinite-dimensional phase space. Then, clearly, the loop splits in that space.

Further changes in the equation parameters may cause additional period-doubling or period-quadrupling relative to the initial plain periodic pulsations. For example, period-four pulsations are observed when the parameter ϵ is increased to 0.793 (see [18]). If the trajectory in the parameter space is chosen correctly, we can have a sequence of an infinite number of period-doubling bifurcations, and this results in chaotic soliton solutions. This route to chaotic soliton solutions is similar to the Feigenbaum route to chaos in finite-dimensional systems or "logistic" maps [19]. However, as we have several equation parameters which can be changed, the route to chaotic solitons can also take different forms.

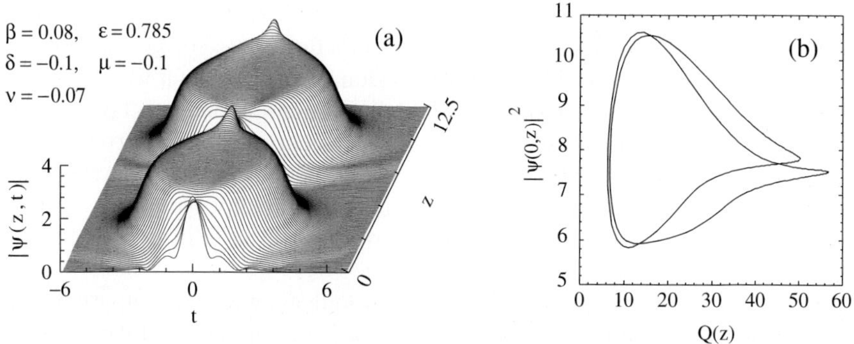

Fig. 4. (**a**) Period-two pulsating soliton solution of the CGLE. (**b**) Square of soliton amplitude $|\psi|^2$ versus soliton energy Q. Here, the single loop of the plain pulsating soliton is split into two. The parameters are: $D = +1$, $\epsilon = 0.785$, $\delta = -0.1$, $\beta = 0.08$, $\mu = -0.1$, and $\nu = -0.07$

9 Chaotic Soliton as a Strange Attractor

Pulsating solitons may become chaotic at certain values of the parameters. These have profiles that evolve in z but never repeat. However, being chaotic, the shape remains localized. This remarkable feature allows us to label them as solitons. A numerical example is shown in Fig. 5a. The profile remains smooth, and is not subject to any major "destruction". A chaotic soliton can be obtained as a result of a sequence of period-doubling bifurcations [5] or as a result of a single bifurcation directly from a pulsating soliton, depending on the way the equation parameters change. Other scenarios for a route to

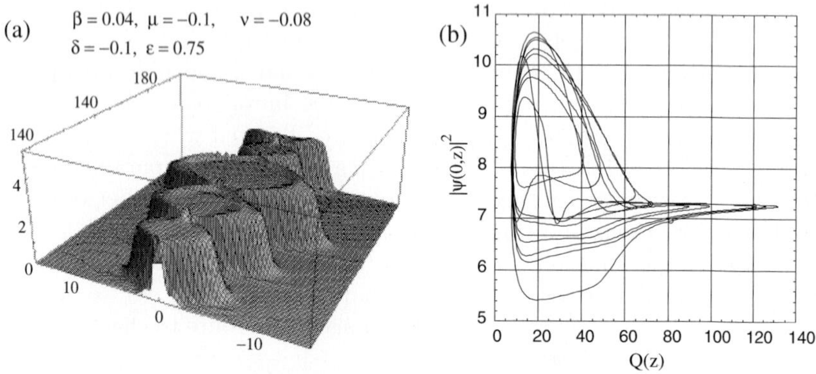

Fig. 5. (**a**) Chaotic soliton solution of the CGLE. The soliton profile never repeats. (**b**) Square of soliton amplitude $|\psi|^2$ versus soliton energy Q. This trajectory, if continued, will densely fill a finite region. The parameters are given within the figure

chaotic motion are also possible. The example of a chaotic soliton shown in Fig. 5 always remains symmetric. In other cases, the soliton profile may evolve so that it is always asymmetric.

The soliton shape is constrained within certain limits, and this means that the solution densely occupies a certain region in a phase space. This can be illustrated if we choose two soliton parameters (i.e. energy and the squared amplitude at the soliton centre), and construct a parametric plot. A sample of such a plot for a finite interval of z values is shown in Fig. 5b. If the trajectory is continued for higher values of z, it will densely cover a finite region in this two-dimensional plot. Similar behaviour will be observed if we choose any other soliton parameters.

A smooth localized initial condition, with parameters near to the point in that region will converge to the chaotic soliton, and the trajectory in the phase space will be attracted to the region. Therefore, we can call this type of soliton solution a "strange attractor", in analogy with this notion in low-dimensional problems.

10 Soliton Explosions

Pulsations that appear in the soliton profile are the external revelations of its internal dynamics. Dissipative solitons can take a variety of shapes, and also show a variety of periodic and chaotic changes. One of the most striking behaviours that can be observed numerically [3, 18] and experimentally [20] is that of "explosions". These are soliton solutions that periodically suffer explosive instabilities, but return to their original shapes after each explosion. Such a solution has intervals of almost stationary propagation, but, over and over again, the instability develops, producing explosions, and then the stationary shape is subsequently recovered.

An example of an exploding soliton is shown in Fig. 6a. It has the following main properties:

(1) Explosions occur intermittently. In our continuous model, they occur more or less regularly, but the period changes dramatically with a change of parameters.
(2) The successive explosions have similar features, but are not identical.
(3) Explosions happen spontaneously, but additional perturbations can trigger them.
(4) One of the basic features of this type of solution is that the recurrence is back to the stationary soliton solution.

These characteristics have been observed both theoretically [3, 18] and experimentally [20]. In some cases, each explosion may occur predominantly on one side of the soliton.

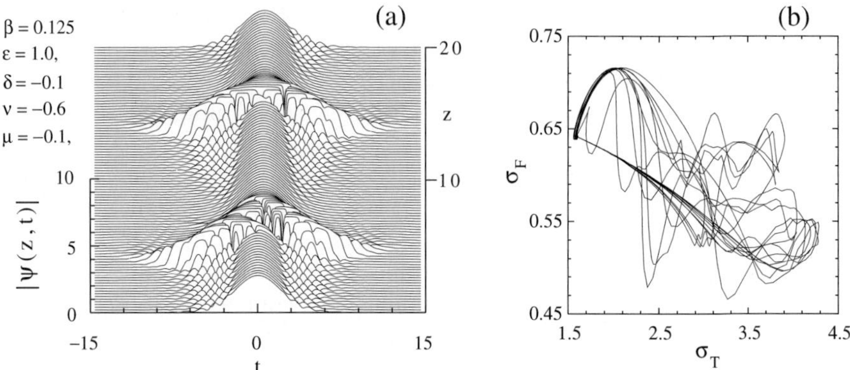

Fig. 6. (**a**) Soliton evolution showing two consecutive explosions. The values of the parameters are given on the l.h.s. of the figure. (**b**) The soliton spectral width σ_F versus soliton temporal width σ_T. This is clearly a chaotic solution. Each trajectory returns to the same point which represents the stationary (laminar) regime of evolution

Among the unusual properties of these localized solutions is the amazingly wide range of parameters of the system where they exist [18]. Hence, exploding solitons can, in principle, be observed in a variety of applications.

Despite the return to the same profile, each exploding soliton solution belongs to the class of chaotic solutions. None of the explosions repeats the previous one. When plotting one of the parameters of the soliton (e.g. soliton spectral width) versus another one (e.g. temporal soliton width), the curves for each explosion will be different. This is clearly seen in Fig. 6b.

Being a chaotic solution, an exploding soliton is another example of a strange attractor. A wide range of initial conditions which do not belong to this object will eventually be transformed into an exploding soliton solution.

11 Spectrum of Eigenvalues

We need a special technique to investigate each chaotic soliton. In the case of exploding solitons, one of the possible approaches for these studies is linear stability analysis of the solution when it is in the laminar regime of its evolution. We note that the soliton is not completely stationary, even during this part of the development. However, there is a stationary solution of the CGLE that serves as a ground state for recoveries after the explosion occurs. The solution approaches this stationary solution with relatively high accuracy before the next instability develops. This stationary solution can be found using, for example, a shooting technique based on z-independent ODEs (see Chap. 13 in [13]). The next step is linear stability analysis of this stationary solution.

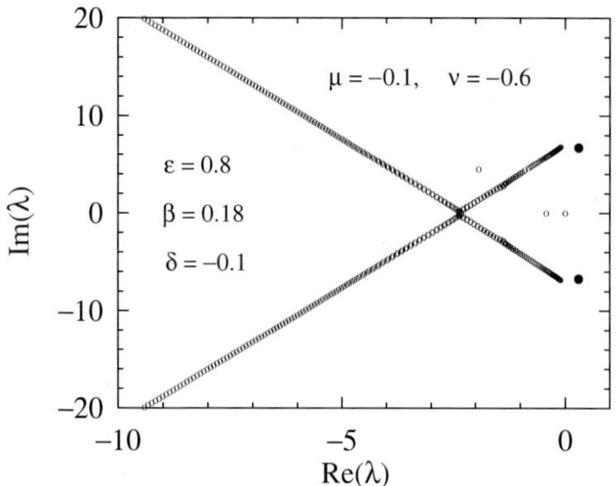

Fig. 7. The spectrum of eigenvalues on the complex plane for an exploding soliton

The complex plane, with the eigenvalues, is shown in Fig. 7. The method for obtaining the eigenvalues is described in [21] and Sect. 5. The full spectrum consists of two complex conjugate eigenvalues with positive real parts and a continuous spectrum of complex conjugate eigenvalues, all with negative real parts. This particular spectrum is obtained for the solution presented in Fig. 6, but it is qualitatively the same for a wide range of parameters where exploding solutions exist.

An important feature of the spectrum is that it has two almost identical pairs of complex conjugate eigenvalues with positive real parts. They differ slightly, but appear to coincide on the scale of Fig. 7. These eigenvalues are responsible for the instability of the soliton. The eigenfunctions corresponding to these eigenvalues are, respectively, even and odd functions of t. They are shown in Fig. 8. Each of these functions is non-zero mainly in the wings of the soliton.

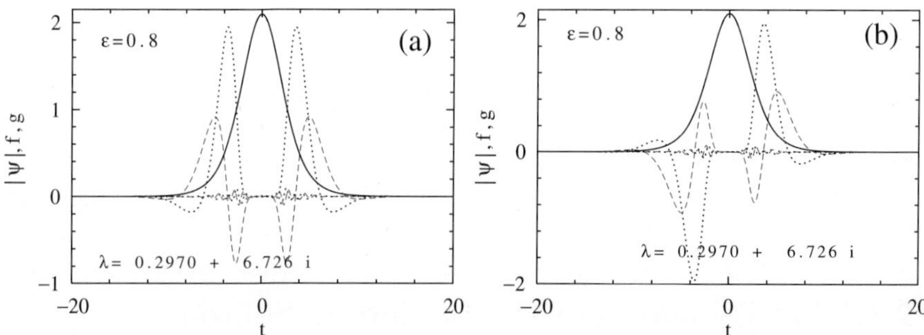

Fig. 8. Real (*dotted line*) and imaginary (*dashed line*) parts of the (**a**) even and (**b**) odd perturbation functions. The *solid* lines in (**a**) and (**b**) show the amplitude of the soliton itself

The whole continuous spectrum of eigenvalues is located on the left half of the complex plane. The corresponding eigenfunctions are much broader than the soliton width. These eigenfunctions are continuous waves of different frequencies and wavenumbers which are perturbed in the central zone by the soliton. In the absence of the soliton, small amplitude radiation waves decay, due to δ being negative. This corresponds to the pair of eigenvalues at the r.h.s edge of the continuous spectrum with real parts exactly equal to -0.1. All other eigenvalues of the continuous spectrum have real parts below -0.1 (i.e. larger than 0.1 in absolute value), due to the influence of spectral filtering on radiation waves of different central frequencies.

This spectrum does not change qualitatively when we change the parameters of the system in the vicinity of the chosen point. The real part of the discrete eigenvalue is shown in Fig. 9 as a function of ϵ. We can see that, when $\epsilon \approx 0.8$, the real part has a maximum, and no other eigenvalues appear around this point. The second eigenvalue appears only when ϵ is below 0.5. Hence, we expect that the qualitative behaviour will be the same for a wide range of ϵ, from 0.5 to 1.4, where this eigenvalue moves to the left half of the complex plane. Exploding solitons also exist when $0.2 < \epsilon < 0.5$, but their behaviour is more complicated during the explosive part of the solution, due to the simultaneous existence of several eigenvalues with positive real parts.

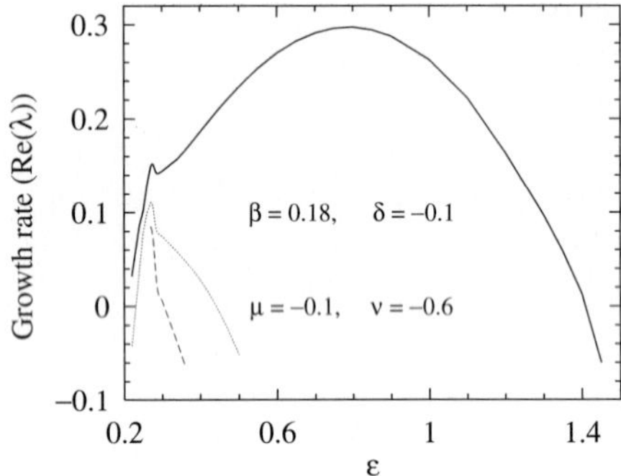

Fig. 9. Real part of the discrete eigenvalue (*solid line*). It is positive in the interval $0.2 < \epsilon < 1.4$. *Dotted* and *dashed* lines show additional eigenvalues of the discrete spectrum that appear when $\epsilon < 0.5$

12 Global Dynamics of the Exploding Soliton

In the presence of eigenvalues with positive real parts, the soliton evolution experiences the following scenario. Suppose, initially, we have the stationary solution with small perturbations. We note that the real parts of the eigenvalues are relatively small, so that perturbations grow slowly. The imaginary parts of the eigenvalues result in oscillations simultaneously with an increase in the perturbations. We also note that the soliton center is not influenced by this instability, because the eigenfunctions are almost zero in the central part of the soliton.

After the initial linear growth of the perturbation, its amplitude becomes comparable with the soliton amplitude, and the dynamics becomes strongly nonlinear. The nonlinearity mixes all perturbations, creating radiative waves. The amplitudes of the radiative waves increase at the expense of the initial perturbation. Consequently, the fraction of the initial perturbation within them becomes small. The solution at this stage appears to be completely chaotic. However, the solution remains localized, both in amplitude and in width, due to the choice of the system parameters. Specifically, due to μ being negative, the total amplitude is limited from above. In addition, a positive β ensures that the total width in the frequency domain also stays finite, provided that the other parameters are within a certain range. It is also important that the stationary soliton shape is fixed, thus providing a point of return.

As all radiative waves have eigenvalues with negative real parts, they decay and quickly disappear, since the eigenvalues for most of them have much larger negative real parts than the initial perturbation. This means that the evolution returns to the state of a stationary soliton with a small perturbation that has an eigenvalue with positive real part. As the real part of the discrete eigenvalue is relatively small, the instability again develops later, thus repeating the whole process of the evolution described above. This pattern is repeated indefinitely along the z axis.

One cycle of this evolution is shown, schematically, in Fig. 10. The fixed point, depicted by a black dot in this figure, corresponds to the stationary soliton solution. It can be classified as a stable-unstable focus, because all the eigenvalues in the stability analysis appear as complex-conjugate pairs. We stress here that our system has an infinite number of degrees of freedom, and

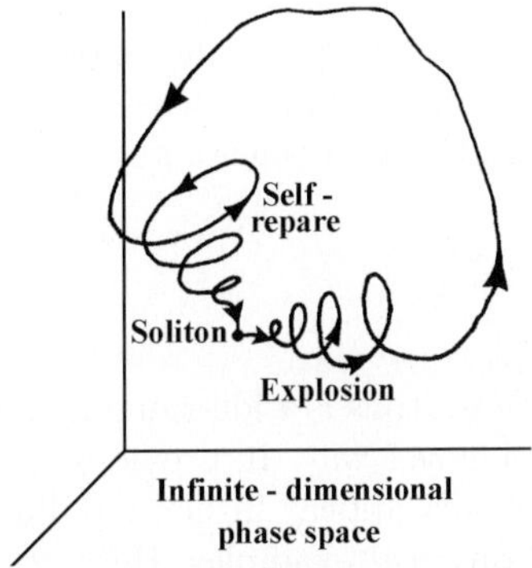

Fig. 10. One cycle of the evolution of an exploding soliton

that the evolution actually unfolds in an infinite-dimensional phase space. It cannot be reduced to a finite-dimensional problem, as all the eigenvalues play essential roles in the dynamics. As the fixed point is unstable, the trajectory leaves it in a direction in the phase space defined by the discrete eigenvalues. This motion is exponential as well as oscillatory. After complicated dynamics in the whole phase space, the trajectory, being homoclinic, returns to the same fixed point, but does so along a different path, as defined by the continuous spectrum. This return is also accompanied by oscillations, as all the eigenvalues in this problem are complex.

This scenario has common features with the one described by Shil'nikov's theorem [22, 23, 24]. The latter is usually applied to a third-order dissipative system or a higher-order one that can be reduced to a third-order system. In a third-order dissipative dynamical system, the stability of the singular points is described by three eigenvalues. Suppose that a given singular point is a saddle-focus. Then this set of eigenvalues consists of two complex conjugate eigenvalues with positive real parts and one real negative eigenvalue that is larger in absolute value than the real parts of the other two eigenvalues. We also suppose that there is a homoclinic orbit based on this singular point. Then, the evolution of the system close to this homoclinic orbit will be chaotic and will periodically leave and then return to this singular point.

Though there is this similarity, the dynamics of an exploding soliton is more involved. Obviously, our system, having an infinite number of degrees of freedom, is much more complicated than any system with three degrees of freedom. Correspondingly, the number of eigenvalues in our case is infinite, rather than just three, and all of them play essential roles in the dynamics. The singular point in our case is a stable – unstable focus rather than saddle-focus. Nevertheless, we have only two nearly identical pairs of eigenvalues with positive real parts responsible for the instability in our system, while the remaining eigenvalues ensure a quick return to the original state. Furthermore, the total phase space in our system is also bounded. As a result, the solution always stays localized, and our system qualitatively behaves in the same way as a system described by Shil'nikov's theorem. In other words, we have a homoclitic orbit that starts and ends at the singular point defined by the exploding soliton, and all nearby trajectories are chaotic. For a mathematical description of similar phenomena, we refer the reader to [25].

13 Conclusion

The concept of dissipative solitons is wide-ranging and allows many phenomena to be explained in a novel way. It provides a new way of undertanding localized structures which appear in non-equilibrium systems, including chemical, biological and physical examples. Here, we have considered a limited number of such structures that appear in optics in (1+1)D geometry. We have also concentrated on those aspects of the problem that can be studied on the basis of a qualitative analysis of nonlinear systems. Other techniques and more complicated cases are presented in other chapters of this book.

Acknowledgements

Our work is supported by the Australian Research Council.

References

1. N. Akhmediev and A. Ankiewicz, Solitons around us: Integrable, Hamiltonian and Dissipative systems, In: *Optical Solitons, Theoretical and Experimental Challenges*, Lecture Notes in Physics, vol. 613, Editors: K. Porsezian and V. C. Kurakose, (Springer, Berlin, 2003), pp. 105–126.
2. A. Ankiewicz and Y. Nagai, Chaos, solitons and fractals **13**, 1345–58 (2002).
3. J. M. Soto-Crespo, N. Akhmediev and A. Ankiewicz, Phys. Rev. Lett. **85**, 2937 (2000).
4. N. Akhmediev and A. Ankiewicz, Solitons of the complex Ginzburg-Landau equation, In *Spatial solitons*, Editors: S. Trillo, W. Torruellas, Springer Series in Optical Sciences, vol. 82, (Springer, Berlin, 2001), pp. 311–339.
5. J. M. Soto-Crespo and N. Akhmediev, Phys. Rev. E **66**, 066610 (2002).
6. K. Maruno, A. Ankiewicz, N. Akhmediev Physica D **176**, 44–66 (2003).
7. W. van Saarlos and P. C. Hohenberg, Phys. Rev. Lett. **64**, 749 (1990).
8. R. J. Deissler and H. Brand, Phys. Rev. Lett. **72**, 478 (1994).
9. P. Kolodner, Phys. Rev. A **44**, 6448 (1991).
10. M. Dennin, G. Ahlers and D. S. Cannell, Phys. Rev. Lett. **77**, 2475 (1996).
11. K. G. Müller, Phys. Rev. A **37**, 4836 (1988).
12. Y. Kuramoto, *Chemical Oscillations, Waves and Turbulence*, (Springer, Berlin, 1984).
13. N. Akhmediev and A. Ankiewicz, *Solitons: Nonlinear Pulses and beams*, (Chapman and Hall, London, 1997).
14. R. Conte, Chapter in Dissipative Solitons.
15. W. Van Saarlos and P. C. Hohenberg, Physica D, **56**, 303 (1992).
16. J. M. Soto-Crespo, N. Akhmediev and K. Chiang, Phys. Lett. A. **291**, 115 (2001).
17. T. Kapitula, Chapter in Dissipative Solitons.
18. N. Akhmediev, J. M. Soto-Crespo and G. Town, Phys. Rev. E, **63**, 056602 (2001).
19. M. J. Feigenbaum, *J. Stat. Phys.* **19**, 25 (1978).
20. S. T. Cundiff, J. M. Soto-Crespo and N. Akhmediev, Phys. Rev. Lett., **88** 073903 (2002).
21. N. Akhmediev and J. M. Soto-Crespo, Phys. Lett. A **317**, 287 (2003).
22. L. P. Shil'nikov, Sov. Math. Doklady **6**, 163 (1965).
23. L. P. Shil'nikov, Math. USSR – Sbornik **10**, 91 (1970).
24. C. P. Silva, Shil'nikov's theorem – a tutorial, *IEEE Transactions on circuits and systems, I: Fundamental theory and applications* **40**, 675 (1993).
25. B. Sandstede, J. Dynamics and Differential Equations **12**, 449 (2000).

Dissipative Magneto-Optic Solitons

A.D. Boardman, L. Velasco, and P. Egan

Joule Physics Laboratory, Institute for Materials Research, University of Salford, Salford, Manchester, M5 4WT, United Kingdom
a.d.bordman@salford.ac.uk
l.velasco@pgt.salford.ac.uk
p.egan@salford.ac.uk

1 Introduction

The study of magneto-optics involves the polarization state of light [1, 2, 3, 4], which is a measure of its vector nature. The displacement vector in a magneto-optic medium is $\mathbf{D} = \epsilon_0[\epsilon \cdot \mathbf{E} + i\mathbf{g} \times \mathbf{E}]$, where ϵ is the relative permittivity that exists in the absence of an applied magnetic field, ϵ_0 is the permittivity of the free space, $\mathbf{g} \propto f(\mathbf{H}_{app})$ is called the gyration vector, and the function $f(\mathbf{H}_{app})$ involves the applied magnetic field $\mathbf{H}_{app}$. The vector $\mathbf{g} \times \mathbf{E}$ is normal to $\mathbf{E}$ and the overall dielectric property tensor of magneto-optic material has off-diagonal elements. For magneto-optic phenomena, as opposed to chiral properties, $\mathbf{g}$ does not depend upon which way the light wave is travelling. This is an important qualification because it means that it continues to act in the same direction, even after the wave has been forced to change direction by a reflection. Hence the physical property described by the form of $\mathbf{D}$ is non-reciprocal. A further generalization is that $\mathbf{g}$ may depend upon the spatial coordinates. It is common practice to write the displacement vector as $\mathbf{D} = \epsilon_0\epsilon \cdot \mathbf{E}$, where ϵ is now the complete magneto-optic tensor. It is also common practice to write the off-diagonal terms of ϵ as $\pm Qn^2$, where n is the refractive index of the un-magnetised dielectric and $Q(\mathbf{r})$ is called the magneto-optic parameter distribution. Some common magneto-optic configurations are shown in Fig. 1, in which the direction of the saturation magnetisation $\mathbf{M}$, relative to the propagation direction, is shown.

The longitudinal case is usually called the Faraday configuration. In the bulk this will cause a rotation about the propagation direction of the electric field carried by a plane wave propagating along $\mathbf{M}$. This propagation can

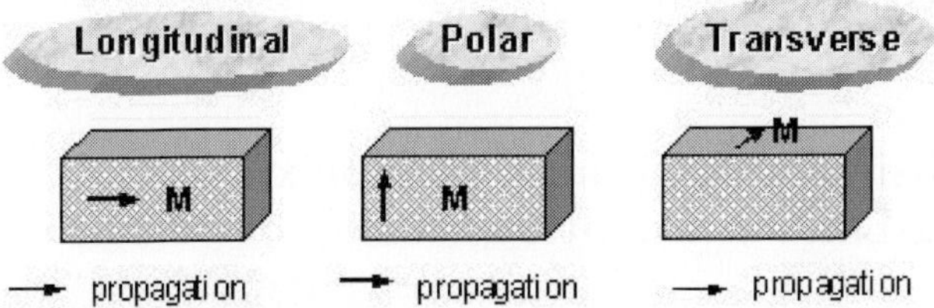

Fig. 1. Common magneto-optic configurations. M is the magnetization

A.D. Boardman, L. Velasco, and P. Egan: *Dissipative Magneto-Optic Solitons*, Lect. Notes Phys. **661**, 19–35 (2005)
www.springerlink.com

be resolved into the propagation of two counter-rotating circularly polarised waves, each seeing a different refractive index. A reversal of the propagation direction reverses these indices and non-reciprocal behaviour occurs. The transverse case maybe called either the Voigt or Cotton-Mouton configuration. This is reciprocal in the bulk and non-reciprocal in an asymmetric waveguide.

Magneto-optics was once described as the stepchild [5] of integrated optics [2]. This impression originated from the once simple desire to insert magneto-optics into known designs rather than address and control the fascinating complexity of the materials. A major task, for example, is using third-order optically nonlinear materials interfaced to magneto-optic materials so that bright solitons can be controlled. This is made possible by the fact that magneto-optic materials have been brought to a high state of readiness by the magnetism community, because the latter maintains a very strong interest in magneto-optic recording media and optically non-reciprocal devices [6]. These include periodic structures and ultra-thin films, the general aim being to exploit modern controllable magnetic properties. Magneto-optic behaviour, however, is a specific, non-reciprocal example of gyrotropic behaviour. In fact the term gyrotropic embraces optical [1, 2, 3, 4] activity and general chiral properties [7] but it is the non-reciprocal behaviour of magneto-optics that is so important for applications. As has been eloquently stated before [8], the vector nature of light leads to robustness and high discrimination levels in applications. The combination of polarisation and nonlinearity is therefore a very powerful mix. Indeed, when coupled to photo-induced Faraday rotation it is possible to discriminate this effect from the background of other nonlinear effects [9]. Non-reciprocal behaviour is characteristic of artificial gyrotropy [2] and can be used in optical isolators and a range of coherence and quantum problems. This is all in sharp contrast to natural gyrotropy, like optical activity. This has been highlighted as being tuned to non-local field-matter interactions and so can be used to probe chiral molecular systems [9]. Actually, it is important to go beyond third-order, Kerr, optical nonlinearity and move towards a saturable model of terms in the polarisation. In this spirit, this chapter seeks to determinate the influence of a magneto-optic presence upon an optically nonlinear material that is modelled by a cubic-quintic form of polarisation. In addition, the coefficients of the envelope equation will be made complex, to take into account both linear and nonlinear damping and cubic gain processes. The emphasis is upon the simulation outcomes, however, rather the applications.

2 The Basic Cubic-Quintic Complex Ginzburg-Landau Equation

The cubic-quintic complex Ginzburg-Landau (CQGLE) equation derives its character not only from an extension of the widely used cubic Kerr nonlinearity to include a quintic contribution, but also from the inclusion of complex coefficients that model loss and gain. The action taken to broaden the scope of the nonlinearity is a step towards acknowledging that the nonlinearity of many materials saturate, as the propagating power increases. The generalisation to complex coefficients permits the modelling of gain and losses, both of a linear and a nonlinear origin. Adding in a magneto-optic material property will modify the basic envelope equation in a rich variety of ways, depending upon whether the full vector character of the propagating electromagnetic waves is simulated.

Optical beams can diffract but this tendency is, broadly speaking, offset by the ability of the material nonlinearity to self-focus the beams. A *scalar* electric field amplitude, $E(x, y, z, \omega)$, associated with the beam satisfies the standard wave equation

$$\frac{\partial^2 E}{\partial z^2} + \nabla_\perp^2 E + \epsilon(\omega)\frac{\omega^2}{c^2}E = 0 \tag{1}$$

where $\nabla_\perp^2 = \frac{\partial^2}{\partial x^2} + \frac{\partial^2}{\partial y^2}$, and the fast time dependence of the electric field varies as $e^{-i\omega t}$ where t is time and ω frequency. The velocity of the light in the vacuum is c and $\epsilon(\omega)$ is the total relative permittivity. The spatial dependence of E can be written as $\Psi(x, y, z)e^{ik_0 z}$, where $k_0 = \frac{\omega}{c}\sqrt{\epsilon_L}$ and ϵ_L is the linear permittivity. The complex amplitude, $\Psi(x, y, z)$, is actually slowly varying, i.e. $\Psi(x, z)$ evolves slowly enough for $|\frac{\partial^2 \Psi}{\partial z^2}| \ll |k_0 \frac{\partial \Psi}{\partial z}|$ to be true. This is a perfectly reasonable assumption as any numerical simulation will reveal. The appropriate form of the wave equation is, therefore,

$$2ik_0\frac{\partial \Psi}{\partial z} + \nabla_\perp^2 \Psi + \left[\epsilon(\omega)\frac{\omega^2}{c^2} - k_0^2\right]\Psi = 0 \tag{2}$$

For a non-dispersive, linear, medium the final term will vanish, since $\epsilon(\omega) = \epsilon_L$ and $k_0^2 = \epsilon_L\frac{\omega^2}{c^2}$. The presence of nonlinearity, creates a small, but significant, nonlinear contribution to the difference $k_0^2 - \epsilon\frac{\omega^2}{c^2}$. The term $\left(\epsilon\frac{\omega^2}{c^2} - k_0^2\right)$ is $\frac{k_0^2}{n_0^2}(\sqrt{\epsilon} + \sqrt{\epsilon_L})(\sqrt{\epsilon} - \sqrt{\epsilon_L})$. This is approximately $\frac{k_0^2}{n_0^2}\left(n_2|\Psi|^2\right)$, after writing the nonlinear index as $n_0^2 + n_2|\Psi|^2$, so that ϵ is $\epsilon_L + 2n_2|\Psi|^2 n_0$. Hence, to a first approximation, the final term in (2) is $\frac{k_0^2}{n_0}\left(n_2|\Psi|^2\right)\Psi$, retaining, for the moment, only cubic nonlinearity. The inclusion of nonlinear effects up to fifth order involves the introduction of another constant, n_4, defined through the envelope equation

$$2ik_0\frac{\partial \Psi}{\partial z} + \nabla_\perp^2 \Psi + 2\frac{k_0^2}{n_0}n_2|\Psi|^2\psi - 2\frac{k_0^2}{n_0}n_4|\Psi|^4\psi = 0 \tag{3}$$

This equation can be usefully scaled by measuring x and y and in the units $\mathbf{D}_0$, equal to the beam width, and z in Rayleigh, or diffraction, lengths $2k_0\mathbf{D}_0^2$. The transformations to the now dimensionless coordinates and the admission of complex coefficients results in the generation of the familiar cubic quintic equation

$$\mathrm{i}\frac{\partial \Psi}{\partial z} + \mathrm{i}\delta\Psi + \left(\frac{1}{2} - \mathrm{i}\beta\right)\nabla_\perp^2\Psi + \left((1-\mathrm{i}\epsilon)|\Psi|^2 - (\nu - \mathrm{i}\mu)|\Psi|^4\right) = 0 \quad (4)$$

The physical interpretation of (4) leads to the conclusion that δ accounts for any linear absorption, β represents diffusion, ϵ is the nonlinear cubic gain, ν is the quintic coefficient that measures the self-defocusing brought on by the negative sign in the last term, and μ is a nonlinear loss term, of quintic origin.

This is a powerful model that has been shown to generate many interesting solutions [10] through the elegant work of Akhmediev and Ankiewicz [11]. The addition of a magneto-optic influence requires some care [12, 13, 14, 15], however, because the choice of applied magnetic field orientation and the choice of the spatial distribution of the magnetisation are both going to be important.

3 Magneto-Optics with Inhomogeneous Magnetisation

This section focuses only upon the influence of the magneto-optics and diffraction upon beam envelope behaviour in classical Faraday and Voigt configurations. The outcomes can be legitimately added later to the effects of nonlinearity. The magneto-optic effect is a perturbation that ranks alongside the perturbation that nonlinearity represents and the combined effect, including diffraction, produces a slowly varying evolution in the electric field amplitude. These are the reasons why magneto-optic behaviour can be studied separately in this section, without adding in the nonlinearity until later.

Assuming the time dependence $e^{-i\omega t}$, Maxwells equations for the Fourier transform of the electric field vector $\mathbf{E}$ lead to

$$\nabla^2\mathbf{E} + \frac{\omega^2}{c^2}\epsilon \cdot \mathbf{E} - \nabla\left(\nabla \cdot \mathbf{E}\right) = 0 \quad (5)$$

For propagation along the z-axis, the permittivity tensor depends upon the direction of an externally applied field $\mathbf{H}_0$. This field produces a magnetisation distribution defined through a function Q that is often used as a constant proportional to the magnitude of the saturation magnetisation. If the applied magnetic field is supplied by a single wire electrode, however, Q is a function of the spatial coordinates. To be specific, if the wire lies along the z-axis then the magnetic field created will be a vector that is tangential to circles in the (x, y) plane, centred upon the wire. There are components of the magnetisation parallel to the x- and y-axes but only the one parallel to the x-axis will be important, and $Q \equiv Q(x)$ models such a magnetisation distribution.

The principal configurations modelled are:

Faraday:

$$\epsilon = \begin{pmatrix} n^2 & -\mathrm{i}Qn^2 & 0 \\ \mathrm{i}Qn^2 & n^2 & 0 \\ 0 & 0 & n^2 \end{pmatrix} \tag{6}$$

Voigt:

$$\epsilon = \begin{pmatrix} n^2 & 0 & 0 \\ 0 & n^2 & -\mathrm{i}Qn^2 \\ 0 & \mathrm{i}Qn^2 & n^2 \end{pmatrix} . \tag{7}$$

Pragmatically, it is assumed, based upon the known properties of most magneto-optic materials, that all the diagonal elements are equal. For a layered structure, this form of $Q(x)$ really is readily provided by a wire electrode arrangement deposited upon the upper planar surface, or buried within the structure. For the Faraday configuration some arrangement of layers, or magnetic domain structure, may be used. Basically, permitting Q to be a function of any of the coordinates is a matter for experimental ingenuity on the one hand, but at the same time it is a feature that adds exciting functionality to the behaviour of optical beams.

The key to magneto-optic behaviour is the fact that $\nabla \cdot \mathbf{E} \neq 0$, and this is characteristic of a forced gyrotropic medium. However, the divergence of the total displacement vector is always zero, so that

$$\nabla \cdot (\epsilon \cdot \mathbf{E}) = 0 \tag{8}$$

Faraday Configuration

In this case, the full electric field in a guiding structure is $\mathbf{E} = (E_x, E_y, E_z) \times \mathrm{e}^{(-\mathrm{i}\omega t)}$. For propagation along the z-axis, which is also the direction of the applied magnetic field, in a typical planar waveguide and $E_z \simeq 0$, in the bulk. For a simple guiding structure consisting of a cladding-(high-index-core)-substrate arrangement, the validity of neglecting E_z will depend upon the size of the refractive-index change across an interface. The answer is that even though E_z rises quite rapidly with this change, it saturates quickly and remains small, compared to the other electric field components. A strategy of neglecting E_z is usually safe, provided that the core-cladding and the core-substrate materials have refractive indices that match to within 1–5%. The neglect of E_z leads to

$$\nabla \cdot \mathbf{E} = \mathrm{i}\frac{dQ(x)}{dx}E_y + \mathrm{i}Q(x)\frac{\partial E_y}{\partial x} - \mathrm{i}Q(x)\frac{\partial E_x}{\partial y} \tag{9}$$

Hence, in the Faraday configuration, linear coupled equations emerge which are

$$\nabla^2 E_x + \frac{\omega^2}{c^2}n_m^2 E_x - iQ\frac{\omega^2}{c^2}n_m^2 E_y + F = 0 \tag{10}$$

$$\nabla^2 E_y + \frac{\omega^2}{c^2} n_m^2 E_y - iQ\frac{\omega^2}{c^2} n_m^2 E_x + G = 0 \tag{11}$$

where

$$F = -\mathrm{i}Q\frac{\partial^2 E_y}{\partial x^2} + \mathrm{i}Q\frac{\partial^2 E_x}{\partial x \partial y} - \mathrm{i}E_y\frac{\partial^2 Q(x)}{\partial x^2} - 2\mathrm{i}\frac{\partial E_y}{\partial x}\frac{\partial Q(x)}{\partial x} + \mathrm{i}\frac{\partial Q(x)}{\partial x}\frac{\partial E_x}{\partial y} \tag{12}$$

$$G = -\mathrm{i}Q(x)\frac{\partial^2 E_x}{\partial y^2} + \mathrm{i}Q(x)\frac{\partial^2 E_y}{\partial x \partial y} - \mathrm{i}Q(x)\frac{\partial^2 E_x}{\partial y^2} \tag{13}$$

The presence of a magnetisation distribution would seem, at this stage of the development to be a major complication but a dimensional analysis of the above equations reveals the situation to be otherwise. First, x- and y-directions can be measured in units of D_0, the beam width. In a planar waveguide structure only x will come into play, but both x and y will feature in the bulk applications described later. For guided waves, E_y is carried as a TM wave, and E_x as a TE wave, and they have slightly different wave numbers. An assumption of phase matching is needed for simplification to be exercised but it is not unreasonable in order to remove this "form" birefringence, and in the bulk this problem does not arise. The z-axis can be measured in terms of diffraction, or Rayleigh, length which is kD_0^2, where k is the bulk wave number, or the average wave number, if the E_x, E_y components are associated with a waveguide. Given this scaling system, the leading terms involving Q have the factor $2kD_0^2n^2$, but none of the Q terms in F and G have this factor. Typically $Q \approx 10^{-4}$ and $\frac{\omega^2}{c^2}D_0^2n^2 \approx 10^{+4}$, so that the effective magneto-optic parameter is a transformation of Q to $Q_1 \approx 1$, which is four orders of magnitude more significant than the Q terms in F and G. Hence F and G may be neglected.

For the bulk or waveguide structure the electric field components will be written as

$$E_x = \Psi_x(x, y, z)\ \exp\left(\mathrm{i}\left(\frac{\omega}{c}\beta\right)\right) \tag{14}$$

$$E_y = \Psi_y(x, y, z)\ \exp\left(\mathrm{i}\left(\frac{\omega}{c}\beta\right)\right) \tag{15}$$

where Ψ_x, Ψ_y are complex and are slowly varying functions of z, $\beta = \frac{\beta_x+\beta_y}{2}$, and $\beta_x = \beta_y$ in the bulk. The adoption of a rotating coordinate system [13] in which $\Psi_+ = \frac{1}{\sqrt{2}}(\Psi_x + i\Psi_y)$, $\Psi_- = \frac{1}{\sqrt{2}}(\Psi_x - i\Psi_y)$, where Ψ_+, Ψ_- are appropriately scaled, uncouples (10) and (11) to give, for example,

$$\mathrm{i}\frac{\partial \Psi_+}{\partial z} + \frac{1}{2}\left(\frac{\partial^2 \Psi_+}{\partial x^2} + \frac{\partial^2 \Psi_+}{\partial y^2}\right) \pm Q_1\Psi_+ = 0 \tag{16}$$

This is the equation that gives the development of a circularly polarised wave. It is satisfying that the transformation shows that an arbitrarily polarised input to a Faraday system evolves, in the linear case, as uncoupled counter-rotating, circularly polarised waves. Equation (16) will be a description of a guided wave if $\frac{\partial^2 \Psi_+}{\partial y^2} = 0$, and a beam in a bulk medium if both and diffraction terms are retained.

Voigt Configuration

To use this effect, a guided wave structure is needed. In fact, for a planar structure TM waves are the only modes that can be deployed. They propagate here along the z-axis with an electric field [15]

$$\mathbf{E} = \Psi \left[\hat{\mathbf{y}}\xi_y(y) + \hat{\mathbf{z}}\xi_z(y)\right] \mathrm{e}^{\mathrm{i}(\omega t - \beta z)} \tag{17}$$

where Ψ is again a slowly varying amplitude, and $\xi_y(y)$, $\xi_z(y)$ are the modal field components. Using the permittivity tensor obtained by (17) the wave equation is

$$\nabla(\nabla \cdot \mathbf{E}) - \nabla^2 \mathbf{E} = \frac{\omega^2}{c^2}(\epsilon \cdot \mathbf{E}) \tag{18}$$

Multiplying (18) by $\mathbf{E}^*$ and integrating over y leads to the conclusions that

$$\int \mathbf{E}^* \cdot \nabla(\nabla \cdot \mathbf{E}) dy = 0 \tag{19}$$

$$\int \mathbf{E}^* \cdot \nabla^2 \mathbf{E} dy = \int \left[-\frac{\omega^2}{c^2} n^2 |\Psi|^2 - 2\mathrm{i}\frac{\omega}{c}\beta\Psi^* \frac{\partial \Psi}{\partial z} \right] (|\xi_y|^2 + |\xi_z|^2) dy \tag{20}$$

$$\int n^2 \mathbf{E}^* \cdot \mathbf{E} dy = |\Psi|^2 \int n^2 \left(|\xi_y|^2 + |\xi_z|^2 \right) dy \tag{21}$$

$$\int \mathbf{E}^* \cdot \mathbf{P}_M dy = -\epsilon_0 \int \left[\xi_y^* \xi_z - \xi_z^* \xi_y \right] n^2 Q dy \tag{22}$$

where $\mathbf{P}_M$ is the part of the polarisation arising from the magnetisation. After further manipulations, equations (19), (20), (21) and (22) yield, neglecting diffraction to make sure that the emphasis is upon the magneto-optic effect for this part of the argument,

$$\mathrm{i}\frac{\partial \Psi}{\partial z} = \frac{\omega}{c}\bar{\epsilon}_{yz}\Psi \tag{23}$$

where $\epsilon_{yz} = -Qn^2$ and

$$\bar{\epsilon}_{yz} = \frac{c}{\omega\beta^2} \frac{\int \epsilon_{yz}\xi_y \left(\frac{\partial \xi_y}{\partial y}\right) dy}{\int \left(|\xi_y|^2 + |\xi_z|^2\right) dy} \tag{24}$$

This integration over a planar waveguide structure vanishes if a symmetric guide is used. A convenient asymmetric system consists of a magneto-optic layer sandwiched between dissimilar substrate and cladding material.

For an application of (23) an asymmetric guiding planar structure will be used, in which the layers will be of infinite extent in the x-direction but any optical beams will be restrained by guiding in the y-direction. This is why the modal fields are needed to average the magneto-optic effect over the system. Adding diffraction of the beam to (23) is achieved in a straightforward manner simply by including the $\frac{\partial^2 \Psi}{\partial x^2}$ term.

Optical nonlinearity can be *added* into (16) and (23), in cubic-quintic form, without having to be concerned with any cross-phase modulation terms. Adopting the Rayleigh length to measure distances along the z-axis, and the beam width as the unit of measurement for distances along the x- and y-axes creates dimensionless envelope equations. The magneto-optic effect can also be reduced to a dimensionless parameter Q_1, and scaling the amplitudes with $\sqrt{L_D/L_{NL}}$, where L_D is the diffraction length and L_{NL} is nonlinear length in the Faraday case, and setting $\Psi = \frac{c}{\omega D_0} \frac{\sqrt{2}}{\sqrt{\beta\chi}}\Psi'$ in the Voigt case, where χ is the cubic nonlinear coefficient averaged over the guiding structure, finally brings the envelope equations into a tidy dimensionless overall form. This action, coupled to the addition of a cubic-quintic nonlinearity, together with complex coefficients, means that the master envelope equations are, replacing Ψ' by Ψ, for convenience:

Faraday Configuration: bulk medium supporting circularly polarised optical vortex

$$\begin{aligned} &\mathrm{i}\frac{\partial \Psi}{\partial z} + \mathrm{i}\delta\Psi + \left(\frac{1}{2} - \mathrm{i}\beta\right)\left(\frac{\partial^2 \Psi}{\partial x^2} + \frac{\partial^2 \Psi}{\partial y^2}\right) \\ &+ (1 - \mathrm{i}\epsilon)|\Psi|^2\Psi - (\nu - \mathrm{i}\mu)|\Psi|^4\Psi + Q_1(x)\Psi = 0 \end{aligned} \tag{25}$$

Voigt Configuration: asymmetric planar guiding structure

$$\begin{aligned} &\mathrm{i}\frac{\partial \Psi}{\partial z} + \mathrm{i}\delta\Psi + \left(\frac{1}{2} - \mathrm{i}\beta\right)\left(\frac{\partial^2 \Psi}{\partial x^2}\right) \\ &+ (1 - \mathrm{i}\epsilon)|\Psi|^2\Psi - (\nu - \mathrm{i}\mu)|\Psi|^4\Psi + Q_1(x)\Psi = 0 \end{aligned} \tag{26}$$

In each system δ is a measure of the linear damping, β represents the possibility of diffusion, ϵ is the cubic gain, μ is a quintic loss and ν is clearly a self-defocusing contribution to the beam evolution. Q_1 is the magneto-optic parameter, which has been made a function of x for the reasons presented earlier.

It must be emphasised that although the length scales are the same in equations (25) and (26), the amplitudes are scaled slightly differently, but this fact has no impact upon the study of the evolution process. Note also that Q_1 is the same order of magnitude in each application but that Q_1 in the Voigt case involves a modal averaging factor arising from the guiding in the asymmetric planar structure. This modal factor is the order of unity and will switch off $Q_1(x)$ if the waveguide is symmetric. Given the master equations (25) and (26) it is now possible to explore the behaviour of Ψ in each case through extensive simulations.

4 Dissipative Solitons in Voigt Configuration

This section concerns one-dimensional spatial solitons. The beam propagation direction is perpendicular to the direction of an applied magnetic field. Guiding takes place in the y-direction and diffraction is permitted in the x-direction. The soliton behaviour is controlled by the complex cubic-quintic envelope equation deploying a parameter selection inspired and informed by the work of Akhmediev and his collaborators. For this reason, the computer experiments use a range of parameters that set the conditions $\delta > 0$, $\beta > 0$, $\epsilon > 0$, $\nu > 0$ and $\mu > 0$ and are given values that permit stable beams to be generated. The input beam for the simulations is an mth-order super-Gaussian beam so that for very large values of m the cross-section of the beam becomes almost rectangular, and has rather sharply sloping sides. This choice of input beam cross-section creates the impression of the presence of an initial constant background. The intrinsic diffraction-nonlinear chirp balancing, coupled to the imposed gain-loss balance can give substantial additional flexibility through the application of an applied magnetic field. It is difficult, however, to manage the parameters, so, as others have done [10, 16], ϵ is selected as a control parameter, whilst keeping the others at fixed values. This is a logical selection, since ϵ provides the nonlinear gain of the system. Also its variance is known to permit discrimination among three main types of evolved stationary soliton beams, so it is interesting to see how the magnetic field impacts upon these beam evolutions, which have been described as plain (bell shaped), composite and moving. A composite beam is interesting because it retains its flat super-Gaussian shape, but it also acquires a kind of "hat", or "cap", that sits on top of this flat region. Clearly this can be technically discussed in terms of source-nonlinear front interaction.

Given this background information, the beam evolutions investigated here deploys the super-Gaussian input beam $\Psi(x, z = 0) = \exp\left(-\left(\frac{x}{15}\right)^8\right)$, where the order 8 is selected to present a sufficiently wide beam on the input plane. Also $\delta = 0.5$, $\beta = 0.5$, $\nu = 0.1$ and $\mu = 1.0$, and these parameters are given fixed values, based upon previous experience. For zero applied magnetic field, and $\epsilon = 2.50$, Fig. 2 shows that a plain beam emerges at $z = 100L_d$. The selected z distance of 100 Rayleigh lengths is considered to be large enough to see the stationary state beam being created. This choice of evolution distance is, of course, only a rule of thumb but it is intuitively acceptable. The stationary beam shape can be understood from the expected tendency of any wide beam to self-focus, or self-trap, until an ordinary, or plain spatial soliton is produced. For this set of parameters the nonlinear gain is not strong enough to prevent this happening. A slight change of ϵ is enough, however, to produce a composite beam, as is clearly shown in Fig. 3.

The optical beams under discussion are contained within a planar layered structure that interfaces nonlinear material to magneto-optic material and is subjected to an applied magnetic field, supplied by placing a current strip

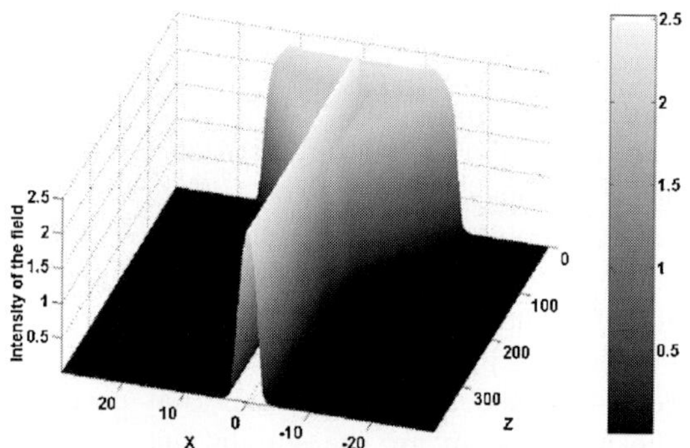

Fig. 2. Formation of a plain beam with $\beta = 0.5$, $\delta = 0.5$, $\nu = 0.1$, $\mu = 1.0$, $\epsilon = 2.50$, $Q_1(x) = 0$

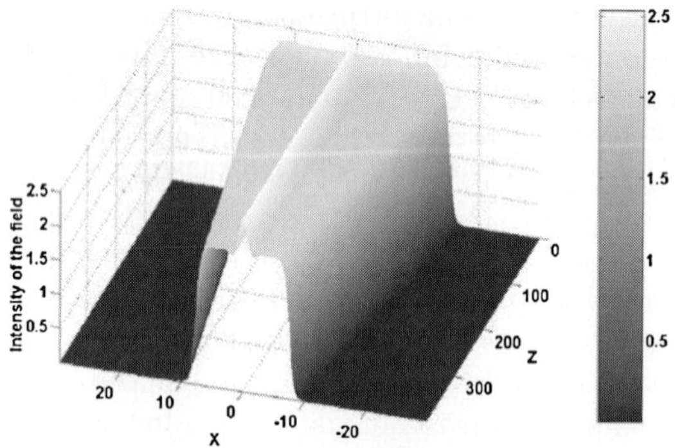

Fig. 3. Formation of composite pulse using $\beta = 0.5$, $\delta = 0.5$, $\nu = 0.1$, $\mu = 1.0$, $\epsilon = 2.52$, $Q_1(x) = 0$

onto the upper surface of the waveguide. This arrangement and waveguide type is sketched in Fig. 4.

A single thin wire carrying a current I is shown and the net effect is to provide access to the magnetic field component directed along the x-axis. The n_i are the, respective, refractive index values of the guide. Only the x-component of the applied magnetic field $\mathbf{H}_0$ is shown.

A very thin strip of conducting material, laid down upon the top surface is quite sufficient to supply the current that will have a typical value of 100 mA, but can be considerably lower. The strip need only be the order of 20 μm wide, so it produces a magnetic field distribution that approximates that of an infinitely thin wire. The magnetic field produced is tangential only directly beneath the wire i.e the magnetic field vector will not in general be parallel to the x-axis. The resultant magnetisation will have a component along both the x- and y-directions. The component along the x-axis gives rise to the transverse, effect but the component along y gives rise to a polar

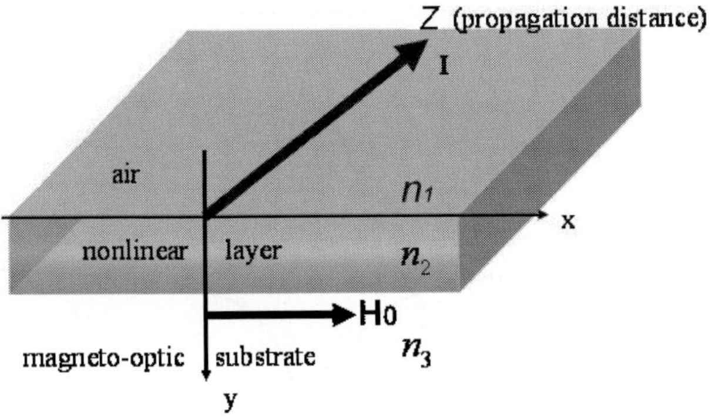

Fig. 4. Formation of composite pulse using $\beta = 0.5$, $\delta = 0.5$, $\nu = 0.1$, $\mu = 1.0$, $\epsilon = 2.52$, $Q_1(x) = 0$

effect that can cause TE-TM coupling. No attempt is being made, however, to phase match the birefringence of this guide, so the role of this polar effect is negligible. It is the component of the magnetisation along the x-axis that counts. Specifically, if the applied magnetic field at a given coordinate position is at some angle ϕ to the y-axis, then the total magnetisation will be multiplied by $\sin(\phi)$ to get the x-component. This action has the effect of confining the magnetisation to a tight region underneath the wire. Typically, such a region would be approximately $-0.25\,\mu\text{m} \leq x \leq 0.25\,\mu\text{m}$ and these distances are where the magnetisation starts to dip below its saturation value, again, for typical magnet-optic materials.

The magnetisation is introduced into the model through a magneto-optic parameter, Q_1, and the arguments given above show that it can be a function of x. Note that a possible functional shape for Q_1 ought to be captured from a straightforward calculation of the magnetization induced by a single current wire. For more complex arrays of wires this task will be more formidable. Even for the simpler case, however, it appears that there is no analytical formula to establish the relationship of the magnetisation to the applied magnetic field. Nevertheless, a simple hyperbolic tangent function is a model that is rather close to experimentally observed dependence of magnetisation upon applied magnetic field. Accordingly, the model $Q_1 = A \tanh\ (BH/H_s)$ can be used in which A, B are just empirical constants that are selected to make $\tanh \to 1$ in the region of high magnetic field. H_s is the saturation magnetic field and is selected to make sure that $Q_1(x)$ acquires its saturation value. It must be emphasised that $Q_1(x)$ can assume whatever shape is desired, depending upon how the magnetisation is created. For a single current wire lying on top of a planar guide a typical shape of $Q_1(x)$ is sketched in Fig. 5. This shows that the magnetisation looks like a hill or valley depending upon the direction of the current and the applied magnetic field direction.

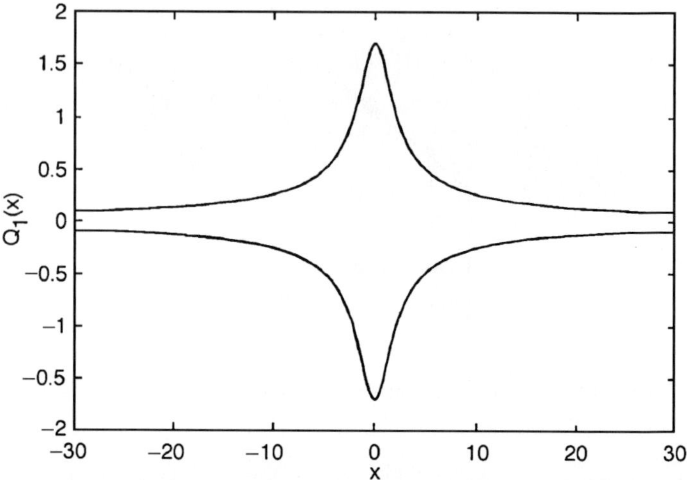

Fig. 5. Typical $Q_1(x)$ distributions for electric currents that flow along $+z$ and $-z$ respectively

Returning now to the full complex cubic-quintic Ginzburg-Landau equation, the influence of the $Q_1(x)$ distribution just discussed will be investigated and the parameter set $\beta = 0.5$, $\delta = 0.5$, $\nu = 0.1$, $\mu = 1.0$, $\epsilon = 2.54$ will be adopted. In the absence of any magnetisation, $\epsilon = 2.54$ is above the threshold for the total parameter set and causes a super-Gaussian input beam to evolve towards a plane wave. The question to address is whether introducing $Q_1(x)$ maintains this dynamic behaviour, or whether the introduced nonreciprocal property creates a significant change to it. Non-reciprocity is expected because $Q_1(x)$ creates forced gyrotropy, but the extent of this needs to be discovered through simulation. The role of $Q_1(x)$ can be seen in qualitative terms from equation (26). Basically, the nonlinearity is provided by a competition between the cubic self-focusing and the quintic self-defocusing nonlinearity i.e. the function $(|\Psi|^2 - \nu|\Psi|^4)\Psi$, leaving aside the dissipative and gain terms, for the moment. The magneto-optic terms are of the form $\pm Q_1\Psi$, so that the plus sign indicates a possible numerical strengthening of the self-focusing term, and the minus sign indicates a strengthening of the self-defocusing term. Admittedly, this is only a qualitative, intuitive, sort of argument but it does indicate the features that should be looked out for in the simulations.

Figure 6 shows how the applied magnetic field influences the soliton development from a super-Gaussian input. Figure 6a shows the outcome for $Q_1(x) = 0$, and sets up the reference behaviour for the parameter set. Figures 6b and 6c show what happens when the magnetisation is given by

$$Q_1(x) = 1.95 \ \tanh\left(\frac{I}{2\pi\sqrt{x^2+b^2}}\frac{1}{23.87}\right)$$

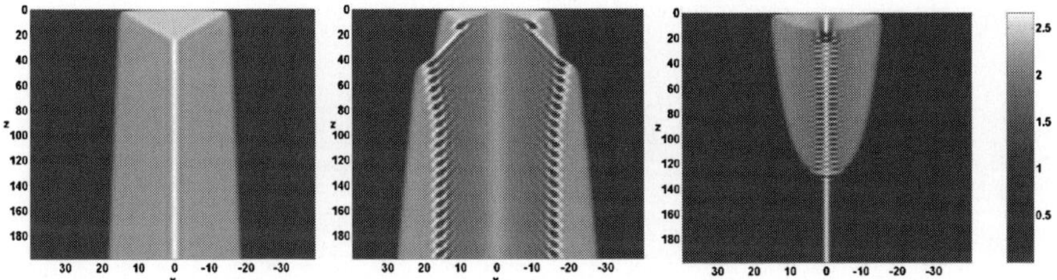

Fig. 6. $\beta = 0.5$, $\delta = 0.5$, $\nu = 0.1$, $\mu = 1.0$, $\epsilon = 2.53$. (**a**) $Q_1(x) = 0$ (**b**) $Q_1(x) < 0$ (**c**) $Q_1(x) > 0$

where the thickness of the nonlinear guide is $\approx 1\,\mu\text{m}$ and the magnitude of the current I is $\approx 200\,\mu\text{m}$. The current is assumed to be positive when flowing along the positive z-direction. It can seen in the figure that the qualitative argument is borne out, and that for the backward flowing (along negative z) current the beam rapidly self-focuses, while for the forward (along position z) flowing current the beam rapidly defocuses. Hence a carefully selected $Q_1(x)$ distribution can be used to control the behaviour of the beams in any way that is desired. The behaviour is also clearly non-reciprocal.

The cubic gain and the quintic loss are crucial to the expected stability of any state that evolves from the super-Gaussian input beam. Indeed, it is possible to discover through computer experiments the broad range of (μ, ϵ) values that will lead to a stable evolution in the $Q_1 = 0$ regime. It has been noted in particular that there are some values of (μ, ϵ) that can drive composite *pulses* to be unstable enough [10] to lead to the spontaneous creation of moving pulses, where none might be expected. It would be an enormous task to investigate the influence of a whole range of $Q_1 > 0$ or $Q_1 < 0$ distributions upon this conclusion but it is possible to select, as a tool, the type of Q_1 used earlier on, and to use sech-type *beams* that are deliberately directed at an angle to the z-axis. This is entirely equivalent, mathematically, to pulses that are moving at a finite velocity with respect to their frame of reference. To launch a sech-profiled beam at an angle to the z-axis requires the input beam to possess a phase factor so that $\Psi(x, z = 0) = 2\,\text{sech}(x - x_0)\,\text{e}^{(-\text{i}\theta(x - x_0))}$, where x_0 is the beam centre, and the proper interpretation of θ is that $\tan^{-1}(\theta)$ is the angle that the propagation direction of the beam makes to the z-axis. Because of the neglect of $\frac{\partial^2 \Psi}{\partial z^2}$ in the envelope equation, θ is small and $x = x_0 + \theta z$. The parameter set selected for the simulations shown in Fig. 7a is $\beta = 0.5$, $\delta = 0.1$, $\nu = 0.1$, $\mu = 0.8$, $\epsilon = 1.868$, $\theta = 4°$ and the reference calculation is once again the $Q_1 = 0$ case.

This $Q_1 = 0$ behaviour can be discussed in terms that associate its behaviour [10] with a combination between a front and a beam. The application of an external magnetic field by placing a current wire at $x = -10$ on the upper surface of the waveguide, and causing the current to flow along the $z < 0$ direction, creates the situations illustrated in Fig. 7b and 7c, respectively. In

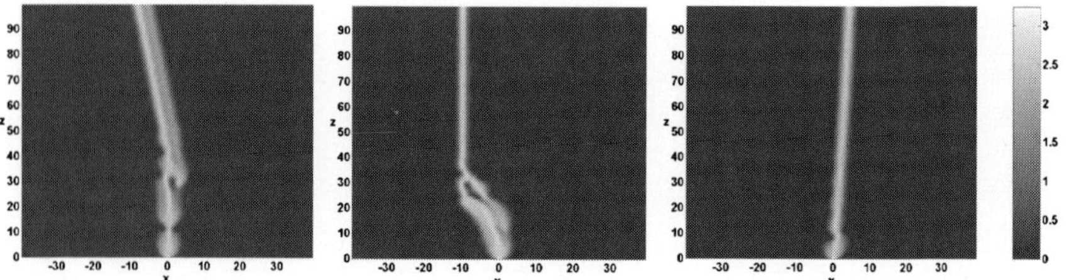

Fig. 7. Evolution of a beam initially directed at a small angle to the z-axis $\beta = 0.5$, $\delta = 0.1$, $\nu = 0.1$, $\mu = 0.8$, $\epsilon = 1.868$, $\theta = 4.75°$ (**a**) $Q_1 = 0$; (**b**) Off-axis beam behaviour for $Q_1 > 0$; (**c**): Off-axis beam behaviour for $Q_1 < 0$

one case, the beam remains on the positive side because a potential barrier created by the magneto-optic properties does not allow it to move from there. Also, the beam has been prevented from becoming established in the region that is clearly accessible when $Q_1 = 0$. In the other case, the beam sets itself along the straight line defined by the current wire direction and once again, the beam has been confined. The magnetic field also stops the formation of the composite beam that is associated with the non-magnetic case.

To investigate the role of the magnetic field in the intersection dynamics an initial input beam $A(x, z = 0) = 2\text{sech}(x-10)\exp(-i\theta(x-10))+2\text{sech}(x+10)$ composed of a beam set an angle to the z-axis and a plain beam, is selected. Each of them is shifted 10 units from the origin and are 20 units apart. Otherwise the parameters are those used above for the single beam.

Figure 8a shows a typical beam intersection for $Q_1 = 0$, in which the beams can coalesce. Note that one of the beams is set at a zero angle to the z-axis and that it remains locked in this position. The introduction of a magneto-optic influence through setting $Q_1 > 0$ or $Q_1 < 0$ creates the effects shown in Figs. 8b and c. In the $Q_1 > 0$ case, the effect of the magnetic field is to prevent an interaction between the two beams, and when $Q_1 < 0$ the two beams merge but yet continue to exist in a stationary single beam state. Once again, the magneto-optic term in the envelope equation acts to create

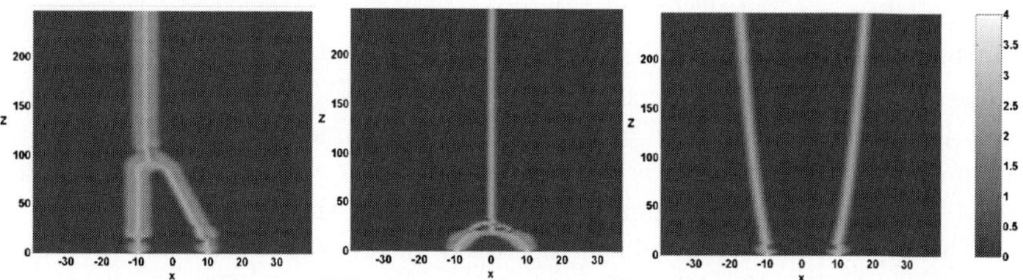

Fig. 8. Interacting beams $\beta = 0.5$, $\delta = 0.1$, $\nu = 0.1$, $\mu = 0.8$, $\epsilon = 1.868$, (**a**) Intersection of two beams creating a fused outcome for $Q_1 = 0$; (**b**) $Q_1 > 0$; (**c**) $Q_1 < 0$

a potential that can act as a barrier or an attractive well. This also can be visualised in terms of the quantity $n_2|\Psi|^2 - n_4|\psi|^4 \pm Q_1(x)$.

5 Optical Singularities in Dissipative Media

The study of optical singularities includes the behaviour of vortices [16] and these will now be investigated for a bulk medium, which makes its presence felt through the occurrence of diffraction along *both* the x- and y-directions. The introduction of magneto-optic behaviour will be through the Faraday effect. This choice is driven by the fact that in the Voigt configuration the magneto-optic effect in bulk media appears only to order Q^2. After scaling this term to the dimensionless form used in the computations then the dimensionless Q_1 used in the waveguide calculations is the order of 1 whilst in the Voigt configuration in the bulk the dimensionless Q_1 is the order of 10^{-4}. In other words it can be neglected. To make the Faraday case simpler to handle only equation (25) will be used which is a description of a circularly polarised beam, arrived at by a transformation of the original Faraday equations to rotating coordinates. The approach to the computations will still be based upon using a broad input beam and numerical experiments that use equation (25), solved with a variable nonlinear gain ϵ.

A circularly polarised beam, containing two optical vortices of topological charge equal to -1, is modelled at the input plane by

$$A(r, z=0) = A_0\left((x+5) + \mathrm{i}(y+5)sgn(m)\right)^{|m|} \exp\left(-\left(\frac{r_1}{r_0}\right)^2\right)$$

$$+A_0\left((x-5) + \mathrm{i}(y-5)sgn(m)\right)^{|m|} \exp\left(-\left(\frac{r_1}{r_0}\right)^2\right)$$

$$r_1 = \sqrt{(x+5)^2 + (y+5)^2}\ ; \quad r_2 = \sqrt{(x-5)^2 + (y-5)^2}\ ; \quad r_0 = 5\ ; \quad m = -1$$

If there is no magnetic field, the pair of optical vortices attract each other until they combine and form a stable optical vortex, centred at $(0, 0)$. Figure 9 shows its intensity plot and Fig. 10 displays the phase and the interferogram. This is an example of the conservation of the angular momentum.

If the magnetic field distribution $Q_1 = 2.34\ \tanh\left(\frac{x}{12}\right) / \cosh\left(\frac{x}{6}\right)$ is applied then, for the same propagation distance, the two vortices disappear and the original light beam splits into two smaller diameter ones.(see Fig. 11). These soliton-like beam are stable, and the magnetic field can control these particles.

A simulation using the same parameters but with the vortices slightly further apart at $(6, 6)$ and $(-6, -6)$ maintain their integrity, as shown in Fig. 12a. The aim of this simulation is to have the vortices far enough apart to avoid union, as in the case $Q_1 = 0$. Figure 12b shows the effect of the

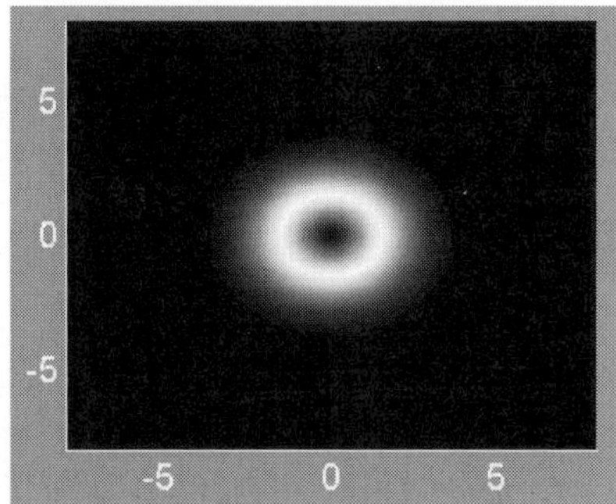

Fig. 9. Intensity at $z = 120$ shows the result of the union of the two initial vortices. $\epsilon = 2.5$, $\beta = 0.5$, $\nu = 0.1$, $\mu = 1$, $Q_1(x) = 0$

Fig. 10. Phase and interferogram for the merged vortex case

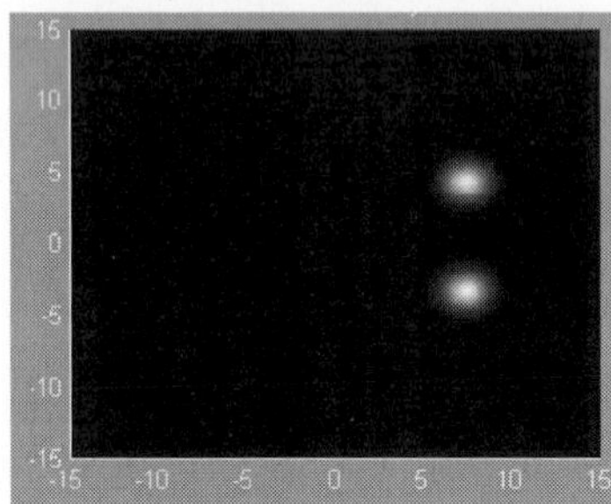

Fig. 11. Propagation to $z = 120$ shows that two soliton-like beams develop

magnetic field. This result is similar to that found when the two vortices were close enough to interact. Clearly the magnetic field has two effects. Firstly, it eliminates the singularity of the beam, transforming it into bright soliton-like beams. Secondly, the bright-solitons can be stimulated to move by the action of the magnetic field.

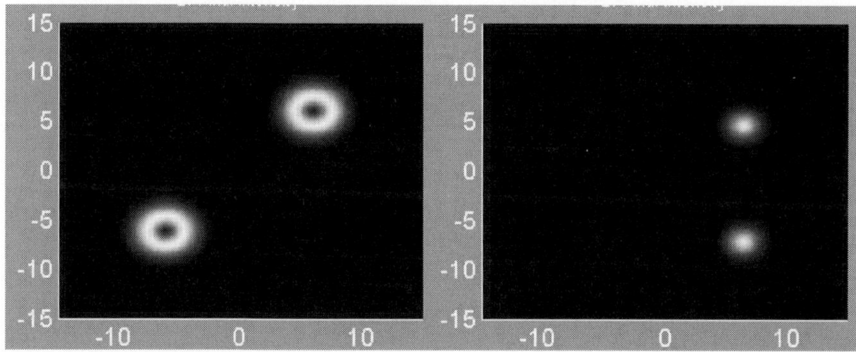

Fig. 12. Intensity at $z = 120$, parameters are $\beta = 0.5$, $\delta = 0.5$, $\nu = 0.1$, $\mu = 1$ and $\epsilon = 2.5$ (**a**) no magnetic field applied, notice the vortex integrity is maintained (**b**) magnetic field is ON

References

1. E. Hecht and A. Zajac, *Optics*, (Addison-Wesley, MA, USA, 1974).
2. J. Petykiewicz, *Wave optics*, (Kluwer, Dordrecht, The Netherlands, 1992).
3. B. E. Saleh and M.C. Teich, *Fundamentals of photonics*, (Wiley, New York, NY, USA, 1991).
4. F. A. Jenkins and H. E. White, *Fundamentals of optics*, (McGraw-Hill, New York, NY, USA, 1976).
5. D. D. Stancil, IEEE J. of Quantum Electronics, **27**, 61 (1991).
6. A. K. Zvezdin and V. A. Kotov, *Modern magneto-optics and modern magneto-optic materials*, (Institute of Physics Publishing, Bristol, UK, 1997).
7. I. V. Lindell, A. H. Sihvola, S. A. Tretyakov and A. J. Viitanen, *Electromagnetic waves in chiral and bi-isotropic media*, (Artech House, Boston, MA, USA, 1994).
8. F. Jonsson and C. Flytzanis, Phys. Rev. Lett, **82**, 1426 (1997).
9. M. Haddad, F. Jonsson, R. Frey and C. Flytzanis, Nonlinear Optics, **25**, 251 (2000).
10. V. Afanasjev, N. Akhmediev and J. M. Soto-Crespo, Phys. Rev. E, **53**, 1931 (1996).
11. N. Akhemediev and A. Ankiewicz, *Solitons nonlinear pulses and beams*, (Chapman & Hall, London, UK, 1997).
12. A. D. Boardman and M. Xie, Spatial solitons, (Springer Series in Optical Sciences, 2002), pp. 417–432.
13. A. D. Boardman and K. Xie, J. Opt. Soc. Am. B, **14**, 3102 (1997).
14. A. D. Boardman and M. Xie, *Magneto-optics: A Critical Review in Introduction to complex mediums for optics and electromgnetics*, W. S. Weiglhofer and A. Lakhtakia (SPIE Press, PM 2003).
15. A. D. Boardman and M. Xie, J .Opt. B: Quantum semiclassical optics, **3**, 5244 (2001).
16. L. C. Crasovan, B. A. Malomed and D. Michalache, Phys. Rev. E, **63**, 016605 (2000).

References

Dissipative Solitons in Semiconductor Optical Amplifiers

E. Ultanir[1], G.I. Stegeman[1], D. Michaelis[2], C.H. Lange[3], and F. Lederer[3]

[1] School of Optics and CREOL, University of Central Florida, 4000 Central Florida Blvd. P.O. Box 162700, Orlando, FL 32816-2700, USA
eultanir@mail.ucf.edu
george@creol.ucf.edu

[2] Fraunhofer Institute of Applied Optics and Precision Mechanics, Albert-Einstein-Strasse 7, 07745 Jena, Germany
dirk@physse.nlwl.uni-jena.de

[3] Friedrich-Schiller-Universität Jena, Max-Wien-Platz 1, 07743 Jena, Germany
c.h.lange@uni-jena.de
pfl@uni-jena.de

Abstract. We present recent experimental and numerical results for dissipative propagating spatial solitons in periodically-patterned semiconductor optical amplifiers (SOAs). These devices are designed to suppress the destabilization of solitons due to the amplification of noise in the soliton tails in uniformly-pumped SOAs. We briefly describe the fabrication of these devices. The zero-parameter and non-local characteristics of the solitons are studied experimentally and compared with simulations. We have also investigated soliton interactions, accounting for the effects of non-locality and zero-parameter properties.

1 Introduction

Until recently, soliton research, especially in the optical domain, has been focused on the now well-known conservative soliton systems where only the diffraction in space, or dispersion in time, needs to be balanced by some robust beam narrowing mechanism [1]. This usually has led to families of solitons which depend, not only on the properties of the physical system, but also on non-trivial soliton parameters which are determined by the excitation conditions. For example, Kerr solitons with different widths or peak intensities can be excited in the same optical system, where broader solitons exhibit smaller peak intensities and vice versa. Quadratic solitons can be regarded as an example of a two-parameter family. Here, the propagation constant and the transverse velocity of the soliton define different characteristics of the family.

A new field which has emerged in the last few years is that of solitons in systems with gain and loss. To achieve these "dissipative" solitons, the systems require loss to be balanced by gain [2, 3], in addition to balancing diffraction with self-focusing. (This is relatively obvious, since stationary soliton solutions are required, i.e. ones whose field distribution does not change

E. Ultanir, G.I. Stegeman, D. Michaelis, C.H. Lange, and F. Lederer: *Dissipative Solitons in Semiconductor Optical Amplifiers*, Lect. Notes Phys. **661**, 37–54 (2005)
www.springerlink.com

with distance.) Such dissipative systems have many unique properties which differ from those of their conservative counterparts. For example, some solutions are "zero-parameter" solitons, i.e. the soliton properties are independent of the intensity or width (or any other optical property) of the input beam. Their properties are completely determined by the external parameters of the optical system, e.g. the gain and its spatial distribution. Their stability is determined by the details of the operating conditions [4, 5, 6]. One approach has been to use multiple round trips in a cavity, or a re-circulating loop [7]. Frequently, multiple discrete elements supply the different functions of gain, loss, nonlinearity and diffraction. Our interests lie in freely-propagating dissipative solitons, so that gain, saturation etc., need to be localized on a single chip. Of the available amplifying media, electrically-pumped semiconductor optical amplifiers (SOAs) are an attractive choice because their technology is very well-developed and their physics is well understood. The principal wavelengths at which amplifiers exist usually coincide with technologically important applications of photonics. This enhances the potential for such solitons, with their special properties, to find their own niches for applications. Moreover, semiconductor active nonlinearities require low powers ($\sim$10 mW), but at the same time have fast relaxation times (ps).

There have been previous reports of spatial solitons in SOAs [8, 9]. In those cases, the gain was current-pumped uniformly along the propagation path. Detailed considerations, described here later in Sect. 2, show that this structure provides only unstable solitons. Although self-trapping was indeed reported in these cases, the distance propagated, in terms of diffraction lengths, was quite limited and the detailed simulations in Sect. 2 show that the beams eventually fall apart into filaments. The key problem is that, in a typical, uniformly-pumped SOA, the gain saturates and the small signal gain experienced in the evanescently-decaying tails of the soliton is then larger than that experienced by the soliton peak. This destabilizes the soliton via preferential amplification of the noise in the tails [10].

In Sect. 3, we will show that using electrical pumping which is periodic along the propagation direction does produce the right conditions for stable solitons. The net response in the soliton tails is negative gain, i.e. loss, so that they cannot grow to destabilize the soliton. Nevertheless, there is sufficient gain supplied to the soliton for stationary propagation when integrated over the complete soliton field distribution. The fabrication of a suitable sample is discussed in Sect. 4. The experimental details of the soliton excitation, the measurement of the soliton properties, i.e. its zero-parameter characteristics, and a comparison between experiment and theory are given in Sect. 5. Section 6 is devoted to soliton interactions, in both experiment and theory. It is shown that the non-local nature of the gain profile and the zero-parameter properties associated with these solitons lead to some unique results for their interactions. Finally, in Sect. 7, we summarize our results and discuss some interesting problems to be addressed in the future.

2 Solitons in Uniformly-Pumped SOAs

An SOA represents a typical nonlinear dissipative system. Dissipative solitons (or auto-solitons) are generated by a balance between linear and nonlinear effects, including loss and/or gain. Their regions of stability can be estimated from plane wave (PW) bifurcation diagrams that describe the steady-state solutions of the equations describing the coupled electric field-carrier density by changing the relevant control parameter, e.g., the system net gain [4, 5, 6]. Such bifurcation diagrams can be plotted, both for the stationary PWs and for the soliton solutions. It has been shown previously that stable solitons may potentially exist around critical points of sub-critical PW bifurcations [4] because the background (potential low-power tails of a soliton) experiences loss, thus suppressing its destabilization. It is this philosophy that we use to establish the conditions under which spatial solitons in SOAs can be potentially stable. This is then verified with beam propagation method (BPM) studies of beam evolution in these structures.

We model the field evolution in a uniformly-pumped SOA by using two coupled equations [11] for the carrier density (N) and the TE-polarized electric field (ψ). The paraxial wave equation, with resonant semiconductor nonlinearities, is coupled to the carrier rate equation which includes carrier diffusion, and spontaneous and Auger recombination. Assuming a linear dependence on the carrier density N' of the induced refractive index and gain (for bulk semiconductors $\sim a(N' - N'_{\mathrm{tr}})$, where a is the gain parameter and N'_{tr} is the carrier density at transparency), the normalized equations for the optical field ψ and normalized carrier density $N = N'/N'_{\mathrm{tr}}$ read:

$$
\begin{aligned}
\frac{\partial \psi}{\partial z} &= \frac{\mathrm{i}}{2}\frac{\partial^2 \psi}{\partial x^2} + \mathrm{i}[-hN - \mathrm{i}(N-1)]\psi - \alpha\psi \\
D\frac{\partial^2 N}{\partial x^2} &+ \pi(z) - N - BN^2 - CN^3 - (N-1)|\psi|^2 = 0
\end{aligned}
\tag{1}
$$

Here h is the linewidth enhancement (Henry) factor, α is the internal loss, which includes scattering losses and the cladding layer absorption, $\pi(z) = \pi_0$ is the uniform current pumping coefficient, D the carrier diffusion coefficient, B the spontaneous recombination coefficient and C the Auger recombination rate. The non-radiative recombination time is scaled to unity. For simplicity, carrier diffusion along the propagation direction z is neglected. We now show analytically that an isolated SOA cannot support stable self-trapped beams via their bifurcation diagrams. We assume $B = C = 0$ because these terms do not change the type of bifurcation and therefore the stability for an amplifying system. For small diffusion, near the bifurcation point (for definition, see below) we obtain a generalized complex Ginzburg-Landau equation:

$$
\frac{\partial \psi}{\partial z} = \mathrm{i}\left(\frac{1}{2} - \mathrm{i}\beta\right)\frac{\partial^2 \psi}{\partial x^2} + \left(\frac{\pi_0 - 1}{1 + |\psi|^2}(-\mathrm{i}h + 1) - \mathrm{i}h\right)\psi - \alpha\psi \tag{2}
$$

Here β accounts for the transverse carrier diffusion and the canonical Ginzburg-Landau equation, frequently used to describe beam propagation in an SOA or a laser diode gain medium, is recovered by expanding the denominator. Fortunately, the solutions of this type of equation are well-known and provide considerable insight, including the stability of the solutions.

To obtain the PW bifurcation behavior of the system, we assume a one-dimensional plane wave solution, i.e., $\psi = \psi_0 \exp(ikz)$, where k is the propagation constant. The small signal net gain of the system, which will serve as a control parameter, is $\delta G = \pi_0 - 1 - \alpha$. Equation (2) has two solutions for $|\psi_0|$, as shown in Fig. 1. The trivial one $|\psi_0| = 0$, and $|\psi_0| = \sqrt{\delta G/\alpha}$ for $\delta \mathrm{G} > 0$, require the dispersion relation $k = -h(\pi + |\psi_0|^2)/(1+|\psi_0|^2) = -h(1+\alpha)$ to be satisfied. Linear stability analysis reveals that the trivial solution destabilizes for $\delta \mathrm{G} > 0$, but shows the non-trivial solution to be stable for homogenous perturbations. Thus, stationary non-zero PW solutions in a conventional SOA bifurcate super-critically ($\partial|\psi_0|^2/\partial\delta \mathrm{G} > 0$) at $\delta \mathrm{G} = 0$ (bifurcation point). In a similar way, one can show that a soliton also bifurcates from the critical point super-critically. The consequence for a soliton or self-trapped beam formation is that a positive net gain is required, but, in turn, this gain will destabilize the soliton tail (growth of amplified spontaneous emission – ASE).

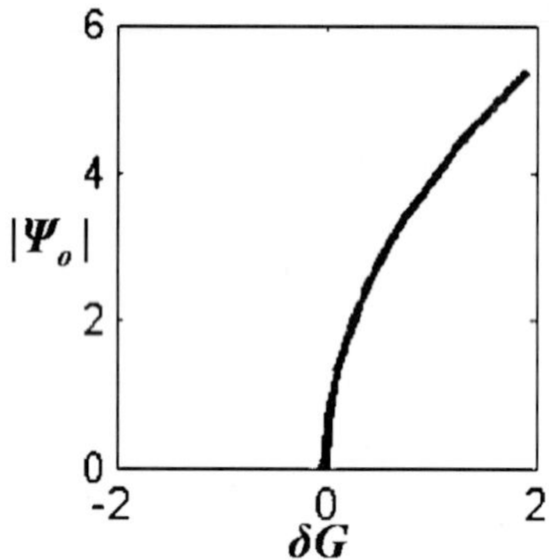

Fig. 1. Super-critical bifurcation behavior in a commercial SOA device

One way to check for the *stability* of these finite beam solutions is to follow their evolution in space with a beam propagation code. This is demonstrated in the simulation of the field evolution in Fig. 2, calculated using (1). An input beam corresponding to the shape of a soliton is used as the initial condition. This soliton solution $\psi(x)\exp(i\lambda z)$ is calculated numerically using a relaxation algorithm, where $\psi(x)$ is the soliton shape and λ is its propagation constant. The distance required before filamentation sets in depends on the noise in the incident optical beam, and the numerical noise generated by the

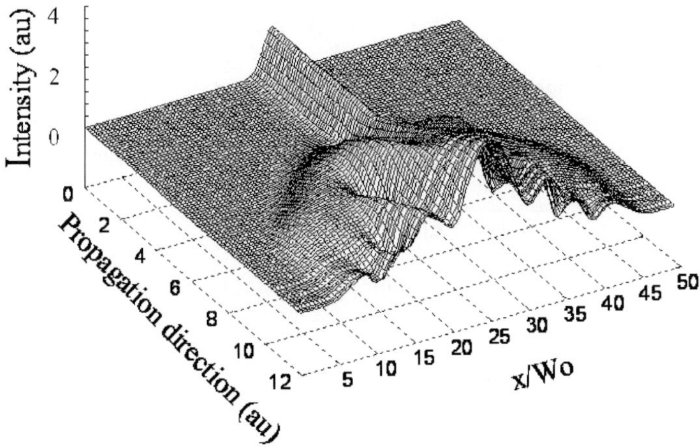

Fig. 2. Beam propagation simulations of the soliton solution $\psi(x)\exp(i\lambda z)$ in a uniformly-pumped SOA

simulator. In conclusion, for δG < 0, the fields are absorbed and for δG > 0, the localized solutions are unstable due to the growth of any noise present.

These types of SOA structures have been analyzed theoretically [12, 13] and experimentally [14] in the past. Previous experimental searches for spatial solitons [8, 9] in such uniformly-pumped SOAs did indeed find self-trapping. Their analysis neglected the gain-loss trade-off, and essentially assumed only a Kerr response. In both cases, the propagation distance, typically one diffraction length, was not sufficient to verify stable soliton formation.

3 Theory of Solitons in Periodically-Pumped SOAs

The key idea, in analogy with the temporal case [7], is to introduce a form of saturable absorption for noise reduction in the system. This can be achieved by introducing periodic losses into an SOA thus forming an SOA/SA module with periodic electrodes. (SA means “saturable absorber”). The structure shown in Fig. 3 is used to control the relative contributions of gain and loss. Here region 1 corresponds to the semiconductor under the pumping electrodes and region 2 to the un-pumped (absorbing) regions. For the Quantum Well (QW) samples actually used in the experiments, the nonlinear gain can be described phenomenologically by [15]

$$f(N) = \ln\left(\frac{NN_{\mathrm{tr}} + N_s}{N_{\mathrm{tr}} + N_s}\right) . \tag{3}$$

Its value, averaged over the pumped and un-pumped regions, is

$$\overline{f}(N_1, N_2) = [f(N_1)w_1 + f(N_2)w_2]/(w_1 + w_2) , \tag{4}$$

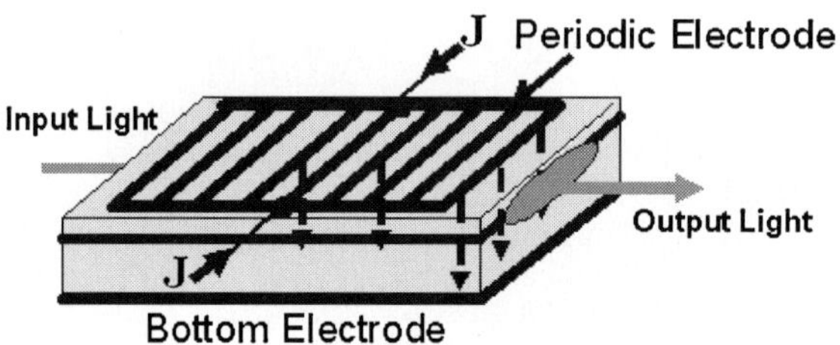

Fig. 3. A periodically-patterned semiconductor optical amplifier (PPSOA). The vertical arrows indicate current flow

where w_1 and w_2 are the widths of the pumped and un-pumped regions, respectively. In a first-order approximation, an averaged evolution of the electric field can be modeled by

$$\begin{aligned} \psi_z &= \frac{\mathrm{i}}{2}\psi_{xx} + \psi[\overline{f}(N_1, N_2)(1 - \mathrm{i}h) - \alpha] \\ DN_{1xx} + \pi_0 - BN_1^2 - CN_1^3 - f(N_1)\,|\psi|^2 &= 0 \\ DN_{2xx} - BN_2^2 - CN_2^3 - f(N_2)\,|\psi|^2 &= 0 \end{aligned} \tag{5}$$

The full system of equations (5) was solved for stationary PWs (no transverse dependence) with periodic pumping ($\pi(z) = 0$ for absorber regions and $\pi(z) = \pi_0$ for amplifier regions). Again, a trivial PW solution, destabilizing at the bifurcation point $\delta\mathrm{G} = 0$, where now $\delta\mathrm{G} = 0.5\pi_0 - 1 - \alpha$, was found, just as for the uniform pumping case. (Here it was assumed that the pumping region had the same length as the un-pumped region, i.e. $w_1 = w_2$.) However, in contrast to the previous case, now the non-trivial solution emanates sub-critically $(\partial|\psi_0|^2\ /\partial\delta\mathrm{G} < 0)$ from $\delta\mathrm{G} = 0$. Thus, in a limited region ($\delta G_{min} < \delta G < 0$), there are two PW solutions and they provide a possible prerequisite for stable soliton formation.

The key for understanding this behavior is the fact that the pumped and the un-pumped regions saturate at different intensity levels, resulting from the nonlinear dependence ($f(N)$, spontaneous recombination $\propto N^2$ and Auger recombination $\propto N^3$) of the saturated carrier density on the field. Fig. 4a shows the difference in the saturation curves for both the pumped and un-pumped regions of a GaAs SOA/SA module that lases at 0.82 µm. For the simulations, we assumed a non-radiative recombination time of 5 ns, a carrier diffusion constant of 33 cm^2/s, a gain parameter of $1.5 \times 10^{-16}\,\mathrm{cm}^2$, a Henry factor ($h$) of 3, a spontaneous recombination coefficient of $1.4 \times 10^{-10}\,\mathrm{cm}^3/\mathrm{s}$, an Auger coefficient of $1.0 \times 10^{-30}\,\mathrm{cm}^6/\mathrm{s}$ and an internal loss (α) of $2.5\,\mathrm{cm}^{-1}$.

Figure 4a shows that, because of saturation, the gain (along the pumped region) and loss (along the un-pumped region) are functions of input intensity. The intensity-dependent net gain (solid line in Fig. 4a) is the sum of these two contributions, including the linear loss α. (Again, each region is

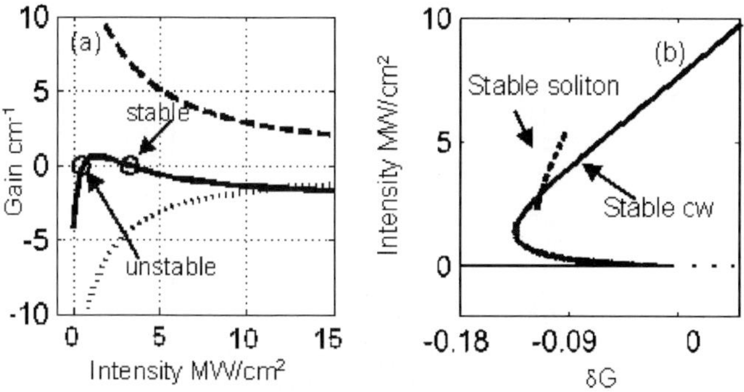

Fig. 4. (**a**) Gain and loss of the system vs. field intensity. The *dotted* curve shows the saturating loss in the non-pumped region, and the *dashed* curve is the saturated gain in the pumped region. The *solid* line represents the sum of these two contributions and the internal linear losses of the system. (**b**) PW and soliton bifurcation diagrams vs. δG. The *thick solid* line shows the PW solutions that sub-critically bifurcate from the trivial solution. The solution with negative slope is unstable between zero gain and the turning point. The *dashed* line indicates the peak intensity of the stable soliton. This branch terminates because of Hopf instabilities. The *dotted* line is the unstable trivial PW solution at positive small signal gain, and the thin *solid* line gives the stable zero PW solutions. The fields shown here are averaged along the pumped and un-pumped segments

assumed to be of equal length). As the pumping current is increased, the net gain vs. intensity curve has two zero-gain crossings, which correspond to two points on the sub-critical curve of Fig. 4b. The crossing at lower intensity corresponds to an unstable stationary solution, since small perturbations will cause it to evolve towards other stable points, i.e., to the trivial solution or the crossing point at high intensity. The latter one is stable because any small perturbation will eventually relax the system back to zero gain. The whole suite of these solutions is displayed in Fig. 4b, i.e. the PW intensity as a function of the small signal net gain that is controlled by the pump current. This PW bifurcation diagram clearly shows the required sub-critical bifurcation.

Three qualitatively different regions can be identified in that plot. For $\delta G < \delta G_{min} \approx 0.134$, the SOA/SA module absorbs any incident PW field. Increasing the pumping current results in $\delta G_{min} < \delta G < 0$, and the system becomes bistable (stable upper and lower branches, unstable negatively sloped branch). If the pumping current is further increased ($\delta G > 0$), the bistability of the PWs is lost because the trivial solution becomes unstable.

General analytic solutions for the dissipative solitons in this system are not possible. We have numerically simulated the stable soliton solutions of the SOA/SA module (1) shown in Fig. 4b by using a beam propagation (BPM) code, and also calculated all the soliton solutions (stable and unstable) by a relaxation code for an InGaAs QW sample in Fig. 5. (The parameters for

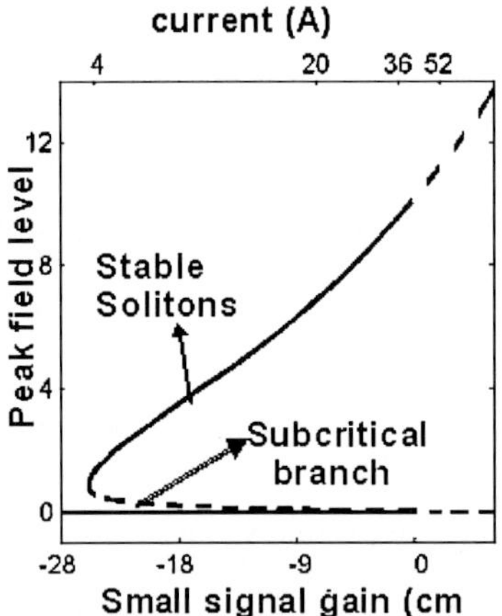

Fig. 5. Plot of peak field level ($\sqrt{\text{mW}/\mu\text{m}}$) of soliton solutions versus small signal gain δG. The *dashed* lines show the unstable solutions, and the *solid* line shows the solitons on a stable background. Current values (× axis-top scale) are calculated assuming 300 µm width contact patterns on a 1 cm long device. (Current $I = \pi q d N_{\text{tr}} A_{contact}/\eta$, where $A_{contact} = 300\,\mu\text{m} \times 1\,\text{cm} \times 0.5$)

the QW simulation are $h = 3$, $\alpha = 0.024$ ($\alpha = \alpha_{\ real}/2\text{g}$) where g is the gain coefficient in QW, $N_{\text{tr}} = 2.5 \times 10^{18}\,\text{cm}^{-3}$, $N_s = 0.6 \times 10^{18}\,\text{cm}^{-3}$, d = 60 Å is the width of the QW region, g = $52\,\text{cm}^{-1}$, $\eta = 0.85$ is internal quantum efficiency, D = 3.311 ($\text{D} = \text{D}_{real} 2\pi \text{ng}/\lambda_0$), B = 3.0 ($\text{B} = \text{B}_{real} N_{\text{tr}}$) and C = 0.219 ($\text{C} = \text{C}_{real} {N_{\text{tr}}}^2$) [15, 16]).

The patterned amplifier was assumed to have 10 µm contact layers with a 20 µm period. It turned out that stable solitons existed near the upper branch of the sub-critically bifurcating PW solution (see Fig. 4b dashed line, the soliton peak intensity curve), since the soliton solution started to breathe, indicating that it may undergo a Hopf bifurcation. In any case, we could numerically verify that, in a certain domain of net losses, stable soliton propagation could be achieved.

These dissipative solitons exhibit some interesting properties. As would be expected from the theory of dissipative solitons [7, 17], the stable SOA spatial solitons exhibit curved (chirped) phase fronts (see Fig. 6a), in contrast to the flat phases associated with conservative solitons. Although the phase of the soliton is highly chirped, BPM simulations have shown that these SOA spatial solitons evolve from a Gaussian-shaped beam with a flat phase input after a relatively short propagation distance of about 1–2 diffraction lengths [see Fig. 6b]. The input power in this case was about 180 mW. This finding is not surprising because zero-parameter dissipative solitons represent strong attractors in parameter space.

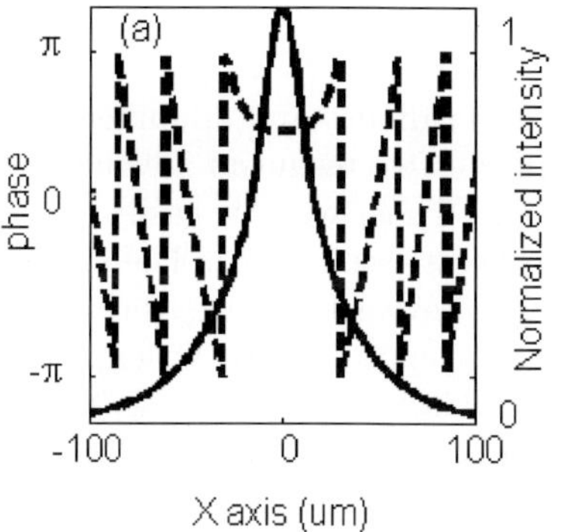

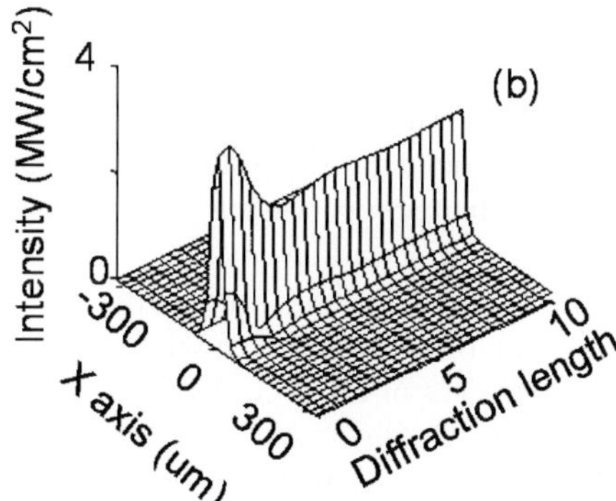

Fig. 6. (**a**) Soliton amplitude (*solid line*) and phase profile (*dashed line*) (**b**) Evolution at $\delta G = -0.106$ of a 179.2 mW total power input beam with a constant phase, sech^2 intensity distribution and 37.5 µm beam waist (FWHM) into a 24.5 µm beam waist (FWHM), 181.4 mW spatial soliton, using the normalized parameters (B = 0.7, C = 0.005, D = 1.56, $\alpha = 0.067$)

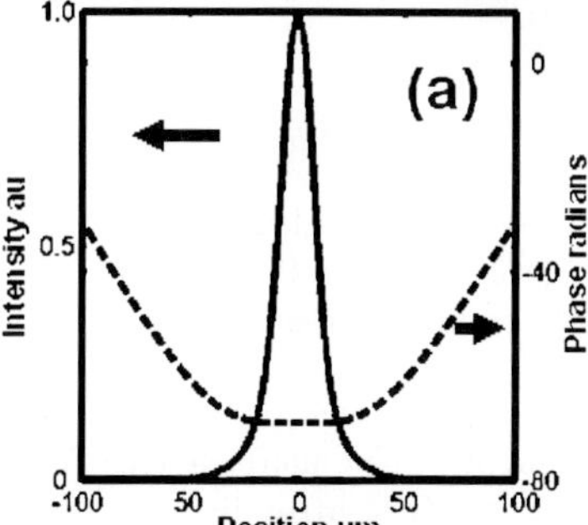

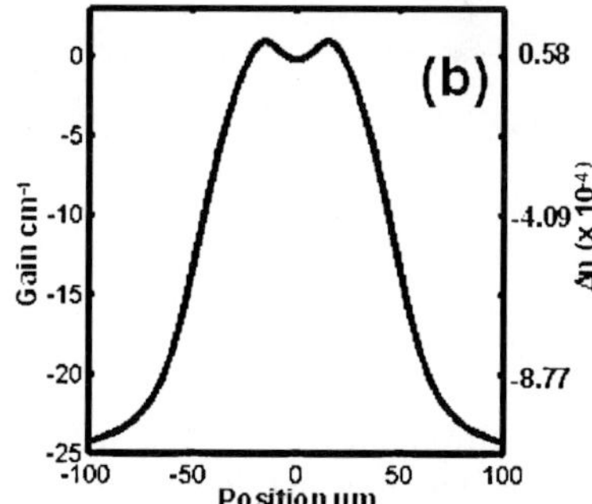

Fig. 7. (**a**) Calculated soliton intensity profile and phase at 4 A (amperes) electrical pumping with the corresponding change (**b**) in the gain and refractive index in the waveguide. (The gain and refractive index changes are calculated with respect to the case when there is no current in the device). Gain is negative around the soliton peak and under the soliton tails, and this stabilizes the soliton and the background. (Current $I = \pi q d N_{\mathrm{tr}} A_{contact}/\eta$, where $A_{contact} = 300\,\mu\mathrm{m} \times 1\,\mathrm{cm} \times 0.5$)

Another important property is the non-locality of the spatial distribution of the gain and the nonlinear index change. This is shown below for a dissipative soliton in Fig. 7.

Note that the optically-induced changes in the optical properties extend 2–3 times further in the medium than the intensity distribution. This will impact on the interactions between dissipative solitons discussed in Sect. 6. Interactions between two solitons can occur even in the absence of significant overlap of their optical fields.

4 Sample Fabrication

Nowadays, the fabrication of semiconductor laser diodes and amplifiers has become a routine procedure, as a result of impressive technical advances in semiconductor processing in the last 20 years. Thus, we can easily fabricate our semiconductor devices in a 3-step process (see Fig. 8). The spatial solitons were investigated experimentally in a device fabricated on a p-n junction diode wafer grown by molecular beam epitaxy. The wafer structure consists of an n-GaAs substrate with an n-GaAs buffer layer, a 100 nm graded n-$Al_xGa_{1-x}As$ (x = 0.2–0.36) layer, a l_m n-$Al_{0.36}Ga_{0.64}As$, and an 80 Å $In_{0.17}Ga_{0.83}As$ layer sandwiched between two symmetric waveguiding layers of 500 nm $Al_{0.2}Ga_{0.8}As$. Then l_m p-$Al_{0.36}Ga_{0.64}As$, 100 nm graded p-$Al_xGa_{1-x}As$ (x = 0.36–0.2) and 400 nm p-GaAs contact layers are deposited on top. The peak electro-luminescence was found to occur at 946 nm at room temperature.

Fig. 8. Sample fabrication: (*left*) p-n junction wafers are cleaved into 1 cm long samples; (*middle*) SiN coating with PECVD; (*right*) lithography of patterned regions, etching SiN regions and metallization of *top* and *bottom* layers

The sample is first coated with an insulator (SiN) using a PECVD machine, and the contact layers are etched on the SiN after lithography. The top and bottom layers of the sample have been metallized with Au-based alloys for ohmic contacts. Finally, the samples are cleaved to 300 μm-wide by 1 cm-long pieces, and mounted on a copper stud for thermal management and light coupling.

5 Experiment Results

The mounted samples are driven by a pulsed current source (400 ns pulses at 1 kHz repetition rate), in order to avoid thermal effects. Stable temperatures (~21°C) for the device in repeated experiments are guaranteed by mounting the samples on a thermo-electrically-cooled copper mount. Light input from a CW Ti:sapphire laser is shaped with a 2 lens elliptical telescope. We used a 40x microscope objective to couple the laser beam into the waveguide. Both focal regions (orthogonal (y-axis) and parallel (x-axis) to the waveguide) are carefully aligned to overlap after the microscope objective at the input facet. This gives maximum coupling of light into the waveguide and a planar phase

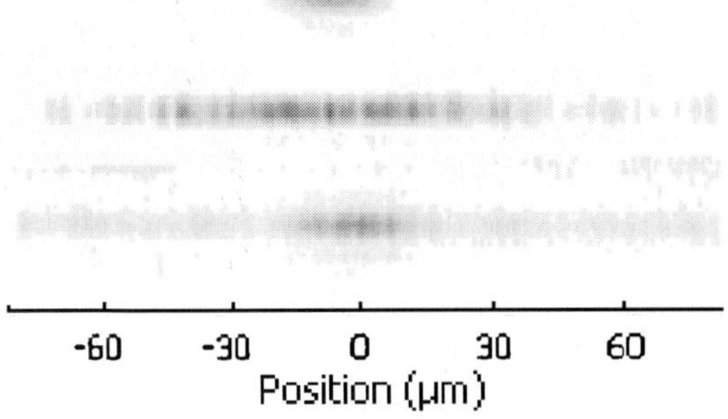

Fig. 9. (*Top*) Input beam profile at $\lambda = 965\,\text{nm}$, Gaussian fitted FWHM = 16.5 µm. (*Middle*) Diffracted beam (62.5 µm FWHM) at the output facet. (*Bottom*) Output beam profile at $\lambda = 950\,\text{nm}$, and 4 A current injection

front at the input. The device is tilted about 2 degrees from normal incidence to avoid any back-reflections. The output of the sample is imaged onto a CCD camera with a microscope objective.

When a 16.5 µm FWHM beam is coupled into the waveguide at a wavelength of 965 nm, where the material is transparent, we observed diffraction to 62 µm FWHM as a result of the 3.6 diffraction lengths of linear propagation. As the wavelength is tuned inside the bandgap ($\lambda < 955\,\text{nm}$), and current is injected into the device, a localized spatial output beam with a FWHM of 21 µm at 950 nm with 4 A of current is seen (see Fig. 9). Figure 10a shows the formation of such a localized beam. The device absorbs all the light until the current reaches 3.8 A. Beyond this point, the system follows the stable sub-critical bifurcation branch (see Fig. 5), and the peak intensity of the localized beam increases. The output profile does not show any formation of noise filaments up to 4.6 A. The striations observed in the experimental pictures were found to be the effect of sample defects inside the waveguide. The sample facets were checked for defects by imaging after cleaving, and only perfectly-cleaved samples have been used. It was found that the stripe locations on the output image move with the sample when the sample is shifted perpendicularly to the laser's propagation axis. Therefore, they are a consequence of localized intrinsic defects in the waveguide. At higher currents, the amplified spontaneous emission increases above the unstable part of the sub-critical branch, and it generates noisy peaks on the sides of the soliton. We calculated the progressive evolution of the light beam with distance for the patterned amplifier device by using a finite difference beam propagation method (FDBPM) based on equation (5) (using $f(N_{1(2)})$ for each individual region, instead of an average nonlinear gain). The simulations also support the bifurcating behavior of the solitons with the corresponding threshold pumping current, Fig. 10b, which indicates the turning point in the bifurcation diagram [18].

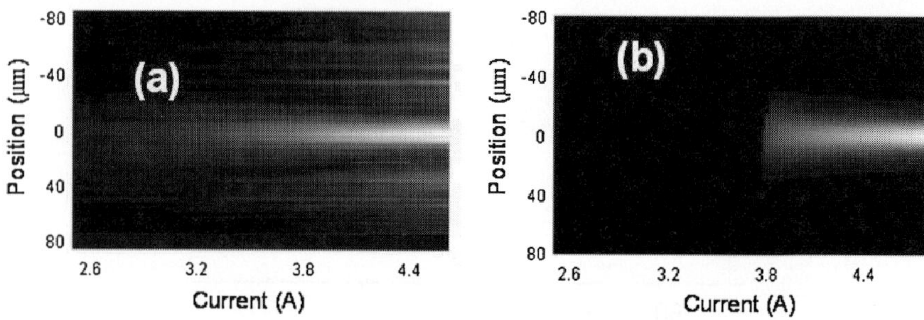

Fig. 10. (**a**) Collage of images (*vertical stripes*) from output facet when the measured input at the focus of the microscope objective is 160 mW & 16.5 μm FWHM. Note that losses, due to Fresnel reflection and mismatch between the Gaussian input beam and the waveguide mode, reduce the power coupled into the waveguide. (**b**) Numerical simulation of the output profile with a 55 mW, 17.5 μm FWHM Gaussian assumed just inside the input facet of the waveguide

Next, we will confirm that the spatial soliton indeed has the properties associated with dissipative solitons. To this end, we will show, for fixed values of the system parameters, that the soliton is a stable attractor and that no families of solitons exist. In the region of attraction, an arbitrary initial input beam will always converge towards one distinct soliton. We increased the input power by using a half-wave plate and a polarizer, and took pictures of the output – see Fig. 11a. Above 60 mW input power, the beam forms a localized structure that does not change its shape or intensity, even though the input power is increased, as expected for dissipative solitons.

Numerical simulations with FDBPM in Fig. 11b show that, once the intensity exceeds the threshold sub-critical branch value for a given pumping current, the soliton does not change its intensity or width, as observed in the experiment. We have also analyzed the change in output beam waist by

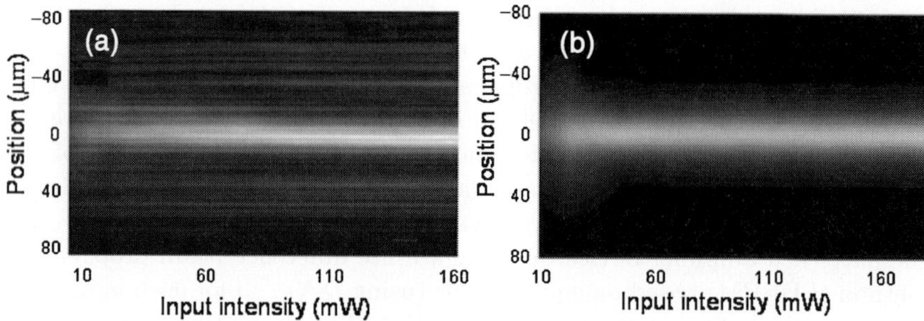

Fig. 11. (**a**) Images from the output facet with 4 A pumping. (**b**) Numerical simulation of solitons at 4 A pumping when the input intensity at the front facet is increased (50% coupling is assumed)

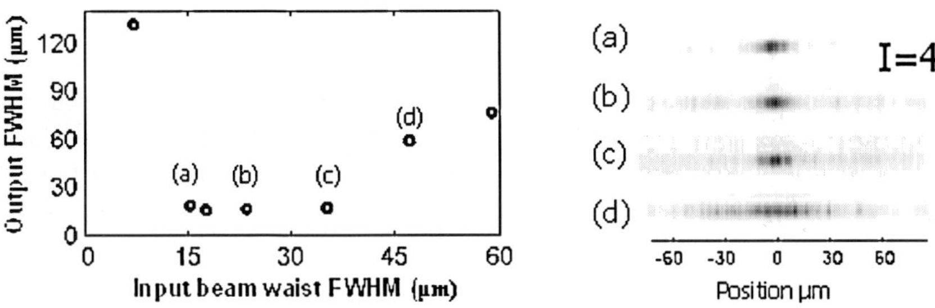

Fig. 12. Output beam waist versus the waist of a Gaussian input beam with 160 mW input power at front facet, and 4 A current injection

increasing the input beam waist from 7 μm FWHM to 60 μm FWHM (Fig. 12). For an input beam waist smaller than 13 μm, the beam diffracts very quickly, and numerical simulations show that it gets totally absorbed if the propagation continues beyond a device length of 1 cm. From 15 μm to 35 μm input FWHM, a stable localized soliton, showing no significant change in its output beam waist, is formed. Still larger input beam waist coupling does not give any soliton for a 1 cm device length, since the diffraction length at these beam waists is greater than 1 cm. However, numerical simulations suggest that for longer propagation distances, formation of multi-solitonic structures is possible.

We investigated 6 different devices from the same wafer. All of them gave essentially the same output images for similar conditions. Thus, we believe these solitons are very robust, since dissipative solitons are strong attractors in these systems [3].

6 Soliton Interactions

The interactions between spatial optical solitons can have very interesting applications, such as all-optical switching, logic operations and soliton junctions [19]. However, the critical issue in implementing any such soliton application is to have a medium in which the solitons can be generated with reasonable input powers (mWs), that the response time be fast (preferably in the picosecond range), and that the signal levels at the output are all restored to a common value (since these solitons belong to a zero-parameter family). Periodically-patterned semiconductor optical amplifiers (PPSOAs) not only support fast switching and low power solitons, but are also compatible with standard semiconductor processing technologies and can be integrated with other semiconductor devices for operation with optical communication networks (fibers, modulators, etc). Therefore, they have significant advantages over solitons in photorefractive media which are slow and require holding beams, or in quadratic or Kerr media which require high powers (>100W).

Here, we describe experiments on the interactions between two coherent dissipative solitons in PPSOAs. We investigate these interactions for pairs of solitons of equal power and width by changing the relative launch angle (collinear and non-collinear interactions), distance and the relative phase between them. Although the interactions exhibit some similarities with interactions in conservative systems, there are also exciting unique features such as soliton "birth" for collinear and non-collinear collisions of two solitons with zero phase difference.

We used the same experimental set-up that we used for observing the soliton (see Sect. 5). Two beams, with their relative phase controlled by a Mach-Zehnder-type interferometer, were incident on the sample. The two beams incident on the input waveguide facet had 15 μm FWHM beam waists in the waveguide plane and 70 mW powers. Spatial solitons, and their interactions in a periodic SOA/SA device, were modeled using the coupled equations for the carrier density (N) and the evolution of the TE-polarized electric field (ψ) (5). As shown in Fig. 7, the calculated soliton beam waist is much smaller than the induced gain (or index) profile associated with the soliton, thus indicating a strong non-locality [20, 21], and the potential for coherent non-local interactions with PPSOA solitons.

Firstly, we examined the interaction between parallel solitons separated by about one beam waist of 15 μm, i.e. in the local limit. The numerical simulations were done using a finite difference beam propagation code with the coupled equations (Equation (5), using the gain function $f(N_{1,2})$, but not the average gain, in each region separately during the propagation). Based on estimates of the coupling losses and 30% loss due to Fresnel reflection, we assumed 50% coupling for the simulations, and used two Gaussian inputs inside the sample, each with a 15 μm FWHM width and 38 mW power. Figure 13(left) shows the calculated output beam profiles versus relative input phase after 1 cm of propagation and 4 A pumping. The solitons fuse and are steered in angle, except where the phase difference between them is π. For π phase, the two solitons repel each other, and we can observe two peaks at the output facet.

Experimentally, the difference between the two separately-excited solitons at the output facet was measured to be 22μm, so the angle between two beams is about 0.04°. The output facet was imaged and data acquired at 200 pictures/30 sec with the piezo-controlled mirror arm of the interferometer moving in every picture. The experimental results in Fig. 13 (left) are in very good agreement with the simulations (Fig. 13(right)). The calibration of the phase difference (x-axis) for the experiment was done using the simulation results as a reference, and we used the same calibration scheme for the experiments that follow, so long as there was no change in voltage control of the piezo. The striations in the output images are as a result of interior defects in the waveguide, as discussed previously [18]. Thus, when the soliton

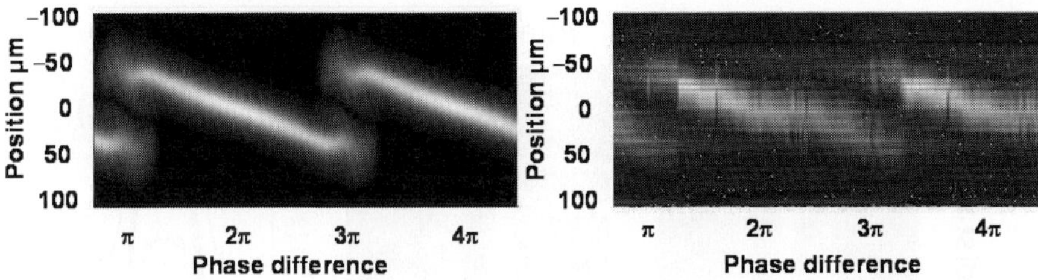

Fig. 13. Output from sample due to soliton interactions after 1 cm propagation. The separation between the input beams, when excited individually, is 15 μm and 22 μm at the output. The *left* picture shows the numerical simulations, while the *right* one shows the experimental results. The experimental results consist of a collage of different outputs (*vertical slices*) obtained as the relative input phase was varied

fields are overlapped strongly, the interaction force depends mainly on the relative phase, just as for Kerr, photo-refractive and quadratic solitons [22].

When the two input solitons are in phase, the total intensity in the area between them increases. Thus there is a local increase in the refractive index, and the beams attract each other. In contrast to this, the intensity and hence the refractive index in the overlap region decreases when the beams are out-of-phase, and the beams repel each other. However, note that, because these are zero-parameter solitons, the soliton outputs all have identical intensity, in contrast to other soliton interactions where the total energy is either conserved (1D Kerr case) or diminished due to radiative losses.

Next, we increased the separation of the beams at the input, while keeping the angle between them at almost 0°. Over a small range of input beam separations, and for zero relative phase, 3 soliton-like outputs were predicted in Fig. 14a, with one well-defined high intensity soliton traveling along the initial propagation axis and the two others diverging. The creation of the central soliton is a consequence of the non-local nature of the nonlinearity (or carrier density profile), so that the addition of the two gain profiles of the input solitons evolves with distance to form another "waveguide" region that could support a soliton (see Fig. 14c). The energy is not conserved, and this is different from conservative soliton collisions. In fact, the total energy of the output solitons is more than the total energy of input ones, a consequence of the gain in the system and the zero-parameter nature of the solitons. Note that the central soliton initially has a higher intensity than either of the satellite solitons due to the different nonlinearity and gain distributions in the center versus the wings, since the central soliton sees a gain profile which is the addition of the two side soliton gain profiles, and this results in the formation of a higher intensity soliton in the middle. Further numerical simulations showed that all 3 solitons stabilize to equal intensities (to zero-parameter solutions) when they propagate long enough, so that the gain profiles no longer overlap significantly. On the other hand, when the

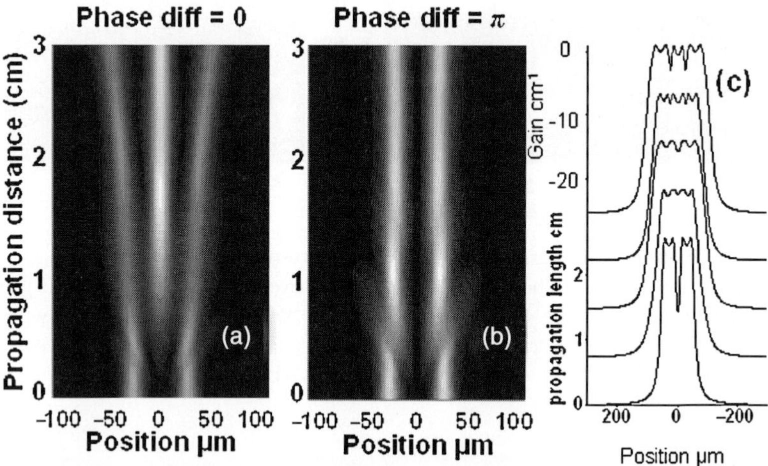

Fig. 14. Numerical beam propagation over 3 cm of soliton interactions (**a**) for 0 and (**b**) π phase difference. The separation between the input beams is 60.2 μm and each has 38 mW power. (**c**) shows the gain profile around the solitons for 0 phase difference at the input

relative input phase difference is around π, only two solitons are obtained at the output (Fig. 14b).

Figure 15 shows the experimental results for the two-soliton interaction with an initial 70 μm separation between them. The solitons are separated by multiple soliton beam waists and therefore there is a non-local nonlinear interaction between the beams. However, the phase in the solitons' tails still affects the interaction, a feature which is different from other non-local Hamiltonian interactions [20, 23]. Note that, at relative phases which are even multiples of 2π, there is evidence of three solitons being present. Further simulations and experiments have shown that non-interacting solitons

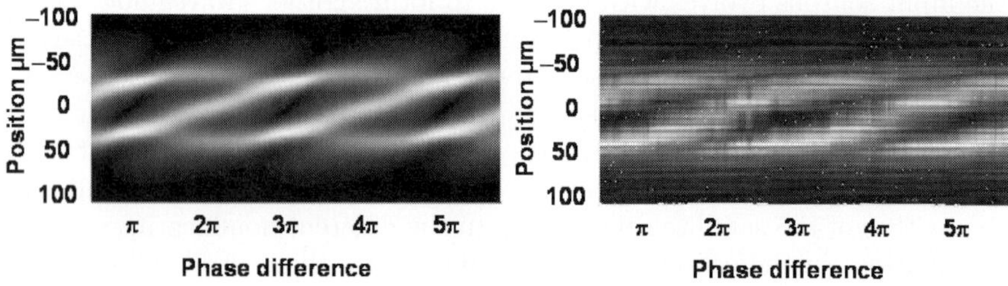

Fig. 15. Soliton interactions after 1 cm propagation. The separation between the beams is 70 μm at the input and 66 μm at the output. The *left* picture shows the numerical simulations, and the *right* one gives the experiment results

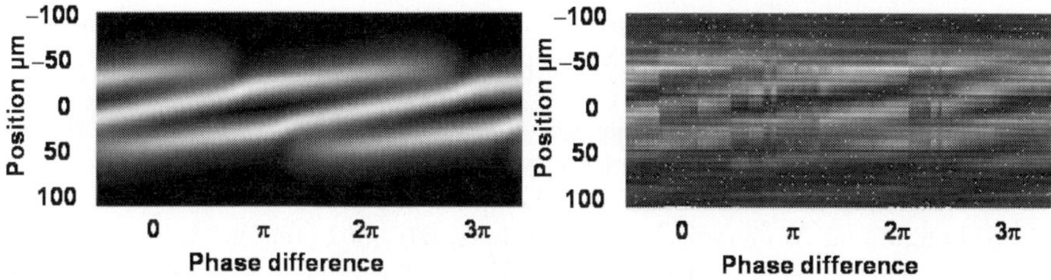

Fig. 16. Soliton interactions after 1 cm propagation. The separation between the input beams is 60.2 μm and the angle between them is 0.5 degrees. The *left* picture shows the numerical simulations, and the *right* one gives the experiment results

only appear when the separation between them is larger than ~100 μm, since the gain profiles no longer overlap significantly.

Another interesting case is the collision between incident solitons which are initially launched with crossing trajectories (i.e. non-collinear collisions). These interactions could be used in all-optically- induced junctions and logic operations [24, 25]. We have investigated the collision for a relative angle of around 0.5°. In contrast to results from non-collinear collisions in conservative systems, the birth of an additional soliton at zero phase difference has been observed in this 1D system, whereas the expected repulsion of the two solitons at π phase difference was obtained. (See Fig. 16).

7 Conclusion

We have discussed the theory and observation of propagating spatial solitons based on semiconductor active nonlinearities. These solitons could have very promising applications due to their interactions, because of their low power requirements, fast switching times and signal level restoration. Preliminary studies of interactions agree very well with the simulations. Thus, these devices could lead to all-optical switching and logic devices based on soliton interactions. We believe that, in the coming years, the spatial solitons in PPSOAs will find an important role in optical computing.

Acknowledgements

This research was supported in the US by an ARO MURI on "Gateless Soliton Computing".

References

1. S. Trillo and W. Torruellas, *Spatial Solitons* (Springer, Berlin, 2001).

2. N. N. Rosanov, *Transverse patterns in wide-aperture nonlinear optical systems*, in Progress in Optics, XXXV ed. by E. Wolf, (Elsevier Science, NORTH-HOLLAND, 1996).
3. N. Akhmediev and A. Ankiewicz, *Solitons: Nonlinear Pulses and Beams* (Chapman and Hall, London, 1997).
4. S. Fauve and O. Thual, Phys. Rev. Lett. **64**, 282 (1990).
5. S. Longhi and A. Geraci, Appl. Phys. Lett. **67**, 3062 (1995).
6. S. V. Federov, A. G. Vladimirov, G. V. Khodova and N. N. Rosanov, Phys. Rev. E **61**, 5814 (2000).
7. Z. Bakonyi, D. Michaelis, U. Peschel, G. Onishchukow and F. Lederer, J. Opt. Soc. Am. B **19**, 487 (2002).
8. G. Khitrova, H. M. Gibbs, Y. Kasamura, H. Iwamura, T. Ikegami, and J. E. Sipe, Phys. Rev. Lett. **70**, 920 (1993).
9. C. Kutsche, P. LiKamWa, J. Loehr and R. Kaspi, Electron. Lett. **34**, 906 (1998).
10. E. A. Ultanir, D. Michaelis, F. Lederer and G. I. Stegeman, Opt. Lett. **28**, 251 (2003).
11. G. P. Agrawal, J. Appl. Phys. **56**, 3100 (1984).
12. D. Mehuys, R. J. Lang, M. Mittelstein, J. Salzaman and A. Yariv, IEEE J. Quant. Elect. **23**, 1909 (1987).
13. H. Paxtion, G. C. Dente, J. Appl. Phys. **70**, 2921 (1991).
14. L. Goldberg, M. R. Surette and D. Mehuys Appl. Phys. Lett. **62**, 2304 (1993).
15. S. Kristjansson, N. Eriksson, P. Modh and A. Larsson, IEEE J. Quantum Electron. **37**, 1441 (2001).
16. D. J. Bossert, D. Gallant, IEEE Photon. Technol. Lett. **8**, 322 (1996).
17. N. N. Akhmediev, V. V. Afanasjev and J. M. Soto-Crespo Phys. Rev. E **53**, 1190 (1996).
18. E. A. Ultanir, D. Michaelis, C. H. Lange and G. I. Stegeman, Phys. Rev. Lett. **90**, 253903 (2003).
19. G. I. Stegeman and M. Segev, Science **286**, 1518 (1999).
20. A. W. Snyder and D. J. Mitchell, Science **276**, 1538 (1997).
21. C. Conti, M. Peccianti and G. Assanto, Phys. Rev. Lett. **91**, 073901 (2003).
22. IEEE J. Quant. Elect. **39** (2003); Feature issue on spatial solitons (Issue 1).
23. M. Peccianti, K. A. Brzdakiewicz and G. Assanto, Opt. Lett. **27**, 1460 (2002).
24. B. Luther-Davies and Y. Xiaoping, Opt. Lett. **17**, 496 (1992).
25. T. T. Shi and S. Chi, Opt. Lett. **15**, 1123 (1990).

Dissipative Solitons in Pattern-Forming Nonlinear Optical Systems: Cavity Solitons and Feedback Solitons

T. Ackemann[1] and W.J. Firth[2]

[1] Institut für Angewandte Physik, Westfälische Wilhelms-Universität Münster, Corrensstraße 2/4, 48149 Münster, Germany
t.ackemann@uni-muenster.de

[2] Department of Physics and Applied Physics, University of Strathclyde, Glasgow G4 0NG, United Kingdom
willie@phys.strath.ac.uk

Abstract. Many dissipative optical systems support patterns. Dissipative solitons are generally found where a pattern coexists with a stable unpatterned state. We consider such phenomena in driven optical cavities containing a nonlinear medium (cavity solitons) and rather similar phenomena (feedback solitons) where a driven nonlinear optical medium is in front of a single feedback mirror. The history, theory, experimental status, and potential application of such solitons are reviewed.

1 Introduction

Spatial optical solitons are beams of light in which nonlinearity counterbalances diffraction, leading to a robust structure which propagates without change of form (Fig. 1a). These intriguing objects are solutions of nonlinear wave equations. In the conservative case, the range of materials is rather limited – they need to be self-focusing – and in many cases the existence of stable spatial solitons is also limited to one-dimensional systems. Such is the case for the simplest soliton medium, one with a Kerr nonlinearity, i.e. a refractive index which changes in proportion to the intensity of the light. However, as this book demonstrates, in the last years it became increasingly clear that more general schemes can support stable soliton-like solutions with lots of intriguing and new properties, if dissipation and feedback are explicitly introduced. Among these *dissipative solitons*, localized bright spots in driven optical cavities (Fig. 1b, c) received particular attention. They share some properties with spatial solitons, and we will refer to them as *cavity solitons* (CS). Actually, a "half-cavity" turns out to be enough, i.e. very similar objects are found in arrangements in which feedback is provided by a single-mirror only (Fig. 1d, "single-mirror feedback scheme"). We will refer to them as *feedback solitons* (FS). Structures such as CS and FS could be natural "bits" for parallel processing of optical information, especially since they can be found in semiconductor micro-resonators.

T. Ackemann and W.J. Firth: *Dissipative Solitons in Pattern-Forming Nonlinear Optical Systems: Cavity Solitons and Feedback Solitons*, Lect. Notes Phys. **661**, 55–100 (2005)
www.springerlink.com

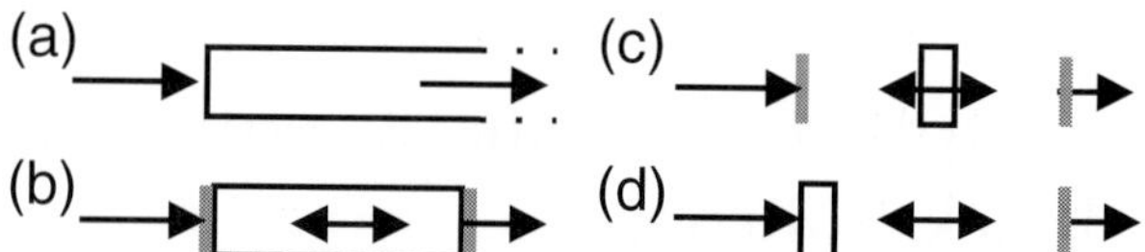

Fig. 1. Nonlinear optical systems supporting pattern formation and solitons (plane mirrors are drawn in *grey*): (**a**) nonlinear beam propagation, (**b**) cavity filled with nonlinear medium, (**c**) cavity with a short medium, (**d**) single-mirror feedback arrangement

In this chapter we briefly review the history and underlying nonlinear optics of cavity and feedback solitons. We then describe in more detail some models which show these soliton-like structures. It is somewhat paradoxical that stable cavity solitons can exist in media with properties different from, even opposite to, those required for Kerr-medium solitons. We therefore discuss candidate physical interpretations of these structures. We examine the perturbation eigenmodes of CS in sample systems, which give information on the soliton's stability and response to external influences such as noise, neighboring solitons or gradients of the holding field. Cavity solitons can be created by localized pulses of light, and could thus be used to capture and process images or information. The ability to control and manipulate these solitons offers potential advantages over competing systems, and we mention some device ideas which might capitalize on these advantages. These theoretical considerations are supplemented by experimental results, mainly on FS. Finally, we comment on similarities and differences of FS and CS.

This article concentrates mainly on the work of ourselves and collaborators. This is primarily for convenience, and serves to illustrate points of general relevance, but we have tried to present a fair and reasonably comprehensive list of relevant references. Further information can be found in earlier reviews and monographies [1, 2, 3, 4, 5, 6, 7, 8, 9, 10]. We mention also a recent feature Section on CS in *IEEE Journal of Quantum Electronics* [11]. Additional material on experiments on CS in semiconductor microresonators can be found in the chapter by Taranenko and Weiss [12] in this book. CS in lasers are covered by Rosanov's chapter [13]. There is a rich phenomenology and literature on CS which rely on the coupling of several *optical fields* through, for example, a $\chi^{(2)}$ nonlinearity. This has been shown in mean-field models to support cavity solitons in both second-harmonic-generation (SHG) [14, 15, 16] and optical parametric oscillator (OPO) [17, 18, 19, 20, 21, 22, 23, 24, 25, 26] configurations. Another example is a two-photon laser [27]. A vector Kerr medium also involves coupled fields, and exhibits *polarised* CS for appropriate parameters [26, 28, 29]. Polarized CS exist also in $\chi^{(2)}$-materials [15, 23] and suitable conditions can be envisaged in certain single-mirror systems [30]. We remark that in these systems CS are often due to a coexistence of two homogeneous states which arise from a symmetry-breaking pitchfork bifurcation.

Thus, they are somehow different from the CS on which this review focuses, and to do them justice would have required a substantial lengthening of this article, and so we have chosen not to deal further with these structures, directing the interested reader to the cited literature. Experimental evidence for such structures was given in intra-cavity four-wave mixing [4, 31, 32] and recently in a single-mirror system [33]. Experimental investigations on CS in self-imaging laser schemes and other oscillators are reported in [4, 34, 35, 36].

2 History

The story of cavity solitons probably began with the seminal paper of Moloney and co-workers [37, 38], who used split-step FFT methods to simulate transverse effects in optical bistability (OB) [39, 40]. The model system was a ring cavity, driven by a gaussian beam, and containing a self-focusing Kerr-like medium. The field was propagated (in z) around the cavity, and added coherently to the driving beam at the input beam-splitter. The transverse simulation was one-dimensional (1D), i.e. the intra-cavity field was described by $E_n(x, z)$ with n counting the cavity round-trips. When the input field was ramped up to exceed the OB switch-up threshold, the beam center switched, and a *switching wave* moved out, switching up most of the beam (Fig. 2). Then something unexpected happened: a new instability. The interface between the "on" and "off" domains spawned what would now be termed a modulational instability (MI) of the "on" region, which broke up into a set of distinct peaks [37, 38]. These were interpreted as a group of spatial solitons circulating in the medium, perturbed by the output coupling losses and sustained by the input field. Such a physical model has been termed "soliton-in-a-box" [41]. Such an interpretation is tenable only for a medium which could sustain solitons, and thus cannot be a general model for CS and FS.

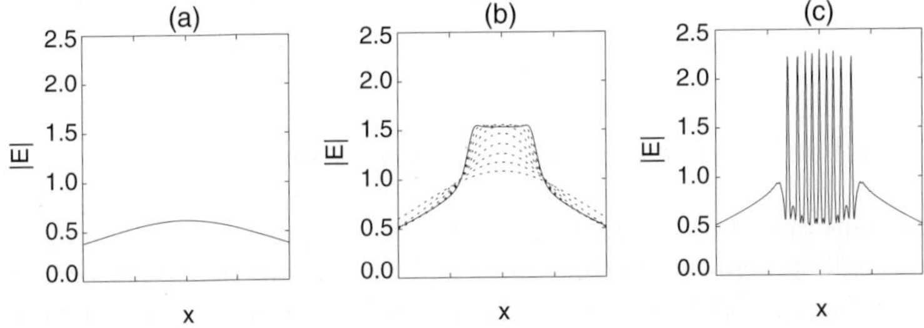

Fig. 2. Gaussian beam switching in an OB cavity. The central section of a smooth broad beam (**a**) switches; a switching wave then forms and moves out (*dotted curves*) in (**b**); the high-intensity central region then breaks up into spikes (after Moloney et al. [37, 38])

In this early cavity soliton work [37, 38] the beam either contains no solitons, or is full of solitons: just two states, and therefore only a one-bit memory in applications terms. McDonald and Firth [42] showed that it was possible to make the individual solitons independently switchable by using a pump beam with a spatially-varying amplitude. They modeled a 20-bit memory of this kind, and also showed that it was possible to switch solitons "off" as well as "on" (with an out-of-phase address pulse) [43].

The other key CS pioneer was Rosanov, who from study of OB switching waves developed the idea of "diffractive autosolitons" [44, 45] in nonlinear optics. Switching waves between co-existent stable states are known in many fields, such as reaction-diffusion systems. Purely diffusive switching waves have monotonic profiles, and the more stable state simply wipes out the less stable, as was shown in an OB model with pure diffusion[46]. When diffraction is present, however, the switching wave typically has ripples. These can trap other switching waves, and thus two can trap each other. Or, in 2D, one switching wave might bend around and close on itself, forming a stable island of one phase surrounded by the other – a diffractive autosoliton (DAS)[44, 45]. Here, then, is a second physical interpretation of cavity solitons: self-trapped switching waves. Note that here there is no requirement for bulk solitons: indeed Rosanov showed that DAS can occur even in a saturable absorber. Nor is OB required: even in its absence one can still have switching waves, e.g. between homogeneous and patterned states.

Rosanov has made many significant contributions to OB and related fields on which excellent reviews are available [1, 9]. His work includes also DAS in lasers. The interested reader is referred to these reviews and to Rosanov's chapter in this book [13].

In terms of early experiments, a 1990 special issue of JOSAB [47] gives a useful snapshot and overview of the field. There were pioneering experiments on driven plano-planar resonators containing a liquid-crystal cell [48, 49] published shortly afterwards. Regarding feedback-mirror experiments, as early as 1988 there was experimental evidence of spatio-temporal structure in sodium vapor coupled to a feedback mirror [50], albeit with a long optical path and low Fresnel number.

3 Mean-Field Models and Cavity Solitons

Analysis and modelling of patterns and solitonic phenomena is much simplified in so-called mean-field cavity models, in which alternation of propagation around the cavity with coherent addition of the input field is replaced by a single partial differential equation with a driving term. In the context of spatial pattern formation, this approach is ascribed to the seminal paper of Lugiato and Lefever[51]. Here we give a heuristic derivation of the Lugiato-Lefever (LL) equation, starting from the well-known Nonlinear Schrödinger Equation (NLS), which describes (conservative) solitons in Kerr media.

The NLS assumes an *infinite* nonlinear medium. Real nonlinear optical media have finite dimensions and (except in glass fibre) solitons can rarely propagate more than a few centimetres before running out of material (Fig. 1a). This perhaps makes it natural to put mirrors around the medium, confining the soliton into a finite slab of material (Fig. 1b). With perfect reflection and zero absorption, one could indeed confine a soliton in such a "box". Real mirrors and materials are lossy, but we can make good the loss by "feeding" the caged soliton with an input field. We are thus led to consider a perturbed NLS:

$$\mathrm{i}\frac{\partial E}{\partial t} + \frac{1}{2}\frac{\partial^2 E}{\partial x^2} + |E|^2 E = \mathrm{i}\varepsilon(-E - \mathrm{i}\theta E + E_{\mathrm{in}}) \tag{1}$$

The terms on the left are standard NLS terms, describing respectively evolution, diffraction and (Kerr) nonlinearity. The three terms on the right side are perturbations of the NLS, all small if ε is. The first is just a linear loss ($\varepsilon > 0$), and the last is the driving field E_{in} needed to sustain E against that loss. Less obvious is the middle term, in θ, but we must remember that coherent light confined between mirrors lies within an optical cavity, and so the response to the driving field will strongly depend on whether or not it is in resonance with the cavity. Hence, therefore, the presence of θ, the *cavity mistuning*. If we ignore the *left* side of (1), then $E = E_{\mathrm{in}}/(1 + i\theta)$, showing that the cavity has a resonance Lorentzian in θ. This is appropriate for high finesse, where just a single longitudinal mode may be considered.

There is one further change from the usual spatial-soliton NLS: propagation (in z) is replaced by evolution (in t). This is natural: the soliton is now in a box, and not going anywhere.

In the limit $\varepsilon \to 0$ equation (1) recovers the NLS, with a soliton solution of sech-profile in x, time independent except for a phase rotation. We might expect, therefore, that for finite ε it has sech-like *cavity soliton* solutions for suitable E_{in}, and indeed it has. It is usual to consider E_{in} to be a plane wave, independent of x (and y in 2D), in which case the soliton sits on a homogeneous non-zero background field being a solution of (1). A finite but relatively broad driving beam supports cavity solitons qualitatively similar to those predicted for the simpler plane-wave input case. We remark that (1) is not the only possible dissipative version of the NLS which might be envisaged. For example, in [52] a damped and parametrically driven version of the NLS is considered which serves as a model for "oscillons", 2D localized states in shaken granular materials [53]. They can be regarded as another manifestation of dissipative solitons.

We now set $\varepsilon = 1$, which is equivalent to a re-scaling. This yields the Lugiato-Lefever (LL) equation, which was originally introduced [51] as a model for pattern formation. Note that in the LL equation time t is scaled to the cavity loss time. The NLS limit is recovered as $\theta \to \infty$. The LL equation is also appropriate as a mean-field model for OB with transverse effects. The term "mean-field" arises because such models are usually derived by assuming a high finesse, so that the cavity field is approximately constant along the cavity axis. The high finesse allows the Airy function response of the cavity to be approximated by a single longitudinal mode, giving the Lorentzian resonance mentioned above. For a plane-wave pump, the plane-wave cavity field obeys $E_0 = E_{\text{in}}/[1 + i(\theta - |E_0|^2)]$, which is three-valued for $\theta \geq \sqrt{3}$, showing that the model exhibits OB[51]. In fact, Lugiato and Lefever showed that E_0 is stable if $|E_0| < 1$, but unstable, usually because of spontaneous pattern formation, above that threshold [51]. In truth Fig. 2 was generated by simulating the LL equation (for $\theta = 2.1$, $E_{\text{in}} = 1.5$ for (a) and 2.5 for (b) and (c)), rather than the original model of Moloney et al. [37, 38]. The strong similarity of the respective results shows how such mean-field models can capture the essential features of a full cavity model while being both cheaper to simulate and easier to analyze.

Such analysis is still by no means simple. Unlike for the NLS, no exact analytical solutions for patterns or solitons are known for the LL equation. The usual approach has been to perturbatively derive amplitude equations or to resort to numerical integration of the model, neither of which is ideal. Perturbative approaches, by their very nature, cannot provide quantitative results and integration gives a very restricted view of the model's bifurcation behavior, by only finding solutions which are dynamically stable. Another technique, which we have used extensively [41, 54, 55, 56, 57] is to use numerical methods to find the system's stationary solutions, their stability and their response to perturbations. We will first apply the method to the LL equation and later to other CS and FS models.

We look for stationary solutions ($\partial/\partial t = 0$) of equation (1), after setting $\varepsilon = 1$ as discussed:

$$0 = -(1 + \mathrm{i}\theta)E + \mathrm{i}|E|^2 E + \mathrm{i}\Delta_\perp E + E_{\text{in}} \tag{2}$$

where the exact form for the transverse Laplacian $\Delta_\perp$ depends on whether we consider 1D, cylindrically symmetric, or fully 2D geometries. We discretize the space variable(s) on N grid points, apply periodic boundary conditions and use a Fast Fourier Transform (FFT) algorithm to evaluate the spatial derivatives. This gives a highly accurate, $O(N)$, set of coupled algebraic equations which can be solved using an iterative Newton method. Given a suitably close initial guess, this method rapidly converges to a stationary solution (not necessarily a stable one) of the original LL equation. These solutions can then be tracked in parameter space, tracing out branches.

The use of a Newton method is advantageous because, as a by-product of this process, it also finds the linearization, in the form of a Jacobian matrix, around the solution found. The resultant eigenvalues, β, give the stability of the solutions and the eigenvectors $\{\boldsymbol{u}\}$ the associated modes.

Figure 3 shows three solution branches tracked in this way. The lower line shows the plane-wave solution discussed already to be stable below $|E_0|^2 = 1$

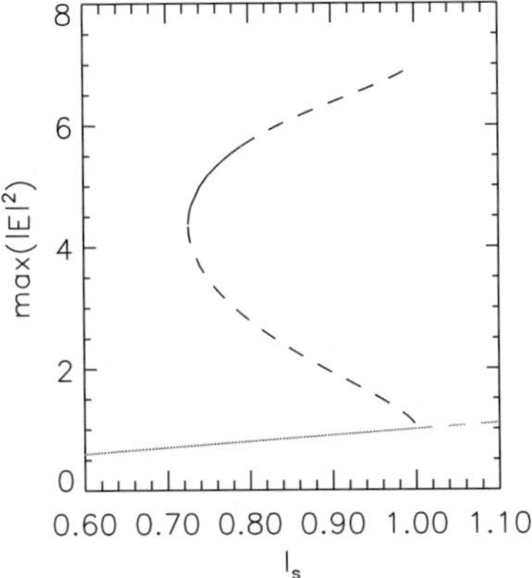

Fig. 3. Branches showing the homogeneous solution and the cylindrically symmetric cavity soliton solution for the 2D Kerr cavity, as a function of the background intra-cavity intensity $I_s = |E_0|^2$. *Solid* curves indicate stable solutions, *dashed* unstable ($\theta = 1.3$)

and unstable above. At this point, a branch of localized, cylindrically symmetric $(\nabla^2\partial^2/\partial r^2+1/r\partial/\partial r)$ solutions bifurcates subcritically before bending around to form a positive-slope branch of finite amplitude. This upper branch is the cavity equivalent of the unstable soliton-like solution to the 2D NLS. This bifurcation structure is typical of cavity solitons in many systems. For values of $|E_{\mathrm{in}}| > E_{\mathrm{in}}^{(\mathrm{sn})}$ the two branches co-exist with the homogeneous background, to which the solitons asymptote at large radius. Note that only for $\theta > \sqrt{3}$ the homogeneous solution is multi-valued and so, over a broad range, any interpretation of CS as self-trapped switching waves must relate to the interface between the homogenous solution and a pattern, rather than simply between homogeneous solutions. We will discuss the relationship between patterns and CS in some detail below.

We now turn to the stability of these cavity soliton solutions. As might be expected, the lower branch of the loop is always unstable but, unlike the 2D NLS case, the upper branch may be stable [58]. The solid portion of the upper branch in Fig. 3 shows where they are stable. We find that the onset of instability is due to the presence of a Hopf bifurcation [58, 59], not, as in bulk Kerr media, collapse. Direct simulation confirms the stability analysis. A perturbed cavity soliton exhibits damped oscillations in the stable domain, which become undamped as the stability boundary is crossed. Inside the instability region, the CS shows periodic oscillations, but still does not collapse, even well beyond threshold. Instead its minimum-amplitude shape becomes very similar to that of the unstable CS solution belonging to the lower branch in Fig. 3. Because that solution has only one unstable eigenmode, it acts as a quasi-attractor for the oscillating CS, which dwells close to it before eventually moving away along its unstable manifold [59].

The stability of 2D cavity solitons in the Lugiato-Lefever mean-field model is a behavior qualitatively different from its bulk-medium equivalent. This encourages exploration of other models which might support CS. Comparing the LL equation with the NLS, we note that the diffraction term and the nonlinearity are not specifically associated with the cavity. We already modified the diffraction term when we considered the 2D Kerr cavity. We can go further and add *dispersion* to make "3D" cavity solitons [60, 61]. Or we can replace diffraction with dispersion, and consider e.g. fibre cavities, where Mitschke and co-workers [62] have found evidence of soliton-like structures in synchronously-pumped fibre loops, and Wabnitz [63] has examined data storage issues. This case is equivalent to diffraction in the 1D geometry of the LL equation, and the existence and stability of CS in 1D is perhaps to be expected.

The nonlinearity also offers considerable scope for variation. Perhaps the obvious generalization from the Kerr nonlinearity is to a two-level atom-like response, which becomes Kerr-like far from the atomic resonance. For exact atomic resonance the medium is just a saturable absorber, with no nonlinear refractive index contribution. It nonetheless supports stable, robust cavity solitons [45, 64, 65].

The simplicity of the saturable absorber makes it a very useful model for CS investigations, and we will use it as illustration of some interesting and general CS phenomena. The spatio-temporal dynamics of the slowly varying amplitude of the electromagnetic field E is modeled by

$$\partial_t E = -E\left(1 + \mathrm{i}\theta + \frac{2C}{1+|E|^2}\right) + \mathrm{i}\Delta_\perp E + E_{\mathrm{in}} \ , \qquad (3)$$

which differs from the Kerr cavity equation (1) only in the nonlinear term, in which C is a scaled atomic density. Since C is real and positive, this a purely dissipative term, which in fact describes the linear as well as nonlinear absorption due to the medium. The field scaling is such that the saturation intensity of the transition corresponds to $|E| = 1$.

The homogeneous solution E_0 of this equation (with plane-wave E_{in}) can be multi-valued if $C > 4$ (absorptive OB). With or without OB, it can become unstable to pattern formation [66], where the wave-vector K of the pattern at threshold is simply given by $K^2 = -\theta$. This obviously requires that the cavity mistuning θ be negative or, in physical terms, that the frequency (and the corresponding wave vector, of course) of the intra-cavity field is higher than the frequency (and wave vector) of the longitudinal (i.e. the on-axis) resonance of the cavity. This has a simple physical interpretation – the unstable mode is off-axis by just enough to compensate for the cavity mistuning, so that this is a so-called "tilted-wave" instability.

Figure 4 gives a demonstration of this scaling behavior. The panels (a) and (b) display the near field intensity distributions observed in a vertical-cavity regenerative amplifier driven electrically above transparency but below the

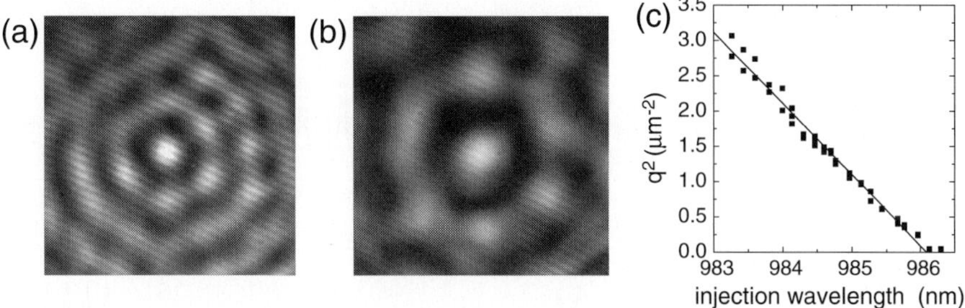

Fig. 4. (**a**, **b**) Near field intensity distribution observed in a vertical cavity regenerative amplifier for an injection detuned by about (**a**) 1.1 nm and (**b**) 0.4 nm (**c**) Square of transverse wavevector in dependence on the injection wavelength (adapted from [67])

self-oscillation threshold (Fig. 1c, [67]). Though the injected external optical field is a smooth on-axis wave, the output shows spontaneous patterning and the scale of the pattern depends on the frequency of the injected field which defines also the oscillation frequency of the intra-cavity field. The patterns become coarser as the injection wave length approaches the longitudinal cavity resonance, as expected from the above consideration. The data depicted in panel (c) demonstrate also very nicely that indeed the square of the transverse wave vector depends linearly on detuning. We note that the simple scaling between the empty cavity detuning and the transverse wave vector holds only for purely absorptive media. In dispersive media (in particular in semiconductor media), the shift in resonance due to the nonlinear refractive index shift induced by the background field needs to be taken into account.

We dwelt on this issue in some detail because it will become clear when we discuss the connection between CS and patterns, and in particular the "pattern element" interpretation of CS, that the scaling properties of the pattern wavelength carry over to the half width of CS and to the minimum allowed distance between them, at least to a great extent. Of course a soliton parameter such as the half width is also affected by diffusion, nonlinearity etc. We remark that such a scaling is a general feature of optical pattern forming systems, since it is a direct consequence of the scaling properties of the paraxial wave equation.

CS in the absorptive model (3) have recently been examined in detail in [56]. We show in Fig. 5a range of such solutions, including not only simple single-peaked CS, but multi-peaked localized solutions in which up to four similar-sized peaks sit on a flat background. The latter four-peak structure invites interpretation as an "island" of pattern set in the homogeneous solution (which is single-valued for these parameters), an interpretation which we will now examine. We should first note that these CS solutions are presented in terms of $A(x)$, where by definition $E = E_0(1 + A)$, so that in terms of A localized solutions sit on a zero background. Second, note that for any

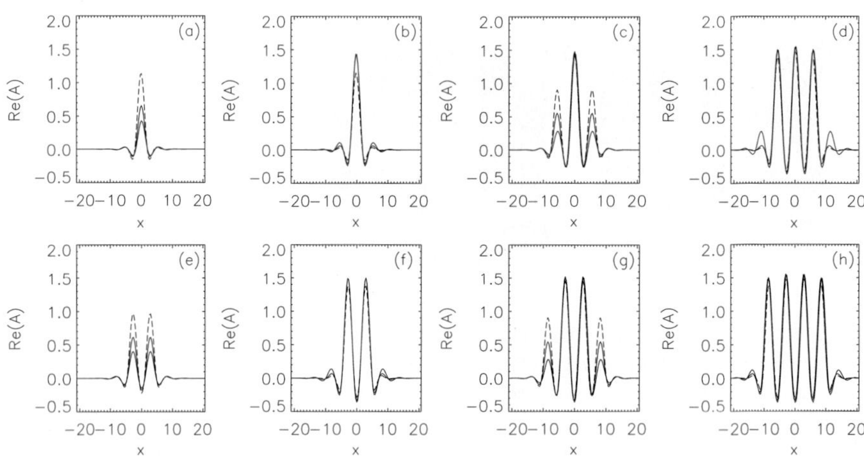

Fig. 5. Sequences of profiles of one-dimensional CS solutions to equation (3) with respectively odd (*above*) and even (*below*) numbers of main peaks. *Dash-dotted*, *solid* and *dashed* lines correspond to solutions at $|E_0|^2 = 1.22$, $|E_0|^2 = 1.33$ and $|E_0|^2 = 1.44$. Other parameters are $\theta = -1.2$ and $C = 5.4$

given input field, all the peaks in these solutions seem to have one of two heights. In particular, there are two single-peak CS of different amplitudes. This is general – the low-amplitude-peak belongs to a branch of unstable CS which bifurcates subcritically from the homogeneous solution at the pattern-formation threshold [65, 56], before bending back in a saddle-node bifurcation to become the high-amplitude-peak branch (which is stable). The Kerr CS discussed show a similar bifurcation structure (see Fig. 3).

Coullet et al [68, 69] have recently offered a general mathematical analysis of this scenario. They consider a situation in which a stable infinite pattern co-exists with a stable homogenous solution. Then one can envisage a configuration in which there is a domain with a pattern and a homogeneous domain, with a *front* at their interface. In general one would expect one or other solution to dominate, such that the front would move, annihilating the weaker solution, until the entire space is filled with the dominant state. One might expect there to be a kind of "Maxwell point" in parameter space, at which the dominant role switches from the pattern to the homogeneous state or vice versa. In fact Pomeau showed that, because motion of the front requires creation or annihilation of pattern cells, the Maxwell point spreads out into a *locking range* of finite width, within which the front is stationary, i.e. the two states can stably coexist in real space as well as in parameter space [70]. Coullet et al. showed, using generic properties of ordinary differential equations, that in parameter space such a locking range is always accompanied by an infinite sequence of localized states. As one enters the locking range from the homogeneous-dominant side, pairs of N-peaked states appear in decreasing

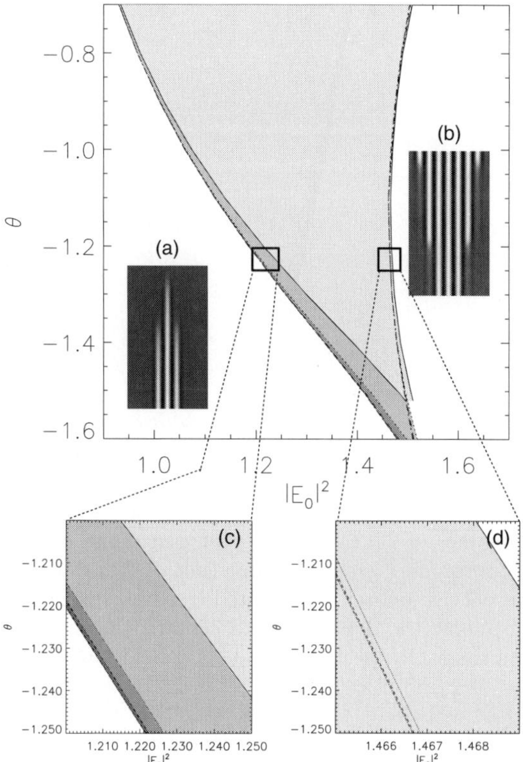

Fig. 6. Existence limits of N-peaked CS structures (*shaded regions*) in two-dimensional parameter space $(|E_0|^2, \theta)$. Structures with $N = 1, 2, 3, 4$ respectively exist between *solid*, *dotted*, *dashed* and *dash-dotted* lines. Panels (**a**) and (**b**) show space-time plots of unlocking behavior, with the transverse coordinate (x) on the horizontal axis and time (t) on the vertical axis. Panel (**c**) shows the fine structure of the locking domain, indicated by the square on the main figure. Parameters: $C = 5.4$

order of N, while as one exits the range on the pattern-dominant side, these states disappear in order, with the $N = 1$ states surviving longest.

The saturable-absorber CS model (3) in one transverse dimension falls within the class of models to which the Coullet et al. theory is relevant, and Fig. 5 is fully consistent with that scenario. McSloy et al. [56] used the stationary-solution approach described above to map out the existence domains of the $N = 1$ to $N = 4$ peaked CS solutions in both one and two dimensions, as a function of the *two* parameters θ and $|E_0|^2$. The results are shown in Fig. 6, and are fully consistent with the analysis of Coullet et al. [68, 69]. In the broad central region of this figure, there is a stable N-peaked localized state (or multi-CS) for any N. (In practice, N will be limited by the size of the holding beam E_{in} relative to the size of the single-peaked

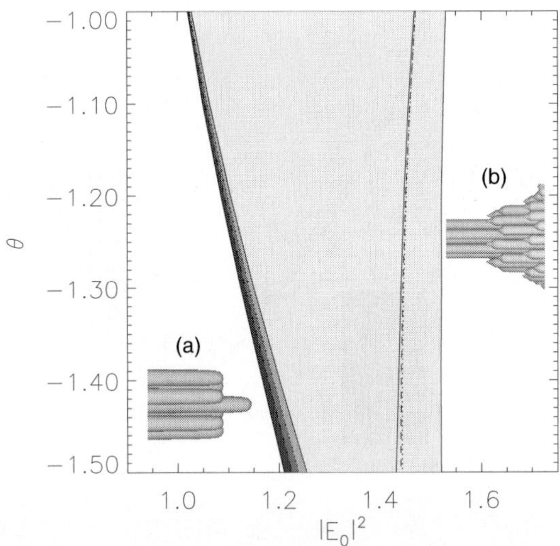

Fig. 7. Locking regimes of 2D *CS* (*solid/light gray*), and of clusters of two (*dotted/gray*), three (*dashed/dark gray*) and four (*dot-dashed/black*) close-packed clusters of CS with respective existence indicated by (*line style/fill shade*) ($C = 5.4$). Panels (**a**) and (**b**) show the dynamics of a centered hexagon structure ($N = 7$) outside the locking range. Respective intra-cavity field intensities are $|E_0|^2 = 1.08$ and 1.45 with $\theta = -1.2$ and $C = 5.4$

CS.) The analysis used by Coullet et al. is not extensible to two transverse dimensions, but in Fig. 7 we show that basically the same phenomena occur [56]. In 2D, multiple peaks transform into clusters, bound states of simple, cylindrically-symmetric, CS [71]. Large clusters, especially symmetric ones, are also referred to as "localized patterns" [64]. In Fig. 7 we plot the existence domains of single CS, and of dipole, triangular, and rhomboidal clusters [56]. In terms both of parameter ranges, and of succession, these 2D CS structures are qualitatively similar to the better-understood 1D scenario. Also shown are plots of the dynamics of the interesting centered-hexagon ($N = 7$) cluster outside the locking range. It "dies" in the left-hand (homogeneous-dominant) region, but grows to form a full hexagonal pattern in the right-hand (pattern-dominant) region.

From this analysis we can conclude that with an appropriate choice of system and parameters we can impress any number of CS up to some aperture-delimited maximum N_{max} on to the output beam of a driven optical cavity containing a suitable nonlinear medium. Such a medium acts as an N_{max}-state memory, assuming that we can create N-clusters of CS with N appropriately-aimed address pulses. We can do better, however (and need to, if such a CS memory is to be economic). Suppose that each of the component CS making up the N_{max} cluster can be independently switched, by selective addressing at the site of each. Then each site acts as a *pixel*, which is "on"

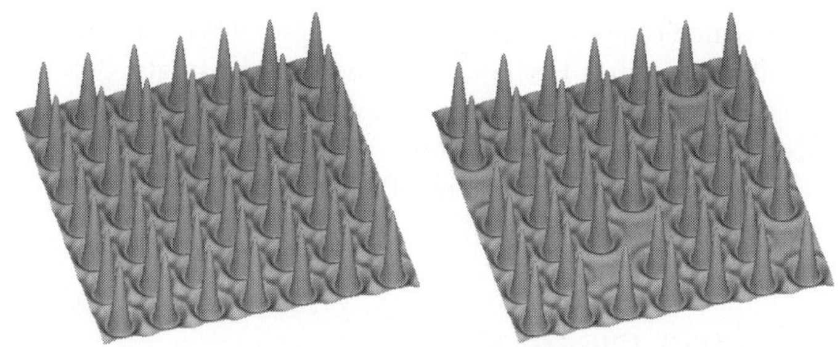

Fig. 8. Regular *vs* irregular patterns. On the *left*, a 7×7 square pattern. On the *right*, the same template, but with five of the peaks (i.e. five individual CS) missing. Both arrays are stable. Assuming that, similarly, any or all of the 49 peaks may be present or absent, there are 2^{49} different co-existent stable states of this array. (Optical cavity with saturable absorber: $\theta = -1.2$ and $C = 5.4$)

when it supports a CS, and "off" when no CS is present at that site. In such a case, we have $2^{N_{max}}$ co-existent states. Some years ago Firth [72] showed that there is a certain equivalence between the condition for the existence of a full set of 2^N coexistent states in a 1D array and the "Smale horseshoe" condition for dynamical chaos. Though he used a simpler model than Coullet et al. [68, 69, 73] (an imposed, not spontaneous, pattern), it seems likely that an analogous development of their approach to consider quasi-random pixel-type patterns will lead to a similar conclusion.

Figure 8 illustrates that this idea can work in practice. Using periodic boundary conditions, a stable square "pattern" exists, as illustrated in the left panel. That this "pattern" can be regarded as an array of independent CS is illustrated in the right panel, where five of the "CS" are missing. This structure is also stable, and indeed this array operates in all respects as a 49-bit pixel array memory, in which the pixel is "on" or "off" according to whether or not there is a CS at the pixel site.

Given that CS are found in both Kerr and saturable absorber media, it may be no surprise that the intermediate case of a two-level atomic response, with mixed absorptive-dispersive nonlinearity, also supports cavity solitons, though more readily on the self-focusing side of resonance [44, 64, 74, 75, 76, 77]. It was also shown that these models reduce in some limit to Swift-Hohenberg-like models [64, 66, 74, 78], a generic class of models in spontaneous pattern formation [79]. An experimental observation of dissipative Kerr-solitons was recently reported for a nonlinear cavity filled with a liquid crystal [80].

A further CS generalization is to consider a nonlinearity mediated by a material excitation. This opens up the further possibility that the medium can have its own dynamics and spatial (usually diffusive) coupling. Semiconductors are particularly interesting among such media, and it has been shown [54, 76, 77, 81, 82, 83] that cavity solitons extend to semiconductor models, even in the presence of such "soliton-antagonistic" effects as diffusion and a measure of self-defocusing. Perhaps even more surprisingly, Michaelis et al. [84] found bright solitons in a cavity model with a purely-defocusing, diffusive saturable Kerr medium, such as is found in semiconductors just below the band edge. Experimental observations of CS in semiconductor microcavities were reported in [85, 86, 87, 88, 89].

4 Self-Propelled Cavity Solitons in Semiconductor Microresonators

Since basic properties of CS in semiconductor microresonators are covered in another chapter of this book [12], and in [86, 90], we will concentrate here on an interesting phenomenon which arises when thermal effects are coupled to the light-carrier interaction equations.

A recent paper [91, 92] demonstrated and analyzed the existence of both bright and dark spontaneously moving CS in a model of a semiconductor microcavity. The motion is caused by temperature–induced changes in the cavity detuning and arises through an instability of the stationary soliton solution when the temperature-tuning coupling is strong enough. The experimental relevance of this coupling is demonstrated by the observation of opto-thermal pulsations in semiconductor amplifiers [93]. Here, we briefly summarize the main features of these phenomena, and detail just a few interesting examples. More details of the phenomena and of models for CS in semiconductor cavities can be found in [82, 91, 92] and references therein.

Before describing the particular model to be looked at, a brief digression on the dynamics of CS is indicated. When the driving field is plane–wave and the system invariant with respect to spatial translations, a given stable CS can exist at any location. This symmetry property is manifest in the eigenvalue spectrum of the soliton (or any stable stationary solution) by the fact that there is a marginal mode with eigenvalue zero, which is connected to the translational degree of freedom. All other eigenvalues than this neutral mode $\boldsymbol{u}_0$ have negative real part, by virtue of the stability of the CS. This means that, as $t \to \infty$, the amplitude a_0 of the neutral mode dominates over all other a_i. Thus the dynamical effect of any perturbation $\boldsymbol{\mathcal{P}}$ on a stationary stable state is primarily determined by its projection onto the neutral mode, which yields the equation[54]:

$$\frac{\mathrm{d}a_0}{\mathrm{d}t} = \frac{1}{\langle \boldsymbol{v}_0 \mid \boldsymbol{u}_0 \rangle} \langle \boldsymbol{v}_0 \mid \boldsymbol{\mathcal{P}} \rangle \tag{4}$$

Here $\boldsymbol{v}_0$ is neutral mode of the corresponding adjoint problem. Because the neutral mode is just the gradient of the CS, physically $\mathrm{d}a_0/\mathrm{d}t$ is the translational velocity of the CS under the influence of the perturbation. Obviously the motion of a such a CS under the influence of an external force is not

Newtonian, but overdamped (Aristotelian). Among the various types of perturbation, three of particular relevance are a phase or amplitude gradient of the driving field [65, 1, 54], and perturbation of one soliton by another (see Sect. 7.1). In the case of a weak phase gradient the input field around a CS at $x = 0$ can be locally approximated by $E_{\text{in}} = E_{0,\text{in}}(1 + ikx)$. Inserting the perturbation part $ikxE_{0,\text{in}}$ into (4) we can calculate the drift velocity of a cavity soliton due to the phase gradient (or, similarly, an amplitude gradient) [54].

By turning a parameter (here, the detuning) into a *dynamical variable* by coupling it to the temperature, a spontaneous transition from stationary to moving solitons becomes possible. We note that the stationary-to-moving CS bifurcation has similarities with the Ising–Bloch transition for *fronts* studied in [94] and more recently in [95, 96]. One particularly interesting feature to emerge is that the spontaneously moving CS have *inertia*, so that their dynamics is quasi-Newtonian. In particular, they can rebound from obstacles (or each other), and can oscillate in a potential well, impossible for Aristotelian particles.

The system under consideration is a semiconductor microcavity (Fig. 1c) consisting of a thin active region sandwiched between two high reflectivity (~99.9%) distributed Bragg reflectors (DBR). The device can be driven by an external pump field E_{in} and, optionally, an external current J (see [12, 67, 85, 86]). We will concentrate on the case $J = 0$, the so-called passive system [82]. The intra-cavity electric field E, carrier density N and temperature difference T between the lattice temperature and the ambient temperature can be described by the following set of partial differential equations [77, 82, 91, 92]:

$$\frac{\partial E}{\partial t} = -(1 + \mathrm{i}\Theta)E + \mathrm{i}\Xi\chi E + E_{\text{in}} + \mathrm{i}\Delta_{\perp} E \tag{5}$$

$$\frac{\partial N}{\partial t} = -\gamma_N \left(N + \beta N^2 - J + (N-1)|E|^2 - D_N \Delta_{\perp} N \right) \tag{6}$$

$$\frac{\partial T}{\partial t} = -\gamma_T \left(T - ZN - PJ^2 - D_T \Delta_{\perp} T \right) \tag{7}$$

where $\Delta_{\perp}$ is the transverse Laplacian in 2D. The strength of the material nonlinearity is parameterized by Ξ. The cavity detuning is denoted by Θ where

$$\Theta = \theta - \alpha T \tag{8}$$

Here θ is the cavity detuning at ambient temperature and α a coupling parameter [97]. Note that if $\alpha = 0$ there is no feedback from heating to the carrier-field dynamics. Thus equation (8) is the driver of the motional instability, because it embodies the most important consequence of heating in semiconductor microcavity devices: namely, a change in the linear refractive index of the semiconductor material and hence a shift in the cavity resonances

[89, 97]. It is thus to be expected that α is the key parameter in the analysis of this instability [92].

Elsewhere in these equations, the term βN^2 describes radiative carrier recombination, carrier and thermal diffusion coefficients [98] are denoted by D_N and D_T, the term ZN describes heating due to nonradiative recombination, and PJ^2 Joule heating by the current (if present).

For the passive MQW device model, the nonlinear susceptibility χ is assumed to be simply a linear function of the carrier density [92]:

$$\chi = \frac{-(\Delta + \mathrm{i})(N-1)}{1+\Delta^2} \tag{9}$$

The parameter Δ represents the band-gap detuning.

Finally, the cavity field decays on a time scale of $\tau \approx 10\,\mathrm{ps}$. This value is used to scale time in (5)–(7). Then, the (normalized) time scale for N and T are $\gamma_N^{-1} \approx 100$ and $\gamma_T^{-1} \sim 10^5$, respectively.

Equations (5)–(7) are known to exhibit plane-wave bistability and to possess stable, stationary CS solutions in the absence of thermal effects ($\alpha = 0$) [54]. We are going to describe structures whose existence is due entirely to the effect of temperature changes on the cavity and thereby on the intra-cavity field. In the passive system, the instability to moving solutions is observed for *dark* solitons.

As intensity minima sitting on a high intensity background, dark solitons are expected to exist on the upper branch of the plane–wave bistability curve. Figure 9 shows an example of a dark soliton branch along with plane–wave and roll (stripe) solutions. This particular roll solution has a wavevector $K \simeq 1.3K_c$ where K_c is the most unstable wavevector at the modulational instability threshold. The homogeneous background field shows the underlying plane–wave bistability. For these parameters, dark solitons exist for $5.542 < E_{\mathrm{in}} < 5.560$ but are always unstable to spontaneous motion.

An example of a moving dark soliton is given in Fig. 10. The physical mechanism is reasonably simple. A stationary dark CS is a region of low intensity, and thus of weak heating, and hence is relatively cold. If the coupling α is positive, that means that the cavity tuning is locally (even) more negative. Without thermal coupling, a CS will move to regions of smaller (absolute) detuning, and so any small displacement of the CS field minimum ("dark spot") from the temperature minimum ("cool spot") will induce it to move further out of the temperature dip, at a speed related to the tuning gradient. This movement is counteracted by the coupling of the temperature to the field and carrier density, which will cause the "cool spot" to follow the "dark spot". If α is small enough, the dark spot will move only slowly, and the temperature will be able to follow any movement of the other fields, and so the CS will be stable against fluctuations. For larger α, however, the dark spot will move faster, and at some point the slow thermal dynamics will prevent the cool spot from keeping up with the dark spot: then there is a transition to a dynamic equilibrium, in which the CS moves at constant

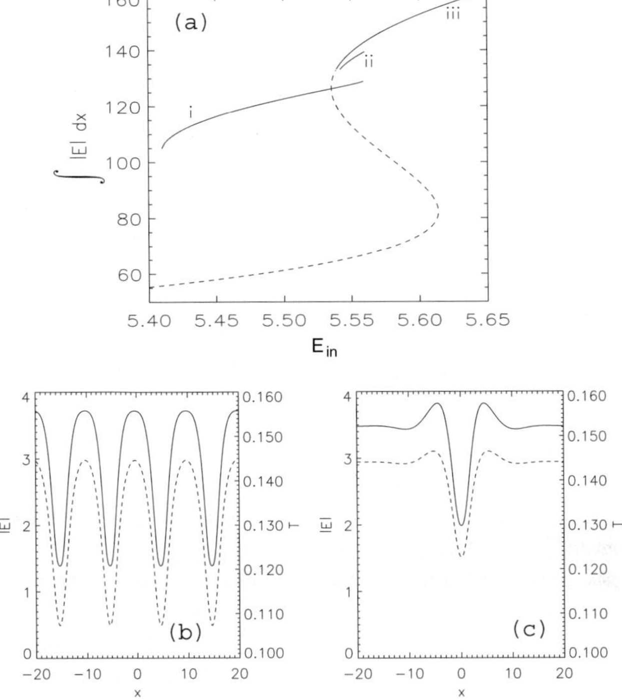

Fig. 9. (**a**) Solution branches for (i) rolls with $K \simeq 1.3K_c$ and (ii) stationary dark CS. Panels (**b**) and (**c**) correspond respectively to points on branches (i) and (ii). *Solid* and *dashed* lines respectively denote $|E|$ and T. The homogeneous solution (iii) is stable on solid parts of the curve and modulationally unstable in the *dashed* region. Parameters are $\alpha = 1$, $D_N = 0.2$, $D_T = 1$, $\gamma_N = 10^{-2}$, $\gamma_T = 10^{-5}$, $\Delta = 10$, $\theta = 0.3$, $\Xi = 80$, $\beta = 1.6$, $Z = 0.172$. (b) $E_{\text{in}} = 5.45$ and (c) $E_{\text{in}} = 5.55$

speed. The moving CS is asymmetric, with the dark spot "ahead" of the cool spot, and so always seeing a tuning gradient. The E and T profiles of a typical moving dark soliton are shown in Fig. 10.

One can perform a weakly nonlinear analysis near the bifurcation point between the stationary and moving solutions [92]. The calculation has similarities to that reported in [96]. Omitting mathematical detail, the result is an equation for the velocity of the moving CS which takes the form:

$$\partial_t v = av + bv^3 . \tag{10}$$

The linear and cubic coefficients a and b in equation (10) are constants which can be determined from the CS solution to the field equations. If they are of the same sign, then $v = 0$ is the only stationary solution, but if they are of opposite sign there are also two constant-velocity solutions $v = \pm\sqrt{-a/b}$. This is typical of a pitchfork bifurcation, with the $v = 0$ solution becoming

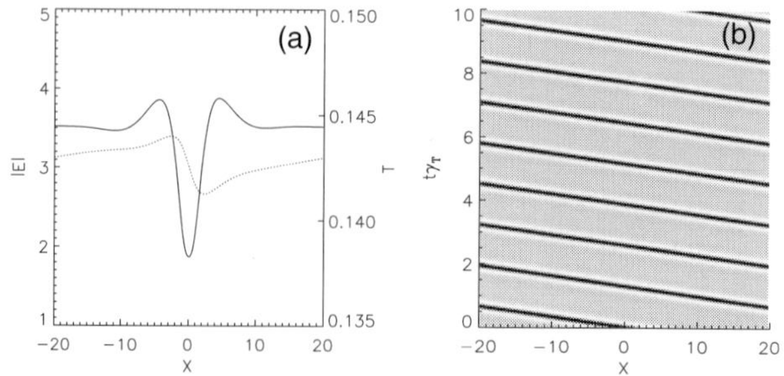

Fig. 10. (**a**) Magnitude of electric field (*solid line*) and temperature (*dotted line*) for a dark, moving CS. (**b**) Space-time plot of $|E|$ for a moving CS (periodic boundary conditions). Parameters are $E_{\rm in} = 5.39$ and $\alpha = 1$. All other parameters as in Fig. 9

unstable when a changes sign. Figure 11 shows a comparison [92] between the speed predicted by this analysis and the results from numerical simulations of equations (5)–(7). The agreement is quite good, even more than three times above threshold.

Note that equation (10) involves the second derivative with respect to time of the CS position, i.e. its acceleration. There is thus a superficial resemblance to a Newtonian force law obeyed by a massive particle. This inertial dynamics is in complete contrast with that of a normally stationary CS. An external perturbation drives the velocity of an otherwise stable stationary soliton, while it drives the acceleration of a spontaneously moving soliton. The latter

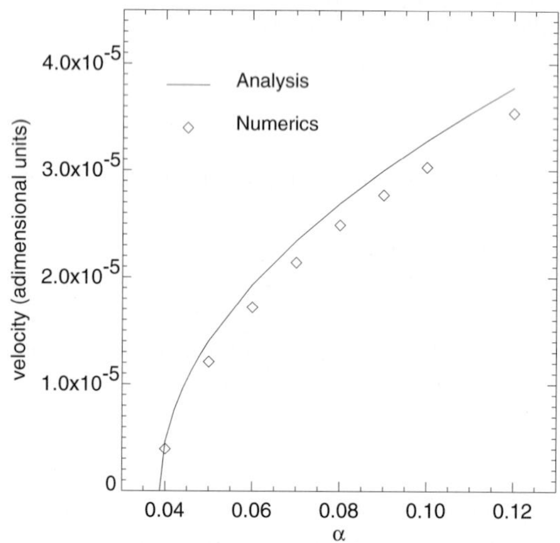

Fig. 11. The velocity of a moving soliton as a function of α: the *diamonds* indicate results from numerical simulations, while the *solid* line comes from weakly nonlinear analysis [92]

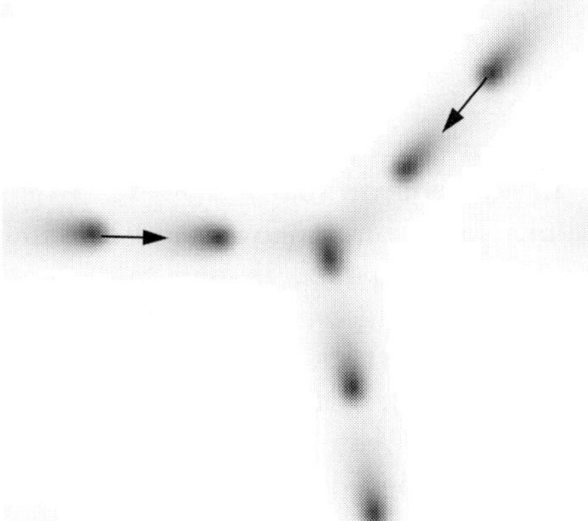

Fig. 12. "Stroboscopic" images of collision and merging of two 2D spontaneously-moving dark CS, showing inertial effects. Arrows indicate initial direction of motion. Shown is the temperature field, obtained in a simplified semiconductor CS model, in which the carrier dynamics is adiabatically eliminated ((21,22) of [92]). *Courtesy A.J. Scroggie*

acquires mass (or rather inertia) because, in essence, the mode which becomes unstable at the bifurcation is *identical* to the neutral mode, giving the CS an extra degree of dynamical freedom [96, 92]. (Note that DAS in lasers also obey a quasi-Newtonian equation of motion [1].)

Figure 12 illustrates the inertia effect in the collision of two self–moving CS in two dimensions. They collide and merge, with the outgoing CS traveling in a direction which, at first glance, looks like that of the mass-center of the incoming CS. One must recall, however, that here there is no "conservation of mass", because the outgoing CS is identical to each of the incoming ones. Nor is there conservation of momentum, in the Newtonian sense, because the *speed* of the outgoing CS is the same of that of the incoming ones, regardless of their directions of motion.

In summary, we have illustrated the existence, in both one and two spatial dimensions, of spontaneously moving cavity solitons in a model of a semiconductor microcavity. These solitons appear through an instability of the stationary solitons arising from localized cavity tuning variations coupled to temperature changes in the semiconductor induced by the light-carrier interaction. Regardless of the details of the present system, the essential ingredients appear to be the existence of stationary soliton solutions and the spontaneous creation of self-sustained parameter gradients. Thus similar phenomena should occur in a variety of optical and other systems. Indeed, spontaneously moving dissipative solitons are found in reaction-diffusion models [99] and were recently identified experimentally in gas discharge systems [100].

5 Solitons in a Single-Mirror Feedback Arrangement

In the preceding sections, it became evident that dissipative solitons are a robust feature of nonlinear cavities if they are operated close to a pattern-forming instability. In addition, in many cases the existence of CS was accompanied by bistability between plane-wave states, though this was not absolutely necessary. Hence it appears to be natural to look for solitons also in other nonlinear optical systems allowing for pattern-formation and/or optical bistability. In the context of pattern formation, the so-called "single-mirror feedback scheme" [101, 102, 103, 6] has emerged as an experimental and theoretical workhorse for studying complex self-organization behavior in space and time and hence it is maybe not too surprising that dissipative solitons can be also found in these systems [104, 105, 106, 107, 108, 109, 110, 111, 112, 113, 114, 115, 116].

5.1 Single-Mirror Feedback Arrangements: Mechanism of Spatial Instability

The basic scheme discussed in the following is depicted in Fig. 1d. It is a thin slice of a nonlinear medium irradiated by a spatially smooth beam (ideally a plane wave with uniform amplitude and phase). A plane feedback mirror is placed at a distance d after the medium to generate a counter-propagating beam in the nonlinear medium [101, 102, 103]. The basic idea in the idealized scheme is the spatial separation of the region in which nonlinearities are at work and the region in which diffraction takes place: If the slice is sufficiently thin, propagation in the slice can be neglected and it will provide just some phase modulation and amplitude modulation. In this way the theoretical treatment is tremendously simplified.

Now consider a weak perturbation in the form of a sinusoidal modulation of the index of refraction in a dispersive nonlinear medium, e.g. a Kerr medium. This modulation will give rise to a weak phase modulation of the transmitted wave. Diffraction during the propagation to the mirror and back will usually convert phase modulation, at least partially, into amplitude modulation. However, in a Kerr medium an amplitude modulation generates a corresponding index modulation, and so the original fluctuation can be enhanced and sustained by positive feedback. The connection between the longitudinal and transverse length scales at which conversion between phase and amplitude modulation takes place is given by the paraxial wave equation. The phase of an off-axis wave with transverse wave vector K evolves like

$$\phi = \frac{\mathrm{i}K^2 z}{2k_0} \tag{11}$$

after propagation over a distance z (k_0 wave vector of light). The periodicity of the phase in (11) in dependency on z is also at the heart of the so-called

Talbot-effect [117, 118], i.e. the periodic recurrence of amplitude and phase modulation along the propagation direction. Equation (11) implies that for a given phase shift the selected wave length scales with the square root of the mirror distance d [101, 102, 103, 6]. Note the similarity to the cavity case: both schemes rely on the differences in phase shift acquired during propagation for different tilt angles. However, in the cavity it is the interference condition provided by the cavity boundary conditions which converts the accumulated phase into amplitude, whereas here it is the interference of the off-axis wave with the on-axis carrier.

For a "dynamic Kerr" nonlinearity, as was assumed in the original proposal [101, 102, 103], the equation of motion for the deviation of the refractive index from its value in thermal equilibrium is

$$\frac{d}{dt}n = -\gamma n + D\Delta_{\perp}n + P \,, \tag{12}$$

where γ is a relaxation rate, $\Delta_{\perp}$ is the transverse part of the Laplacian modeling some nonlocal coupling within the medium, D is the corresponding "diffusion" constant, P denotes the pump rate being proportional to the sum of the intensities of the forward (E_f) and the backward (E_b) field, whose amplitudes are suitably scaled. The transmitted field is given by

$$E_t = E_{\text{in}} \exp\left(-\mathrm{i}\chi k_0 L/2\right) \,, \tag{13}$$

where L is the length of nonlinear medium and the dielectric susceptibility for the model Kerr medium assumed here is $\chi = 2n$. The backward field is calculated from the transmitted one by taking into account its propagation in free space and the reflection from the mirror with reflectivity R:

$$E_b = \sqrt{R}\mathrm{e}^{-\mathrm{i}d\Delta_{\perp}/k_0} E_t \,, \tag{14}$$

Note that E_t and hence E_b and P depend on n. Thus (12) is a nonlinear, nonlocal partial differential equation.

The refractive index change n could have many different physical origins. It might be due to highly off-resonant excitation in an atomic medium, due to a change of a carrier concentration in a semiconductor (at weak excitation) or due to a field-induced reorientation of the director orientation in a liquid crystal. Indeed, soon there were plenty of realizations of the proposed scheme using a variety of different nonlinear media. We refer to review articles and the references given therein for an overview [3, 6]. Typically, the first structures bifurcating at threshold are hexagons [102, 103, 3, 6] and in at least two cases it was shown explicitly shown that they may bifurcate subcritically from the homogeneous state [119, 120]. Hence, already one of the prerequisites identified above for the formation of solitons is fulfilled.

We will now concentrate on the case where the nonlinear medium is sodium vapor driven in the vicinity of the D_1-line. The use of an atomic vapor has the advantage of combining a rather high all-optical nonlinearity with a good optical quality and the possibility of a well established description of the light-matter interaction in terms of the semiclassical density matrix formalism. Very interesting experiments on localized states and solitons have also been performed in the so-called liquid-crystal light-valves [104, 105, 108, 109, 112, 113, 115, 116, 121].

5.2 Experimental Setup

The experimental setup (Fig. 13) consists of a cell containing sodium vapor in a nitrogen atmosphere and a plane feedback mirror at some distance d behind the medium. The cell is heated over a length of 15 mm, while the ends are cooled. The cell temperature is varied in the range around 320°C, yielding a particle density of $10^{19}\,\mathrm{m}^{-3}$ to $10^{20}\,\mathrm{m}^{-3}$. The buffer gas (pressure typically 200–300 hPa) reduces the diffusion, provides a homogeneous linewidth masking the Doppler effect and the hyperfine splitting of the ground state and it quenches the excited state in order to prevent diffusion of radiation. The cell is placed in a magnetic field **B** which has a non-vanishing longitudinal (z-axis) and transverse component (x-axis) with respect to the direction of the laser beam.

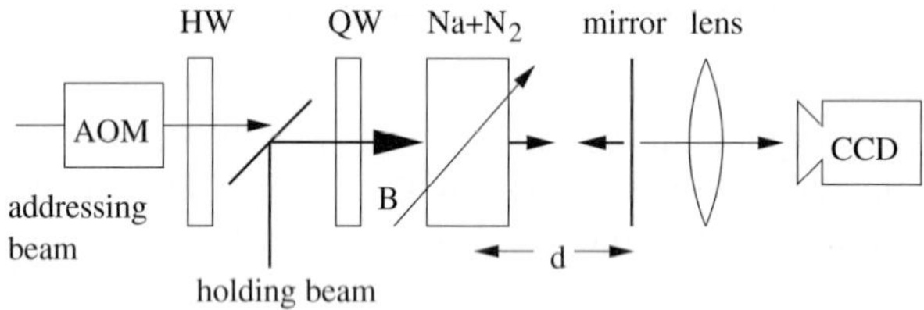

Fig. 13. Scheme of the experimental setup. AOM: acousto-optic modulator, QW: quarter-wave plate, $Na+N_2$: sodium-cell, CCD: charge-coupled device camera

The output beam of a cw dye laser, which is stabilized in frequency and intensity, is carefully spatially filtered by transmitting it through a single mode fiber. After the fiber, the beam is collimated ($w_0 = 1.5\,\mathrm{mm}$, radius at $1/e^2$–point of intensity) and injected into the sodium cell. We will refer to this beam as the "holding" or "background" beam. It is circularly polarized and the detuning Δ of its frequency with respect to the resonance of the Na–D_1 is chosen to be about 10 GHz to the blue side. Under these conditions the nonlinearity has both a dispersive and an absorptive contribution.

A small amount of light is split off the main beam and serves as an "addressing beam" which "ignites" the solitons. It can be gated by an acousto-optical modulator (AOM). The radius ($1/\mathrm{e}^2$) of the addressing beam is 0.17 mm. Its polarization state and frequency is discussed in more detail below. The power in the background beam within the sodium vapor is in the order of 100 mW and in the address beam about 1 mW.

The feedback mirror has a reflectivity of 90 ... 99%. The transmitted part of the light passes a lens which can focus more or less arbitrary planes of the beam path onto the image plane of a CCD camera. If not stated otherwise the imaged plane is at a distance $2d$ behind the sodium cell, i.e. the recorded intensity distribution corresponds to the reflected and re-entrant field E_b.

5.3 Optical Pumping Nonlinearity in Alkali Metal Vapors

For the conditions of the experiment, the optical nonlinearity on the D_1–line can be described in the framework of the homogeneously broadened $J = 1/2 \rightarrow J' = 1/2$ transition depicted in Fig. 14a. Angular selection rules allow the creation of a population difference between the two Zeeman sublevels of the ground state. If we assume e.g. excitation with σ_+-light, then only the $m_J = -1/2$–substate of the ground state absorbs, but both are repopulated by the relaxation processes (spontaneous emission and quenching of the excited state by collisions with nitrogen molecules). The relaxation is even isotropic for the conditions of the experiment since the populations in the sublevels of the excited state are rapidly equalized by the collisions with the buffer gas. This process is referred to as *optical pumping* [122] and leads to a rather efficient optical nonlinearity.

The existence of a population difference between magnetic sublevels implies the existence of a macroscopic magnetic moment of the sample. This will precess in an external magnetic field. If the direction of propagation of the laser beam is used as the axis of quantization (z-axis), then the presence of a transverse component B_x of the magnetic field causes changes in the z-component of the magnetic moment, i.e. "spin-flips" between the Zeeman sublevels occur (Fig. 14b). The mechanism is most effective if the longitudinal component B_z vanishes. In a quantum mechanical description the light-induced level-shift has to be taken into account (Fig. 14b). It acts like an extra contribution to the z-component of the magnetic field and is proportional to the intensity times the detuning of the laser with respect to the atomic resonance.

In a formal description it is convenient to introduce the Bloch vector $\mathbf{m} = (\mathbf{u}, \mathbf{v}, \mathbf{w})$ which is proportional to the ground–state orientation of the sodium vapor [123]. Its equation of motion is given by [124]:

$$\partial_t \mathbf{m} = -(\gamma - D\Delta_\perp)\mathbf{m} + \hat{\mathbf{e}}_z P - P\mathbf{m} - \mathbf{m} \times \boldsymbol{\Omega} \, . \tag{15}$$

The first three terms are identical to the ones we know already from the Kerr case, (12). Here, they are interpreted as collision-induced relaxation, thermal motion of the sodium atoms in the buffer gas atmosphere, and optical pumping. The next term describes the saturation of the optical nonlinearity and would be also present in models for two-level atoms and semiconductors. The

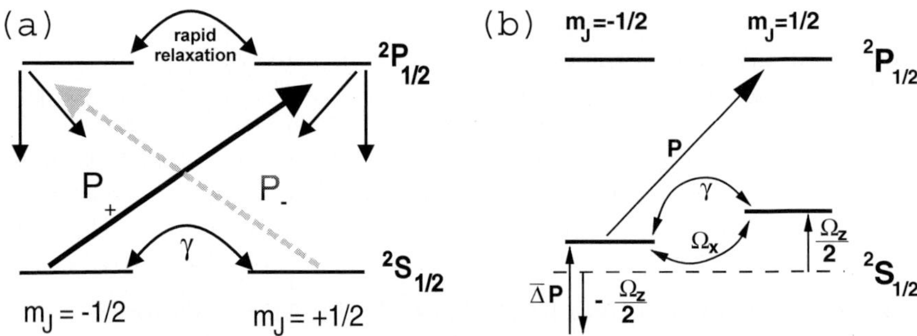

Fig. 14. (**a**) Kastler diagram of a J = 1/2→J′ = 1/2 transition driven by σ_+-light (*solid black arrows*) and/or σ_--light (*dashed grey arrows*). (**b**) Modified Kastler diagram for blue detuned excitation with σ_+-light in the presence of an oblique magnetic field. For simplicity, the splitting of the excited state is not shown. The *dashed* line indicates the position of the ground state of the bare atom

components of the vector $\boldsymbol{\Omega} = (\Omega_x, 0, \Omega_z - \bar{\Delta}P)$ are the Larmor frequencies produced by B_x and B_z, respectively, with the latter being modified by a light-shift term [123]. $\bar{\Delta}$ is the detuning between the incident field and the atomic transition, normalized to the relaxation constant Γ_2 of the polarization of the medium. The complex susceptibility of the vapor depends on the longitudinal component w of the orientation through [123]

$$\chi = -\frac{N|\mu|^2}{2\hbar\epsilon_0\Gamma_2}\frac{\bar{\Delta}+\mathrm{i}}{\bar{\Delta}^2+1}(1-w) \equiv \chi_{lin}(1-w) , \tag{16}$$

where N is the sodium particle density. The equations for the transmitted and re-entrant field distributions were given in (13) and (14).

The homogeneous solution for the steady state of w is given by

$$w_0 = \frac{P_0}{\gamma + P_0} \quad \frac{(\Omega_z - \bar{\Delta}P_0)^2 + (\gamma + P_0)^2}{(\Omega_z - \bar{\Delta}P_0)^2 + (\gamma + P_0)^2 + \Omega_x^2} , \tag{17}$$

where $P_0 = P_{\mathrm{in}}(1 + R|\exp(-\mathrm{i}k_0 l\chi_{lin}(1-w_0)/2)|^2)$. Equation (17) has the interesting property that it has a a resonance–like dependence on the term $\Omega_z - \bar{\Delta}P_0$, i.e. there can be a pronounced minimum of the orientation for a well–defined finite light intensity (see Fig. 15). This is a manifestation of a "light-shift induced level-crossing" produced by the combined action of Zeeman splitting and light-shift (Fig. 14b, [124, 6]). In the presence of optical feedback the corresponding characteristic curve describing the homogeneous solution of the orientation can become very steep (see Fig. 15) or even bistable [124, 111]. There is some analogy between the z-component of the magnetic field in the feedback system and the detuning parameter in a cavity, although we warn that one should not take this analogy too seriously: Ω_z as well as θ determine where (and whether) on the pump axis OB occurs. In addition,

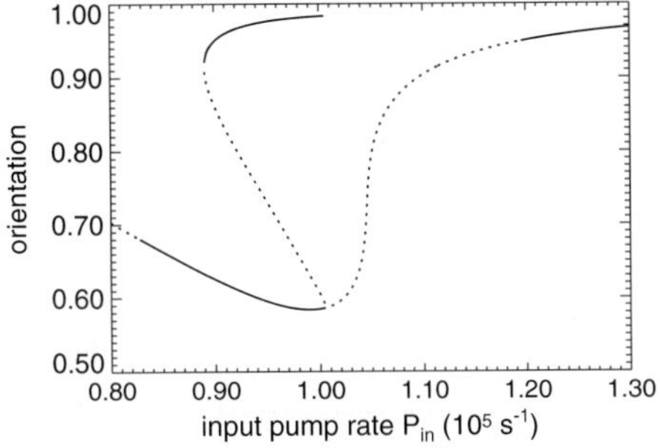

Fig. 15. Solutions and their stability properties, both obtained from semi-analytical calculations. The sigmoid curve extending over the whole range of pump parameter depicts the homogeneous, steady–state solution (orientation versus input pump rate). The *dashed* parts are unstable against periodic perturbations at a finite transverse wave number. The additional curve in the central range of pump parameter depicts the maximum orientation of a branch of feedback solitons (FS) bifurcating subcritically from the homogeneous solution at $P_0 = 100500\,\mathrm{s}^{-1}$. Again the *dashed* part is unstable. Parameters: $\Omega_x = 1.2 \times 10^5\,\mathrm{rad/s}$, $\Omega_z = 9.0 \times 10^5\,\mathrm{rad/s}$, $\Delta = 10.0\,\mathrm{GHz}$, $d = 70\,\mathrm{mm}$, $N = 3.0 \cdot 10^{13}\ \mathrm{cm}^{-3}$, $D = 237\,\mathrm{mm}^2/\mathrm{s}$, $\gamma = 1.5\,\mathrm{s}^{-1}$, $\Gamma_2/(2\pi) = 1.6\,\mathrm{GHz}$, $L = 15\,\mathrm{mm}$, $R = 0.915$ (from [114])

a nonlinear resonance might occur in both cases, if Ω_z or θ are ramped at constant pumping. However, there is no continuous dependence of the selected spatial scale on Ω_z as it is on θ in the cavity case.

In the following, we will investigate the behavior of the system in the vicinity of the minimum of the characteristic curve in a situation in which the homogeneous characteristic is very steep, i.e. close to being bistable ("nascent optical bistability", Fig. 15).

6 Basic Results

6.1 Switch-On and -Off Procedure

Figure 16 illustrates the process of igniting and erasing a FS. The system is prepared in an "unstructured" state by slowly increasing the power of the background beam from low values until the working point close to the minimum of the transmission curve is reached (leftmost image in Fig. 16). Then the polarization of the addressing beam is adjusted by means of a half–wave plate such that it is the same as that of the background beam after recombination. This means that the two beams are circularly polarized with the same helicity after the quarter–wave plate. When the addressing beam is then switched on, a bright spot emerges at the position of the addressing beam (second image from the left). This spot remains stable after the addressing beam is switched off (central image of Fig. 16). The shape of the central lobe deviates from a Gaussian and it is surrounded by weak dark and bright diffraction fringes (see also Figs. 19, 18), i.e. it deviates from the Gaussian profile of the addressing beam. This indicates that this bright peak is an attractor of the dynamics.

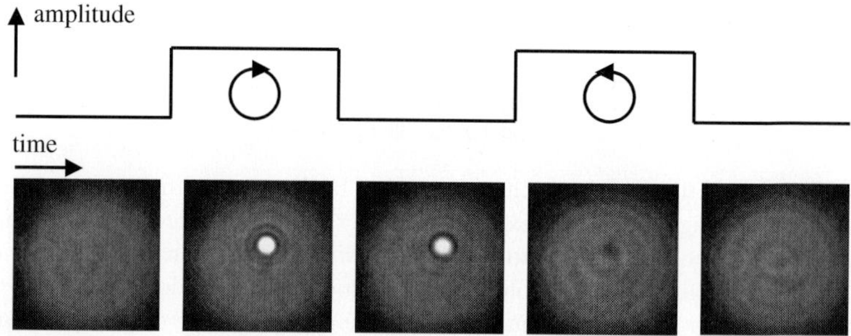

Fig. 16. Ignition and erasure of a localized state with the addressing beam. The upper part of the figure shows schematically the amplitude of the addressing beam, the lower row the near field intensity distribution of the beam reentering the medium. Parameter: $d = 70\,\mathrm{mm}$, $\Delta = 8.2\,\mathrm{GHz}$, $B_\perp = 2.59\,\mu\mathrm{T}$, $B_z = 19.2\,\mu\mathrm{T}$, cell temperature 312.3°C, $p_{\mathrm{N}_2} = 197\,\mathrm{hPa}$, $P_{in} = 100\,\mathrm{mW}$. The images are plotted in a linear grey–level scale with white denoting high intensity. The absolute scale was adjusted such that the background beam is always clearly visible; therefore the center of the FS is overexposed. The frame size is $2.6\,\mathrm{mm}\times 2.6\,\mathrm{mm}$ (from [111])

Now the polarization of the addressing beam is rotated by 90° by turning the half–wave plate; therefore the background beam and the addressing beam are still circularly polarized in the medium but with *opposite helicity.* If the addressing beam is switched on again at the position of the peak, a dark hole appears (second image from the right in Fig. 16). The hole disappears after the addressing beam is switched off again. Thus the bright peak is erased and the homogeneous state is reached again.

This procedure can be repeated over and over again. This observation proves that there is bistability between a state with one bright peak on an unstructured background and the unstructured background alone. This is the expected behavior for a FS or a CS.

The remaining question needed to be answered is why we can utilize "polarization" of the opposite helicity to erase a FS, whereas we stated in the first sections that an addressing beam of opposite phase should be used. The reason becomes immediately clear by looking at Fig. 14a. Circular polarization components of opposite helicity pump antagonistically. Hence, it is possible to mimic destructive interference by using the opposite polarization component (see, e.g., (7) of [111] for the complete equations). This has the additional advantage that there is no need for interferometric stability between the beam pathes of the holding and the addressing beam, since light of opposite polarization does not interfere. Indeed, the frequency of the addressing beam was shifted from the frequency of the holding beam by 140 MHz due to the presence of the AOM. Hence, even the switch-on is created by an incoherent superposition of the holding and the addressing beam again eliminating the need for interferometric path length stability. We remark that, of course, switching of FS is also possible by using constructive and destructive interference of coherent beams of the same circular polarization, however this scheme is considerably less robust.

These observation obviously motivate considerations on whether polarization degrees of freedom might be utilized for phase–insensitive control of CS in semiconductor microcavities. We remark that the phase–insensitive control of all-optical flip–flops based on small–area semiconductor amplifiers was achieved by injecting beams of very different wavelength (several tens of nm) which perturb incoherently the carrier concentration [125]. For the case of CS, interesting studies exist using beams of orthogonal linear polarization [85]. However, the mechanisms of incoherent switching are unclear and a complete control of ignition and erasure was not achieved, yet.

In general, the FS are not stable at the positions at which they are ignited, but they start to drift after the addressing beam is switched off. This is due to a drift motion of the FS in intensity and phase gradients of the light field as discussed in the preceding sections for CS. In a rotationally symmetric holding beam, the FS will be trapped either in beam center or on a ring with a certain distance from the center, since these are the points equivalent by symmetry. For the present situation, the amplitude gradients pull the FS towards the center, whereas the phase gradients push them out. As a result, there is a stable equilibrium at a finite distance from beam center. This is very apparent where more than one FS exist on the beam (Fig. 17). This issue is discussed in more detail in [114, 110].

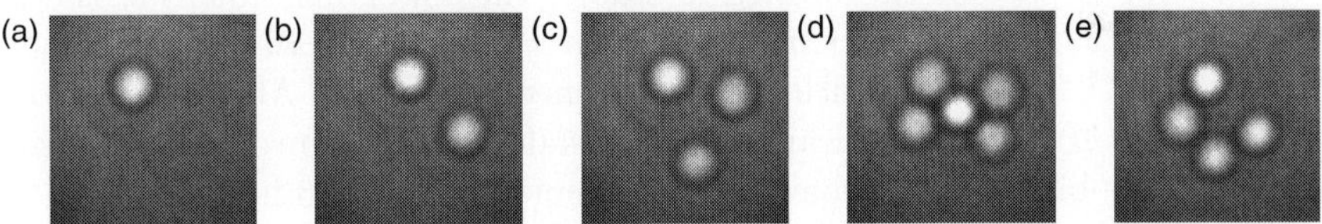

Fig. 17. Stable clusters of localized structures. The images are overexposed in order to emphasize the diffraction fringes surrounding each localized structure. Parameters: $B_\perp = 0.78\,\mu\mathrm{T}$, $B_\parallel = 14.40\,\mu\mathrm{T}$, $\Delta = 18.6\,\mathrm{GHz}$, $d = 70\,\mathrm{mm}$, $p = 310\,\mathrm{mbar}$, $T = 315^\circ$ C. Power: (**a**) 131 mW, (**b**) 133 mW, (**c**) 133 mW, (**d**) 135 mW and (**e**) 138 mW. The transverse size of the images is 2.6 mm (from [107])

6.2 Existence of FS: Theory

Figure 15 shows the calculated homogeneous solution for the orientation versus the input pump rate for a typical situation in which the experiments with FS have been carried out. After the minimum, the homogeneous characteristic has a sigmoid form, which corresponds to the situation of "nascent optical bistability" for which many semi-analytical studies of CS were done [64, 74]. The sections of the homogeneous solution that are drawn dashed are linearly unstable against periodic perturbations at a finite wave number, i.e. there is a pattern forming modulational instability (MI) [126]. We will investigate this in more detail below (see Sect. 7.2).

For a calculation of the soliton branches a method similar to the one discussed above for CS is utilized, i.e. we will analyze in the following the system which is obtained from (15) by setting the time derivative to zero. In order to simplify the calculations, radial symmetry is assumed. Then, the stationary states are found by discretizing the differential equation in space and searching for solutions of the resulting large set of coupled equations with a Newton method. The fact that in the feedback problem, (14), the Laplacian operator appears in an exponential whereas it is only first order in the cavity case, (1), constitutes a technical difficulty which is solved by evaluating the derivatives by a Hankel transform method in Fourier space [127].

The calculations show that a branch of FS emerges from the homogeneous state at the point where the homogeneous state becomes modulationally unstable near the minimum of the characteristic ($P_{\rm in} = 1.005 \cdot 10^5 {\rm s}^{-1}$, Fig. 15). The branch bifurcates backwards, turns around at $P_{\rm in} = 0.891 \cdot 10^5 {\rm s}^{-1}$ and terminates again at a point which coincides with the MI point with a relative deviation of less than 10^{-3}. Here, the FS becomes unstable against the formation of spatially extended patterns. With the present accuracy of the program code, it is not sure, whether the two points coincide or not.

The upper part of the branch is stable, the lower unstable, i.e. the bifurcation to the FS is subcritical. Between $P_{\rm in} = 0.891 \cdot 10^5\,{\rm s}^{-1}$ and $P_{\rm in} = 1.005 \cdot 10^5\,{\rm s}^{-1}$ FS coexist with the homogeneous solution. At the limit point $P_{\rm in} = 0.891 \cdot 10^5\,{\rm s}^{-1}$ they disappear via a saddle–node bifurcation. The similarity to the bifurcation behavior of CS depicted in Fig. 3 is obvious.

Figure 18a shows the calculated profile of a FS on a plane–wave background. It consists of a high–amplitude central peak which is surrounded by oscillating tails. This matches the experimental observation. For $x \to \pm\infty$ it approaches the homogeneous solution. Numerical simulations of the full two-dimensional equations show that it can exist anywhere in the transverse plane, if a plane wave input beam and periodic boundary conditions are used. Furthermore several FS can exist at the same time at different positions of the transverse plane. These properties are completely identical to those of CS (cf. Fig. 5b for the profile of a corresponding CS). In the experimental realization there is – as already discussed – no complete arbitrariness of positioning since

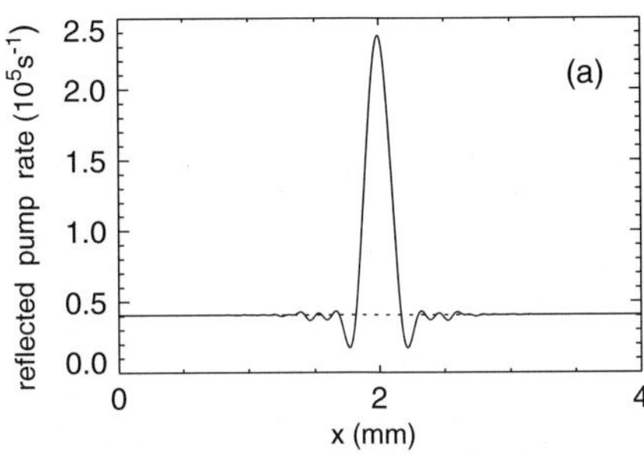

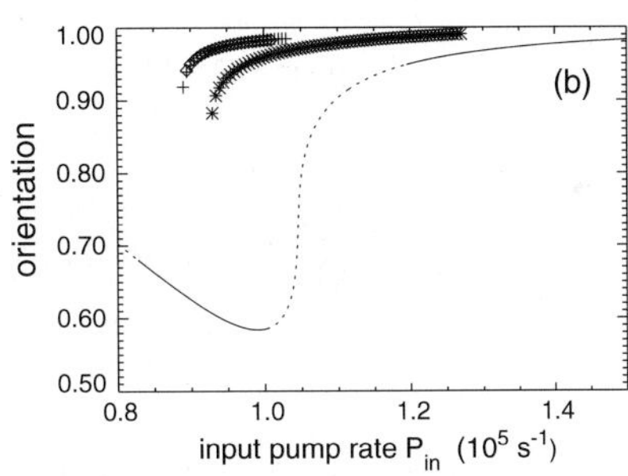

Fig. 18. (**a**) Transverse cut through the center of a calculated FS for a plane–wave background beam for $P_{\rm in} = 0.96 \cdot 10^5 {\rm s}^{-1}$ (other parameters as in Fig. 15). The graph shows the pump rate of the reflected field versus one transverse coordinate. (**b**) Bifurcation diagram for spatially extended periodic patterns. Shown is the maximum orientation versus the pump rate. The lower curve corresponds to the homogeneous solution. It is unstable in the *dotted* part. The wavenumbers of the patterns are: Double crosses (*lower right curve*) 17.1 $\rm mm^{-1}$, crosses 10.2 $\rm mm^{-1}$, rhombs 7.9 $\rm mm^{-1}$ (from [114])

the symmetry of the system is reduced due to the radial gradients induced by the Gaussian input beam.

Since the bifurcation to FS is subcritical one can create a FS by a large–amplitude perturbation (sometimes called "hard" perturbation) in the bistable region before the bifurcation point. This is the region in which the experiment described above was performed. The unstable branch will be very important in the turn-on experiments discussed below since it defines the separatrix in phase space.

6.3 Mechanism of Stabilization

As mentioned in the description of the experimental setup, the plane imaged onto the camera can be varied. Thus the change of the intensity distribution in the light field during its propagation in the free space between the nonlinear medium and the feedback mirror and also behind the mirror can be monitored. The latter corresponds to the intensity distribution in the backward beam, of course. Corresponding results are shown in Fig. 19a. At the position of a FS, there is an intensity peak at the exit window of the cell. During the propagation of the beam, however, the central intensity increases strongly and simultaneously the width of the peak decreases. This behavior is possible if the main effect of the FS is a phase modulation that creates locally a concave deformation of the wavefront which makes the beam locally converge to a focus. The central intensity reaches a peak and then declines, but when the beam reenters the medium after being reflected it still has a very pronounced intensity peak, surrounded by wiggles which obviously correspond to diffraction fringes. The behavior is reproduced well in simulations

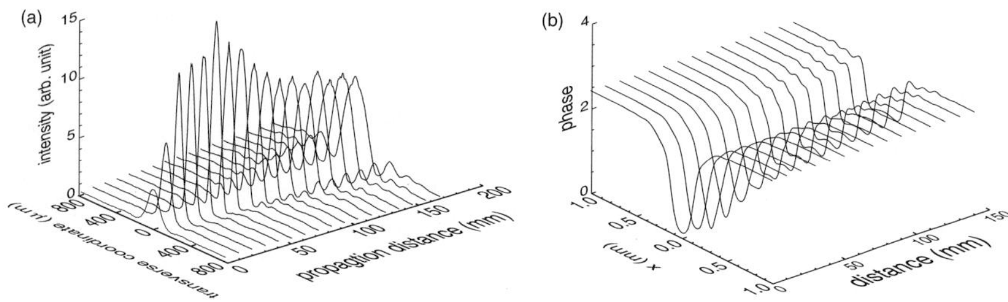

Fig. 19. (**a**) Observed profiles of the intensity distribution produced by a FS during propagation from the sodium cell to the mirror and back. Parameters: $B_x = 4.17\,\mu\mathrm{T}$, $B_z = 8.23\,\mu\mathrm{T}$, $\Delta = 13.5\,\mathrm{GHz}$, $d = 80\,\mathrm{mm}$, $P_{las} = 107\,\mathrm{mW}$, $T = 340.0°\mathrm{C}$, $p_{N_2} = 216\,\mathrm{hPa}$ (from [107]). (**b**) Numerically calculated phase distribution produced by a FS during the propagation between the sodium cell and the mirror and back. Parameters: $\Omega_x = 1.2{\cdot}10^5\,\mathrm{rad/s}$, $\Omega_z = 9.0{\cdot}10^5\,\mathrm{rad/s}$, $\Delta = 10.0\,\mathrm{GHz}$, $\gamma = 1.5\,\mathrm{s}^{-1}$, $D = 237\,\mathrm{mm}^2\mathrm{s}^{-1}$, $\Gamma_2 = 1.0{\cdot}10^{10}\,\mathrm{s}^{-1}$, $N = 0.3{\cdot}10^{20}\,\mathrm{m}^{-3}$, $d = 70\,\mathrm{mm}$, $P_{\mathrm{in}} = 96000\,\mathrm{s}^{-1}$ (from [110])

(Fig. 5a of [110]). These also confirm that there is a strong local curvature of the phase surface introduced by the FS (Fig. 19b).

It can be concluded that the mechanism of the formation of the FS is the following: the address beam locally introduces a distribution of the orientation which acts like a focusing lens. This lens introduces a phase encoding which makes the diameter of the local field distribution shrink until the focus is reached. The process works best if there is a strong dependence of the orientation on the intensity, as in the case of nascent OB. Indeed in the corresponding intensity range the medium is self-focussing, while it is self-defocusing at the intensity level of the background beam which corresponds to the descending slope of the plot (Fig. 15). It will be of importance in Sect. 7 that strong diffraction fringes occur in the process.

We mention that the stabilization of localized states by self-induced lensing was suggested first, to our knowledge, in [128, 49] for plano-planar cavities with a liquid-crystal nonlinearity. It can be shown that even in a simple lensing model the scaling of the size of a FS with wavelength and mirror distance is the same as the scaling for the pitch of a pattern due to the Talbot effect, (11), [129]. We commented on the generality of this scaling behavior already before in the CS case (see also [49]).

This interpretation of a FS seems to correspond best to the "soliton-in-a-box"-picture of CS, i.e. in the "half-cavity box" formed by the medium and the mirror a self-consistent localized state forms due to the interplay of nonlinear refraction (lensing) and diffraction. We mention that the analogy can be made even closer by regarding the medium as an "active mirror" [130, 131], although to our knowledge this idea was not worked out for localized states. However, we will see that the "part-of-a-pattern"-interpretation can also be applied in the FS case.

7 Interaction Behavior

7.1 Interaction Between Two Solitons

An analysis of images containing several FS (e.g., Fig. 17) suggests that certain distances between single FS are preferred by the system. For a detailed analysis we recorded a large number of images for fixed experimental parameters at a constant time interval of 200 ms. Within this time, typically spontaneous transitions between different configurations of FS occur. The observed configurations varied in the number of constituents (from 1 to 3) and in the position of the FS on the preferred ring around the beam center. These different configurations are interpreted as manifestations of the high degeneracy of the situation because of multistability due to symmetry and due to the coexistence of several states. Random transitions between these states occur due to noise. For the histogram of distances displayed in Fig. 20g the subset of images containing only two FS has been considered.

A preference for three discrete values of the distance is clearly visible from Fig. 20. By weighted averaging over the three humps we find the values 445 µm, 573 µm and 717 µm. Figure 20 a–c shows typical examples of these configurations. The peaks for larger distances between two FS are broader than the first peak; the reason for this result is probably that the configurations with a larger distance are more sensitive to noise or parameter fluctuations. The images of the bound states depicted in Fig. 20a and d are the analogue to the cavity case depicted in Fig. 5f.

For studying the interaction behavior calculations are carried out under the assumption of a plane wave background beam. In this way the drift of the FS in a gradient of the background is eliminated. It turns out that – depending

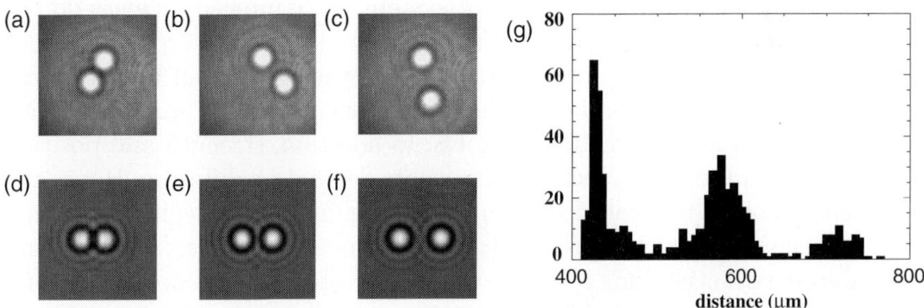

Fig. 20. Stable configurations of two localized structures. (**a**)–(**c**): Experiment (overexposed in order to visualize the diffraction fringes), parameters: $B_\perp = 2.36\,\mu\text{T}$, $B_\parallel = 10.98\,\mu\text{T}$, $\Delta = 13.9\,\text{GHz}$, $d = 63\,\text{mm}$, $p = 200\,\text{mbar}$, $T = 316°\text{C}$. (**d**)–(**f**): Numerical simulation, parameters $B_\perp = 3.94\,\mu\text{T}$, $B_\parallel = 15.72\,\mu\text{T}$, $\Delta = 9.5\,\text{GHz}$, $d = 63\,\text{mm}$, sodium particle density $N = 0.3 \cdot 10^{14}\,\text{mm}^{-3}$. The transverse size of the images is 2020 µm. (**g**) Histogram of distances between two localized structures; 669 images are evaluated (from [107])

on the initial distance between two FS – they either attract or repel each other until a stable distance is achieved. The resulting histograms are much sharper than the experimental one proving that the stable configurations belonging to bound states of FS are indeed discrete. Figure 20d–f show the configurations for the three smallest stable distances between two FS. The corresponding distances are 393 μm, 533 μm and 691 μm, i.e. they show a rather close correspondence to the experiment. We find even larger distances than these three, but they are rather unstable with respect to noise.

As explained in the sections on CS, stationary solitons have a neutral mode related to the translational symmetry. Consider now the perturbation of a soliton with respect to another soliton. The oscillations of the phase and amplitude of the soliton field as it dies away into the background field will provide a perturbation to the other soliton. This perturbation induces a drift motion with some velocity. We can expect to find equilibrium positions where the relative velocity is zero, and that these define stable or unstable bound states of two solitons. Because the solitons are surrounded by several diffraction fringes, a large number of equilibrium positions can be expected between two solitons. Only the smallest distances, however, are observed in the limited beam. Furthermore we observe numerically that the stability of a bound state of FS against perturbations decreases with increasing distance, which is consistent with the fact that the amplitude of the fringes decreases. The pitch of the diffraction fringes and the corresponding distances between the FS depend on the mirror distance d with roughly the same scaling as for extended patterns (11). They can be also changed to some extent by introducing a Fourier filter in the feedback loop [112, 127].

Obviously, it is worthwhile to strive for analytical insight into the FS interaction, especially on the establishment of a concrete relationship between the pitch of the diffraction fringes and the mutual distances. This is however quite difficult to perform in the present system, since intensity fringes do not interact directly, but the interaction is mediated rather by the spatial profiles in the three different components of the Bloch vector induced by the intensity profile. The spatial profile of these components can be quite different, e.g. u has a hole at the positions of a FS. In addition, strength and position of the ripples in the Bloch vector components depend on diffusion. Hence the situation is more complicated than in a Kerr medium (with negligible or only small diffusion), in which there is a 1:1 correspondence between intensity and the induced refractive index. Nevertheless, our observations are backed up, complemented and extended by recent nice experiments on FS interaction in liquid-crystal light-valves with a Kerr-like nonlinearity [112, 115]. This establishes a rather universal behavior of FS. A corresponding interaction behavior is also found for CS [21, 54, 71, 75, 81, 84] for nonlinearities based on a intensity dependence of the refractive index as well as on saturable absorption. For CS, semi-analytical studies were performed revealing quantitatively the interaction potential [21, 54, 71, 132]. Actually, the universality applies not only for FS and CS but even for dissipative solitons in non-optical systems [99] (and Refs. in [107]). However, in some sense this is not too surprising since the qualitative aspects of solitons with oscillatory tails should be fixed by symmetry considerations.

An analysis of the clusters with more than two constituents in Fig. 17 reveals that most of the distances observed between the constituents appear already in the bound states of two single FS for identical parameters; the deviations are characteristic for the type of cluster. It should be kept in mind, however, that in the experiment the shape of the clusters is not only a result of the interaction of FS with each other, but also of the interaction of the FS with the inhomogeneous holding beam – favoring positions on ring segments.

7.2 Connection Between Solitons and Extended Patterns

The results of a linear stability analysis of the homogeneous state reported in Fig. 15 show that the FS branch arises from a point in which the homogeneous solution becomes unstable versus spatially periodic perturbations, i.e. pattern formation. Indeed, it was also established experimentally, that the existence region of FS is at the boundary of the existence region of spatially extended patterns (Fig. 4 of [111]).

Figure 18b shows three branches of perfect periodic hexagonal pattern existing in the parameter regime discussed in Fig. 15. These patterns differ by their length scale. The branches of two of these (with a wavenumber of $10.2\,\mathrm{mm}^{-1}$ and $7.9\,\mathrm{mm}^{-1}$) are very similar to each other and to the FS branch depicted in Fig. 15 except for the fact that the pattern branches extend in the MI region. The third branch has a wave number of $17.1\,\mathrm{mm}^{-1}$ and has a different amplitude and existence range. Interestingly, the next-neighbor distance between two spots of the hexagonal patterns corresponds nearly exactly to the distance between the FS in bound states of two FS for these parameters. This holds for all three patterns (distances in bound states: 434 µm, 720 µm, 937 µm; next neighbor distances in patterns: 424 µm, 711 µm, 918 µm [127]) This suggests a "FS = part of a pattern" interpretation like in the case of CS discussed above. Using suitable initial conditions also a great variety of clusters and of patches of hexagonal patterns surrounded by voids ("localized patterns") can be found (Fig. 21a–c). It is also possible to create patterns in which single or multiple constituents are missing (Fig. 21d, e), [133]). Corresponding states were discussed above for CS (see also [56, 64, 71, 73, 87]).

We warn that the perfect patterns mentioned above as well as the closely packed localized patterns form only, if they are seeded. Otherwise spatially irregular states with clusters of FS and voids in between them emerge, if simulations are started from noisy initial conditions above or below the MI point (Fig. 19 of [114]). These show a great similarity to the "cracking hexagons" obtained for CS [55].

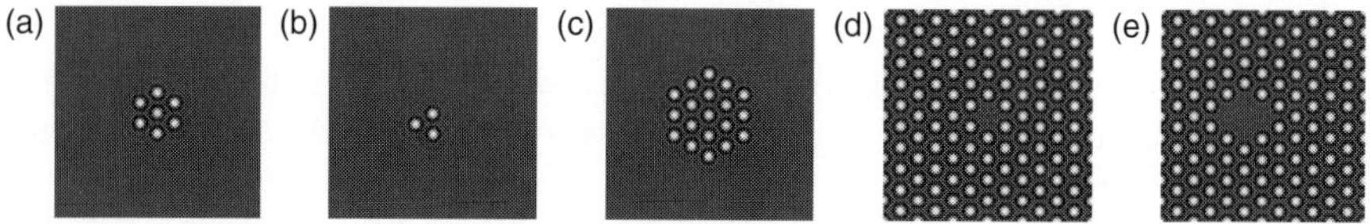

Fig. 21. Localized patterns with the same pitch as the extended pattern with $10.2\,\mathrm{mm}^{-1}$ in Fig. 18b). $P_0 = 90000\,\mathrm{s}^{-1}$ (from [114])

The results on bound states of FS or CS indicate the intriguing possibility to build up nearly arbitrary states from (weakly) interacting FS or CS which serve as "quasi-particles". The perfect patterns might be also interpreted as being due to weakly interacting FS (see the discussion above for CS, see also [84, 87]). The interpretation of the pattern with a wavenumber of $17.1\,\mathrm{mm}^{-1}$ is not quite clear, since there the constituents differ significantly from the shape of a single FS.

Finally, we note that obviously bistability between plane wave states is not a prerequisite for FS (see Fig. 15). We obtained this result for CS before (see also [57, 132]). What is needed is bistability between a plane wave state and a spatially extended pattern. Nevertheless, we find numerically that FS only exist, if the slope of the homogeneous characteristic is "sufficiently high", i.e. if one is "sufficiently close" to nascent OB (see Fig. 6 of [114]). We are not aware of a rigorous analysis of the situation, but it seems to be reasonable that large-amplitude structures like FS or CS are somehow favored by the existence of the second high-amplitude branch. For patterns, modifications of the standard bifurcation in the vicinity of bistability between plane wave states is known [134, 133, 132].

8 Applications

Arrays of cavity solitons may have applications in parallel information processing [135]. They can be created at any location by a suitable address pulse, and are non-diffracting and dynamically stable, all of which makes them suitable "bits" for image or data capture, storage and processing. This has been demonstrated numerically in [65] and confirmed in the prototype experiments discussed above and in [104, 105, 85, 88, 86]. Cavity and feedback solitons also offer functionalities which are beyond any micro-structured material array, whether optical or electronic. In particular, we have seen that they can be optically manipulated, e.g. by imposing a spatial phase profile on the driving field [65, 77, 54]. Processing schemes which take advantage of their unique properties may avert the unequal competition with silicon which has plagued other all-optical processing schemes.

Any perturbation to the pump field which has a finite gradient at the soliton location will couple to the "neutral mode" identified above, and cause the soliton to move. This has implications for the response of CS and FS to noise and to any stray gradients, and also for interaction between them, as is illustrated by the experimental results on FS discussed above. This coupling can also be used positively, to *control* the motion and location of the solitons through the spatial phase profile of the pump field. One can think of the solitons in a "landscape" determined by the phase of the holding (control) field, a landscape in which they move in response to phase gradients. Thus a simple memory array [65] consists of a regular landscape of "hills" and "valleys", with the solitons attracted to the peaks. Unlike one formed from machined pixels, however, this landscape is reconfigurable by changing the phase profile of the control field. This allows these "soliton bits" to be manipulated, by either global or local reconfigurations of the control field. This plasticity opens up possibilities for novel processing functions and applications [135]. In the following, we will discuss two prototype experiments which give proofs of principle of these ideas for FS in sodium vapor.

First, we show how "FS bits" can be pinned to a fixed position in the transverse plane by spatial modulations of the holding beam. For the experiment shown in Fig. 22 this proposal has been realized in a very simple manner by introducing a quadratic aperture into the input beam and thus inducing a quadratic diffraction pattern. The small amplitude modulations appearing in the background beam can be seen in Fig. 22a. In addition, there will be phase modulation. We are able to create a 2×2–Array of FS: When the input power is increased, the FS pop up spontaneously on the four positions prescribed by the diffraction pattern. Probably because of small misalignments in the experimental setup we did not succeed in producing bistability between the states with and without FS on all four positions simultaneously. In addition to the pinning of FS, this feature would be necessary for an optical memory.

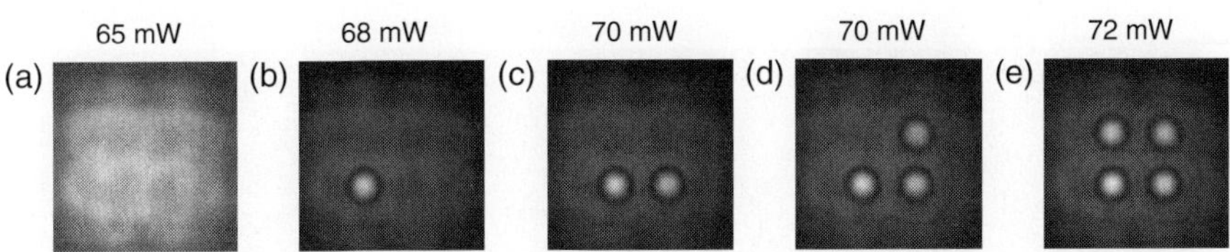

Fig. 22. Pinning of FS on positions that are prescribed by modulations in the background light field. Parameters: $B_\perp = 2.38\,\mu\text{T}$, $B_z = 7.55\,\mu\text{T}$, $\Delta = 12.0$ GHz, $d = 70\,\text{mm}$, $T = 303.0°\text{C}$ (from [110])

Another possible application that uses FS as optical bits is demonstrated in Fig. 23. Here the feedback mirror has been slightly tilted in order to initiate a drift motion of the FS. In the experiment shown in Fig. 23 the direction of the drift motion is upwards. For $t = 0\,\mu\text{s}$ there is one FS in the upper part of the background beam and a second FS is being created with the addressing beam in the lower part. Therefore in the following the system is in a state

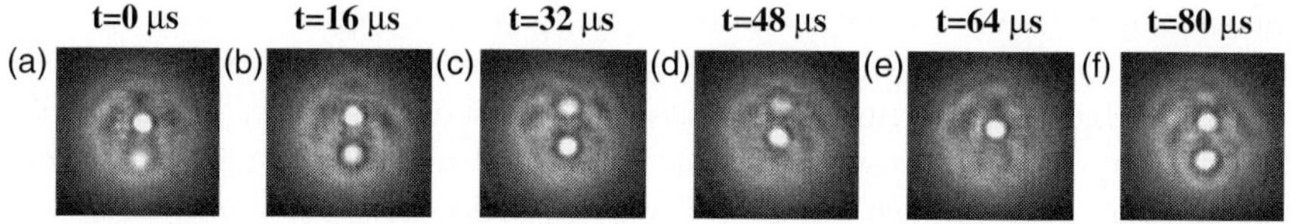

Fig. 23. Optical buffer register. The FS are drifting upwards since the feedback mirror is tilted. New FS are created in the lower part of the image by the addressing beam with a frequency of 14 kHz. Parameters: $B_\perp = 0\,\mu\mathrm{T}$, $B_z = 14.41\,\mu\mathrm{T}$, $P_0 = 115\,\mathrm{mW}$, $\Delta = 11.9\,\mathrm{GHz}$, $d = 70\,\mathrm{mm}$, $T = 329.5°\mathrm{C}$ (from [110])

where two storage positions are occupied with a FS. After 64 μs the upper FS has disappeared, at $t = 80\,\mu\mathrm{s}$ a new FS has been created in the lower part. An empty storage location can be created simply by omitting the addressing beam in one cycle.

Such an all–optical buffer register could be used in optical telecommunications for the translation of serial data into parallel data, or could serve as a buffer for the temporary storage of data [136], which is considered to be a key-element of future all-optical communication and processing networks. Since the drift velocity of the FS depends on the tilt angle of the feedback mirror, the delay of the buffer register can be adjusted easily. In the situation considered here, the total number of FS is rather limited due to the small beam radius. However, in principle much larger systems are accessible, if the material questions are solved. Figure 24 shows a simulation of an optical buffer memory using drifting CS (saturable absorber model) for a homogeneous, but phase-tilted, input beam. Note that such schemes depend on the non-diffracting property of a CS or FS: an ordinary beam created by a bit-pulse would diffract as it drifted, and the information would be lost.

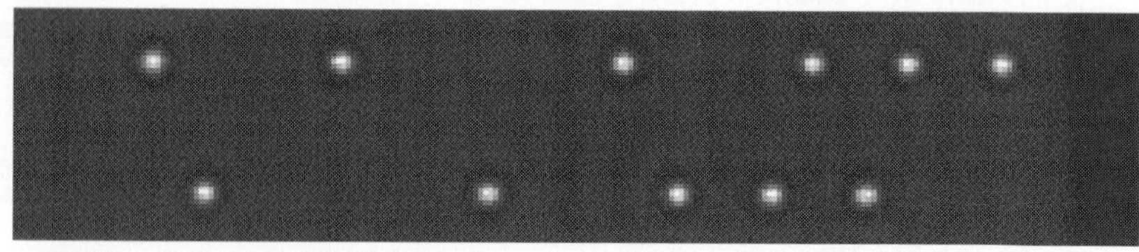

Fig. 24. Two frames from a soliton buffer memory simulation in the saturable-absorber cavity (3). CS are written at the left by "1" pulses of an optical bit-stream, and drift rightwards on a uniform phase gradient (from [135])

For applications, it is very important to determine the response and switching times of FS and CS. Hence, we next discuss the transient evolution of the ignition process of a FS. The power of the background beam is adjusted such that the system is in the unstructured branch within the bistable region. The switching process is observed with a fast photodetector placed in the near field image plane at the location of the addressing beam.

The ignition pulse used in the experiment has a width of 1.7 µs (dashed curve in Fig. 25a). Obviously the detector monitoring the transmitted light has to show a corresponding signal during the time interval in which the addressing pulse is present (solid curve in Fig. 25a). Afterwards the transmission drops again to nearly the initial value, but the switching process has been started. In the example shown it takes 25.4 µs (using a 50%–criterion) until the FS is finally switched up. The success of the ignition process depends on the amplitude I_a as well as on the duration of τ_a of the addressing pulse. For zero-dimensional bistable systems it was found that for switching pulses that are short compared to the relaxation of the medium, only the pulse area, i.e. the product $F = I_a \cdot \tau_a$ in the case of a rectangular pulse, is the decisive factor [137, 138, 139, 140]. Although we did not perform the systematic studies necessary to establish a pulse area law strictly, it will be seen in the following that our experimental results for various amplitudes and widths are in accordance with such a law.

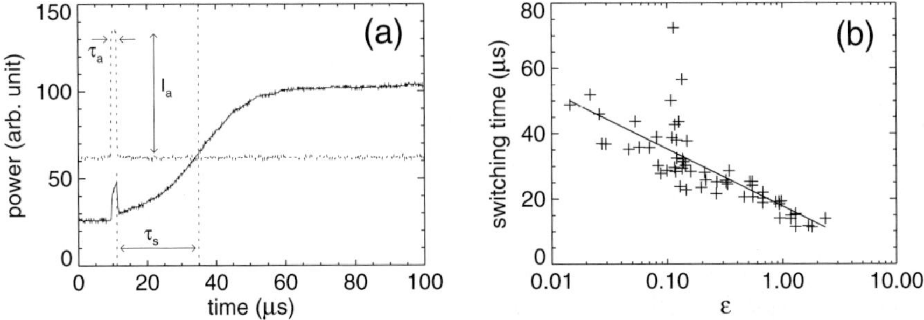

Fig. 25. (**a**) Temporal evolution of transmitted power during ignition of a FS (*solid line*) after an addressing pulse is applied (*dashed line*). (**b**) Distribution of switching times in dependence on the control parameter ϵ (see text). The spread in switching times which is apparent around $\epsilon \approx 0.1$ is probably related to perturbations of parameters. Parameters: $d = 70\,$mm, $\Delta = 8.9\,$GHz, $B_\perp = 2.35\,\mu$T, $B_z = 15.1\,\mu$T, cell temperature 313.2°C, $p_{N_2} = 200\,$hPa, $P_{in} = 76\,$mW (from [111])

Figure 25b shows the distribution of switching times versus a control parameter ϵ, which is related by $\epsilon = (F_a - F_c)/F_c$ to the pulse area F_a and the critical pulse area F_c. The switching time increases logarithmically, if ϵ is decreased. The increase of the switching time observed in the experiment is called *noncritical slowing down* and has been established for zero-dimensional bistable systems [138, 139, 140]. It may be useful to call the interpretation of the phenomenon to mind: the action of the short addressing pulse is integrated over its duration and puts the system in phase space beyond the separatrix. ϵ parametrizes the distance to the separatrix. After the addressing pulse relaxational dynamics towards the stable fixed point takes place. If one starts close to the separatrix (unstable fixed point) the dynamics is governed by the exponential growth of deviations from this fixed point, which explains the observed logarithmic dependence of the switching time on the (excess) pulse area of the addressing beam.

Non-critical slowing down was discussed before numerically as well as analytically for CS [43, 54, 56] and was experimentally observed for LS in a liquid crystal light valve [104, 105] with switching times in the regions of some hundreds of milliseconds. Obviously, these switching times as well as the ones obtained here, are not very attractive for applications. However, the switching times scale somehow with the relevant relaxation rates, though, of course the problem of the divergence at the separatrix always exists. For semiconductor systems, switching times in the nanosecond-range are found [88, 12, 90] after first experiments were limited by thermal time scales [85]. These investigations demonstrate clearly that the knowledge of unstable states can be very significant in understanding and control of spatial structures in nonlinear optical systems. This makes the analysis of stationary solutions an even more valuable technique in the study of model systems.

9 Conclusion

We have discussed a class of stable soliton-like structures predicted to exist in pattern-forming nonlinear optical systems containing any of a wide variety of nonlinear materials. This class includes semiconductor micro-resonators, which is promising for possible applications of these cavity and feedback solitons. They can be formed into two-dimensional arrays of information bits which can be written, stored, read, erased, and spatially manipulated in various ways. They can thus act as the basis of a new kind of all-optical parallel processor, with functionalities not available to other processing and storage devices in information technology.

Apart from this applicative aspect, their formation, stabilization and interaction involves a lot of interesting physics. Particularly intriguing is their "quasi-particle" nature and their interaction properties. Although obviously numerous details remain to be worked out, these might open the intriguing possibility to construct a parallel to the chain of complexity known from equilibrium physics (atom – molecule/cluster – crystal/solid state lattice) in non-equilibrium physics: single CS/FS – clusters/molecules – localized patterns – spatially extended periodic patterns.

Another intriguing aspect is the astonishing parallels between CS and FS in shape, bifurcation characteristics and interaction behavior. This illustrates that the existence of dissipation and feedback creates a universality transcending the differences in geometry between the two systems and that both should be regarded as representatives of the very general class of dissipative optical solitons. Nevertheless, it is of course instructive and important to understand and interpret the mechanism of localization in each specific system and we commented on several aspects of that.

Acknowledgements

This work is based on research which was partially supported by ESPRIT project 28235 PIANOS and EPSRC grant GR/M 19727. We thank our colleagues and collaborators for many helpful discussions and insights. We especially thank Andrew Scroggie for important contributions to this work. T.A. is grateful to Burkhard Schäpers and Wulfhard Lange for the fruitful collaboration on the subject. Their work was supported by the Deutsche Forschungsgemeinschaft.

References

1. N. N. Rosanov, *Transverse patterns in wide-aperture nonlinear optical systems.* Progress in Optics **XXXV**, 1–60 (1996).
2. L. A. Lugiato, M. Brambilla, and A. Gatti, *Optical pattern formation.* Adv. Atom. Mol. Opt. Phys. **40**, 229–306 (1999).
3. F. T. Arecchi, S. Boccaletti, and P. L. Ramazza. *Pattern formation and competition in nonlinear optics.* Phys. Rep. **318**, 1–83 (1999).
4. C. O. Weiss, M. Vaupel, K. Staliunas, G. Slekys, and V. B. Taranenko. *Solitons and Vortices in lasers.* Appl. Phys. B **68**, 151–168 (1999).
5. S. Trillo and W. E. Torruellas, editors. *Spatial Solitons*, volume 82 of *Springer Series in Optical Sciences.* (Springer, Berlin, 2001).
6. T. Ackemann and W. Lange. *Optical pattern formation in alkali metal vapors: Mechanisms, phenomena and use.* Appl. Phys. B **72**, 21–34 (2001).
7. W. J. Firth. *Theory of Cavity Solitons.* In A. D. Boardman and A. P. Sukhorukov, editors, *Soliton-Driven Photonics*, pages 459–485 (Kluwer Academic Publishers, London, 2001).
8. W. J. Firth and C. O. Weiss. *Cavity and feedback solitons.* Opt. Photon. News **13**(2), 54–58 (2002).
9. N. N. Rosanov. *Spatial hysteresis and optical patterns.* Springer Series in Synergetics. Springer, Berlin (2002).
10. U. Peschel, D. Michaelis, and C. O. Weiss. *Spatial solitons in optical cavities.* IEEE J. Quantum Electron. **39**, 51–64 (2003).
11. L. A. Lugiato. *Introduction to the feature section on cavity solitons: An overview.* IEEE J. Quantum Electron. **39**(2), 193–196 (2003).
12. V. B. Taranenko, G. Slekys, C. O. Weiss. *Dissipative Solitons.*
13. N. N. Rosanov. *Dissipative Solitons.*
14. C. Etrich, U. Peschel, and F. Lederer. *Solitary Waves in Quadratically Nonlinear Resonators.* Phys. Rev. Lett. **79**, 2454–2457 (1997).
15. U. Peschel, D. Michaelis, C. Etrich, and F. Lederer. *Formation, motion, and decay of vectorial cavity solitons.* Phys. Rev. E **58**, R2745–R2748 (1998).

16. D. Michaelis, U. Peschel, C. Etrich, and F. Lederer. *Quadratic Cavity Solitons – The Up-Conversion Case.* IEEE J. Quantum Electron. **39**, 255–268 (2003).
17. S. Longhi. *Localized structures in optical parametric oscillation.* Physica Scripta **56**, 611–618 (1997).
18. K. Staliunas and V. J. Sánchez-Morcillo. *Localized structures in degenerate optical parametric oscillators.* Opt. Commun. **139**, 306–312 (1997).
19. G.-L. Oppo, A. J. Scroggie, and W. J. Firth. *From domain walls to localized structures in degenerate optical parametric oscillators.* J. Opt. B: Quantum Semiclass. Opt. **1**, 133–138 (1999).
20. M. Le Berre, D. Leduc, E. Ressayre, and A. Tallet. *Striped and circular domain walls in the DOPO.* J. Opt. B: Quantum Semiclass. Opt. **1**, 153–160 (1999).
21. D. V. Skryabin and W. J. Firth. *Interaction of cavity solitons in degenerate optical parametric oscillators.* Opt. Lett. **24**, 1056–1059 (1999).
22. M. Tlidi, M. Le Berre, E. Ressayre, A. Tallet, and L. Di Menza. *High-intensity localized structures in the degenerate optical parametric oscillator: Comparisom between the propagation and the mean-field model.* Phys. Rev. A **61**, 043806 (2000).
23. G. Izús, M. San Miguel, and M. Santagiustina. *Bloch domain walls in type II optical parametric oscillators.* Opt. Lett. **25**, 1454–6 (2000).
24. G. L. Oppo, A. J. Scroggie, and W. J. Firth. *Characterization, dynamics and stabilization of diffractive domain walls and dark ring cavity solitons in parametric oscillators.* Phys. Rev. E **63**, 066209 (2001).
25. C. Etrich, D. Michaelis, and F. Lederer. *Bifurcations, stability, and multistability of cavity solitons in parametric downconversion.* J. Opt. Soc. Am. B **19**, 792–801 (2002).
26. D. Gomila, P. Colet, M. San Miguel, A. Scroggie, and G. L. Oppo. *Stable droplets and dark-ring cavity solitons in nonlinear optical devices.* IEEE J. Quantum Electron. **39**, 238–244 (2003).
27. R. Vilaseca, M. C. Torrent, J. García-Ojalvo, E. Brambilla, and M. San Miguel. *Two-photon cavity solitons in active optical media.* Phys. Rev. Lett. **87**, 083902 (2001).
28. R. Gallego, M. San Miguel, and R. Toral. *Self-similar domain growth, localized structures, and labyrinthine patterns in vectorial Kerr resonators.* Phys. Rev. E **61**, 2241–4 (2000).
29. V. J. Sánchez-Morcillo, I. Pérez-Arjona, Silva F., G. J. Valcárcel, and E. Roldán. *Vectorial Kerr cavity solitons.* Opt. Lett. **25**, 957–959 (2000).
30. E. Große Westhoff, V. Kneisel, Yu. A. Logvin, T. Ackemann, and W. Lange. *Pattern formation in the presence of an intrinsic polarization instability.* J. Opt. B: Quantum Semiclass. Opt. **2**, 386–392 (2000).
31. V. B. Taranenko, K. Staliunas, and C. O. Weiss. *Pattern formation and localized structures in degenerate optical parametric mixing.* Phys. Rev. Lett. **81**, 2236–2239 (1998).
32. V. B. Taranenko, M. Zander, P. Wobben, and C. O. Weiss. *Stability of localized structures in degenerate wave mixing.* Appl. Phys. B **69**, 337–339 (1999).
33. M. Pesch, E. Große Westhoff, T. Ackemann, and W. Lange. Vectorial solitons and higher-order localized states in a single-mirror feedback system. In *Nonlinear Guided Waves and Their Applications.* Toronto, Canada, March 28–31, 2004. Paper TuC24 (2004).

34. V. Yu. Bazhenov, V. B. Taranenko, and M. V. Vasnetsov. *Transverse optical effects in bistable active cavity with nonlinear absorber on bacteriorhodopsin.* Proc. SPIE **1840**, 183–193 (1992).
35. M. Saffman, D. Montgomery, and D. Z. Anderson. *Collapse of a transverse-mode continuum in a self-imaging photorefractively pumped ring resonator.* Opt. Lett. **19**, 518–520 (1994).
36. V. B. Taranenko, K. Staliunas, and C. O. Weiss. *Spatial soliton laser: localized structures in a laser with a saturable absorber in a self-imaging resonator.* Phys. Rev. A **56**, 1582–1591 (1997).
37. D. W. McLaughlin, J. V. Moloney, and A. C. Newell. *Solitary waves as fixed points of infinite-dimensional maps in an optical bistable ring cavity.* Phys. Rev. Lett. **51**, 75–78 (1983).
38. J. V. Moloney and A. C. Newell. *Nonlinear Optics.* Addison-Wesley, Redwood City (1992). Fig. 5.16, p. 225 and associated text.
39. H. M. Gibbs. *Optical Bistability: Controlling Light with Light.* Academic Press, Orlando (1985).
40. L. A. Lugiato. *Theory of optical bistability.* Progress in Optics XXI pages 70–216 (1984).
41. W. J. Firth and G. K. Harkness. *Cavity solitons.* Asian J. Phys. **7**, 665–677 (1998).
42. G. S. McDonald and W. J. Firth. *Spatial solitary-wave optical memory.* J. Opt. Soc. Am. B **7**, 1328–1335 (1990).
43. G. S. McDonald and W. J. Firth. *Switching dynamics of spatial solitary wave pixels.* J. Opt. Soc. Am. B **10**, 1081–1089 (1993).
44. N. N. Rosanov and G. V. Khodova. *Autosolitons in nonlinear interferometers.* Opt. Spectrosc. **65**, 449–450 (1988).
45. N. N. Rosanov and G. V. Khodova. *Diffractive autosolitons in nonlinear interferometers.* J. Opt. Soc. Am. B **7**(6), 1057–65 (1990).
46. W. J. Firth and I. Galbraith. *Diffusive transverse coupling of bistable elements – switching waves and crosstalk.* IEEE J. Quantum Electron. **21**, 1399–1403 (1985).
47. N. B. Abraham and W. J. Firth. *Overview of transverse effects in nonlinear-optical systems.* J. Opt. Soc. Am. B **7**, 951–961 (1990).
48. M. Kreuzer, H. Gottschilk, Th. Tschudi, and R. Neubecker. *Structure formation and self-organization phenomena in bistable optical elements.* Mol. Cryst. Liquid Cryst. **207**, 219–230 (1991).
49. R. Neubecker and T. Tschudi. *Self-induced mode as a building element of transversal pattern formation.* J. Mod. Opt. **41**, 885–906 (1994).
50. G. Giusfredi, J. F. Valley, R. Pon, G. Khitrova, and H. M. Gibbs. *Optical instabilities in sodium vapor.* J. Opt. Soc. Am. B **5**, 1181–1191 (1988).
51. L. A. Lugiato and R. Lefever. *Spatial dissipative structures in passive optical systems.* Phys. Rev. Lett. **58**, 2209–2211 (1987).
52. I. V. Barashenkov, N. V. Alexeeva, and E. V. Zemlyanaya. *Two- and three-dimensional oscillons in nonlinear Faraday resonance.* Phys. Rev. Lett. **89**, 104101 (2002).
53. P. B. Umbanhowar, F. Melo, and H. L. Swinney. *Localized excitations in a vertically vibrated granular layer.* Nature **382**, 793–796 (1996).
54. T. Maggipinto, M. Brambilla, G. K. Harkness, and W. J. Firth. *Cavity solitons in semiconductor microresonators: Existence, stability, and dynamical properties.* Phys. Rev. E **62**(6), 8726–8739 (2000).

55. G. K. Harkness, W. J. Firth, G. L. Oppo, and J. M. McSloy. *Computationally Determined Existence and Stability of Transverse Structures: I. Periodic Optical Patterns.* Phys. Rev. E **66**, 046605 (2002).
56. J. M. McSloy, W. J. Firth, G. L. Oppo, and G. K. Harkness. *Computationally Determined Existence and Stability of Transverse Structures: II. Multi-Peaked Cavity Solitons.* Phys. Rev. E **66**, 046606 (2002).
57. T. Maggipinto, M. Brambilla, and W. J. Firth. *Characterization of stationary patterns and their link with cavity solitons in semiconductor microresonators.* IEEE J. Quantum Electron. **39**, 206–215 (2003).
58. W. J. Firth, A. Lord, and A. J. Scroggie. *Optical bullet holes.* Phys. Scr. **T67**, 12–16 (1996).
59. W. J. Firth, G. K. Harkness, A. Lord, J. M. McSloy, D. Gomila, and P. Colet. *Dynamical properties of two-dimensional Kerr cavity solitons.* J. Opt. Soc. Am. B **19**(4), 747–752 (2002).
60. K. Staliunas. *Three-dimensional Turing structures and spatial solitons in optical parametric oscillators.* Phys. Rev. Lett. **81**, 81–84 (1998).
61. M. Tlidi and P. Mandel. *Three-dimensional optical crystals and localized structures in cavity second harmonic generation.* Phys. Rev. Lett. **83**, 4995–4998 (1999).
62. G. Steinmeyer, A. Schwache, and F. Mitschke. *Quantitative characterization of turbulence in an optical experiment.* Phys. Rev. E **53**, 5399–5402 (1996).
63. S. Wabnitz. *Suppression of interactions in a phase-locked optical memory.* Opt. Lett. **18**, 601–603 (1993).
64. M. Tlidi, P. Mandel, and R. Lefever. *Localized structures and localized patterns in optical bistability.* Phys. Rev. Lett. **73**, 640–643 (1994).
65. W. J. Firth and A. J. Scroggie. *Optical bullet holes: robust controllable localized states of a nonlinear cavity.* Phys. Rev. Lett. **76**, 1623–1626 (1996).
66. W. J. Firth and A. J. Scroggie. *Spontaneous pattern formation in an absorptive system.* Europhys. Lett. **26**, 521–526 (1994).
67. T. Ackemann, S. Barland, J. R. Tredicce, M. Cara, S. Balle, R. Jäger, P. M. Grabherr, M. Miller, and K. J. Ebeling. *Spatial structure of broad-area vertical-cavity regenerative amplifiers.* Opt. Lett. **25**, 814–816 (2000).
68. P. Coullet, C. Riera, and C. Tresser. *Stable static localized structures in one dimension.* Phys. Rev. Lett. **84**, 3069–3072 (2000).
69. P. Coullet, C. Riera, and C. Tresser. *Qualitative theory of stable stationary localized structures in one dimension.* Prog. Theor. Phys. Suppl. **139**, 46–58 (2000).
70. Y. Pomeau. *Front motion, metastability and subcritical bifurcations in hydrodynamics.* Physica D **23**, 3–11 (1986).
71. A. G. Vladimirov, J. M. McSloy, D. V. Skryabin, and W. J. Firth. *Two-dimensional clusters of solitary structures in driven optical cavities.* Phys. Rev. E **65**, 046606 (2002).
72. W. J. Firth. *Optical Memory and Spatial Chaos.* Phys. Rev. Lett. **61**, 329–332 (1988).
73. P. Coullet, C. Riera, and C. Tresser. *A new approach to data storage using localized structures.* Chaos **14**, 193–198 (2004).
74. M. Tlidi and P. Mandel. *Spatial patterns in nascent optical bistability.* Chaos, Solitons & Fractals **4**, 1475–1486 (1994).
75. M. Brambilla, L. A. Lugiato, and M. Stefani. *Interaction and control of optical localized structures.* Europhys. Lett. **34**, 109–114 (1996).

76. M. Brambilla, L. A. Lugiato, F. Prati, L. Spinelli, and W. J. Firth. *Spatial soliton pixels in semiconductor devices.* Phys. Rev. Lett. **79**, 2042–2045 (1997).
77. L. Spinelli, G. Tissoni, M. Brambilla, F. Prati, and L. A. Lugiato. *Spatial solitons in semiconductor microcavities.* Phys. Rev. A **58**, 2542–2559 (1998).
78. M. Tlidi, M. Georgiou, and P. Mandel. *Transverse patterns in nascent optical bistability.* Phys. Rev. A **48**, 4605–4609 (1993).
79. M. C. Cross and P. C. Hohenberg. *Pattern formation outside of equilibrium.* Rev. Mod. Phys. **65**, 851–1112 (1993).
80. S. Hoogland, J. J. Baumberg, S. Coyle, J. Baggett, M. J. Coles, and H. J. Coles. *Self-organized patterns and spatial solitons in liquid-crystal microcavities.* Phys. Rev. A **66**, 055801 (2002).
81. G. Tissoni, L. Spinelli, M. Brambilla, I. Perrini, T. Maggipinto, and L. A. Lugiato. *Cavity solitons in bulk semiconductor microcavities: dynamical properties and control.* J. Opt. Soc. Am. B **16**, 2095–2105 (1999).
82. L. Spinelli, G. Tissoni, M. Tarenghi, and M. Brambilla. *First principle theory for cavity solitons in semiconductor microresonators.* Eur. Phys. J. D **15**, 257–266 (2001).
83. S. Barbay, J. Koehler, R. Kuszelewicz, T. Maggipinto, I. M. Perrini, and M. Brambilla. *Optical patterns and cavity solitons in quantum-dot microresonators.* IEEE J. Quantum Electron. **39**, 245–254 (2003).
84. D. Michaelis, U. Peschel, and F. Lederer. *Multistable localized structures and superlattices in semiconductor optical resonators.* Phys. Rev. A **56**, R3366–R3369 (1997).
85. V. B. Taranenko and C. O. Weiss. *Incoherent optical switching of semiconductor resonator solitons.* Appl. Phys. B **72**(7), 893–895 (2001).
86. S. Barland, J. R. Tredicce, M. Brambilla, L. A. Lugiato, S. Balle, M. Giudici, T. Maggipinto, L. Spinelli, G. Tissoni, T. Knödel, M. Miller, and R. Jäger. *Cavity solitons as pixels in semiconductors.* Nature **419**, 699–702 (2002).
87. V. B. Taranenko, C. O. Weiss, and B. Schäpers. *From coherent to incoherent hexagonal patterns in semiconductor resonators.* Phys. Rev. A **65**, 13812 (2002).
88. V. B. Taranenko, F. J. Ahlers, and K. Pierz. *Coherent switching of semiconductor resonator solitons.* Appl. Phys. B **75**, 75–77 (2002).
89. I. Ganne, G. Slekys, I. Sagnes, and R. Kuszelewicz. *Precursor forms of cavity solitons in nonlinear semiconductor microresonators.* Phys. Rev. E **66**, 066613 (2002).
90. X. Hachair, S. Barland, L. Furfaro, M. Giudici, S. Balle, J. Tredicce, M. Brambilla, T. Maggipinto, I. M. Perrini, G. Tissoni, and L. Lugiato. *Cavity solitons in broad-area vertical-cavity surface-emitting lasers below threshold.* Phys. Rev. A **69**, 043817 (2004).
91. L. Spinelli, G. Tissoni, L. Lugiato, and M. Brambilla. *Thermal effects and transverse structures in semiconductor microcavities with population inversion.* Phys. Rev. A **66**, 023817 (2002).
92. A. J. Scroggie, J. M. McSloy, and W. J. Firth. *Self-Propelled Cavity Solitons in Semiconductor Microresonators.* Phys. Rev. E **66**, 036607 (2002).
93. S. Barland, O. Piro, M. Giudici, J. R. Tredicce, and S. Balle. *Experimental evidence of van der Pol–Fitzhugh–Nagumo dynamics in semiconductor optical amplifiers.* Phys. Rev. E **68**, 036209 (2003).
94. P. Coullet, J. Lega, B. Houchmanzadeh, and J. Lajzerowics. *Breaking chirality in nonequilibrium systems.* Phys. Rev. Lett. **65**, 1352–1355 (1990).

95. D. Michaelis, U. Peschel, F. Lederer, D. V. Skryabin, and W. J. Firth. *Universal criterion and amplitude equation for a nonequilibrium Ising-Bloch transition.* Phys. Rev. E **63**, 066602 (2001).
96. D. V. Skryabin, A. Yulin, D. Michaelis, W. J. Firth, G.-L. Oppo, U. Peschel, and F. Lederer. *Perturbation theory for domain walls in the parametric Ginzburg-Landau equation.* Phys. Rev. E **64**, 056618 (2001).
97. C. Degen, I. Fischer, W. Elsäßer, L. Fratta, P. Debernardi, G. Bava, M. Brunner, R. Hövel, M. Moser, and K. Gulden. *Transverse modes in thermally detuned oxide-confined vertical-cavity surface-emitting lasers.* Phys. Rev. A **63**, 23817 (2001).
98. T. Rössler, R. A. Indik, G. K. Harkness, and J. V. Moloney. *Modeling the interplay of thermal effects and transverse mode behavior in native-oxide confined vertical-cavity surface-emitting lasers.* Phys. Rev. A **58**, 3279–3292 (1998).
99. M. Bode. *Pattern formation in dissipative systems: A particle approach.* Adv. in Solid State Phys. **41**, 369–381 (2001).
100. H. U. Bödeker, M. C. Röttger, A. W. Liehr, Frank. T. D., R. Friedrich, and H. G. Purwins. *Noise-covered drift bifurcation of dissipative solitons in a planar gas-discharge system.* Phys. Rev. E **67**, 056220 (2003).
101. W. J. Firth. *Spatial instabilities in a Kerr medium with single feedback mirror.* J. Mod. Opt. **37**, 151–153 (1990).
102. G. D'Alessandro and W. J. Firth. *Spontaneous hexagon formation in a nonlinear optical medium with feedback mirror.* Phys. Rev. Lett. **66**, 2597–2600 (1991).
103. G. D'Alessandro and W. J. Firth. *Hexagonal spatial pattern for a Kerr slice with a feedback mirror.* Phys. Rev. A **46**, 537–548 (1992).
104. M. Kreuzer, A. Schreiber, and B. Thüring. *Evolution and switching dynamics of solitary spots in nonlinear optical feedback systems.* Mol. Cryst. Liq. Cryst. **282**, 91–105 (1996).
105. A. Schreiber, M. Kreuzer, B. Thüring, and T. Tschudi. *Experimental investigation of solitary structures in a nonlinear optical feedback system.* Opt. Commun. **136**, 415–418 (1997).
106. B. A. Samson and M. A. Vorontsov. *Localized states in a nonlinear optical system with a binary-phase slice and a feedback mirror.* Phys. Rev. A **56**, 1621–1626 (1997).
107. B. Schäpers, M. Feldmann, T. Ackemann, and W. Lange. *Interaction of localized structures in an optical pattern forming system.* Phys. Rev. Lett. **85**, 748–751 (2000).
108. P. L. Ramazza, S. Ducci, S. Boccaletti, and F. T. Arecchi. *Localized versus delocalized patterns in a nonlinear optical interferometer.* J. Opt. B: Quantum Semiclass. Opt. **2**(3), 399–405 (2000).
109. M. G. Clerc, S. Residori, and C. S. Riera. *First-order Frédericksz transition in the presence of light-driven feedback in nematic liquid cyrstals.* Phys. Rev. E **63**, 060701(R) (2001).
110. B. Schäpers, T. Ackemann, and W. Lange. *Characteristics and possible applications of localized structures in an optical pattern–forming system.* Proc. SPIE **4271**, 130–137 (2001).
111. B. Schäpers, T. Ackemann, and W. Lange. *Robust control of switching of localized structures and its dynamics in a single-mirror feedback scheme.* J. Opt. Soc. Am. B **19**(4), 707–715 (2002).

112. P. L. Ramazza, E. Benkler, U. Bortolozzo, S. Boccaletti, S. Ducci, and F. T. Arecchi. *Taloiring the profile and interactions of optical localized structures.* Phys. Rev. E **65**, 066204 (2002).
113. B. Gütlich, M. Kreuzer, R. Neubecker, and T. Tschudi. *Manipulation of solitary structures in a nonlinear optical single feedback experiment.* Mol. Cryst. Liq. Cryst. **375**, 281–289 (2002).
114. B. Schäpers, T. Ackemann, and W. Lange. *Properties of feedback solitons in a single-mirror experiment.* IEEE J. Quantum Electron. **39**(2), 227–237 (2003).
115. B. Gütlich, R. Neubecker, M. Kreuzer, and T. Tschudi. *Control and manipulation of solitary structures in a nonlinear optical single feedback experiment.* Chaos **13**, 239–246 (2003).
116. S. Rankin, E. Yao, and F. Papoff. *Traveling waves and counterpropagating bright droplets as a result of tailoring the transverse dispersion relation in a multistable optical system.* Phys. Rev. A **68**, 013821 (2003).
117. W. H. F. Talbot. *Facts relating to optical science. No. IV.* Philos. Mag. **9**(Third series), 401–407 (1836).
118. E. Ciaramella, M. Tamburrini, and E. Santamato. *Talbot assisted hexagonal beam patterning in a thin liquid crystal film with a single feedback mirror at negative distance.* Appl. Phys. Lett. **63**, 1604–1606 (1993).
119. T. Ackemann, B. Giese, B. Schäpers, and W. Lange. *Investigation of pattern forming mechanisms by Fourier filtering: properties of hexagons and the transition to stripes in an anisotropic system.* J. Opt. B: Quantum Semiclass. Opt. **1**, 70–76 (1999).
120. S. G. Odoulov, M. Yu. Goulkov, and O. A. Shinkarenko. *Threshold behavior in formation of optical hexagons and first order optical phase transition.* Phys. Rev. Lett. **83**, 3637–3640 (1999).
121. S. A. Akhmanov, M. A. Vorontsov, V. Yu. Ivanov, A. V. Larichev, and N. I. Zheleynykh. *Controlling transverse-wave interactions in nonlinear optics: generation and interaction of spatiotemporal structures.* J. Opt. Soc. Am. B **9**, 78–90 (1992).
122. A. Kastler. *Optical methods of atomic orientation and of magnetic resonance.* J. Opt. Soc. Am. **47**, 460–465 (1957).
123. F. Mitschke, R. Deserno, W. Lange, and J. Mlynek. *Magnetically induced optical self-pulsing in a nonlinear resonator.* Phys. Rev. A **33**, 3219–3231 (1986).
124. T. Ackemann, A. Heuer, Yu. A. Logvin, and W. Lange. *Light-shift induced level crossing and resonatorless optical bistability in sodium vapor.* Phys. Rev. A **56**, 2321–2326 (1997).
125. D. N. Maywar, G. P. Agrawal, and Y. Nakano. *Robust optical control of an optical-amplifier-based flip-flop.* Opt. Express **6**, 75–80 (2000).
126. W. Lange, Yu. A. Logvin, and T. Ackemann. *Spontaneous optical patterns in an atomic vapor: observation and simulation.* Physica D **96**, 230–241 (1996).
127. B. Schäpers. *Lokalisierte Strukturen in einem atomaren Dampf mit optischer Rückkopplung.* Phd thesis, Westfälische Wilhelms-Universität Münster (2001).
128. M. Kreuzer. *Grundlagen und Anwendungen von Füssigkristallen in der optischen Informations- und Kommunikationstechnologie.* PhD thesis, Darmstadt (1994).
129. T. Ackemann (2002). Unpublished.
130. G. Grynberg, A. Petrossian, M. Pinard, and M. Vallet. *Phase-contrast mirror based on four-wave mixing.* Europhys. Lett. **17**, 213 (1992).

131. G. Grynberg. *Roll and hexagonal patterns in a phase-contrast oscillator.* J. Phys. III **3**, 1345–1355 (1993).
132. M. Tlidi, A. G. Vladimirov, and P. Mandel. *Interaction and Stability of Periodic and Localized Structures in Optical Bistable Systems.* IEEE J. Quantum Electron. **39**, 216–226 (2003).
133. Yu. A. Logvin, B. Schäpers, and T. Ackemann. *Stationary and drifting localized structures near a multiple bifurcation point.* Phys. Rev. E **61**, 4622–4625 (2000).
134. S. Mêtens, G. Dewel, P. Borckmanns, and R. Engelhardt. *Pattern selection in bistable systems.* Europhys. Lett. **37**, 109–114 (1997).
135. W. J. Firth. *Processing Information with Arrays of Spatial Solitons.* Proc. SPIE **4016**, 388–394 (2000).
136. E. Lugagne Delpon, J. L. Oudar, and H. Lootvoet. *Operation of a* 4×1 *optical register as a fast access optical buffer memory.* Electron. Lett. **33**, 1161–1162 (1997).
137. P. Mandel. *Scaling properties of switching pulses.* Opt. Commun. **55**, 293–296 (1985).
138. B. Segard, J. Zemmouri, and B. Macke. *Noncritical slowing down in optical bistability.* Opt. Commun. **63**, 339–343 (1987).
139. J. Y. Bigot, A. Daunois, and P. Mandel. *Slowing down far from the limit points in optical bistability.* Phys. Lett. A **123**, 123–127 (1987).
140. F. Mitschke, C. Boden, W. Lange, and P. Mandel. *Exploring the dynamics of the unstable branch of bistable systems.* Opt. Commun. **71**, 385–392 (1989).

Solitons in Laser Systems with Saturable Absorption

N.N. Rosanov

Research Institute for Laser Physics Birzhevaya Liniya 12, Saint Petersburg, Russia
nrosanov@yahoo.com

Abstract. A review is presented of the main features of localized structures – dissipative solitons – in optical systems with nonlinear amplification and absorption, without driving (holding) radiation, including cases with and without feedback. The focus is on two-dimensional laser solitons. For the case of cylindrically- symmetric intensity distributions, there is a discrete set of such solitons with different values of topological charge and different numbers of oscillations in the radial profiles of their amplitude and phase, within certain intervals of the system parameters. Even these simplest dissipative solitons have certain internal structures that become apparent in the distribution of the radiation energy flows. For weakly-coupled solitons whose tail overlap is only small, the distribution of energy flows in the vicinity of each constituent soliton is topologically similar to the distribution for individual solitons. In addition to weakly-coupled solitons, strongly coupled states exist in parameter domains overlapping with those for symmetric solitons. Symmetric structures are motionless, while asymmetric structures move and rotate. Strongly-coupled soliton states are characterized by asymmetric multi-humped intensity distributions. Rotation occurs even in the absence of radiation wavefront dislocations. We present examples of bifurcations of the phase portrait of energy flows during the transient process of the rotating structure formation, as well as different rotating chains of localized strongly-coupled laser vortices. Under conditions of modulation instability of homogeneous field distributions, new regimes arise, including localized structures with simultaneous rotation and pulsation, and "bio-solitons" with initial growth of structure like a labyrinth, with periodic separation of the fragments; then the fragments repeat the stages of growth and separation of new generations of fragments.

1 Introduction: Definitions, Examples, History

By "dissipative soliton", or "autosoliton", we usually mean a stable localized field structure in a homogeneous non-conservative (with essential energy exchange) medium or system. The field can be an electromagnetic one or can represent the spatio-temporal distribution of medium temperature, concentration of particles of different types, etc. At the soliton periphery, the soliton field approaches certain background values – *viz.* a constant or weakly-modulated function (the latter in the presence of modulation instability of the field homogeneous distributions). In the soliton's central part, the field is considerably greater (for bright solitons) or lower (for dark solitons) than the background values.

N.N. Rosanov: *Solitons in Laser Systems with Saturable Absorption*,
Lect. Notes Phys. **661**, 101–130 (2005)
www.springerlink.com

For solitons, any spreading of the field structure during its evolution is prevented by the features of the nonlinear medium. In certain cases, the field structure is not nonlinearly confined in all three spatial directions (dimensionality $D = 3$), but only in one ($D = 1$) or two ($D = 2$) of them, whereas in the other $3 - D$ directions the structure is confined by the medium linear inhomogeneity, or it is not confined at all (e.g.,in a cigar-shaped structure; these structures must be stable against perturbations dependent on the "other" co-ordinates). Thus, stable localized structures with the *nonlinear confinement* in D directions will be referred to as D-dimensional solitons (e.g., $D = 1$ in single-mode optical fibres). A further generalization of the soliton notion arises in the case of periodic or quasi-periodic spatial modulation of the medium characteristics. Examples in optics are "discrete solitons" (where the modulation is across the direction of radiation propagation, see [1]) and "solitons with dispersion management" (with modulation along the radiation propagation direction) [2].

A standard example of a dissipative soliton is a pulse of fire propagating along a long strip of grass. When the leading edge of fire approaches, the grass is heated and blazes up. After the combustion of the grass portion, the fire pulse moves to the next portion, the soil cools due to heat transfer to the environment, and new grass grows. Therefore the initial values of the temperature and grass density are restored behind the trailing edge of the fire pulse – a moving dissipative soliton.

In contrast to the more familiar *conservative solitons* in nonlinear media without energy dissipation, the energy sources and sinks are integral parts of a medium or system where dissipative solitons exist. Dissipative solitons were first observed by M. Faraday in 1831 in experiments with fine powder (e.g., sand) upon a vibrating surface [3]. Faraday called them *little heaps*; in a recent, more detailed, study of granular matter on vibrating surfaces, these solitons are termed *oscillons* [4]. We note that Faraday had performed these experiments a few years before the first observation of conservative solitons by R. Scott (1834) [5].

Dissipative solitons occur in many non-equilibrium physical, chemical, and biological systems. An important example of biological dissipative solitons is the propagation of electric pulses along nerve and cardiac muscle fibres [6, 7]. Such solitons arise in chemical reactions involving different agents – *activators* promoting an increase in concentration of an agent, and *inhibitors* which prevent this process – accompanied by diffusion. They are known in gases, plasmas, and semiconductors under electric discharge. Numerous examples, and the theory of corresponding reaction-diffusion systems, are given by Kerner and Osipov [8]; these authors also proposed the term "autosoliton".

The fundamental difference between conservative and dissipative solitons is in the actual type of spectrum of their main characteristics, for example the field maximum at the soliton's centre or the soliton's width. For conservative solitons, this spectrum is *continuous*: there are families of solitons with continuously-varying characteristics. But this spectrum is *discrete* for dissipative solitons because of the additional requirement of energy balance. (This requirement is clearly inapplicable for conservative solitons). Thus, dissipative solitons are "calibrated". If fluctuations arise, they induce continuous variations of the characteristics of conservative solitons, and, even after weak fluctuations, these characteristics can differ radically from their initial values. However, the characteristics of dissipative solitons are restored in this case because of the discrete nature of the spectra of these solitons – a feature which is very important for a number of applications. More precisely, fluctuations, or noise can induce spontaneous switching between solitonic and background states which are stable in the absence of the noise. Therefore, dissipative solitons only have a limited lifetime. However, the noise intensity is extremely low in systems of practical interest. Fluctuation switching occurs with a noticeable probability only when a system's parameters are in close proximity to the soliton stability boundaries, and we usually have the condition that noise does not significantly exhibit itself.

Modulation instability of the medium homogeneous states is not a prerequisite for dissipative soliton existence, as they are known both in the presence and absence of this type of pattern formation. Their excitation has a threshold, and small initial perturbations dissolve in the process of evolution, whereas fairly large perturbations initiate dissipative soliton formation. This feature is also important in applications to noise suppression.

The *mechanism of transport phenomena* is important for dissipative solitons with an essential energy exchange. In many cases, especially for non-coherent fields, this mechanism is *diffusion*. In optics, a different mechanism is typical – *viz. diffraction* or *dispersion* accompanied by interference phenomena in the case of coherent light. Optical dissipative solitons with a diffractive mechanism of transport phenomena were first found in the theoretical study of driven nonlinear wide-aperture interferometers under conditions of classical bistability, when two homogeneous states of the field inside the interferometer are stable [9, 10], and under conditions of modulation instability of one or both of these states [11]. They were first demonstrated experimentally in a liquid-crystal valve system with spatial filtering in the feedback [12, 13]. Another example of a similar "driven" passive optical system in which dissipative solitons were found experimentally is a single-mirror feedback system containing a cell with Na vapour [14]. In the strict sense, dissipative solitons in driven optical systems are "oscillons", because the optical field oscillates at (very high) frequency. However, in terms of the field envelope, these solitons are stationary. Driven systems include mirrors or other sharp medium inhomogeneities at least in one direction (the longitudinal one, along the radiation main propagation direction), so, strictly speaking, only $1D$ and $2D$-solitons are possible with them.

Only active optical systems without an external signal, such as wide-aperture *lasers with saturable absorbers* will be considered henceforth.

Dissipative solitons in such systems were predicted in [15, 16] (see also [17]). They differ from driven optical systems in the absence of the external signal; optical feedback elements like mirrors are then not necessary. This gives laser solitons much more freedom, including arbitrariness of the phase of their oscillations. Note also that $1D$, $2D$ and $3D$ solitons ("laser bullets", see [18]) have been demonstrated in numerical simulations with laser systems. Since many features of laser solitons were recently reviewed in [18] (see also [19, 20]), below we will summarize only briefly the features of $1D$ and $3D$ laser solitons, and our main attention will be paid to the extremely rich variety of $2D$ laser solitons in media with fast nonlinearities.

2 Model and Evolution Equation

When formulating a theoretical model of active (laser) systems with infinite dimensions in one or more directions, certain precautions are necessary. In fact, in these cases, if the small-signal amplification is larger then the small-signal absorption, localized structures cannot be stable. The instability arises because the spontaneous emission will be amplified in these directions at the localized structure periphery, where the radiation intensity is small. To prevent this instability, it is sufficient to satisfy the conditions of a lasing hard (threshold-like) excitation (there is the opposite relation between small-signal amplification and absorption). Practically, it is possible to insert an anisotropy of gain or absorption (dichroism), or an angular selector, into the system.

Further, the following laser systems will be considered: Firstly, in systems without feedback, e.g. Fig. 1, the medium has nonlinear amplification and absorption; additionally, for temporal solitons (with radiation pulses which do not spread in the longitudinal direction), the presence of frequency dispersion is important. This means we have a frequency dependence of the linear refractive index and a finite width of the spectral contours of amplification and absorption. Then, in active nonlinear single-mode fibres (Fig. 1a), $1D$ temporal dissipative solitons exist. In a planar waveguide having a single mode with respect to the vertical direction and being filled with a medium with nonlinear amplification and absorption, there are $1D$ spatial (with time-independent envelope, Fig. 1b), $1D$ temporal (Fig. 1c), and $2D$ spatio-temporal solitons, with envelopes depending on time and on the longitudinal co-ordinate (Fig. 1d). Finally, in continuous media with nonlinear amplification and absorption and frequency dispersion (Fig. 1e), $3D$ laser bullets are possible, in addition to all the previous types of $1D$ and $2D$-solitons listed. Secondly, the systems shown in Fig. 2 consist of cavities formed by plane mirrors and an intracavity medium possessing nonlinear (intensity-dependent) amplification and absorption. Here $1D$ and $2D$ spatial (with time-independent envelope) solitons exist under certain conditions.

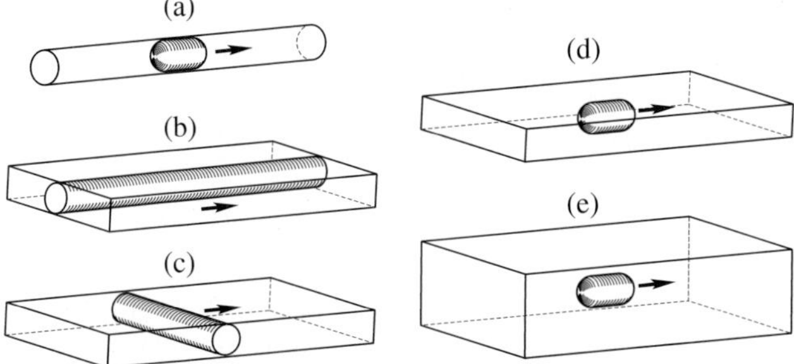

Fig. 1. Laser systems without feedback: active nonlinear fibre (**a**), active planar waveguide (**b**, **c**, **d**) and a continuous medium with frequency dispersion and nonlinear amplification and absorption (**e**); temporal $1D$-solitons are possible in the systems (**a**, **c**, **e**), spatial $1D$-solitons exist in the systems (**b**, **e**), spatio-temporal solitons are possible in the systems (**d**, **e**), spatial $2D$- and spatio-temporal $3D$-solitons can be found in the system (**e**)

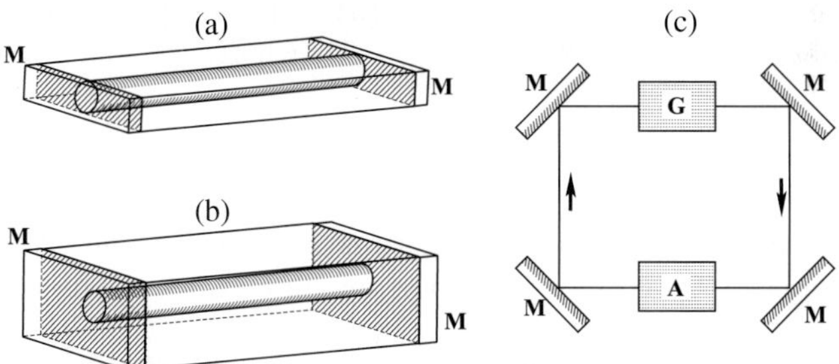

Fig. 2. Two-mirror (**a**, **b**) and ring (**c**) schemes for a laser with a saturable absorber; M indicates the cavity mirrors, G is the medium with gain, A is an absorber

The evolution equation describing the kinetics of all these systems is the generalized Ginzburg–Landau equation for the slowly-varying electric field complex envelope E [18]:

$$\frac{\partial E}{\partial \zeta} = (\mathrm{i} + d)\Delta_D E + f(|E|^2)E . \tag{1}$$

Here, the field E is scalar, and this corresponds to a fixed (e.g., linear) field polarization and the paraxial approach (non-paraxial effects are considered in [18]). All the variables and values are dimensionless. In most cases, we will neglect detunings between central frequencies of radiation and of the spectral contours of amplification and absorption. Then, the function f is real and describes nonlinear gain and linear (non-resonant) and nonlinear resonant losses. This function does not depend on co-ordinates explicitly, if the medium or system does not include inhomogeneities and aperture confinement.

The sense of other variables differs for systems with and without a cavity. In the second case (see Fig. 1), the evolution equation (1) is given in the co-ordinate frame moving along the longitudinal co-ordinate z with the light group velocity v_{gr}, where in the dimension form

$$\frac{1}{v_{\mathrm{gr}}} = \frac{\mathrm{d}k}{\mathrm{d}\omega}, \tag{2}$$

k is the linear wavenumber, and the derivative over frequency ω is taken at the carrier frequency of light. Then the evolution equation variable $\zeta = z$. The Laplacian Δ_D is taken over variables x_n that include the transverse Cartesian co-ordinates and the retarded time $\tau = t - z/v_{\mathrm{gr}}$:

$$\Delta_D = \nabla_D^2 = \sum_{n=1}^{D} \partial^2/\partial x_n^2 . \tag{3}$$

It describes diffraction and frequency dispersion. In the dimension form, there are coefficients before the derivatives in (3). For continuous media, the coefficient before the temporal derivative is $-D_2$ where

$$D_2 = \frac{1}{2}\frac{\mathrm{d}^2k^2}{\mathrm{d}\omega^2} - \frac{1}{v_{\mathrm{gr}}^2} . \tag{4}$$

Note that contrary to the case of conservative solitons, the sign of the dispersion coefficient D_2 is not very important. In fact, for the fiber $1D$-system, Fig. 1a, and for zero frequency detunings, (1) is invariant to the D_2 sign change with the simultaneous replacement $E \to E^*$. This means that if a dissipative soliton exists for anomalous dispersion, $D_2 < 0$, then it also exists for normal dispersion, $D_2 > 0$. However, for the case of normal dispersion, the coefficient of the temporal derivative has the opposite sign as compared with the coefficients in front of the spatial derivatives, and their combination cannot be represented by a Laplacian. Therefore, for the sake of simplicity, we consider further the case of anomalous frequency dispersion, where (3) is correct. The value d in (1) reflects the finite width of the spectral lines of the medium amplification and absorption. It should be non-negative, so $d \geq 0$, otherwise modulation instability occurs, even for a weak field. In a more detailed description of cavity systems, the diffusion coefficient d depends on the radiation intensity. However, this dependence does not qualitatively change the soliton features [18].

In the cavity systems (see Fig. 2), the evolution equation is obtained from the initial paraxial equation, like (1), by averaging over its longitudinal co-ordinate z, using the boundary conditions on the mirrors [21, 18]. This mean-field approximation is valid if the envelope changes during one radiation round-trip over the cavity are small. In this case, the variable ζ is the time t, the term $\mathrm{i}\Delta_D E$ on the right-hand- side of (1) represents diffraction, the frequency dispersion is not usually important, and the Laplacian Δ_D can be taken over the transverse co-ordinates x_n only. Note that, in the mean-field approximation, two-mirror (Fig. 2a, b) and ring (Fig. 2c) cavities are described by the same equation.

Equation (1) has an evident solution $E = 0$ corresponding to the non-lasing regime. It is stable against small perturbations if

$$\mathrm{Re}\, f_0 < 0\,, \quad f_0 = f(0)\,. \tag{5}$$

There are also stationary solutions with the envelope harmonic dependence on the evolution variable ζ and homogeneous distribution of the intensity $I = |E|^2$:

$$E = \sqrt{I}\exp\left(-\mathrm{i}\alpha\zeta\right)\,. \tag{6}$$

The intensity is determined as a root of the equation

$$\mathrm{Re}\left[f(I)\right] = 0\,. \tag{7}$$

The number of the roots depends on the system's parameters. The regime is mono-stable with the only stationary homogeneous envelope distribution $E = 0$ (non-lasing regime), if $\mathrm{Re}\left[f(I)\right] < 0$ for any intensity value within the region of interest.

For bright localized structures, the intensity $I = |E|^2 \to 0$ at the structure's periphery. Therefore a necessary condition of their stability is inequality (5). Let us introduce the radiation power, or energy (if $D = 3$), by the relation

$$W(\zeta) = \int I \mathrm{d}\boldsymbol{r}\,, \tag{8}$$

where $\boldsymbol{r} = (x_1, \ldots, x_D)$. The rate of variation of power W with ζ can be found using the evolution equation (1), replacing D-fold integrals by $(D-1)$-fold ones with the help of Green's formulae and neglecting outside-integral terms containing field amplitudes which are small when far from the structure centre. Then we have

$$\frac{1}{2}\frac{\mathrm{d}W}{\mathrm{d}\zeta} = \int I \;\; \mathrm{Re}\left[f(I)\right]\mathrm{d}\boldsymbol{r} - d\int |\nabla E|^2 \mathrm{d}\boldsymbol{r}\,. \tag{9}$$

Taking into account (5) and the inequality $d > 0$, we conclude that the power W decreases with ζ monotonically under conditions of mono-stability, when Re $f(I) < 0$ for any intensity I. Therefore, a change of sign of the function Re $f(I)$ is necessary for the existence of localized structures – not only motionless, but also moving and rotating ones. This coincides with the condition of bistability.

3 Stationary Symmetric Solitons

Let us start with the case of stationary structures where the field envelope depends on the evolution variable harmonically:

$$E = F \exp(-\mathrm{i}\alpha\zeta) \,. \tag{10}$$

and the complex amplitude F depends only on D co-ordinates $(x_1, \ldots, x_D)$. Then, it follows from (1) that:

$$(\mathrm{i} + d)\Delta_D F + [\mathrm{i}\alpha + f(|F|^2)]F = 0 \,. \tag{11}$$

Next, we consider symmetric structures with amplitude F depending only on the radius $r = \sqrt{x_1^2 + \ldots + x_D^2}$. In this case

$$\frac{\mathrm{d}^2 F}{\mathrm{d}r^2} + \frac{D-1}{r}\frac{\mathrm{d}F}{\mathrm{d}r} + \frac{\mathrm{i}\alpha + f(|F|^2)}{\mathrm{i} + d} F = 0 \,. \tag{12}$$

For the bright localized structures in which we are interested, the intensity $I = |F|^2 \to 0$ when $r \to \infty$. Then it follows from (12) that

$$F = F_\infty \exp(-pr) \,, \quad p^2 = \frac{\mathrm{i}\alpha + f_0}{\mathrm{i} + d} \,, \quad \mathrm{Re}\, p > 0 \,. \tag{13}$$

For $r \to 0$, the power series for localized solutions of (12) is

$$F = F_0(1 + F_2 r^2 + \ldots) \,, \quad F_2 = -\frac{\mathrm{i}\alpha + f(|F_0|^2)}{2D(\mathrm{i} + d)} \,, \tag{14}$$

where $F_0 = F(0)$.

Let us introduce the real amplitude $A(r)$ and phase $\Phi_0(r)$, so that $F = A \exp(\mathrm{i}\Phi_0)$. We obtain the following real ordinary differential equations for them:

$$\begin{aligned} &\frac{\mathrm{d}^2 A}{\mathrm{d}r^2} + \frac{D-1}{r}\frac{\mathrm{d}A}{\mathrm{d}r} - A\left(\frac{\mathrm{d}\Phi_0}{\mathrm{d}r}\right)^2 + A\,\mathrm{Re}\left[\frac{\mathrm{i}\alpha + f(A^2)}{\mathrm{i} + d}\right] = 0 \,, \\ &\frac{\mathrm{d}^2 \Phi_0}{\mathrm{d}r^2} + \frac{D-1}{r}\frac{\mathrm{d}\Phi}{\mathrm{d}r} + \frac{2}{A}\frac{\mathrm{d}A}{\mathrm{d}r}\frac{\mathrm{d}\Phi}{\mathrm{d}r} + \mathrm{Im}\left[\frac{\mathrm{i}\alpha + f(A^2)}{\mathrm{i} + d}\right] = 0 \,. \end{aligned} \tag{15}$$

The phase space corresponding to (15) is 4-dimensional. This dimensionality can be reduced to 3 by taking into account the fact that the phase itself, $\Phi_0(r)$, does not appear explicitly in (15), and introducing, additionally, the following new unknown functions to A

$$Q = \frac{\mathrm{d}\Phi_0}{\mathrm{d}r} \,, \quad K = \frac{1}{A}\frac{\mathrm{d}A}{\mathrm{d}r} \,. \tag{16}$$

Then, it follows from (12) that

$$\begin{aligned} &\frac{\mathrm{d}A}{\mathrm{d}r} - KA = 0 \,, \\ &\frac{\mathrm{d}Q}{\mathrm{d}r} + 2QK + \frac{D-1}{r}Q + \mathrm{Im}\left[\frac{\mathrm{i}\alpha + f(A^2)}{\mathrm{i} + d}\right] = 0 \,, \\ &\frac{\mathrm{d}K}{\mathrm{d}r} + K^2 - Q^2 + \frac{D-1}{r}K + \mathrm{Re}\left[\frac{\mathrm{i}\alpha + f(A^2)}{\mathrm{i} + d}\right] = 0 \,. \end{aligned} \tag{17}$$

The asymptotic behaviour of $A(r)$, $Q(r)$ and $K(r)$ for $r \to 0$ and $r \to \infty$ follows from (14) and (13). Now, the radial dependence of functions A, Q and K for the symmetric localized structure corresponds to the trajectory $\mathrm{A_0L}$ in the phase space of (17), Fig. 3. The trajectory starts at a point on the axis A with co-ordinates $(A_0, 0, 0)$ (at the centre of the structure, $r = 0$, the value A_0 is unknown) and should finally reach the point L on the QK-plane with known co-ordinates $(0, -\mathrm{Re}\,p, -\mathrm{Im}\,p)$ (the non-lasing regime, $r \to \infty$). Let us recall that a trajectory in the phase space is determined completely by its initial point. To come to the final point, L, two conditions should be met at the point of the trajectory intersection with plane QK: $Q = -\mathrm{Re}\,p$ and $K = -\mathrm{Im}\,p$. To achieve this, it is not sufficient just to choose a proper value of A_0. Such trajectories are possible only for certain discrete values of the eigenvalue α that corresponds to the *discrete spectrum* of localized dissipative structures.

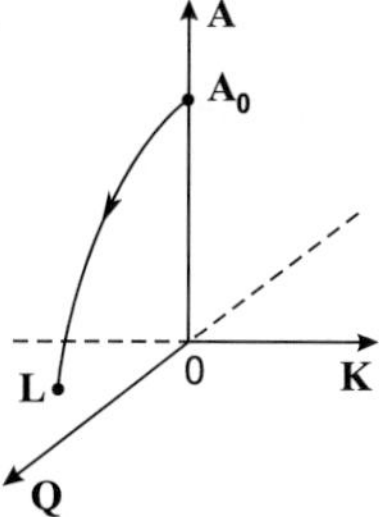

Fig. 3. Phase space of system (17) and trajectory $\mathrm{A_0L}$ (*line with arrow*) representing the radial profile of a symmetric localized structure. Point $\mathrm{A_0}$ corresponds to the structure centre ($r = 0$), and point L corresponds to the non-lasing regime at the structure periphery ($r = \infty$)

Accepting the two-level model, both for amplification and absorption, we can specifically define the form of the nonlinear function f. Neglecting the frequency detunings, it is

$$f(|E|^2) = -1 + \frac{g_0}{1 + |E|^2} - \frac{a_0}{1 + b|E|^2} \,. \qquad (18)$$

Here g_0 and a_0 are small-signal gain and absorption coefficients, respectively, and b is the ratio of the saturation intensities for gain and absorption. The non-resonant absorption coefficient is re-scaled to unity by the evolution variable normalization. In the numerical examples below, we fix the following values of parameters: $a_0 = 2$ and $b = 10$.

In Fig. 4, the dependence of the localized structure spectral parameter – eigenvalue α – on the small-signal gain coefficient g_0 is shown for all three possible values of dimensionality D. The dependence is actually a spiral-type curve. For $D = 1$, the spiral has an infinite number of increasingly narrow

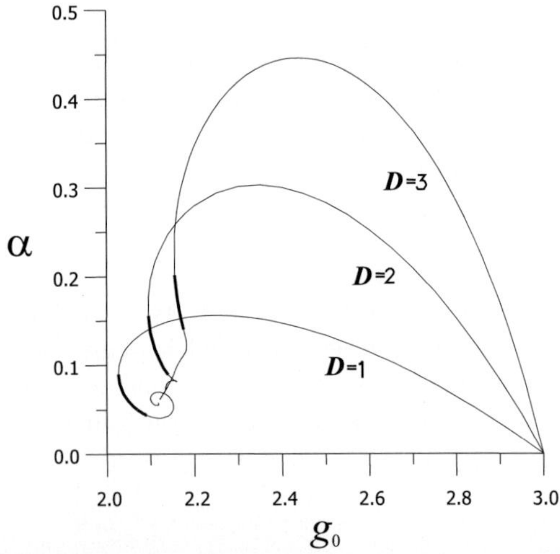

Fig. 4. Dependence of the spectral parameter α on the small-signal gain coefficient g_0 [22] for stable (*solid lines*) and unstable (*dashed lines*) localized structures of laser radiation with dimensionality $D = 1$, 2 and 3

coils [22]. In the general case, the shape of this curve is more complex and includes self-intersections (see below Fig. 5). Only certain parts of the spiral coils correspond to stable localized structures, i.e., solitons. The linear stability analysis will be described below. Outside the ranges of soliton stability, there are localized structures which pulsate periodically or chaotically. We re-direct the reader to [18, 23] for the cases $D = 1$ and 3, so we will restrict ourselves to a consideration of two-dimensional structures only.

4 Two-Dimensional Laser Solitons

We now set the system dimensionality to $D = 2$ in the evolution equation, (1), and consider, to be definite, a wide-aperture laser with a saturable absorber ($\zeta = t$). The simplest case of stationary localized solutions with cylindrically-symmetric distributions of intensity will be our initial interest here. In polar co-ordinates r, φ ($x = r\cos\varphi$, $y = r\sin\varphi$) the electric field envelope for such structures is

$$E = F(r)\exp(\mathrm{i}M\varphi)\exp(-\mathrm{i}\alpha t)\ . \tag{19}$$

The spectral parameter α has the meaning of a nonlinear frequency shift, and the integer $M = 0,\ \pm1,\ \pm2, \ldots$ is the topological charge, or azimuthal index. Note that the total envelope is symmetric only if $M = 0$, just as for the case of two-dimensional structures in Sect. 3. Now, the complex amplitude F is determined by equation

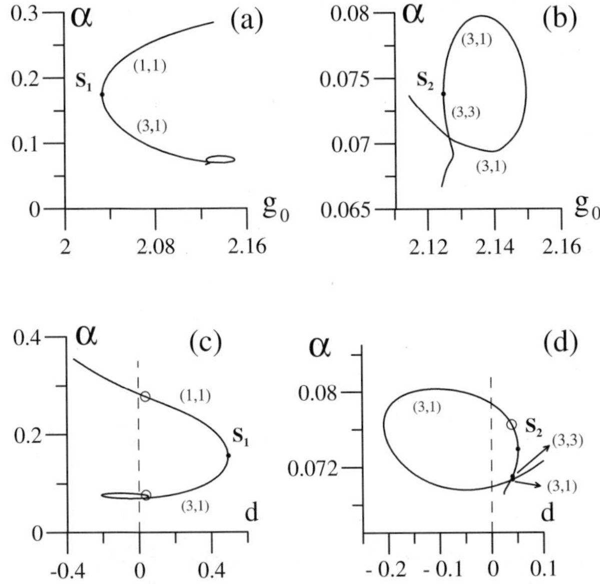

Fig. 5. Dependence of the spectral parameter α on the small-signal gain coefficient g_0 (**a**, **b**; $d = 0$) and on the diffusion coefficient d (**c**, **d**; $g_0 = 2.13$) for localized structures of laser radiation with topological charge $M = 1$. In **b** and **d** the region of the second loop of the spiral is shown in more detail. Near the spiral loops, the values (N_Q, N_K) are given as they change at transition through points S_1, S_2 with vertical tangents. Unstable localized structures are marked by circles. Points with arrows near the spiral self-intersection [(**d**), $d = 0.04$] denote fundamental and excited solitons; their radial profiles are shown in Fig. 6c, d [24]

$$\frac{\mathrm{d}^2 F}{\mathrm{d}r^2} + \frac{1}{r}\frac{\mathrm{d}F}{\mathrm{d}r} - \frac{M^2}{r^2}F + \frac{\mathrm{i}\alpha + f(|F|^2)}{\mathrm{i} + d}F = 0 \, . \tag{20}$$

The asymptotic behaviour at $r \to \infty$ is the same as in Sect. 3 [it does not depend on M – see (13)]. However, this is not so for $r \to 0$, as then

$$F = F_0 r^{|M|}(1 + F_2 r^2 + \ldots) \, , \quad F_2 = -\frac{\mathrm{i}\alpha + f^{(0)}}{4(|M|+1)(\mathrm{i}+d)} \, , \tag{21}$$

where $f^{(0)} = f(|F_0|^2)$ for $M = 0$ and f_0 for $M \neq 0$.

It is helpful to re-introduce the real amplitude $A(r) = |F|$ and phase $\Phi = \arg F$. The total phase Φ depends on the radial and angular (for $M \neq 0$) co-ordinates

$$\Phi = \Phi_0(r) + M\varphi \, . \tag{22}$$

Note that the “radial phase” Φ_0 is a linear function of radius when $r \to \infty$:

$$\Phi_0 = \mathrm{const} - \mathrm{Im}\,(p)\; r \, , \tag{23}$$

and is a quadratic function of radius when $r \to 0$:

$$\Phi_0 = -\frac{r^2}{4(|M|+1)} \mathrm{Im}\left[\frac{\mathrm{i}\alpha + f^{(0)}}{4(|M|+1)(\mathrm{i}+d)}\right] . \tag{24}$$

Below, we give the results of numerical simulations for the nonlinear function (18) for parameters $a_0 = 2$, $b = 10$. As a control (varying) parameter, we can use the small-signal gain coefficient g_0 or the diffusion coefficient d. The dependence of the eigenvalue α on g_0 for $M = 0$ was given in Fig. 4. Such dependencies – "spirals" – are shown in Fig. 5 for various M. Different spiral coils are characterized by a different numbers of zeros, N_Q and N_K, of functions $Q(r)$ and $K(r)$, respectively, for $r > 0$. The numbers N_Q and N_K change by one (at $M = 0$) or by two (at any M) when passing along a spiral through points S_1 and S_2 with a vertical tangent.

Now let us consider the stability of the localized structures. The perturbed envelope includes small amplitudes a and b of perturbations with angular harmonics $m = 0, 1, 2, \ldots$:

$$E = [F(r) + \delta F] \exp(\mathrm{i}M\varphi) \exp(-\mathrm{i}\alpha t) , \tag{25}$$

where

$$\delta F = a(r) \exp(\mathrm{i}m\varphi + \gamma t) + b^*(r) \exp(-\mathrm{i}m\varphi + \gamma^* t) . \tag{26}$$

The localized solution is stable if there are no eigenvalues γ with positive real part. At the boundary of the stability range, we have max (Re γ) = 0, where the maximum is chosen from all the eigenvalues. The case of a boundary with Im $\gamma \neq 0$ corresponds to an Andronov-Hopf bifurcation. For $m = 0$, this bifurcation leads to temporal oscillations of the perturbed structure. For $m \neq 0$, (26) can be rewritten in the form

$$\delta F = a(r) \exp\left[\mathrm{i}m(\varphi - \Omega t)\right] + b^*(r) \exp\left[-\mathrm{i}m(\varphi - \Omega t)\right] , \tag{27}$$

which corresponds to a rotation of the perturbed structure with an angular velocity $\Omega = -\mathrm{Im}(\gamma)/m$. According to (25) and (26), the intensity, $I = |E + \delta F|^2$, of the rotating structure exhibits $|m|$ maxima as the angle φ is varied (a multi-humped structure) with r and t fixed.

After linearization of the evolution equation (1) with respect to a and b, we obtain a linear equation

$$\boldsymbol{L\Psi} = \gamma \boldsymbol{\Psi} , \quad \boldsymbol{\Psi} = (a, b)^{\mathrm{T}} \tag{28}$$

(where T denotes transposition) with the following form of the linear operator $\boldsymbol{L}$:

$$\boldsymbol{L} = \begin{pmatrix} (d+\mathrm{i})\Delta_+ + \mathrm{i}\alpha + F_0', & (f_0' F^2) \\ (f_0' F^2)^*, & (d-\mathrm{i})\Delta_- - \mathrm{i}\alpha + F_0'^* \end{pmatrix} . \tag{29}$$

Here $\Delta_\pm = \mathrm{d}^2/\mathrm{d}r^2 + (1/r)\mathrm{d}/\mathrm{d}r - (M \pm m)^2/r^2$ and $F_0' = f_0 + f_0'|F|^2$. Under conditions of bistability, when the non-lasing regime is stable, the continuous spectrum of γ is situated on the left half-plane of the complex plane $\gamma = \mathrm{Re}\,\gamma + \mathrm{iIm}\,\gamma$ and has no connections with unstable regions. Therefore, it is the discrete spectrum that determines soliton stability. The "neutral modes" – eigenfunctions of (28) with zero eigenvalue ($\gamma = 0$) – correspond to the symmetry of the evolution equation (1) with respect to the phase shift ($m = 0$) and to shifts of co-ordinates ($m = 1$).

The discrete spectrum of localized perturbation eigenvalues γ can be found numerically, and the results of calculations show the following. For $M = 0$, the outer coil of spirals similar to these shown in Fig. 5, up to the point S_1, corresponds to unstable localized solutions only. Stable solitons appear for parameters in the parts of the following coils where $N_Q = 1$ and $N_K = 0$. For $M = 1$ and 2, we have $N_Q = N_K = 1$ in the outer coil, and there are no stable solitons there either. Stable solitons appear in those parts of the following coils where $N_Q = 3$ and $N_K = 1$ (*fundamental soliton*) and $N_Q = 3$ and $N_K = 3$ (*excited soliton*). In Fig. 5d, the points with arrows show two nearby values of the soliton eigenvalue α for the same small-signal gain g_0 that belong to two different coils of the spiral near its self-intersection. These points characterize fundamental and excited solitons with radial profiles of amplitude and phase shown in Fig. 6c, d. Indeed, we are mainly interested in stable localized structures. For them, the amplitude decays monotonically with radius in the case $M = 0$, and has one or more maxima if $M \neq 0$. The phase radial profile includes oscillations whose number depends on $|M|$ – see

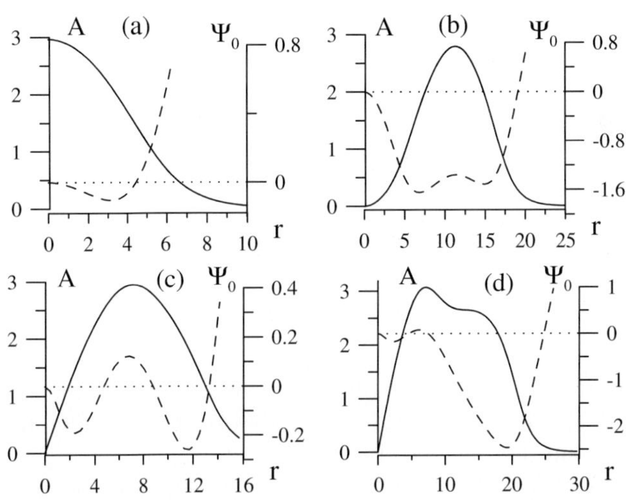

Fig. 6. Radial profiles of amplitude A (*solid lines*) and radial phase Φ_0 (*dashed lines*) for solitons with topological charges $M = 0$ (**a**), $M = 1$ [fundamental (**c**) and excited (**d**) solitons], and $M = 2$ (**b**) [24]

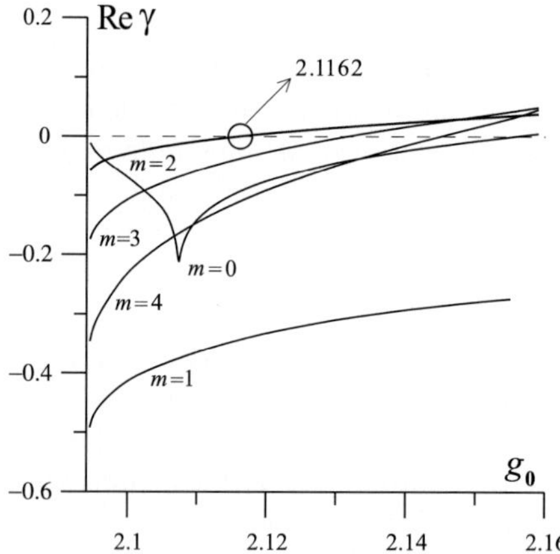

Fig. 7. Growth rate, Reγ, of perturbations with angular harmonic m versus the small-signal gain for a fundamental laser soliton, $M = 0$. The *circle* indicates the instability threshold [18]

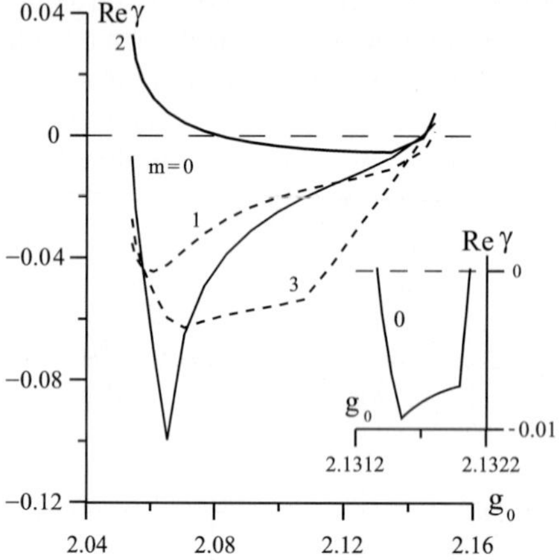

Fig. 8. Growth rate, Re γ, of perturbations with angular harmonic m from 0 to 3 versus the small-signal gain g_0 for a vortex laser soliton ($M = 1$). Inset shows the maximum growth rate of perturbations with $m = 0$ for the excited soliton in the narrow range of its stability. Diffusion coefficient $d = 0.1$ [25]

Fig. 6. Note that these profiles differ only quantitatively for the cases of a stable soliton in the fundamental and excited states (for different spiral coils in Fig. 5, compare Fig. 6c and d).

More detailed results of the linear stability analysis for localized structures with topological numbers $M = 0$, 1 and 2 are given in Figs. 7, 8, and 9a, respectively. The fundamental soliton ($M = 0$) is stable even for zero diffusion ($d = 0$). Perturbations with $m = 2$ are responsible for its destabilization

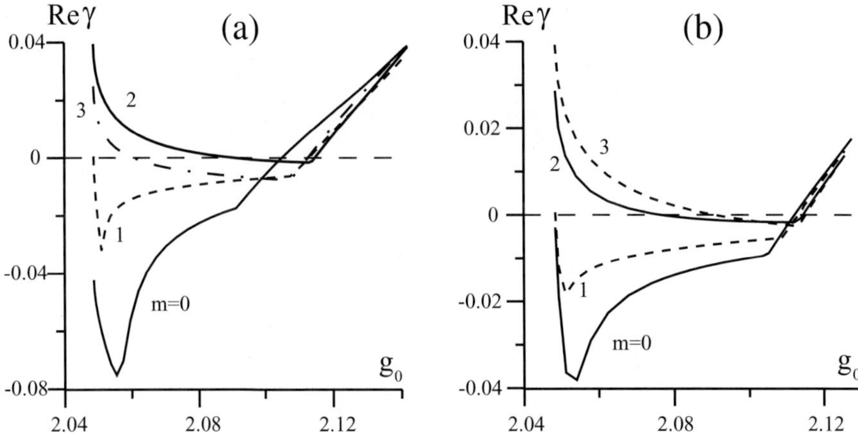

Fig. 9. Growth rate, $\mathrm{Re}\,\gamma$, of perturbations with angular harmonic m from 0 to 3 versus the small-signal gain coefficient g_0 for a vortex soliton of (**a**) the second order ($M = 2$) and (**b**) the third order ($M = 3$). Here $d = 0.1$ [25]

with an increase in gain – see Fig. 7. Calculations show that localized vortices ($M \neq 0$) are stable only for positive diffusion coefficient. With an increase in gain, vortex solitons are destabilized by perturbations with $m = 0$. With a decrease of gain, for $M = 0$ the bifurcation is a saddle-node (a merging of stable and unstable localized solutions; only transversely homogeneous regimes are possible for a lower gain). For $M = 1$ and 2, solitons are destabilized by perturbations with $m = 2$, and solitons with $M = 3$ are destabilized by perturbation harmonics with $m = 3$. These instabilities correspond to Andronov-Hopf bifurcations and are connected to the formation of rotating structures.

5 Numerical Simulations of Asymmetric Solitons

The results of the semi-analytical approach, presented in the previous sections, indicate the possibility of asymmetric structures arising beyond the ranges of stability of the symmetric solitons. However, this approach is practically incapable of describing the final stage of such structures. Therefore, to study them, we need to perform direct numerical simulations and solve the evolution equation (1). Here we present the results of such simulations, based on the standard splitting method and algorithm of the fast Fourier transform [18].

Let us recall that destabilization of a fundamental soliton ($M = 0$) above the high-gain boundary of its stability range is associated with asymmetric perturbation harmonics with $m = 2$ (see Fig. 7). Numerical calculations confirm the subcritical (with hysteresis jump) character of the

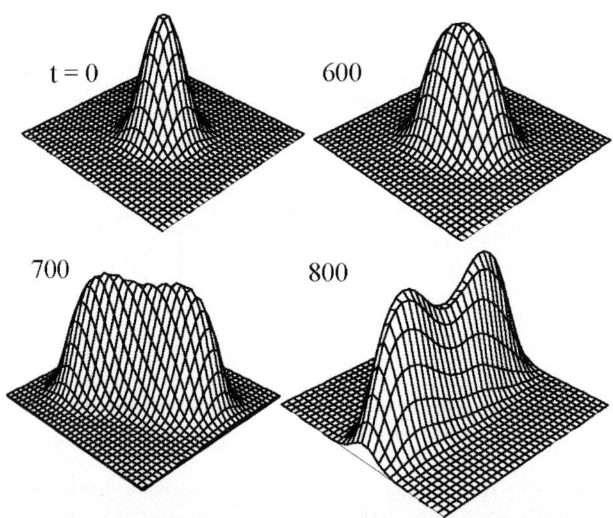

Fig. 10. Transverse intensity distributions illustrating transformation of the initially cylindrically-symmetric soliton (time instant $t = 0$) into an asymmetric rotating soliton with rapid change in the small-signal gain coefficient. Here $d = 0$ [25]

corresponding Andronov-Hopf bifurcation. With an increase in gain, an initially cylindrically-symmetric soliton (Fig. 10, $t = 0$) transforms into a soliton rotating with a constant angular velocity. The latter soliton is cylindrically asymmetric and has two intensity maxima (it is a "two-humped" structure, Fig. 10, $t = 800$). More precisely, a stationary rotating soliton can be formed if the small-signal gain coefficient g_0 first increases above the stability threshold and then decreases to a value within the range of this soliton stability – see below Fig. 13. Note that there are no wavefront vortices in this case, so the rotation is not related to the structure topological charge. Such rotating laser solitons were first found in [26]. Below, we will give new examples of rotating solitons and study their bifurcations. Note that Fig. 10 corresponds to a fast gain increase. If it is slow, then at the first stage a metastable non-rotating localized structure arises and this oscillates quasi-periodically. Then, at the second stage, the structure transforms into a stable asymmetric rotating soliton [25]. For the parameters of Fig. 10, there is hysteresis between stationary fundamental solitons and asymmetric rotating solitons – see below Fig. 13.

For localized laser vortices ($M \neq 0$), both high- and low-gain boundaries of the stability range correspond to Andronov-Hopf bifurcations (see Figs. 8, 9). Numerical simulations of laser vortex destabilization give a wide spectrum of different stable localized structures. It would appear reasonable to say that an Andronov-Hopf bifurcation results in the formation of a rotating soliton with the number of maxima being equal to the value m of the perturbation responsible for the instability. However, this conclusion is not universally

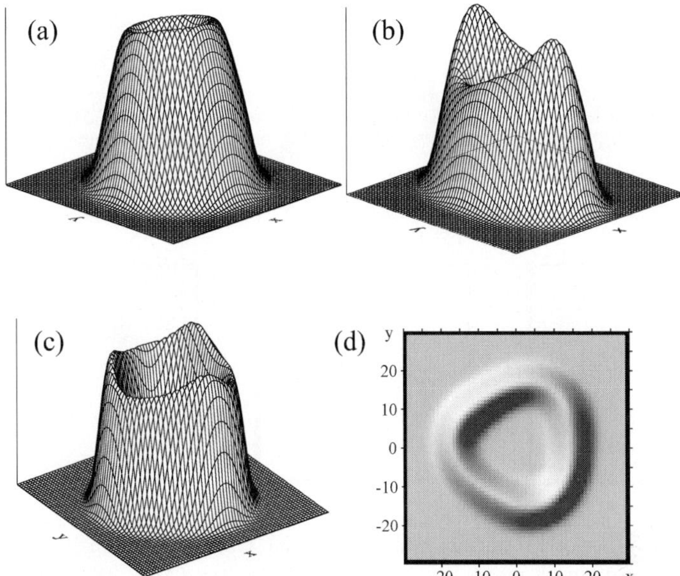

Fig. 11. Instantaneous transverse intensity distributions for rotating one- (**a**), two- (**b**), and three-humped [(**c**, **d**), wired surface and surface relief, respectively] vortex solitons with topological charges $M = 1$ (**a**, **b**) and 3 (**c**, **d**). Here $d = 0.06$ [25]

valid, because linear stability analysis does not describe the nonlinear stage of perturbation growth. For example, the symmetric vortex of the first order ($M = 1$) is associated with two types of stable rotating asymmetric solitons: *viz.* two-humped solitons existing in the low-gain region, and one-humped solitons in the high-gain region – see Fig. 11a, b. Here, the formation of the one-humped soliton cannot be explained by the results of the linear stability analysis – see Fig. 8.

Bifurcations of higher-order localized vortices are similar, or even more complex. In this case, stable rotating and simultaneously oscillating solitons (Fig. 12, $M = 2$) exist for a fairly wide range of gain. An example of a bifurcation of a third-order vortex ($M = 3$) is given in Fig. 11c, where the stable state of a rotating asymmetric vortex with three intensity maxima is presented. The number of the maxima corresponds to the value $m = 3$ of the perturbation associated with the bifurcation of the symmetric laser vortex that occurs with a decrease in gain (see Fig. 9b).

Co-existence of symmetric and asymmetric solitons, and hysteresis jumps between them with gain variations, are shown in Fig. 13, where $M = 1$. With an increase in gain, the initially symmetric localized vortex loses its symmetry and transforms into a rotating one-humped soliton. With further gain increase, this soliton also loses its stability. Then, after an intermediate

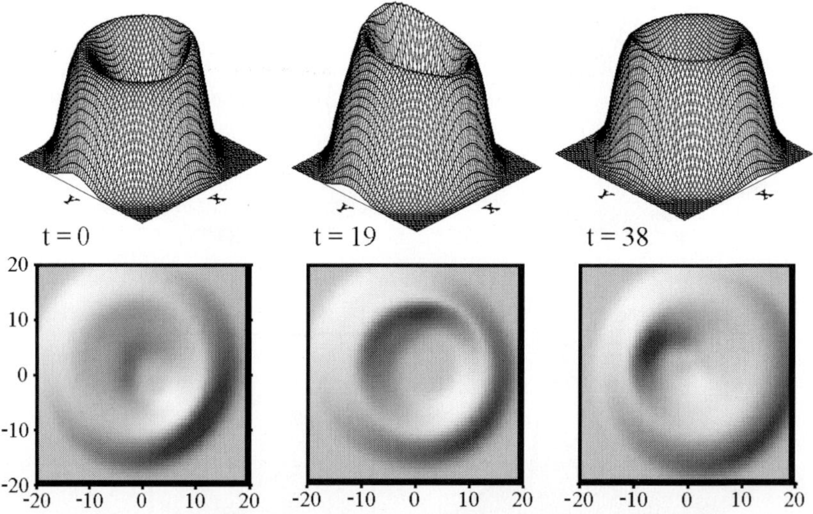

Fig. 12. Quasi-periodic temporal variation of the transverse intensity distribution of a rotating and simultaneously oscillating laser vortex ($M = 2$) – wired surfaces (*upper figures*) and surface reliefs (*lower figures*). Here $d = 0.06$ [25]

metastable regime of oscillations, the lasing zone progressively widens in the form of a cylindrical switching wave [18].

With a decrease in gain, the one-humped rotating soliton transforms into a vortex soliton with an axially-symmetric intensity distribution. There exists an extremely narrow hysteresis range where both types of solitons coexist. The hysteresis range is fairly wide in the low-gain region where symmetric vortices co-exist with two-humped solitons (see Fig. 13). Two-humped solitons cannot be formed from symmetric vortices with a slow gain variation in this case, but they can be excited by a sufficiently strong asymmetric perturbation. A further decrease in gain results in the splitting of both symmetric and asymmetric (rotating) vortex solitons into two fragments without wavefront dislocations. Then these fragments disappear.

There is also a different manner of formation of rotating soliton structures. Let us consider two single solitons with a fairly large distance L between their centres relative to the individual soliton width. Then the overlap of the solitons is small, and the instantaneous envelope of the total field can be written in the form

$$E = A(r_1)\exp\left[\mathrm{i}\Psi_0(r_1) + \mathrm{i}M_1\varphi_1\right] + A(r_2)\exp\left[\mathrm{i}\Psi_0(r_2) + \mathrm{i}M_2\varphi_2 + \mathrm{i}\vartheta\right] \quad (30)$$

(for the case of two-soliton structures with topological charges $|M_1| = |M_2|$). Here $(r_n,\ \varphi_n)$ are the two sets of polar co-ordinates whose centres coincide with the centres of n-th soliton, $n = 1$ and 2:

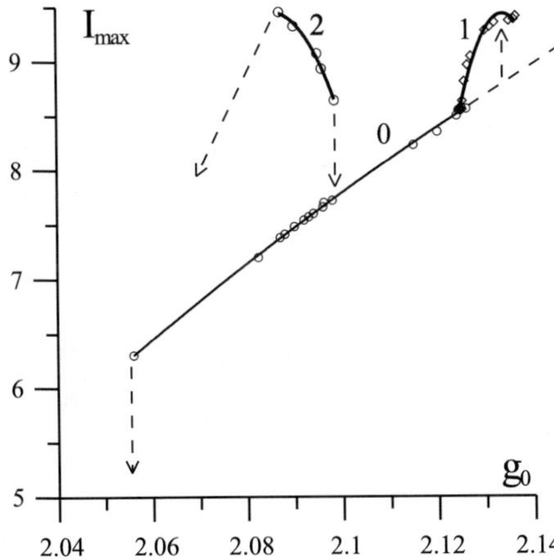

Fig. 13. Maximum intensity, $I_{\max}$, of stable localized vortices with topological charge $M = 1$ versus the small-signal gain coefficient g_0. Curves 0, 1, and 2 correspond to symmetric, asymmetric one-humped, and two-humped vortices, respectively. *Vertical dashed* lines represent hysteresis jumps with gain variation. The oblique *dashed* line indicates a jump to a regime of soliton splitting into two decaying fragments. Here $d = 0.06$ [25]

$$r_n^2 = \left(x \pm \frac{L}{2}\right)^2 + y^2 , \quad \cos\varphi_n = \frac{x \pm \frac{L}{2}}{r_n} , \quad \sin\varphi_n = \frac{y}{r_n} . \tag{31}$$

Here ϑ is a constant phase difference which influences the character of the interaction of solitons.

An example of a stable motionless symmetric pair of coupled fundamental solitons (topological charges $M_1 = M_2 = 0$) is presented below in Fig. 16a. There, the soliton overlap is fairly small, so it is an example of weakly-coupled solitons. However, for certain conditions, the two solitons approach each other over time and can form a strongly-coupled asymmetric structure rotating with a certain angular velocity. An example of a corresponding two-humped rotating structure is shown below in Fig. 16c. Other examples of strongly-coupled asymmetric rotating structures are given in Fig. 14. Different chain links there correspond to fundamental or excited solitons with topological indices $M = \pm 1$.

6 Energy Flows and Soliton Internal Structure

To gain a more penetrating insight into the nature of weakly and strongly coupled dissipative soliton structures, let us consider flows of energy inside them. In the paraxial approximation used for the derivation of the evolution equation (1), and for quasi-monochromatic radiation with a fixed polarization, e.g., close to the linear one, the Poynting vector **S** averaged over the

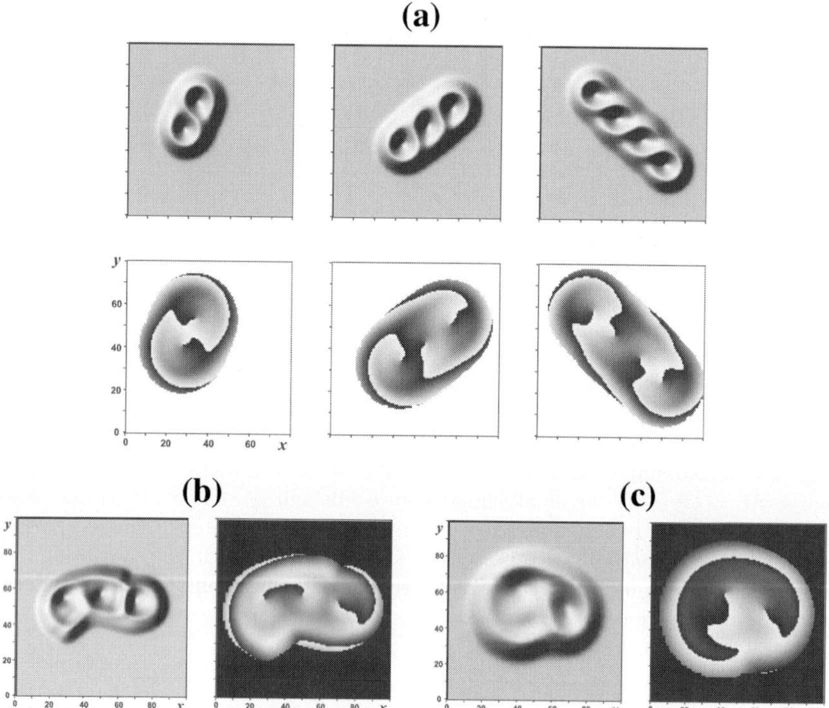

Fig. 14. Intensity [(**a**), *upper* figures; (**b**) and (**c**), *left* figures] and phase [(**a**), *lower* figures; (**b**) and (**c**), *right* figures] transverse distributions for rotating linear chains of solitons with the same topological charge, $M = 1$ (**a**), a chain with charges $M_1 = M_2 = 1$, $M_3 = -1$ (**b**), and an excited pair of solitons with charges 1 and -1 (**c**) [24]

optical period is related to the electric field complex envelope E, real amplitude A, and real phase Φ by the following relations:

$$\boldsymbol{S} = A^2 \nabla \Phi = \mathrm{Im}(E^* \nabla E) . \tag{32}$$

In the systems considered, the longitudinal co-ordinate is separated, and the main power flow is directed along the longitudinal co-ordinate. However, in the evolution equation (1), the longitudinal dependencies are averaged, and we are interested in smaller transverse components of the Poynting vector. For a fixed time instant t, they are

$$\boldsymbol{S}_\perp(x, y) = A^2 \nabla_\perp \Phi = \mathrm{Im}(E^* \nabla_\perp E) . \tag{33}$$

Let us introduce lines of flow of the radiation energy at time instant t. These are curves in the plane (x, y) whose tangents have the same directions as the transverse Poynting vector, $\boldsymbol{S}_\perp = (S_x, S_y)$, at any point (x, y). This condition can be formulated as the following equation:

$$\frac{\mathrm{d}x}{S_x(x,y)} = \frac{\mathrm{d}y}{S_y(x,y)} \; . \tag{34}$$

It is convenient to re-write (34) by introducing a parameter τ which varies along the length of the curve (compare with ray tracing in geometrical optics). Then

$$\frac{\mathrm{d}x}{\mathrm{d}\tau} = S_x(x,y) \; , \quad \frac{\mathrm{d}y}{\mathrm{d}\tau} = S_y(x,y) \; . \tag{35}$$

Transverse distributions of the Poynting vector and of the intensity give the full description of the electric field (to within a non-essential phase constant).

Equations of type (35) are studied in detail in the theory of nonlinear oscillations [27]. Note that, for dissipative systems, $\mathrm{div}_\perp \boldsymbol{S}_\perp \neq 0$, which is related to the energy exchange between the field and the amplifying and absorbing media, whereas $\mathrm{div}_\perp \boldsymbol{S}_\perp = 0$ for the degenerate case of conservative systems.

Our next problem is to separate the phase plane into (a finite number of) cells with topologically equivalent trajectories inside the cells. To solve this problem, it is first necessary to find the *fixed points*, with co-ordinates (x_0, y_0), where simultaneously $S_x(x_0, y_0) = S_y(x_0, y_0) = 0$. Second, limit cycles and saddle separatrices should be found.

Taking into account (33), the Poynting vector becomes zero in two cases: when the field amplitude (intensity) or the phase gradient is zero. Correspondingly, there are two types of fixed points. In the first case, the envelope expansion, as a series in the vicinity of the fixed point, begins with terms linear in the deviation of the co-ordinates x and y from (x_0, y_0):

$$E(x,y) = E_{(x)}x + E_{(y)}y + \frac{1}{2}E_{(xx)}x^2 + E_{(xy)}xy + \frac{1}{2}E_{(yy)}y^2 + \dots \; , \tag{36}$$

where derivatives $E_{(x)}$, etc., are calculated at the fixed point. Then

$$S_x = qy \; , \quad S_y = -qx \; , \quad q = \mathrm{Im}(E_{(x)}E^*_{(y)}) \; . \tag{37}$$

Now, in the vicinity of the fixed point, (35) takes the form

$$\frac{\mathrm{d}x}{\mathrm{d}\tau} = qy \; , \quad \frac{\mathrm{d}y}{\mathrm{d}\tau} = -qx \; . \tag{38}$$

The lines of energy flow correspond in this case to the obvious solution of (38), *viz.* concentric circles of arbitrary radius R with the centre being the fixed point:

$$x^2 + y^2 = R^2 \; . \tag{39}$$

In this case, the fixed point is a centre. This means it is degenerate and can change its type due to even small subsequent terms in the series (36). Further, we will see that, for laser solitons, this point usually corresponds to a focus.

The envelope expansion near a fixed point of the second type includes a constant term, $E_0 \neq 0$:

$$E(x,y) = E_0 + E_{(x)}x + E_{(y)}y + \frac{1}{2}E_{(xx)}x^2 + E_{(xy)}xy + \frac{1}{2}E_{(yy)}y^2 + \dots , \quad (40)$$

The condition that the Poynting vector becomes zero at the fixed point is $\mathrm{Im}(E_0^* E_{(x)}) = \mathrm{Im}(E_0^* E_{(y)}) = 0$. The lines of energy flow are determined by the equations

$$\frac{\mathrm{d}x}{\mathrm{d}\tau} = \mathrm{Im}(E_0^* E_{(xx)})x + (q+p)y \,, \quad \frac{\mathrm{d}y}{\mathrm{d}\tau} = (p-q)x + \mathrm{Im}(E_0^* E_{(yy)})y \,, \quad (41)$$

where $p = \mathrm{Im}(E_0^* E_{(xy)})$. The fixed point type is determined by the features of the two roots of the quadratic equation

$$\lambda^2 + \sigma\lambda + \Delta = 0 \,, \quad (42)$$

where

$$\sigma = -\mathrm{Im}(E_0^* E_{(xx)}) - \mathrm{Im}(E_0^* E_{(yy)}) \,,$$
$$\Delta = \mathrm{Im}(E_0^* E_{(xx)})\mathrm{Im}(E_0^* E_{(yy)}) - p^2 + q^2 \,. \quad (43)$$

For $\sigma \neq 0$ and $\Delta \neq 0$, the fixed point is non-degenerate, and is of the type node, focus or saddle. Nodes can be "quasi-stable" or "quasi-unstable" with respect to perturbation evolution in the framework of (35) in the limit $\tau \to +\infty$. Note that it is not real temporal stability, because the time instant is fixed in (35). In our calculations, foci are "quasi-stable".

With a variation of the system parameters, initial conditions, or just time, the lines of energy flow change. Of special interest are *bifurcations* when the phase plane separation into cells changes topologically. For equations of type (35), all the types of bifurcations are known in non-degenerate cases, resulting in the problem being simplified.

For stationary solitons with cylindrically-symmetric intensity distributions [see (19)], the Poynting vector in polar co-ordinates with unit vectors $\boldsymbol{e}_r, \boldsymbol{e}_\varphi$ is

$$\boldsymbol{S}_\perp = A^2(r)\left(\frac{\mathrm{d}\Phi_0}{\mathrm{d}r}\boldsymbol{e}_r + \frac{M}{r}\boldsymbol{e}_\varphi\right) . \quad (44)$$

Therefore, the equations for the lines of energy flow are

$$\frac{\mathrm{d}r}{\mathrm{d}\tau} = A^2\frac{\mathrm{d}\Phi_0}{\mathrm{d}r} \,, \quad \frac{\mathrm{d}\varphi}{\mathrm{d}\tau} = \frac{M}{r^2}A^2 \,. \quad (45)$$

For the trajectories we have

$$\frac{\mathrm{d}\varphi}{\mathrm{d}r} = \frac{M}{r^2 \, \mathrm{d}\Phi_0/\mathrm{d}r} \,, \quad \varphi - \varphi_0 = M\int_{r_0}^{r} \frac{\mathrm{d}r}{r^2 \, \mathrm{d}\Phi_0/\mathrm{d}r} \,. \quad (46)$$

At $\tau \to \pm\infty$, the radius $r(\tau)$ approaches a constant value, R_0, which is a root of the equation $\frac{\mathrm{d}\Phi_0}{\mathrm{d}r} = 0$. One of these roots is $R_0 = 0$; it corresponds to the soliton centre and is a quasi-stable node (if $M = 0$) or focus (if $M \neq 0$).

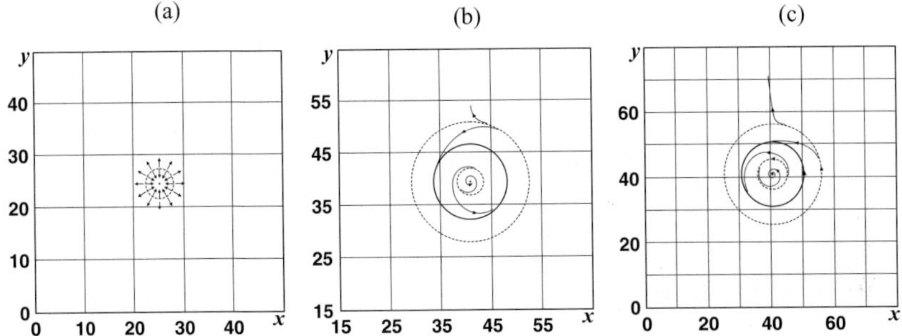

Fig. 15. Phase portrait of energy flow for single symmetric solitons with topological charge $M = 0$ (**a**), 1 (**b**), and 2 (**c**). At the centres ($r = 0$), there are fixed points – a node (**a**) or focus (**b**, **c**). Phase plane cells with trajectories of different character are separated by circles corresponding to quasi-stable (*solid thick lines*) and quasi-unstable (*solid dashed lines*) limit cycles or to a set of fixed points (degenerated case, $M = 0$). *Thin lines with arrows* indicate the direction of the transverse Poynting vector [24]

Let us recall that the Poynting vector becomes zero at fixed points. According to Fig. 6, the number of other roots (R_0) is odd and depends on the topological charge M; it is one for $M = 0$ and 3 for $M = 1$ and 2. For $M = 0$, the trajectories are radial rays which start on the circle of radius R_0. For $M \neq 0$, these circles are limit circles, and the affix (representative point) rotates around them with a constant angular velocity $\Omega = MA^2(R_0)/R_0^2$. The limit cycle closest to the centre is quasi-unstable; it separates trajectories attracted to the neighbouring quasi-stable fixed point and the next quasi-stable limit cycle. The last cycle is quasi-unstable, separating trajectories which are attracted to the previous quasi-stable cycle from those which approach infinity where the Poynting vector becomes zero. The case $M = 0$ is degenerate, because then the only circle with $R_0 \neq 0$ is not a limit cycle (a trajectory), but a set of separate fixed points. The phase plane separation into cells for $M = 0$, 1, and 2 is shown in Fig. 15. It is easy to find similar figures for $M < 0$ because of the symmetry of the evolution equation (1) with respect to inversion of one of the Cartesian co-ordinates, say $y \to -y$.

Now let us consider the energy flow for cylindrically-asymmetric soliton structures. In Fig. 16a, b the transverse intensity distribution and the energy diagram are shown for the case of a stationary stable symmetric pair of weakly-coupled solitons with zero topological charge. Comparing Fig. 16b with Fig. 15a, one can see that the degeneration of the circles around the individual and almost independent solitons is now lifted: each of the two circles transforms into four arcs between two nodes and two saddles. In the region between the solitons, a new saddle arises, and its two separatrices approach infinity. The phase plane separation into cells is clear from Fig. 16b.

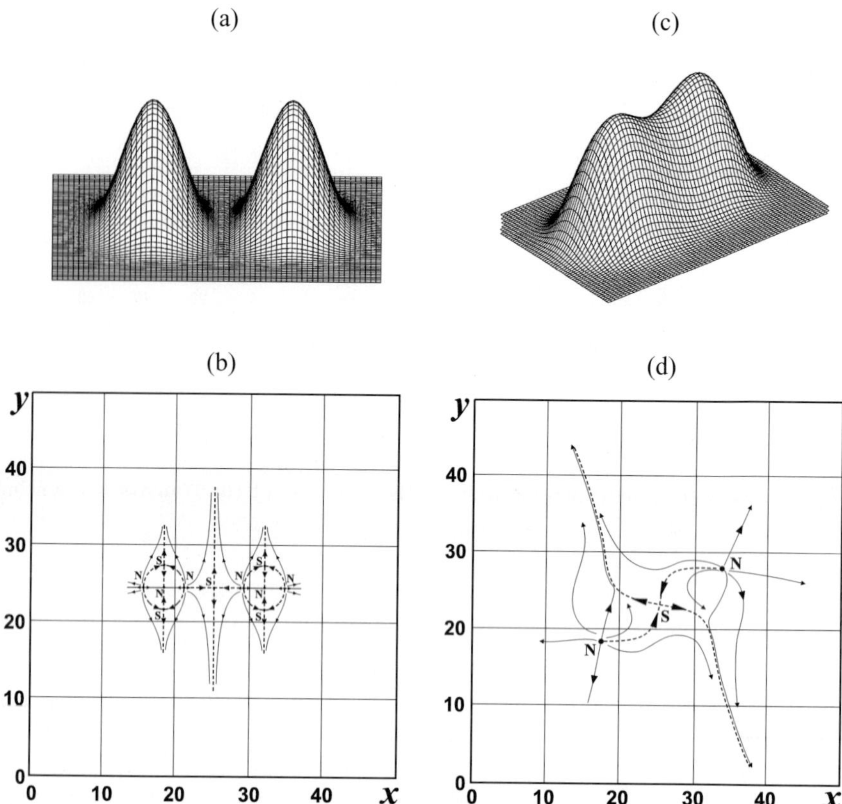

Fig. 16. Transverse intensity distributions (**a**, **c**) and phase planes of energy flow (**b**, **d**) for stable pairs of weakly (**a**, **b**) and strongly (**c**, **d**) coupled solitons with topological charges $M_1 = M_2 = 0$. Separatrices of saddle S are shown by *thick dashed* lines, N are nodes, and arrows show the direction of the Poynting vector [24]

In Fig. 16c, d, we show the transverse intensity distribution and the energy diagram for the radically different case of an asymmetric rotating two-humped soliton structure with zero topological charge. It can be interpreted as a strongly-coupled state of two fundamental solitons. In comparison with Fig. 15a, one can see that closed lines, typical for the individual solitons, are now absent, and almost all trajectories approach infinity as $\tau \to +\infty$. It is evident that the phase plane differs topologically from the previous case. Let us recall that, in this case, the structure rotation is not related to any wavefront dislocation. This means that the main condition for rotation is asymmetry of the field distribution.

The dynamics of bifurcations *in time* of energy flow is illustrated in Fig. 17 for the case of a two-vortex structure. The initial state, Fig. 17a, corresponds

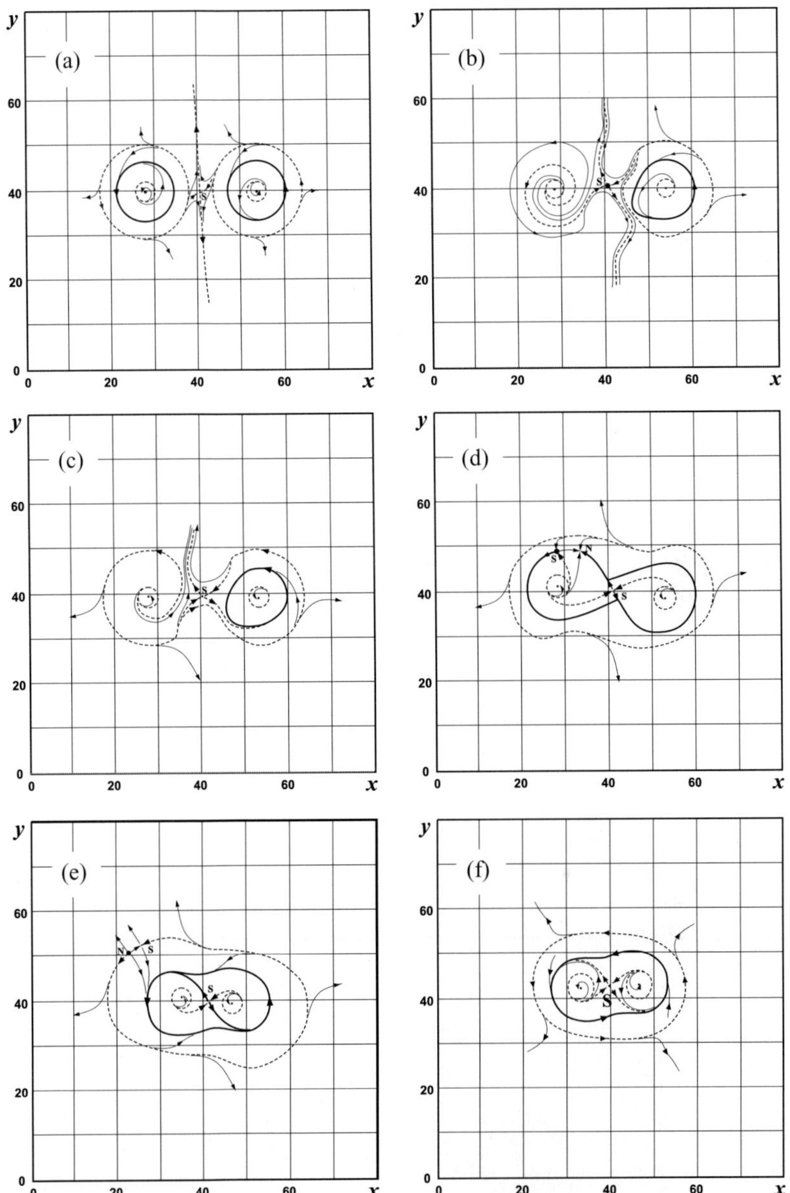

Fig. 17. Phase portraits of energy flow during the transient process of the formation of a strongly coupled rotating pair of solitons with topological indices $M_1 = M_2 = 1$ at time instants $t = 0$ (**a**), 5.6 (**b**), 6 (**c**), 30 (**d**), 59 (**e**), and $t = \infty$ (**f**) [24]

to the superposition (30) of fields of two vortex solitons with topological charges $M_1 = M_2 = 1$. The two vortices correspond to the left and right fixed points, which are quasi-stable foci. Initially, there are three individual limit cycles around each of the two foci. Finally, as $\tau \to +\infty$, an asymmetric rotating two-humped structure establishes itself with only one quasi-unstable limit cycle around each focus (Fig. 17f). During the transient between these situations, a number of bifurcations occur; only some of them are shown in Fig. 17. The first bifurcation corresponds to the merging of the left quasi-stable and outer quasi-unstable limit cycles – compare Fig. 17a and b. The last bifurcation corresponds to the merging and disappearance of two fixed points, node N and neighbouring saddle S in Fig. 17e, resulting in the establishment of the final structure with two common (quasi-stable and quasi-unstable) limit cycles enclosing the rest of the individual elements – two foci and two quasi-unstable limit cycles (Fig. 17f).

The phase portrait of soliton pairs with topological charges of opposite signs, $M_1 = 1$ and $M_2 = -1$ is somewhat similar, but is more complicated. In Fig. 18a, b, we show the initial state corresponding to weak coupling of two such solitons. The centre of each of them – a quasi-stable focus – is

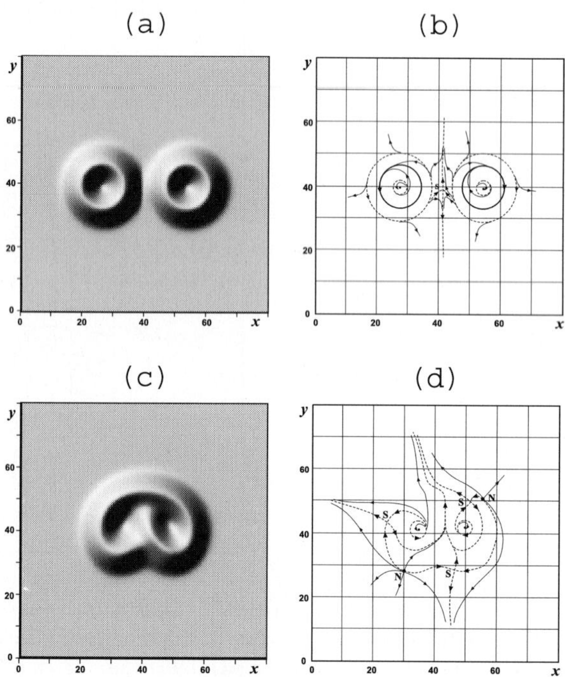

Fig. 18. Initial (**a**, **b**) and final (**c**, **d**) transverse intensity distributions (**a**, **c**) and phase planes of energy flow (**b**, **d**) for a pair of solitons with topological indices $M_1 = 1$ and $M_2 = -1$ [24]

surrounded by three individual limit cycles, corresponding to only slightly distorted cycles of single vortex solitons (see Fig. 15b). After a sufficiently long time interval, an asymmetric rotating structure appears (Fig. 18c, d). It can be interpreted as a strongly-coupled pair of two vortex solitons. Note that each of the two foci (centres of vortices) is finally surrounded by only one individual quasi-unstable limit cycle. A number of new saddles arise, and some of their separatrices approach each other asymptotically.

We do not present here any phase portraits for the other asymmetric rotating structures shown in Fig. 14, because they are more complicated.

7 Effect of Frequency Detunings and Bio-Solitons

In the presence of frequency detunings Δ_{g} and Δ_{a} between the central frequencies of the spectral lines of amplification and absorption and the frequency of an empty cavity mode, the nonlinear function $f(|E|^2)$ in the evolution equation (1), with the dimensionless form of Δ_{g} and Δ_{a}, is complex:

$$f(|E|^2) = -1 + \frac{(1 - \mathrm{i}\Delta_{\mathrm{g}})g_0}{1 + |E|^2} - \frac{(1 - \mathrm{i}\Delta_{\mathrm{a}})a_0}{1 + b|E|^2} \ . \tag{47}$$

The additional control parameters, Δ_{g} and Δ_{a}, admit even more diverse structures, briefly discussed below. The appearance of different types of instability, including the modulation instability of homogeneous field distributions, is typical. An example of a localized structure existing under these conditions, with simultaneous oscillations and rotation, is given in Fig. 19.

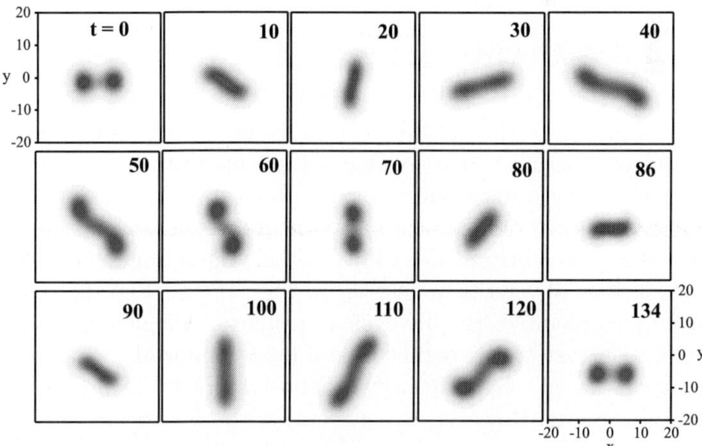

Fig. 19. Transverse intensity distributions for periodically oscillating and simultaneously rotating localized structure at time instants indicated near corresponding Figs. [28]

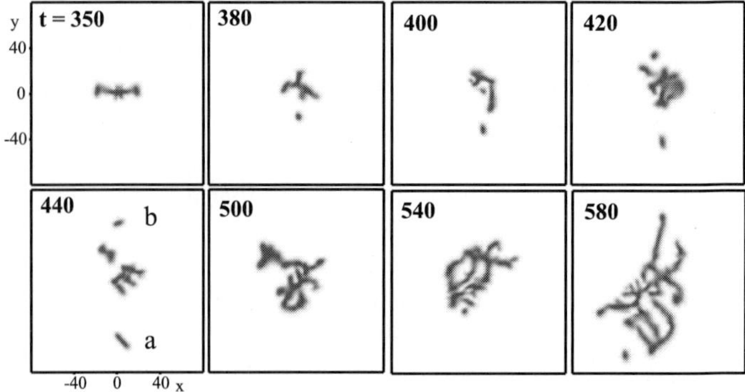

Fig. 20. Transverse intensity distributions for a "bio-soliton" arising near the stability boundary of stationary solitons under conditions of modulation instability; a and b, at time instant $t = 440$, indicate two local fragments separating from the main structure [28]

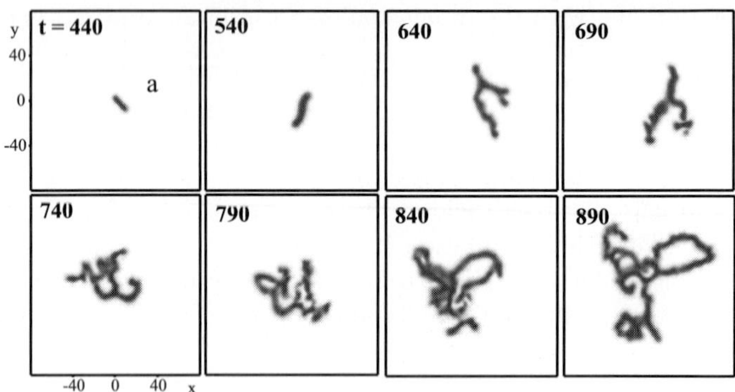

Fig. 21. Evolution of the fragment a of Fig. 20 [28]

With a variation of the detunings, this regime also loses its stability, and more complicated structures arise. One of them is illustrated in Fig. 20. Here, an initially round field distribution evolves with time into a labyrinth structure with spontaneous detachment of fragments which were initially small (a and b in Fig. 20, time instant $t = 440$). Further, these fragments grow, transforming into new labyrinths, as shown in Fig. 21 and Fig. 22. This regime of "bio-soliton" resembles the life cycle of primary biological animals such as amoebae, in contrast to the regime of stable – "immortal" – solitons.

The various laser solitons described above seem to be promising for information processing. However, experiments with two-dimensional laser solitons have been performed, until recently, with extremely slow absorbers [19]. More rapid operations are possible when using semiconductor vertical cavity surface-emitting lasers (VCSELs) with saturable absorbers, which can be integrated in the same semiconductor device.

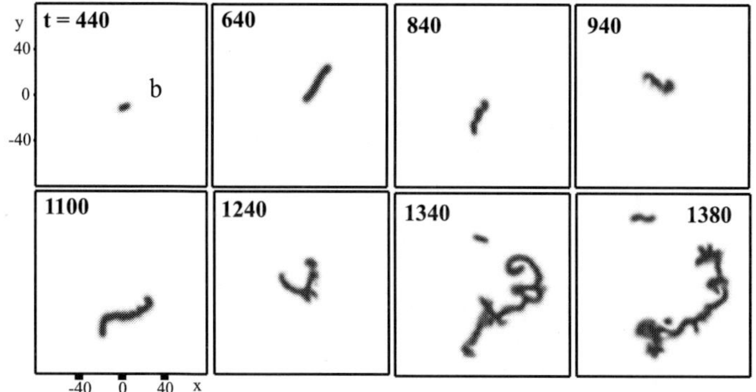

Fig. 22. Evolution of the fragment b of Fig. 20 [28]

I am very grateful to S.V. Fedorov, A.N. Shatsev and A.M. Kokushkin for joint work and for their help.

References

1. N. K. Efremidis and D. N. Christodoulides, see the chapter in this book.
2. U. Peschel, D. Michaelis, Z. Bakonyi, G. Onishchukov, and F. Lederer, see the chapter in this book.
3. M. Faraday, Philos. Trans. R. Soc. London **122** Part II, 299 (1831).
4. P. Umbanhowar, F. Melo, H.L. Swinney, Nature (London) **382**, 793 (1996).
5. R. J. Scott, Report on waves, in Rep. 14th Meeting of the British Assoc. for the Advancement of Science, London, John Murray (1844).
6. A. L. Hodgkin, A. F. Huxley, J. Physiol. **116**, 449 (1952).
7. D. Noble: J. Physiol. **162**, 317 (1962).
8. B. S. Kerner, V. V. Osipov, *Autosolitons: A New Approach to Problems of Self-Organization and Turbulence* (Kluwer, Dordrecht, 1994).
9. N. N. Rosanov, G. V. Khodova, Opt. Spectrosc. **65**, 449 (1988).
10. N. N. Rosanov, A. V. Fedorov, G. V. Khodova, Phys. Status Solidi B **150**, 499 (1988).
11. N. N. Rosanov, G. V. Khodova, Opt. Spectrosc. **72**, 1394 (1992).
12. A. N. Rakhmanov, Opt. Spectrosc. **74**, 701 (1993).
13. A. N. Rakhmanov, V. I. Shmalhausen, Proc. SPIE **2108**, 428 (1993).
14. B. Schapers, M. Feldmann, T. Ackemann, W. Lange, Phys. Rev. Lett. **85**, 748 (2000).
15. N. N. Rosanov, S. V. Fedorov, Opt. Spectrosc. **72**, 1394 (1992).
16. S. V. Fedorov, G. V. Khodova, N. N. Rosanov, Proc. SPIE **1840**, 208 (1992).
17. V. Yu. Bazhenov, V. B. Taranenko, M. V. Vasnetsov, Proc. SPIE **1840**, 183 (1992).
18. N. N. Rosanov, *Spatial Hysteresis and Optical Patterns* (Springer, Berlin, 2002).
19. C. O. Weiss, M. Vaupel, K. Staliunas, G. Slekys, V. B. Taranenko, Appl. Phys. B: Laser Opt. **68**, 151 (1999).

20. C. O. Weiss, G. Slekys, V. B. Taranenko, K. Staliunas, R. Kuszelewicz, Spatial Solitons in Resonators, In: *Spatial Solitons*. ed. by S. Trillo, W. E. Torruellas (Springer, Berlin, 2001), pp. 393–414.
21. A. F. Suchkov, Sov. Phys. JETP **22**, 1026 (1966).
22. A. G. Vladimirov, N. N. Rosanov, S. V. Fedorov, G. V. Khodova, Quantum Electron. **27**, 949 (1997).
23. N. Akhmediev and A. Ankiewicz, see the chapter in this book.
24. N. N. Rosanov, S. V. Fedorov, A. N. Shatsev, JETP, **98**, 427 (2004).
25. S. V. Fedorov, N. N. Rosanov, A. N. Shatsev, N.A. Veretenov, A.G. Vladimirov, IEEE J. Quantum Electron. **39**, 197 (2003).
26. N. N. Rosanov, A. V. Fedorov, S. V. Fedorov, G. V. Khodova, Opt. Spectrosc. **79**, 795 (1995).
27. A. A. Andronov, A. A. Vitt, S. E. Khaikin: *Theory of Oscillations* (Pergamon, Oxford, 1966).
28. N. N. Rosanov, S. V. Fedorov, A. N. Shatsev: Opt. Spectrosc. **95**, 843 (2003).

Spatial Resonator Solitons

V.B. Taranenko, G. Slekys, and C.O. Weiss

Physikalisch-Technische Bundesanstalt, 38116 Braunschweig, Germany
Victor.Taranenko@ptb.de
g.slekys@ptb.de
carl.weiss@ptb.de

Abstract. Spatial solitons can exist in various kinds of nonlinear optical resonators with and without amplification. In past years, different types of these localized structures, such as vortices, bright, dark solitons and phase solitons, have been experimentally shown to exist. Many links appear to exist with fields separate from optics, such as fluids, phase transitions or particle physics. These spatial resonator solitons are bistable, and, due to their mobility, suggest systems of information processing not possible with the fixed bistable elements which form the basic ingredients of traditional electronic processing. The recent demonstration of the existence and manipulation of spatial solitons in semiconductor microresonators represents a step in the direction of such optical parallel processing applications. We review pattern formation and solitons in a general context, show some proof-of-principle soliton experiments in slow systems, and describe in more detail the experiments on semiconductor resonator solitons which are aimed at applications.

1 Introduction: A Multi-Disciplinary View of Pattern Formation and Solitons

Patterns or "structures" are known to form in non-equilibrium nonlinear systems of all kinds. Their properties seem to be determined by a maximum dissipation principle. A system without structure is only able to dissipate as much energy as given by the microscopic dissipation processes. Then, if a system with structure (compatible with the system equations, parameters, boundary conditions and possibly initial conditions) has a higher dissipation, that structure will appear spontaneously, e.g. through a modulational instability. This has been discussed at length in relation to Bénard convection, but is easily found to apply in other fields.

In optics, if a medium is pumped sufficiently, then microscopic relaxation processes do not provide the highest possible dissipation. As is well-known, a coherent, structured and well-ordered optical laser field which gives the system a much higher dissipation can form. If the pumping is increased, the laser emission may convert to a pulsed instead of a continuous emission; this has the form of regular or chaotic pulsing, as described by the Lorenz Model. This pulsed laser emission again increases the system's capability to dissipate over the continuous emission. Such pulsing means a modulational instability

V.B. Taranenko, G. Slekys, and C.O. Weiss: *Spatial Resonator Solitons*,
Lect. Notes Phys. **661**, 131–160 (2005)
www.springerlink.com

in the temporal domain. Equally, modulations or patterns appear in space, such as a hexagonal structure of the field in the emission cross-section [1].

If noise sources act, the "coherent" structures permitting the high dissipation can appear stochastically. Examples are the appearance of vortices in wave turbulence [2], vortex-antivortex generation in a noise driven Ginzburg Landau equation [3], and, evidently, phenomena such as earthquakes and stock market crashes [4]. The mentioned vortex-antivortex generation would seem to us to be a good picture for the spontaneous appearance of particle-antiparticle pairs out of the vacuum in quantum physics and we would think that it is also the proper model for the enigmatic phenomenon of $1/f$ noise [5], particularly since this description contains the two best accepted explanations of "self-induced criticality" [6] and the often-mentioned coupled oscillator model as special cases.

"Coherence", "structure" or "order" in space or time thus occurs in non-equilibrium systems to maximize the energy dissipation of the system. If one accepts such a principle (and, to our knowledge, no counter-examples are known), one can see this world as driven by a "pressure to dissipate" (where the pump evidently is the sun). Water pressed through a porous material or through rocks with cracks will form highly complex flow paths through the material, and in a similar way, one can visualize, by analogy, complex pattern-formation in nature, culminating in biological structures, as forming under the pressure to dissipate and in the cracks and pores given by the boundary conditions in the world.

In this picture there is no need for a "plan" or "intention" to generate the complex structures which we find developing under the irradiation of the sun (i.e. an absence of "teleological" elements). Everything is governed by a simple extremum principle, which is "blind" in the sense that it can only find the nearest potential minimum. (And the system will directly head there even if "death" lurks in this minimum. There are good examples for such "suicidal" systems in laser physics [7].) A "teleological" element or "planning" and an "intention" appears only with the appearance of a brain, which can simulate the world and find out that there are several minima, of which the nearest may not be the deepest (or that at the nearest minimum, "death" may lurk). Even then, it appears that the teleological element seems to be the exception. Even within the most complex structure, the human species, most decisions are made unconsciously.

It is, evidently, one of the goals of nonlinear physics to clarify and understand the origin of life and of biological structures. Although the path from simple periodic patterns, appearing as modulational instabilities, and spatial solitons, to structures as complex as biological ones would seem exceedingly long at a first glance, this may not be so at a second glance.

A simple periodic pattern such as a hexagonal field structure in the cross section of a nonlinear resonator can disintegrate into spatial resonator solitons [8]. A closer examination of these structures shows many common traits with biological structures [9]. Spatial solitons in nonlinear dissipative resonators

1. are mobile (allowing them to move to the point in space where they are most stable i.e. find the best "living conditions" (hence, they can search around for food));
2. they possess a "metabolism": they have internal energy flow and spatially-varying dissipation, which stabilize them;
3. they can "die": when the field sustaining them ("food") is reduced, they will self-extinguish (die). And even a later increase of the field, to a strength which allowed the solitons to exist before, will not re-ignite them, because of their bistability (death is irreversible).
4. they can multiply ("self-replicate") by repeated splitting [10].

There are further common traits of dissipative resonator solitons and biological structures, among them the fact that both exist only at medium nonlinearities and not at large or small nonlinearities (see below).

Consequently, one could very well conjecture that the simplest biological ("live") structures are not viruses (as commonly taught), but dissipative solitons. This, by the way, sheds a completely new light on the question of what biological life is: life is usually ascribed to chemistry-based structures (with carbon as the central compound). Whereas, if one takes the analogies above seriously, "life" is not bound to chemistry but constitutes itself as organizational "forms" of matter or fields. Thus the probability of finding "live" structures is much larger than finding "life" only in its form which is bound to carbon or chemistry.

These analogies of dissipative solitons with biological structures are one of the reasons for which one might want to study spatial solitons. Furthermore, these solitons are all, by definition, bistable i.e. they constitute natural carriers of information. In contrast to the information-carrying elements in electronic information processing, they are mobile and would therefore perhaps permit processing functions beyond the reach of electronic computing. We have recently studied the "aggregation" of spatial solitons and again find traits in common with biology. Considering that these "aggregations" of information carriers constitute something like a "brain" (a collection of information-carrying elements, bound loosely to one another, capable of exchanging information among the elements...) there is another motivation for the study: just as there are analogies between solitons and simple biological ("live") structures, one might hope to find operating principles of the brain by studying the structure and properties of "aggregations" of spatial solitons.

Optics, and particularly resonator optics, suggest themselves for the investigations, because of possible photonic applications, such as parallel processing, already urgently awaited e.g. in telecommunications. For this reason, in the past we have conducted a series of "proof-of-existence" experiments on resonator solitons, as well as on their manipulation and properties [11].

At present, spatial resonator solitons in semiconductor micro-resonators are investigated, as they are most likely best-suited for applications.

2 Proof-of-Existence Experiments on Resonator Solitons with Slow Materials

Spatial resonator solitons are self-formed localized structures, which are free to move around. We have investigated several types.

Vortices occur in resonators with a gain medium without phase preference, such as laser resonators. They are helical defects in the wavefront and therefore possess, strictly speaking, "tri-stability": right- as well as left-handed vortices co-exist with the wave without defect. We have given an overview of laser vortices in [11], including examples of their dynamics in Internet films [12]. Vortices were, until now, not studied with regard to their usefulness as binary information carriers (except for some basic experiments relating to pattern recognition with lasers [13]).

Bright solitons are more reminiscent of the original idea of optical information processing, namely the use of optically-bistable resonators. It was shown theoretically in [14] that such solitons can exist in the simplest kind of nonlinear resonator: one filled with a medium with Kerr-nonlinearity. We believe that this prediction was confirmed in experiments [15] which used a liquid crystal material (which should essentially behave like a Kerr medium) in a resonator. Bright solitons have therefore been considered more closely for information processing tasks. In order to convince oneself that such bright solitons in bistable resonators exist widely, we conducted experiments with nonlinear media with slow response – for the purpose of being able to observe the phenomena (which entail $2D$ space-time dynamics) on a convenient time scale. As opposed to semiconductors, which have characteristic times of ps to ns, photorefractive media and nonlinear absorbers like bacterio-rhodopsin [16] have time constants of 10 ms to 1 s, and so are quite suitable for recording with ordinary video equipment. In general, a resonator containing only such a non-linear absorber is too lossy for bistability. Therefore, we used resonators additionally containing gain elements to compensate for the losses. A system like a laser with a nonlinear absorber results, and it is well-suited to study these transverse effects, such as spatial modulational instabilities, patterns and solitons.

The first such experiments yielded evidence of the existence of spatial solitons [17]. In refined experiments, we showed the existence, the bistability and development of bright solitons in this system [18]. Figure 1 shows that the solitons can be "ignited" anywhere in the cross-section of the resonator. Figure 2 shows the motion of a bright soliton in a phase gradient, and Fig. 3 shows the "trapping" of a bright soliton in a "phase-trough". These are functions that are required for applications. Figure 4 shows that a large numbers

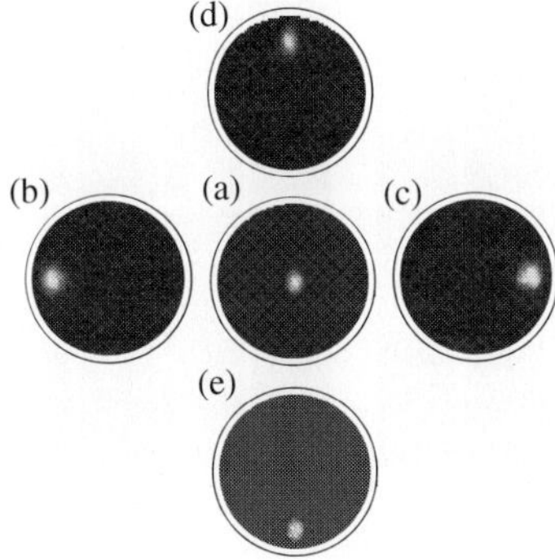

Fig. 1. In a laser with an internal nonlinear absorber (bacterio-rhodopsin), a soliton can be written in arbitrary locations in the resonator cross-section. The resonator used is degenerate for all transverse modes, in order to allow an arbitrary field configuration to be resonant

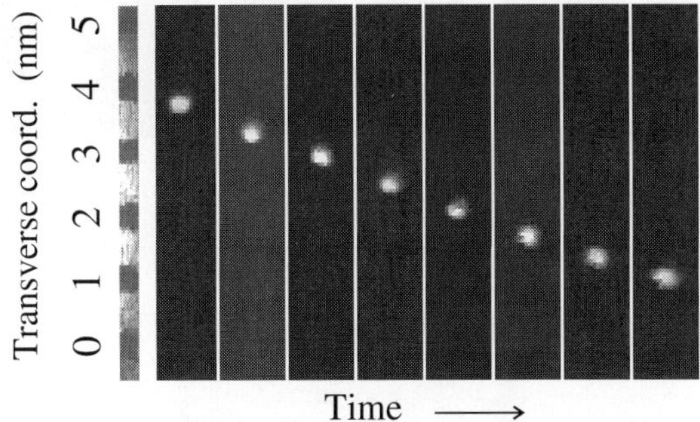

Fig. 2. A soliton can be made to move across the resonator cross-section by using a phase gradient in the resonator

of these bright solitons can exist at the same time [10], a point which is equally necessary for technical applications.

An interesting feature of these solitons is that they are found to have quantized velocity magnitudes (including velocity of magnitude zero) [19]. Although the velocity magnitudes of such solitons are fixed, their direction of motion is completely free, and it varies under the action of noise, for example.

It may be mentioned that in resonators with phase-sensitive gain, such as gain supplied by 4-wave-mixing or degenerate parametric mixing, solitons which relate only to the phase structure of the field (phase-solitons) were predicted [20] and experimentally observed [21]. See Fig. 5.

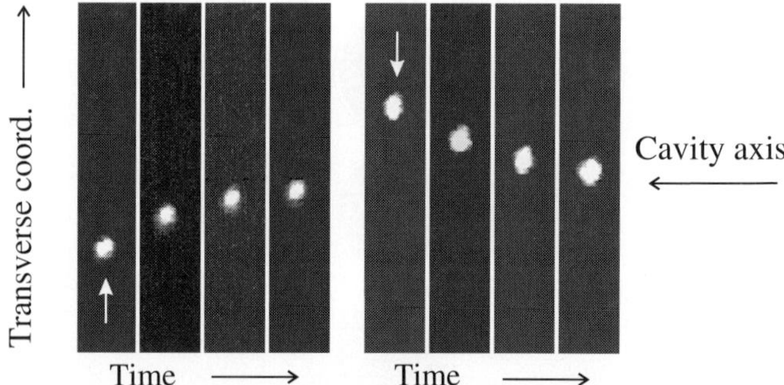

Fig. 3. A soliton is drawn from any side to the axis of a resonator with a "phase trough" at its center, and is trapped there

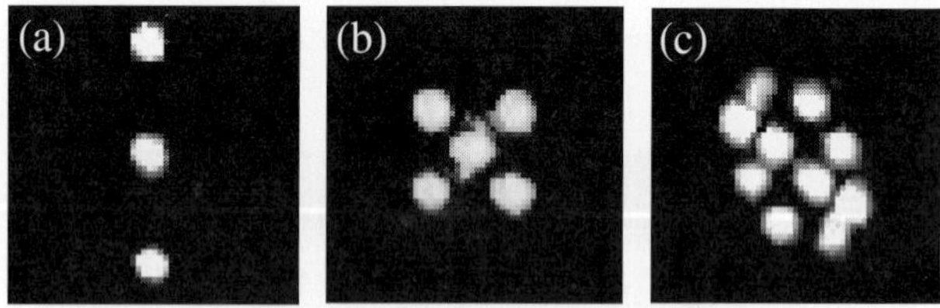

Fig. 4. Co-existence of a large number of stationary bright solitons

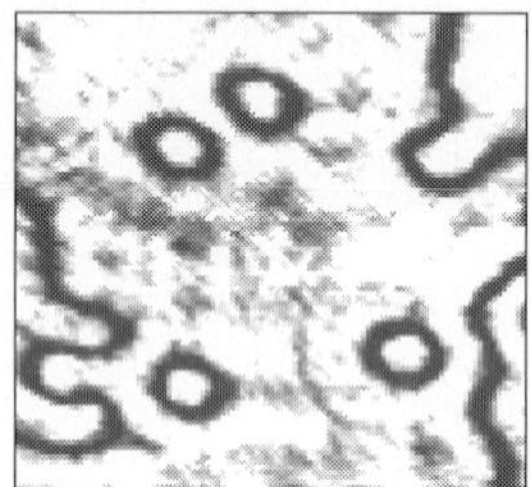

Fig. 5. Phase solitons co-existing with non-stationary phase fronts. The phase of the field changes by 180 degrees at the fronts and around the soliton. By interference, this produces the dark ring surrounding the solitons and the dark fronts at the phase-domain boundaries

3 Semiconductor Resonator Solitons

To make bright solitons suitable for technical applications, fast materials have to be used. We found that the optimum system concerning speed of response and nonlinearity is the semiconductor micro-resonator. This is a resonator approximately 1 wavelength long which provides the shortest conceivable resonator response time of 100 fs – 1 ps. The finesse of these resonators has to be around 100–200 to provide bistability, together with a nonlinear medium. For the latter, a semiconductor slice of the length of the resonator ($\sim 1\lambda$) is suitable, and we stress that the response time of semiconductor material is well-matched to that of the resonator. Importantly, it makes no sense to use materials "faster" than semiconductors in such resonators. Since the small length of the resonator provides a response time which is at the conceivable limit for optical resonators, a faster nonlinear material would not decrease the system response time. However, since higher speed comes at the cost of a smaller nonlinearity, a faster material would require unnecessarily high light intensities. Thus the 1λ microresonator, combined with a semiconductor material as the nonlinear component, represents the optimum in terms of response time and required light intensity. This structure is incidentally the structure of VCSELs (vertical cavity surface emitting lasers) which have already been developed to a certain degree of perfection during recent years. Absorbing ("passive"), as well as population-inverted ("active"), material is suitable to allow the existence of spatial solitons in such resonators. Thus a direct connection with VCSEL technology exists.

Figure 6 shows the structure of the nonlinear resonator. Its length is $\sim 1\lambda$ while the transverse size is typically 5 cm. The short length and wide area of this micro-resonator permit only one longitudinal mode (Fig. 7a), while allowing an enormous number of transverse modes, so that a very large number of spatial solitons can co-exist. The resonator is obviously of the plane mirror type, implying frequency degeneracy of all transverse modes, and thus allowing arbitrary field patterns to be resonant inside the resonator. This is another pre-requisite for the existence and ability to manipulate spatial solitons.

Soliton existence in a resonator is closely linked with plane wave resonator bi-stability (Fig. 7b) caused by longitudinal nonlinear effects, *viz.* the nonlinear changes of the resonator length (due to nonlinear refraction changes) and finesse (due to nonlinear absorption changes) [22]. The longitudinal nonlinear effects, combined with transverse nonlinear effects (such as self-focusing), can balance diffraction and form resonator solitons. Generally, these nonlinear

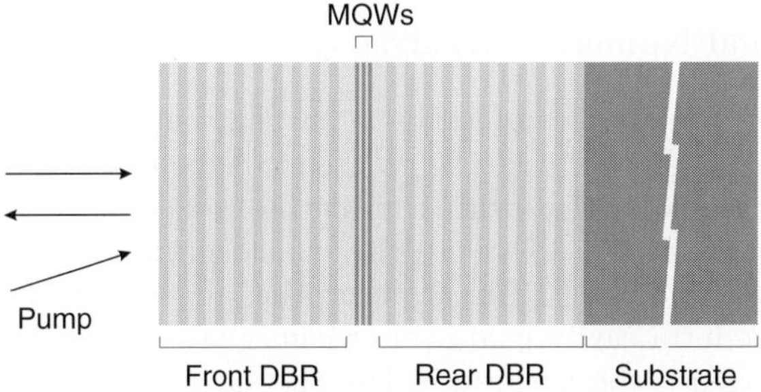

Fig. 6. Schematic of semiconductor micro-resonator consisting of two plane distributed Bragg reflectors (DBR) and multiple quantum wells (MQW). The observation is in reflection

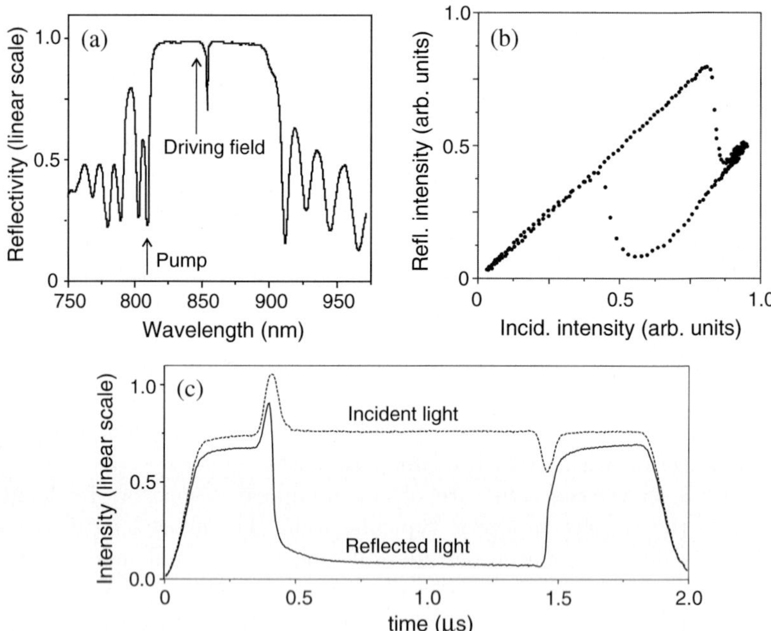

Fig. 7. (**a**) Semiconductor micro-resonator reflectance spectrum, (**b**) typical bistability loop in reflection and (**c**) dynamics of switching. Arrows mark the driving field that is detuned from the resonator resonance and the pump field that is tuned to be coupled into the resonator through one of the short-wavelength interference notches of the resonator reflectance spectrum. (**c**) shows that the resonator can be switched from one branch of the bistability loop to the other by using short positive (negative) pulses

effects can co-operate or counteract each other, with the consequence of reduced soliton stability in the latter case.

3.1 Model and Numerical Analysis

As a guide for the experiments, we use a phenomenological model of a driven wide area multi-quantum well (MQW)-semiconductor micro-resonator similar to [23, 24]. The optical field A inside the resonator is described in the mean-field approximation [25]. The driving incident field A_{in} is assumed to be a stationary plane wave. Nonlinear absorption and refractive index changes induced by the intra-cavity field in the vicinity of the MQW-structure band edge are assumed to be proportional to the carrier density N (normalized to the saturation carrier density). The equation of motion for N includes optical pumping P, carrier recombination and diffusion. The resulting coupled equations describing the spatio-temporal dynamics of A and N have the form:

$$\frac{\partial A}{\partial t} = A_{in} - \{[1 + C\ Im(\alpha)(1 - N)] + \mathrm{i}(\theta - C\ Re(\alpha)N - \nabla_\perp^2)\}\sqrt{T}\ A$$
$$\frac{\partial N}{\partial t} = P - \gamma\left[N - |A|^2(1 - N) - d\nabla_\perp^2 N\right] , \tag{1}$$

where C is the saturable absorption scaled to the resonator transmission T. (T is assumed to be small, since the mirror reflectivity is typically $\geq$0.995). $Im(\alpha)N$ and $Re(\alpha)N$ describe the absorptive and refractive nonlinearities, respectively, θ is the detuning of the driving field from the resonator resonance, γ is the photon lifetime in the resonator normalized to the carrier recombination time, d is the diffusion coefficient scaled to the diffraction coefficient, and $\nabla_\perp^2 = \partial^2/\partial x^2 + \partial^2/\partial y^2$ is the transverse Laplacian.

The linear effects in the resonator are spreading of light by diffraction, and carrier diffusion (terms with $\nabla_\perp^2$ in (1)). The material nonlinearity that can balance this linear spreading can do it in various ways. It has a real (refractive) and imaginary (dissipative) part, and can act longitudinally and transversely. The nonlinear changes of the resonator finesse (due to nonlinear absorption change) and length (due to nonlinear refractive index change) constitute longitudinal nonlinear effects, also known by the name *nonlinear resonance* [20]. The transverse effects of the nonlinear refractive index can be self-focusing (favourable for bright, and unfavourable for dark solitons) and self-defocusing (favourable for dark, and unfavourable for bright solitons). Absorption (or gain) saturation (bleaching), leading to *nonlinear gain guiding* (in laser terminology), is a transverse effect. Longitudinal and transverse effects can counteract each other, or cooperate.

There are two principal external control parameters – the driving field intensity $|A_{in}|^2$ and the resonator detuning θ. Figure 8 shows typical existence domains (in coordinates θ, $|A_{in}|^2$) for all possible structures (patterns, bright and dark spatial solitons), and plane-wave bistability as calculated from (1) for the case of a mixed absorptive/dispersive nonlinearity. For large resonator detuning, the intra-cavity field is transversely homogeneous and stable. If, in this case, the driving beam has a Gaussian profile (Fig. 9b) and its maximum amplitude exceeds the switch-on threshold, then the part of the beam cross-section which is limited by the so-called Maxwellian intensity [26] is switched, thus forming a dark switched domain in reflection (Fig. 9a).

For small resonator detuning and for driving intensities not quite sufficient for reaching the resonance condition for the whole resonator area, the system "chooses" to distribute the light intensity in the resonator in isolated spots where the intensity is then high/low enough to reach the resonance condition, thus forming bright/dark patterns (Fig. 10). Instead of saying "the system chooses", one could express this point more mathematically by describing it as a modulational instability. The detuned plane wave field, without spatial structure and with an intensity which is not sufficient to reach the resonance condition, is unstable against structured solutions. According

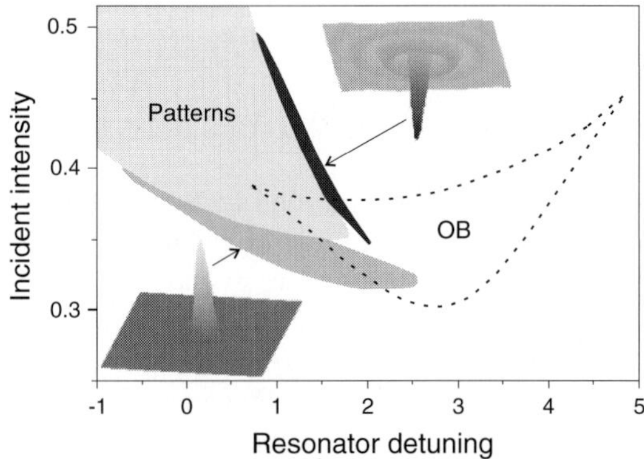

Fig. 8. Numerical solutions of (1) for the unpumped and mixed absorptive/self-focusing case. The area within the *dashed* lines is the optical bistability domain for plane waves. The *shaded* areas are domains of stability for *bright/dark* solitons and patterns. The insets are *bright* and *dark* solitons in a 3D representation

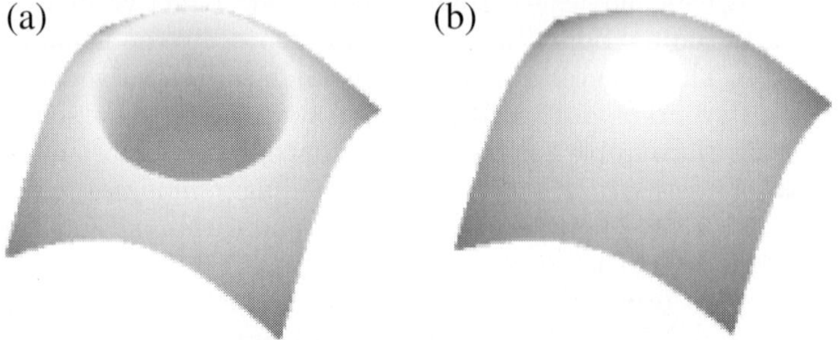

Fig. 9. (**a**) Maxwellian (see [26]) switched domain in reflection, numerically-calculated for Gaussian profile of the driving beam which is shown in (**b**). Parameters in this simulation correspond to the plane-wave OB area in Fig. 8

to our numerical solutions of (1), a large number of such structured solutions co-exist and are stable (see, for example, the patterns in Fig. 18).

The bright/dark soliton structures (inset in Fig. 8) can be interpreted as small circular switching fronts, connecting two stable states: a low transmission and a high transmission state. Such a front can, in $2D$, surround a domain of one state. When this domain is comparable in diameter to the "thickness" of the front, then each piece of the front interacts with the piece on the opposite side of the circular small domain, which can lead, particularly if the system is not far from a modulational instability (see Fig. 8, Patterns), to a stabilization of the diameter of the small domain. In this case, the small domain is an isolated self-trapped structure or a dissipative resonator soliton.

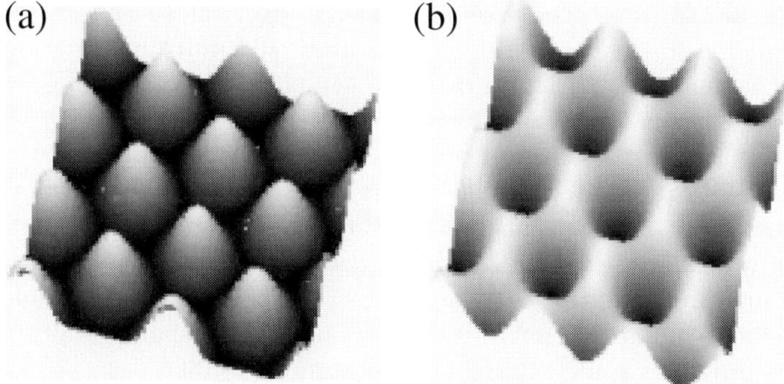

Fig. 10. (**a**) *Bright* and (**b**) *dark*-spot-hexagons, numerically-calculated for parameters corresponding to the Pattern existence domain of Fig. 8. The driving light intensity increases from (a) to (b)

3.2 Experimental Arrangement

Figure 11 shows the optical arrangement of the semiconductor soliton experiments, and in particular for their switching on or off. The semiconductor micro-resonator consists of a MQW (GaAs/AlGaAs or GaInAs/GaPAs) structure sandwiched between high-reflectivity ($\geq$0.995) DBR-mirrors (Fig. 6) used at room temperature. The micro-resonator structures were grown on GaAs substrates by a molecular beam epitaxy technique that allows

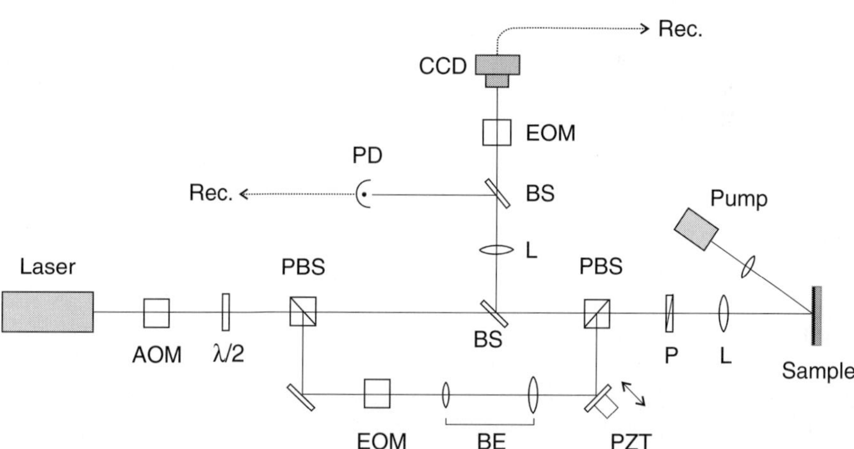

Fig. 11. Experimental setup. Laser: Ti:Sa (or diode) laser, AOM: acousto-optic modulator, $\lambda/2$: halfwave plate, PBS: polarization beam spliters, EOM: electro-optical amplitude modulators, BE: beam expander, PZT: piezo-electric transducer, P: polarizer, L: lenses, BS: beam splitters, PD: photodiode. In some cases, an optical pump is used ("Pump") – see text

growth of MQW structures with small radial layer thickness variation. The best sample used in our experiments has only ∼0.3 nm/mm variation of the resonator resonance wavelength over the sample cross section of 5 cm diameter.

The driving light beam was generated by either a tunable (in the range 750–950 nm) Ti:Sa laser or a single-mode laser diode (∼854 nm), both emitting continuously. For experimental convenience and to limit thermal effects, the experiments are performed within a few microseconds, by admitting the light through an acusto-optical modulator. A laser beam of suitable wavelength is focused onto the micro-resonator surface in a spot of ∼50 μm diameter, thus providing quite a large Fresnel number (≥100).

Part of the laser light is split away from the driving beam and is superimposed with the main beam in a Mach-Zehnder interferometer arrangement, to serve as a writing/erasing (address) beam. This beam is tightly focused and directed to some particular location in the illuminated area to create or destroy a spatial soliton. The switching light is opened for a few nanoseconds using an electro-optic modulator. For the case of incoherent switching, the polarization of the address beam is set perpendicular to that of the main beam to avoid interference. For the case of coherent switching, the polarizations are parallel and a phase control of the switching field is always needed, both for switching a soliton on, and switching it off. One of the interferometer mirrors can be moved by a piezo-electric element to control the phase difference between the background light and the address light.

Optical pumping of the MQW-structures was achieved by a multi-mode laser diode or a single-mode Ti:Sa laser. To couple the pump light into the micro-resonator, the pump laser wavelength was tuned to the short wavelength reflection minimum, as shown in Fig. 7a. The observations are made in reflection (because the GaAs substrate is opaque) by a CCD camera combined with a fast shutter (another electro-optic modulator), which allows us to take nanosecond snapshots of the illuminated area on the resonator sample at a given time. Recording of "movies", on a nanosecond time scale, is also possible. To follow the intensity in time at certain points (e.g. at the location of a soliton), a fast photodiode can be imaged onto arbitrary locations within the illuminated area.

3.3 Results and Discussions

Switched (dark) domains exist for small negative resonator detuning in the vicinity of the absorption band edge. The switching front surrounding these domains is defined by the spatial profile of the driving beam. The switching front is located where the driving intensity equals the Maxwellian intensity. Figure 12 shows snapshots of various switched domains, demonstrating that they have the same shapes as the driving beams. When changing the intensity of the driving beam in time, the fronts around the switched domains obviously follow an equal-intensity contour of the incident light (Fig. 13).

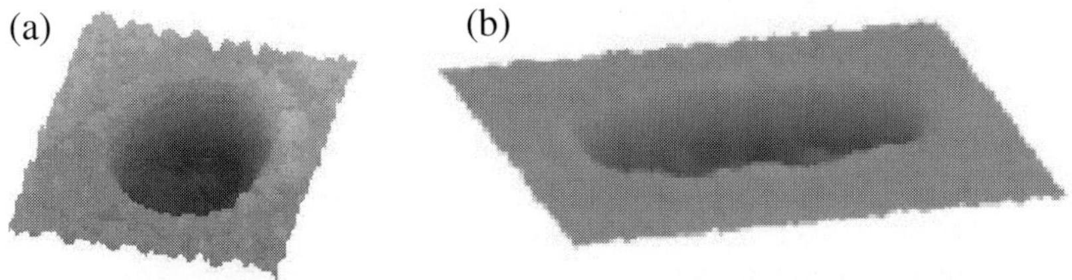

Fig. 12. Dark switched domains observed at small (negative) resonator detuning for (**a**) *round* and (**b**) *oval* driving beam shapes

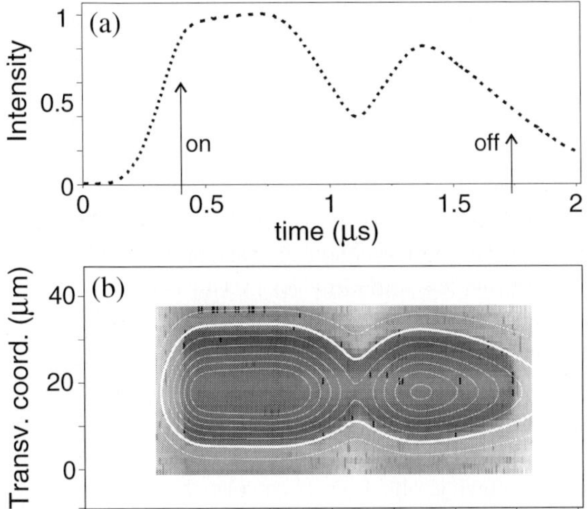

Fig. 13. (**a**) Incident driving light intensity at center of Gaussian beam and (**b**) reflectivity of sample (on diameter of circular driving beam cross-section). Equi-intensity contours of incident light in (**b**), demonstrate that the borders of the dark switched area follow one of the equi-intensity contours (*heavy line*)

In contrast to these dark Maxwellian switched domains, spatial patterns and solitons are self-sustained objects, independent of boundary conditions or a beam profile, and can be both bright and dark. Figure 14 shows snapshots of bright and dark small (∼10 μm) round spots (solitons), at large (negative) resonator detuning, where the shapes/sizes are independent of the shapes/intensities of the driving beam [27]. This independence from boundary conditions allows us to distinguish between patterns and solitons on the one hand, and Maxwellian switched domains on the other hand.

Resonator solitons can be written and erased by focused optical (coherent) pulses, independently from other bright (dark) spots (solitons). Figure 15 shows how solitons can be switched by light which is coherent with the background light [28]. Figure 15a shows switching a bright soliton on. The driving

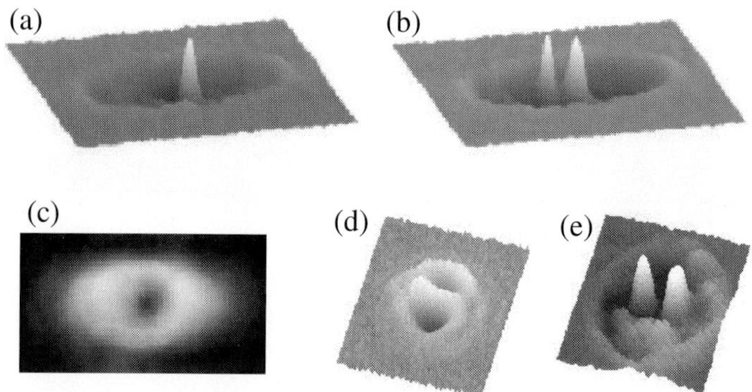

Fig. 14. *Bright* and *dark-spot* switched structures observed experimentally at large (negative) resonator detuning. The fact that the structures are independent of the driving beam shape shows that they are self-localized (patterns or spatial solitons)

light intensity is chosen slightly below the spontaneous switching threshold. At $t \approx 1.2\,\mu s$, the writing pulse is applied. It is in phase with the driving light, as visible from the constructive interference.

A bright soliton results, showing up in the reflected intensity time trace as a strong reduction of the intensity. Figure 15b shows switching a soliton off. The driving light is increased to a level where a soliton is formed spontaneously. The address pulse is then applied in counter-phase to the driving light, as visible from the destructive interference. The soliton then disappears, showing up in the reflected intensity time trace as a reversion to the incident intensity value. For clarity, the Fig. 15 insets show $2D$ snapshots before and after the switching pulses. Thus, depending on conditions, we find switched domains, patterns and bright and dark resonator solitons. Periodic patterns can be distinguished from collections of solitons by the mutual independence of the latter.

To find the most stable resonator solitons for applications, one can play with (a) the nonlinear (absorptive/dispersive) material response by choice of the driving field wavelength and intensity, (b) the resonator detuning, and finally, (c) the carrier population inversion (when using pumping). We recall that all nonlinearities change their sign at transparency i.e. at the point where the valence- and the conduction-band populations are equal. In moving from below transparency (absorption) to above transparency (population inversion, producing light amplification), the nonlinear absorption changes to nonlinear gain, self-focusing changes to self-defocusing and vice versa, and a decrease of optical resonator length with intensity changes to an increase (and vice versa). The population of the bands can be controlled by pumping (P in (1)), i.e. transferring electrons from the valence band to the conduction band. We do this by optical excitation [29], with radiation of a wavelength shorter than the band edge wavelength. If the structures were suitable for

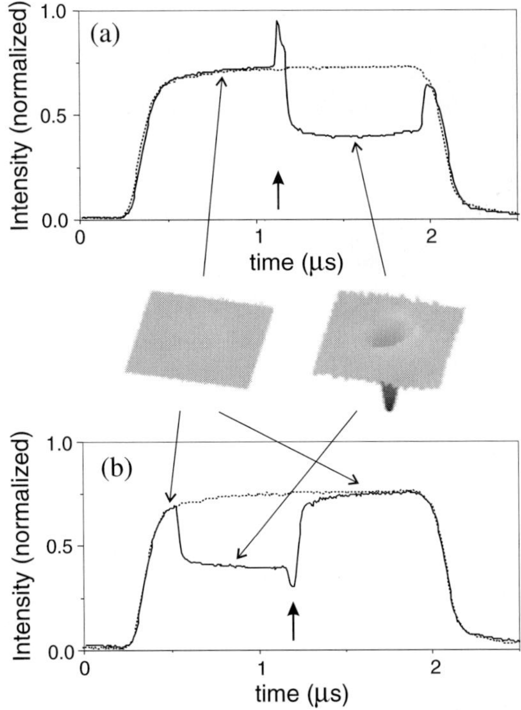

Fig. 15. Recording of coherent (**a**) switching-on and (**b**) switching-off of a bright soliton. The *heavy* arrows mark the application of switching pulses. *Dotted* traces: incident intensity; solid traces: reflected intensity at center of soliton. The insets show intensity snapshots, namely the un-switched state (*left*) and soliton (*right*). For details, see text

supporting electrical currents (i.e. if it were a real VCSEL-structure), pumping could evidently be effected by electrical excitation.

Illumination Below Band-Gap (Defocusing Nonlinearity)

Working well below the band-gap (in the dispersive/defocusing limit), with the driving field wavelength ~30 nm longer than the band edge wavelength, we observe spontaneous formation of hexagonal patterns (Fig. 16). The hexagon period scales linearly with $\theta^{-1/2}$ [30] indicating that the hexagons are formed by the tilted-wave mechanism [31], which is the basic mechanism for resonator pattern formation [32]. Dark-spot hexagons (Fig. 16a) convert to bright-spot hexagons (Fig. 16b) when the driving intensity increases. This is in qualitative agreement with numerical simulations (Fig. 10).

At high driving intensity, we find that the bright spots in such hexagonal patterns can be switched independently of one another by focused optical

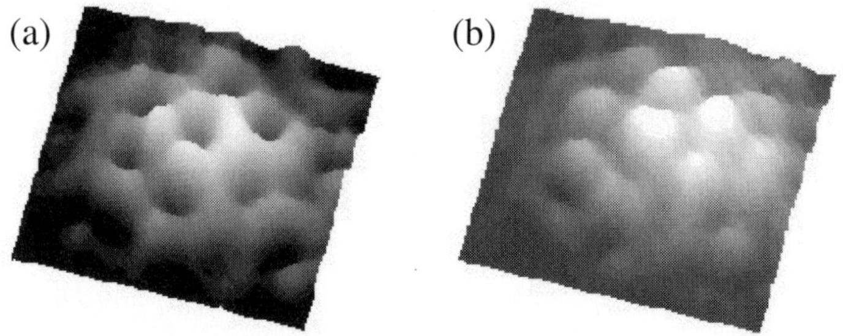

Fig. 16. (**a**) Bright (*dark in reflection*) and (**b**) dark (*bright in reflection*) hexagonal patterns for the dispersive/defocusing case. The driving light intensity increases from (**a**) to (**b**)

(incoherent) pulses [30]. Figure 17 shows the experimental results. Figure 17a shows the hexagonal pattern formed. The focused light pulse can be aimed at individual bright spots, such as the ones marked "1" or "2". Figure 17b shows that after the switching pulse was aimed at "1", spot "1" was switched off. Figure 17c shows the same for spot "2". We note that, in these experiments, we speak of true logic switching – the spots remain switched off after the switching pulse (if the energy of the pulse is sufficient, otherwise the bright spot reappears after the switching pulse). These observations of local switching indicate that these hexagonal patterns are not coherent patterns. The individual spots are spatial solitons – they are independent, even with this dense packing, where the spot separation is about equal to the spot size.

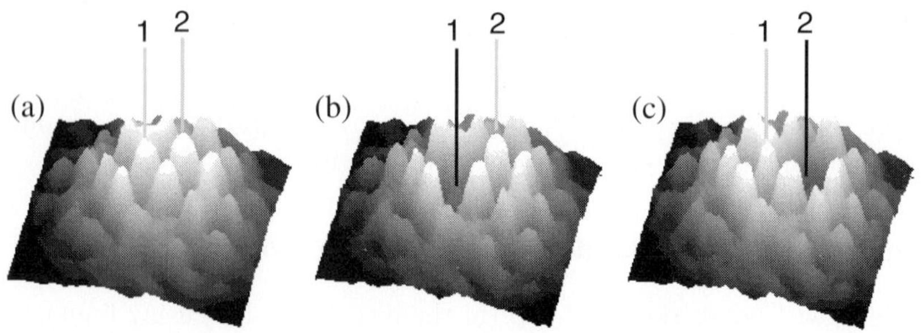

Fig. 17. Switching-off of individual spots of a hexagonal structure with address pulses aimed at different bright spots (marked 1 and 2) of the pattern

These experimental findings can be understood in the frame of the model (1) [8]. Figure 18 shows the bistable plane wave characteristic of the semiconductor resonator for conditions roughly corresponding to the experimental conditions. At the intensities marked (a) to (d), patterned solutions exist. The pattern period in Fig. 18a scales again linearly with $\theta^{-1/2}$ due to the

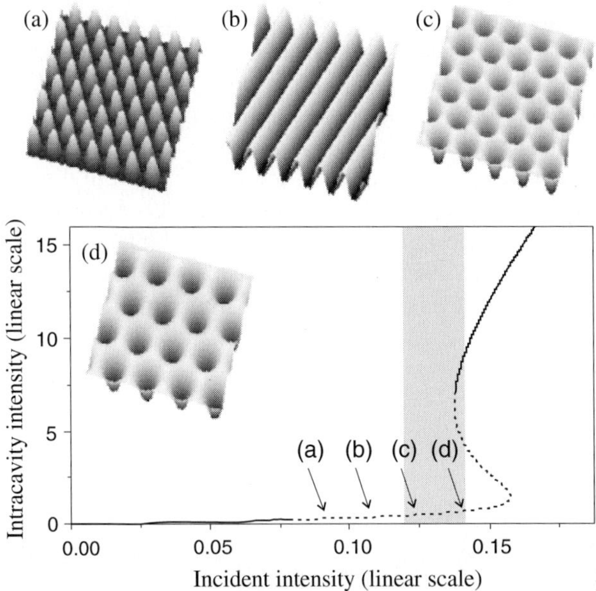

Fig. 18. Numerical solutions of (1) for intra-cavity light intensity as a function of incident intensity: (**a**–**d**) homogeneous solution (*dashed* line marks modulationally-unstable part of the curve) and patterns. The *shaded* area denotes the existence range for *dark-spot* hexagons

tilted wave mechanism. When the driving field is detuned, the resonance condition of the resonator cannot be fulfilled by plane waves travelling exactly perpendicularly to the mirror plane. However, the resonance condition can be fulfilled if the wave plane is somewhat inclined with respect to the mirror plane (the tilted wave mechanism [31]). The system "chooses" therefore to support resonant, tilted waves. Fig. 18a is precisely the superposition of six tilted waves that support each other by (nonlinear) 4-wave-mixing. The pattern period corresponds to the resonator detuning as in the experiment for structures Fig. 16a. In this pattern (Fig. 18a), the bright spots are not independent. Individual spots cannot be switched as in the experiment of Fig. 17.

In the high intensity pattern Fig. 18d, the pattern period is remarkably different from Fig. 18a, even though the (external) detuning is the same. This is an indication that the internal detuning is smaller and means that the resonator length is nonlinearly changed by the intensity-dependent refractive index (nonlinear resonance). From the ratio of the pattern periods in Fig. 18a and d, one sees that the nonlinear change of detuning is about half of the external detuning. That means the nonlinear detuning is by no means a small effect. This, in turn, indicates that the detuning can be varied substantially in the resonator cross-section by a spatial variation of the resonator field intensity. In other words, the resonator has, at the higher intensity, a rather wide freedom to (self-consistently) arrange its field structure. One can expect that this would allow a large number of possible stable patterns which the system can choose from – or which are chosen by the initial conditions.

Figure 19 shows that, at the high intensity corresponding to Fig. 18d, the model (1) allows us to reproduce the experimental findings on switching individual bright spots. Figure 19a is the regular hexagonal pattern, at high intensity. Figure 19b shows the field with one bright spot switched off as a stable solution and Fig. 19c shows also a triplet of bright spots switched off as a stable solution, just as observed in the experiments [8].

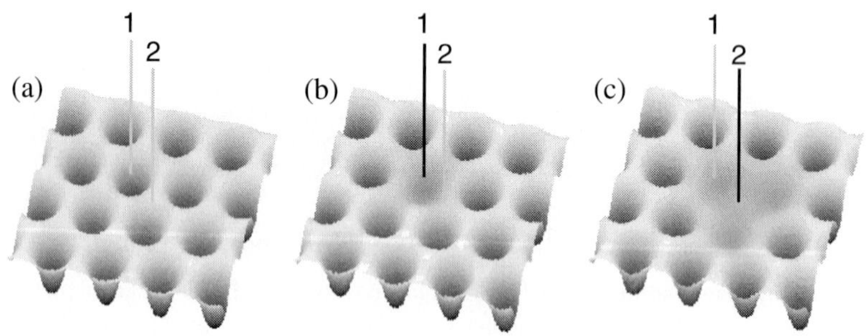

Fig. 19. Stable hexagonal arrangements of dark spatial solitons: (**a**) without defects, (**b**) with single-soliton defect, (**c**) with triple-soliton defect. The parameters are the same as in Fig. 18 (**d**). Compare with the experiment Fig. 17 and [30]

Thus, while Fig. 18a is a completely coherent space-filling pattern, Fig. 18d is really a cluster of (densest packed) individual dark solitons. The increase of intensity from (a) to (d) allows the transition from extended "coherent" patterns to localized structures, by means of the increased nonlinearity, and this gives the system an additional internal degree of freedom. We note that the transition from the coherent low intensity pattern to the incoherent higher intensity structure proceeds through stripe patterns, as shown in Fig. 18b [8]. For the intensity of Fig. 18c, the individual spots are still not independent, corresponding to the experiment Fig. 16b.

Illumination Near Band-Gap (Absorptive/Defocusing Nonlinearity)

Working at wavelengths close to the band edge, we find bright and dark solitons (Fig. 20), as well as collections of several spots (Fig. 14b, d, e) (all independent on the illumination beam profile). The nonlinearity of the MQW structure near the band edge is predominantly absorptive. To compare with calculations, we can therefore, to a first approximation, neglect the refractive part of the complex nonlinearity in the model equations (1) and describe

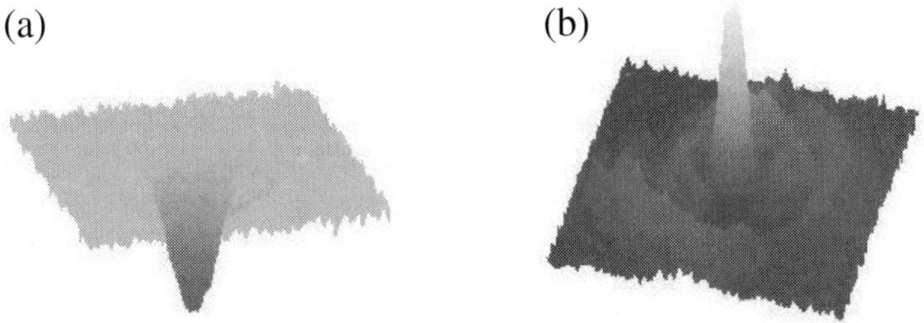

Fig. 20. (**a**) *Dark* and (**b**) *bright* solitons experimentally observed near band-gap

the nonlinear medium as a saturable absorber. Numerical simulations for this case (Fig. 21) confirm the existence of both bright and dark resonator solitons as they are observed in the experiment (Figs. 14, 20). We can contrast these purely dissipative resonator spatial solitons with propagating spatial solitons (in a bulk nonlinear material) [33] – the latter cannot be supported by saturable absorption.

Figure 22 shows details of the spontaneous formation of such bright solitons as occur in Fig. 14c and Fig. 20a. As discussed in [27], in this case material heating leads to a slow spontaneous formation of solitons, associated with the shift of the semiconductor band edge with temperature [34]. In Fig. 22, at $t \approx 1.3\,\mu\mathrm{s}$ (arrow), the resonator switches to high transmission (small reflection). The switched area then contracts slowly to the stable structure Fig. 22b, which exists after $t \approx 3.0\,\mu\mathrm{s}$.

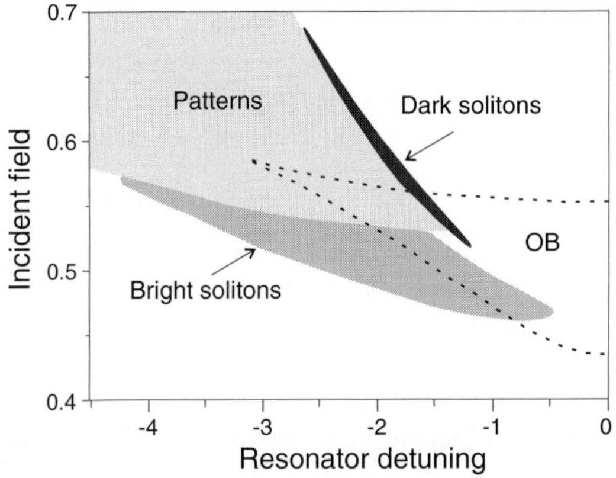

Fig. 21. Numerical solutions of (1) for unpumped ($P = 0$), absorptive ($Re(\alpha) = 0$) case. The area within the *dashed* lines is the optical bistability domain for plane waves. The *shaded* areas are the domains of existence of bright and dark solitons

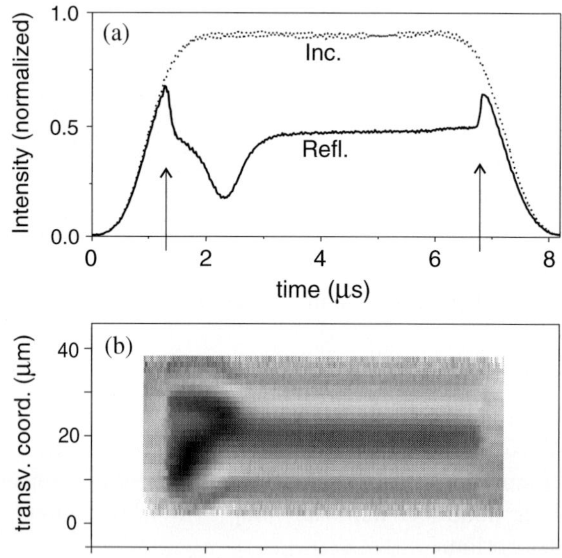

Fig. 22. Bright soliton formation below the band-gap. (**a**) Intensity of incident (*dotted*) and reflected (*solid*) light at the center of the soliton, as a function of time. (**b**) Reflectivity on a diameter of the illuminated area, as a function of time. The arrows mark the switch-on and off. For details, see text

After the resonator has switched to low reflection, its internal field is high, and so is the dissipation. A rising temperature ΔT decreases the band gap energy [34] ($E_g(\Delta T \approx E_g^0 - \alpha\Delta T)$) and therefore shifts the bistable resonator characteristic towards higher intensity. Thus the basin of attraction for solitons, which is located near the plane wave switch-off intensity (see locations of the existence domains for the bright solitons and the plane wave bistability in Fig. 21), is shifted to the incident intensity, whereupon a soliton can form. Evidently, for different parameters, the shift can be substantially larger or smaller than the width of the bistability loop, in which case no stable soliton can appear. As a consequence, we note that, in the absence of thermal effects (good heat-sinking of sample), solitons would not appear spontaneously, but would have to be switched on by a local pulsed light injection.

Apart from the coherent switching mentioned in Sect. 3.3 (Fig. 15), bright solitons can also be switched by light incoherent with the background field. Under these conditions, the switching occurs due to change of carrier density alone. Figure 23a shows the incoherent switch-on of a bright soliton, where a perpendicularly-polarized switching pulse of ≈10 ns duration is applied at $t = 4$ μs. As is apparent, a soliton forms after this incoherent light pulse. The slow formation of the soliton is apparent in Fig. 23a (taking roughly the time from $t = 4$ to $t = 4.5$ μs), indicating again the influence of material heating.

It should be emphasized that this heating is not the mechanism for switching a soliton on, it only accompanies the soliton formation and slows it. However, it allows switching a soliton off incoherently [35]. This is shown in Fig. 23b, where the driving light is initially raised to a level at which a soliton forms spontaneously. Slow soliton formation, due to the heating, is again apparent. The incoherent switching pulse is then applied and this leads to disappearance of the soliton.

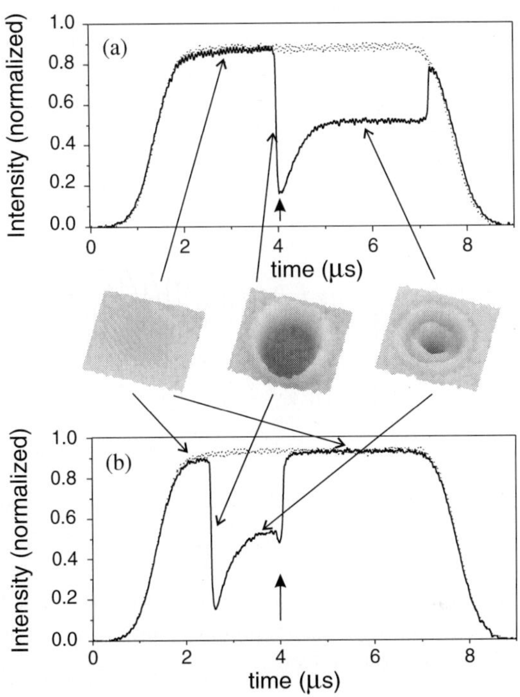

Fig. 23. Recording of incoherent (**a**) switching-on and (**b**) switching-off of a soliton. Snapshot pictures show un-switched state (*left*), circular switched domain (*center*) and a soliton (*right*). The dotted trace gives the incident intensity, while the solid trace gives the reflected intensity at the center of soliton. Arrows mark the time of the 10 ns switching pulse

The soliton can thus be switched on and also off by an incoherent pulse. The reason for the latter is a thermal effect. Initially the material is "cold", and then a switching pulse leads then to the creation of a soliton. Dissipation in the material at the location of the soliton raises the temperature and the soliton is slowly formed. At the raised temperature, the band edge (and with it the bistability characteristic and the existence range of solitons) is shifted, so that a new pulse brings the system out of the range of existence of solitons. Consequently the soliton is switched off.

Thus switching on a soliton is possible incoherently with the "cold" material and switching off is possible incoherently with the "heated" material. When the driving intensity is chosen to be slightly below the spontaneous switching threshold, the nonlinear resonator is cold. An incoherent pulse increases the carrier density locally and can switch the soliton on, thus causing local heating. Another incoherent pulse, aimed into the heated area, can then switch the soliton off and thereby return the resonator to its initial temperature, so that the soliton can be switched on/off again.

This thermal effect, combined with electronic nonlinearity, can cause spatial and temporal instabilities. A bright soliton switching on spontaneously in the cold material will then heat locally and can thereby destroy the condition for its existence, so that it switches off. After the material has cooled, the soliton switches on again, etc. Regenerative pulsing of the soliton results, and an example of this is shown in the observation in Fig. 24.

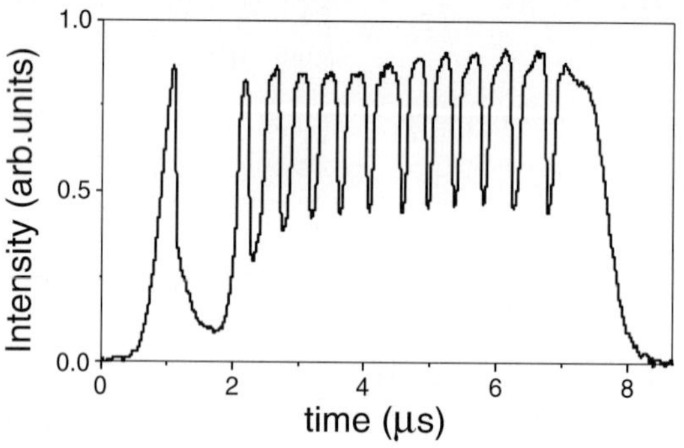

Fig. 24. Reflected light intensity measured at the center of a soliton. This shows regenerative pulsing (i.e. repeated switching on and off of a soliton), resulting from combined thermal and electronic effects

As opposed to the bright soliton that heats the material locally, a dark soliton cools the material locally. This results in a shift of the band edge and causes the switching characteristic to shift oppositely to the bright soliton case. A consequence is that the dark soliton moves laterally to places of uncooled material. Here, however, it cools the material again, so that a continuous motion results, analogous to what we have described as the "restless vortex" in [36]. With regard to material heating effects , dark solitons therefore tend to be non-stationary. Such motion of a dark soliton was observed e.g. in [27] and is shown in Fig. 25.

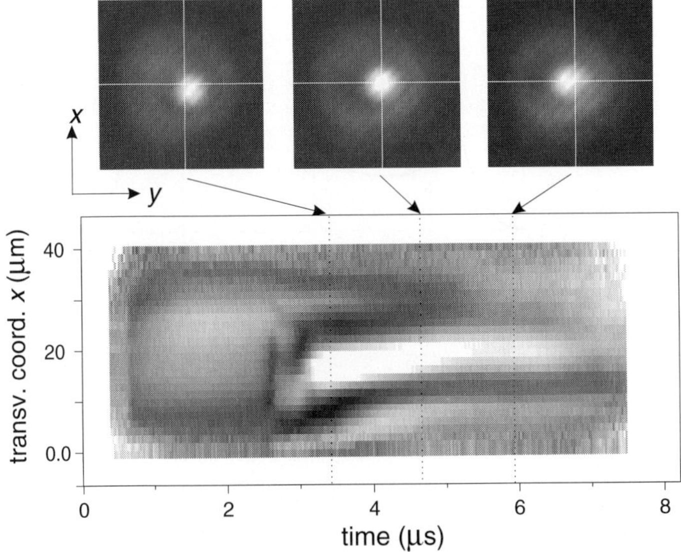

Fig. 25. Formation of a dark soliton (compare with Fig. 22 and caption) and its motion. The motion visible in the streak picture (*bottom*) is shown in 2D snapshots (*top*) for clarity

Illumination Above Band-Gap (Absorptive/Focusing Case)

For excitation above the band-gap, bright solitons form [37], and these have the same appearance as the bright solitons below or near the band gap (Figs. 14c and 20a).

Figure 26 shows the dynamics of the bright soliton formation for excitation above band-gap. The difference between solitons existing above and below the band-gap can be understood from the model. From (1), we obtain the reflected intensity as a function of the incident intensity for wavelengths above the band-gap ($Re(\alpha) > 0$), as well as below the band-gap ($Re(\alpha) < 0$), for plane waves (Fig. 27). One sees that the bistability range is large below the band-gap and small above it. Solving (1) numerically, a typical bright soliton (top of Fig. 27) is found co-existing with the homogeneous intensity solutions in the shaded regions of Fig. 27a, b.

Above the band-gap, Fig. 26 shows that the soliton is switched on "immediately" without the slow thermal process described above. Figure 27a shows why. The plane wave characteristic of the resonator above band-gap is either bistable in a very narrow range, or even mono-stable (due to the contribution of the self-focusing reactive nonlinearity [38]), but still has bistability between the soliton state (not plane wave) and the unswitched state. In this case, the electronic switching leads directly to the basis of attraction for solitons and the switch-on of the soliton is purely electronic and fast. The

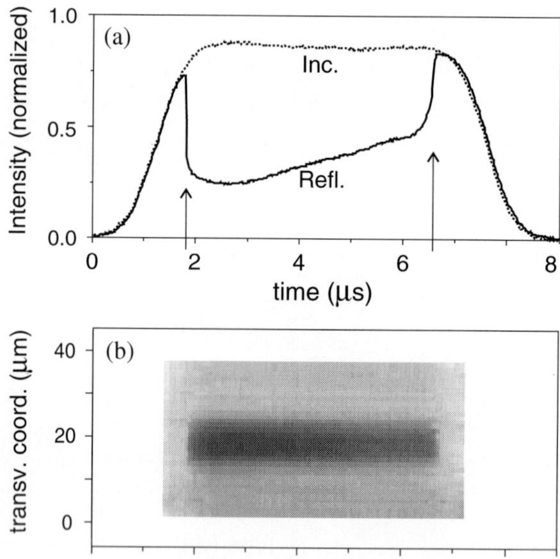

Fig. 26. Bright soliton formation above the band-gap. (**a**) Intensity of incident (*dotted*) and reflected (*solid*) light, at the center of the soliton, as a function of time. Arrows mark the switch-on and off. (**b**) Reflectivity on a diameter of the illuminated area, as a function of time. The soliton forms abruptly (without mediation by a thermal effect as in Fig. 22)

widths of the bistability characteristics observed experimentally [37] correspond to the calculated ones Fig. 27. Conversely, the wide bistability range below band-gap requires a (slow) thermal shift of the characteristic to reach the soliton existence range.

Nonetheless, above the band-gap, there is also strong dissipation after the switch-on. The associated temperature rise influences a soliton, and can even destabilize it. The destabilization effect can be seen in Fig. 26a. Over a time of a few μs after the soliton switch-on, the soliton weakens (reflectivity increases slowly), presumably because of the rise of temperature and the associated shift of the band-gap. At 6.5 μs, the soliton switches off, although the illumination has not yet dropped.

Thus, while the dissipation does not hinder the fast switch-on of the soliton, it can finally destabilize the soliton. After the soliton is switched off, the material cools and the band-gap shifts back, so that the soliton can switch on again.

Optical Pumping

The thermal effects discussed above result from the local heating caused by the high intra-cavity intensity within the bright soliton. They limit the

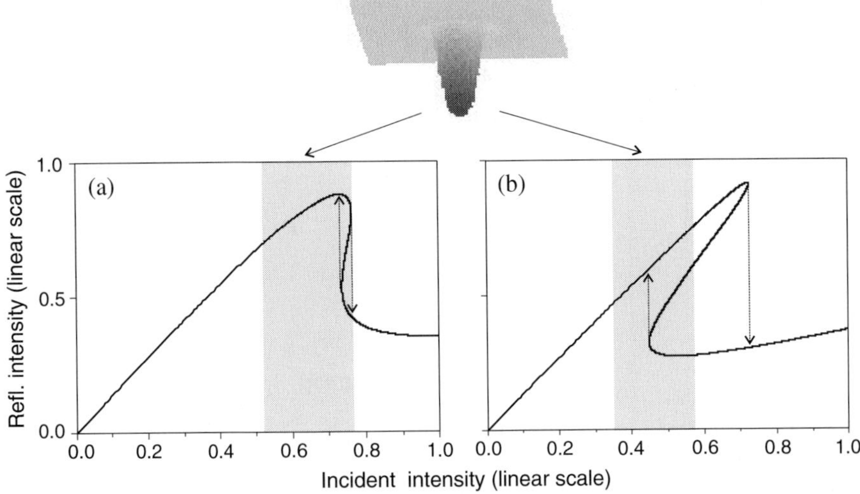

Fig. 27. Steady-state plane wave solution of (1) (**a**) above band-gap and (**b**) below band-gap. The soliton solution shown exists for incident intensities corresponding to the *shaded* areas, and co-exists with the homogeneous solution. If the temperature increases, the characteristics shift together and the soliton existence range moves to higher incident intensities. Reflected and incident intensities are normalized to the same value. Note the different widths of the bistability ranges above and below the band-gap

switching speed of solitons and they will also limit the speed at which solitons could be moved around, thus limiting applications. The picture is that a soliton carries with it a temperature profile, so that the temperature becomes a dynamic and spatial variable influencing the soliton stability.

On the other hand, spatially-uniform heating will not cause such problems, as it shifts parameters but does not constitute a variable in the system. The unwanted heating effects are directly proportional to the light intensity sustaining a soliton. For this reason, and quite generally, it is desirable to reduce the light intensities required for sustaining solitons.

Conceptually, this can be expected if part of the power sustaining a resonator soliton could be provided incoherently with the driving field, e.g. by means of optical or electrical pumping. Pumping of the MQW structure generates carriers and allows conversion from absorption to gain. If pumping is strong enough, the semiconductor micro-resonator can emit light as a laser [39].

Figure 28 shows the variation of the plane-wave bistability domain with the pump, as calculated from (1). The increase of the pump intensity leads to a shrinking of the bistability domain for plane waves (Fig. 28, pump intensities from P_1 to P_3) and the resonator solitons" existence domain, while reducing the light intensity necessary to sustain the solitons. This reduction

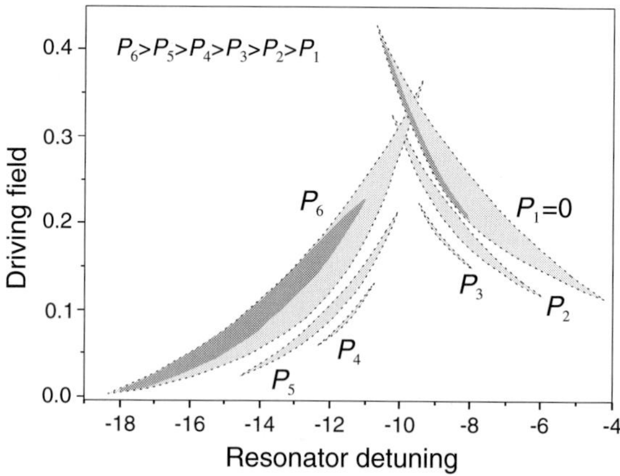

Fig. 28. Calculated plane-wave bistability domains as a function of the pump intensity. *Dark* shaded areas show bright soliton existence domains for two extreme cases: without pump ($P_1 = 0$), and with pump near lasing threshold (P_6)

of the sustaining light intensity was observed experimentally (Fig. 29) [29]. When pumping below the transparency point of the material, and with the driving laser wavelength near the semiconductor MQW structure band edge, bright and dark solitons form in a similar fashion to the unpumped case (Fig. 20). As the pump reduces the sustaining light intensity of the solitons, the heating effects become weak and the solitons switch on quickly, being unmediated by heating [29].

When the pump intensity approaches the transparency point of the semiconductor material, the resonator solitons' domain of existence disappears (Fig. 28, pump intensities between P_3 and P_4 and Fig. 29). It reappears above the transparency point (Fig. 28, pump intensities from P_4 to P_6) and expands with the pump intensity (Fig. 30).

Switched structures observed below the lasing threshold are shown in Fig. 31. Figure 32 shows structures observed slightly above the lasing threshold when a driving field is used simultaneously with the laser emission. The structures of Fig. 32 are reminiscent of the solitons in electrically-pumped resonators [40]. It becomes clear that these structures (Fig. 32) must be soliton-collections or patterns, and not (linear) mode patterns, when looking at Fig. 32c. Here, three bright spots are visibly separated by darker lines. In a "mode-pattern" the phase change from one bright spot to the next is π (resulting in black lines due to destructive interference separating the bright spots). Therefore a "flower" mode pattern must always have (and has) an even number of "petals", whereas here we observe an odd number, inconsistent with phases in mode patterns. Thus we can conclude that there is no phase change between the bright spots and that the latter are formed

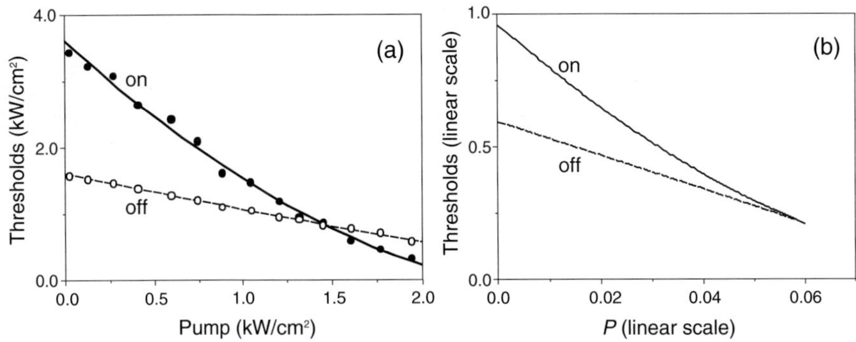

Fig. 29. (**a**) Measured and (**b**) calculated switch-on and switch-off intensities for the driving light as a function of the pump intensity. The unphysical crossing of the on and off curves is an artifact from material heating

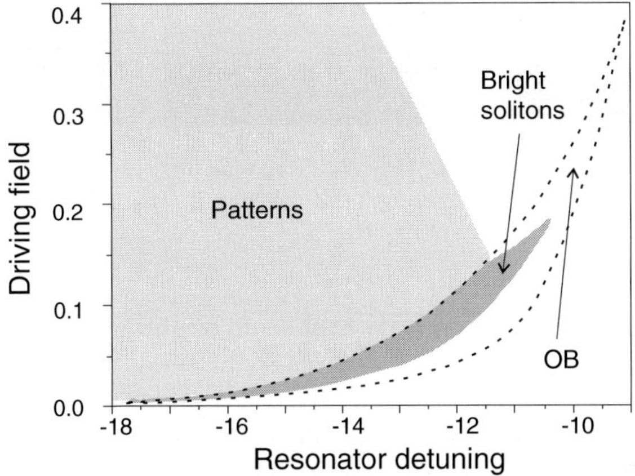

Fig. 30. Results of numerical simulations of below band-gap solitons using the model (1) for a micro-resonator pumped close to the lasing threshold. *Shaded* areas are domains of existence of bright resonator solitons and patterns. The area within the *dashed* lines is the optical bistability domain for plane waves

by self-localization, i.e. they are solitons or patterns in a resonator above transparency of the nonlinear material.

We note that optically-pumped resonators allow more homogeneous pumping conditions than electrical pumping [41]. This suggests that optical pumping lends itself more readily to localization and motion control of solitons than electrical pumping.

There is a difference between resonator solitons below band-gap in pumped and unpumped material. The nonlinear resonance mechanism of soliton formation [20] uses a defocusing nonlinearity below transparency and a focusing nonlinearity above transparency. Defocusing nonlinearity stabilizes dark

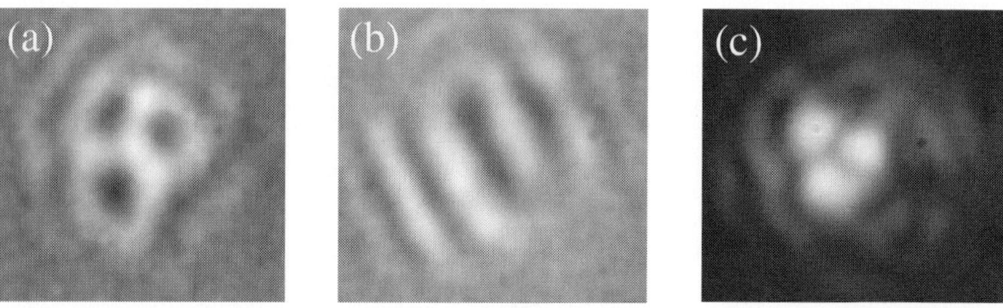

Fig. 31. Snapshots of typical optical structures at optical pump intensities slightly below lasing threshold. (The driving light intensity increases from (**a**) to (**c**)). With an increase in driving intensity, the dark spot patterns change to bright spot patterns (reminiscent of the changes in Figs. 16 and 18)

Fig. 32. Snapshots of typical optical structures at optical pump intensities slightly above lasing threshold. (The driving light intensity increases from (**a**) to (**c**)). These structures are always bright spot patterns

solitons and focusing nonlinearity stabilizes bright solitons. It follows that dark solitons should prevail for unpumped material and bright solitons for pumped material. Figure 30 shows typical examples of calculated resonator solitons for a pumped semiconductor micro-resonator. Bright solitons have a large existence range in the pumped case (Fig. 28, at P_6), while dark solitons exist, but with a smaller range of stability, in the unpumped case (Fig. 28, at P_1).

Thus, pumped semiconductor resonators are well-suited for sustaining solitons below band-gap. We note that (i) the background light intensity necessary to sustain and switch resonator solitons is substantially reduced by the pumping and therefore destabilizing thermal effects are minimized, and (ii) the nonlinear resonance effect and the transverse nonlinear effect (self-focusing) co-operate to stabilize bright solitons, so the domain of existence of (purely dispersive) bright solitons below band-gap can be quite large.

4 Conclusion

We have shown in these experiments that optical solitons generally exist in nonlinear optical resonators. They can be vortices, phase solitons, or bright and dark solitons. For technical applications, the experiments on semiconductor micro-resonators have shown the existence of bright and dark solitons. One can experimentally distinguish switched areas, patterns and solitons from each other and from mode-fields (i.e. fields whose structures are standing transverse waves, resulting from boundaries). Thermal effects can lead to spatial and temporal instabilities. They can be controlled in various ways, e.g. by working above band-gap or by pumping. Furthermore, pumping allows us to control the magnitude and the sign of the material nonlinearities, and thus to maximize the stability ranges for bright (dark) solitons. Coherent periodic patterns can disintegrate, with increasing driving fields, into independent spatial solitons.

Optical resonator solitons show properties reminiscent of simple biological structures, and one may speculate that aggregations of resonator solitons could allow information processing resembling brain functions.

Acknowledgement

This work was supported by Deutsche Forschungsgemeinschaft under grant We743/12–1. The similarities between dissipative solitons and simple organic structures was mentioned to us first by N. Akhmediev

References

1. W. J. Firth, A. J. Scroggie, Europhys. Lett. **26**, 521 (1994).
2. U. Frisch, *Turbulence.* The Legacy of A. N. Kolmogorov, (Cambridge University Press, 1995).
3. K. Staliunas, Int. Journ. of Bifurcation and Chaos **11**, 2845 (2001).
4. K. Staliunas, Anticorrelation and Subdiffusion in Financial Systems; *xxx.lanl.gov*(*cond.mat*/0203591), subm. Phys. Rev. E, (2002).
5. For a review see e.g. P. Dutta, P. M. Horn, Rev. Mod. Phys. **53**, 497 (1981).
6. P. Bak, Phys. Rev. Lett. **59**, 381 (1987).
7. See as examples: A. Jaques, P. Glorieux, Opt. Comm. **40**, 455 (1982) and S. Rushin, S. M. Bauer, Appl. Phys. **24**, 45 (1981).
8. V. B. Taranenko, C. O. Weiss, B. Schaepers, Phys. Rev. A **65**, 013812 (2002).
9. N. Akhmediev mentioned this likening of dissipative solitons to "animals" and of conservative solitons to "dead" things to my knowledge first.
10. G. Slekys, K. Staliunas, C. O. Weiss, Opt. Comm. **149**, 113 (1998).
11. C. O. Weiss, M. Vaupel, K. Staliunas, G. Slekys, V. B. Taranenko, Appl. Phys. B: Lasers Opt. **68**, 151 (1999).
12. See the supplementary electronic material to [11]: http://link.springer.de/jounals/apb
13. C. P. Smith, Y. Dihardja, C. O. Weiss, L. A. Lugiato, F. Prati, P. Vanotti, Opt. Comm. **102**, 105 (1999).
14. G. S. Mc Donald, W. J. Firth, Journ. Opt. Soc. Amer. **B7**, 1328 (1990) G. S. Mc Donald, W. J. Firth, J. Mod. Opt. **37**, 613 (1990).

15. M. Kreuzer, H. Gottschling, T. Tschudi, R. Neubecker, Mol. Cryst. Liq. Cryst. **207**, 219 (1991).
16. D. Oesterhelt, W. Stoekenius, Nature **233**, 149 (1971).
17. V. Yu. Bazhenov, V. B. Taranenko, M. V. Vasnetsov, Proc. SPIE **1840**, 183 (1992).
18. V. B. Taranenko, K. Staliunas, C. O. Weiss, Phys. Rev. A **56**, 1582 (1997).
19. K. Staliunas, V. B. Taranenko, G. Slekys, R. Viselga, C. O. Weiss, Phys. Rev. A **57**, 359 (1998).
20. K. Staliunas, V. J. Sanchez-Morcillo, Opt. Comm. **139**, 306 (1997).
21. V. B. Taranenko, K. Staliunas, C. O. Weiss, Phys. Rev. Lett. **81**, 2226 (1998).
22. H. M. Gibbs, textitOptical Bistability – Controlling Light with Light, (Academic Press, 1985).
23. L. Spinelli, G. Tissoni, M. Brambilla, F. Prati, L. A. Lugiato, Phys. Rev. A **58**, 2542 (1998).
24. D. Michaelis, U. Peschel, F. Lederer, Phys. Rev. A **56**, R3366 (1997).
25. L. A. Lugiato, R. Lefever, Phys. Rev. Lett. **58**, 2209 (1987).
26. N. N. Rosanov, Prog. Opt. **35**, 1 (1996).
27. V. B. Taranenko, I. Ganne, R. Kuszelewicz, C. O. Weiss, Appl. Phys. B: Lasers Opt. **72**, 377 (2001).
28. V. B. Taranenko, F.-J. Ahlers, K. Pierz, Appl. Phys. B: Lasers Opt. **75**, 75 (2002).
29. V. B. Taranenko, C. O. Weiss, W. Stolz, Opt. Lett. **26**, 1574 (2001).
30. V. B. Taranenko, I. Ganne, R. Kuszelewicz, C. O. Weiss, Phys. Rev. A **61**, 063818 (2000).
31. P. K. Jacobsen, J. V. Moloney, A. C. Newell, R. Indik, Phys. Rev. A **45**, 8129 (1992).
32. W. J. Firth, A. J. Scroggie, Europhys. Lett. **26**, 521 (1994).
33. G. I. Stegeman, M. Segev, Science **286**, 1518 (1999).
34. T. Rossler, R. A. Indik, G. K. Harkness, J. V. Moloney, C. Z. Ning, Phys. Rev. A **58**, 3279 (1998).
35. V. B. Taranenko, C. O. Weiss, Appl. Phys. B: Lasers Opt. **72**, 893 (2001).
36. C. O. Weiss, H. R. Telle, K. Staliunas, M. Brambilla, Phys. Rev. A **47**, R1616 (1993).
37. V. B. Taranenko, C. O. Weiss, W. Stolz, J. Opt. Soc. Am. B **19**, 8129 (2002).
38. S. H. Park, J. F. Morhange, A. D. Jeery, R. A. Morgan, A. Chavez-Pirson, H. M. Gibbs, S. W. Koch, N. Peyghambarian, M. Derstine, A. C. Gossard, J. H. English, W. Weidmann, Appl. Phys. Lett. **52**, 1201 (1988).
39. V. B. Taranenko, C. O. Weiss, Spatial solitons in an optically pumped semiconductor microresonator, xxx.lanl.gov. (nlin.PS/0204048 (2002)).
40. S. Barland, J. R. Tredicce, M. Brambilla, L. A. Lugiato, S. Balle, M. Guidici, T. Maggipinto, L. Spinelli, G. Tissoni, T. Knodl. M. Miller, R. Jaeger, Nature **419**, 699 (2002).
41. W. J. Alford, T. D. Raymond, A. A. Allerman, J. Opt. Soc. Am. B **19**, 663 (2002).

Dynamics of Dissipative Temporal Solitons

U. Peschel[1], D. Michaelis[2], Z. Bakonyi[3], G. Onishchukov[3], and F. Lederer[4]

[1] Friedrich-Schiller-Universitä Jena, Institute of Condensed Matter Theory and Optics, Max-Wien-Platz 1, 07743 Jena, Germany
p6peul@uni-jena.de
[2] Fraunhofer Institute of Applied Optics and Precision Mechanics, Albert-Einstein-Strasse 7, 07745 Jena, Germany
dirk@physse.nlwl.uni-jena.de
[3] Friedrich-Schiller-Universitä Jena, Institute of Applied Physics, Max-Wien-Platz 1, 07743 Jena, Germany
zbakonyi@iap.uni-jena.de
George.Onishchukov@uni-jena.de
[4] Friedrich-Schiller-Universität Jena, Max-Wien-Platz 1, 07743 Jena, Germany
pfl@uni-jena.de

1 Introduction

The properties and the dynamics of localized structures, frequently termed solitary waves or solitons, define, to a large extent, the behavior of the relevant nonlinear system [1]. Thus, it is a crucial and fundamental issue of nonlinear dynamics to fully characterize these objects in various conservative and dissipative nonlinear environments. Apart from this fundamental point of view, solitons (henceforth we adopt this term, even for localized solutions of non-integrable systems) exhibit a remarkable potential for applications, particularly if optical systems are considered. Regarding the type of localization, one can distinguish between temporal and spatial solitons. Spatial solitons are self-confined beams, which are shape-invariant upon propagation. (For an overview, see [2, 3]). It can be anticipated that they could play a vital role in all-optical processing and logic, since we can use their complex collision behavior [4]. Temporal solitons, on the other hand, represent shape-invariant (or breathing) pulses. It is now common belief that robust temporal solitons will play a major role as elementary units (bits) of information in future all-optical networks [5, 6]. Until now, the main emphasis has been on temporal and spatial soliton families in conservative systems, where energy is conserved. Recently, another class of solitons, which are characterized by a permanent energy exchange with their environment, has attracted much attention. These solitons are termed dissipative solitons or auto-solitons . They emerge as a result of a balance between linear (delocalization and losses) and nonlinear (self-phase modulation and gain/loss saturation) effects. Except for very few cases [7], they form zero-parameter families and their features are entirely fixed by the underlying optical system. Cavity solitons form a prominent type. They appear as spatially-localized transverse peaks in transmission or reflection, e.g. from a Fabry-Perot cavity. They rely

U. Peschel, D. Michaelis, Z. Bakonyi, G. Onishchukov, and F. Lederer: *Dynamics of Dissipative Temporal Solitons*, Lect. Notes Phys. **661**, 161–181 (2005)
www.springerlink.com

strongly on the nonlinear interaction of forward and backward propagating fields, where cavity (radiation) losses are compensated for by gain or an external holding beam. Cavity solitons have been observed in Fabry-Perot cavities and single mirror feedback setups (for an overview see [8]). In contrast, propagating dissipative solitons consist of forward-propagating fields only. Recently, the existence of both stable temporal [9] and spatial propagating dissipative solitons [10] was experimentally demonstrated. It is evident that the study of dissipative solitons is of great practical relevance, because most real optical systems are dissipative by nature.

Here, we deal exclusively with temporal dissipative solitons. The identification of existence domains of stable dissipative propagating solitons is by no means a straightforward task. This subject has attracted much interest within the last decade in the theoretical literature. The underlying physical system usually includes a semiconductor optical amplifier (SOA). A fast system is frequently realized by a nonlinear optical loop mirror [11], while a slow one is usually realized by a SOA with an ion-implanted output facet saturable absorber (SA). The SA is introduced to absorb the amplified spontaneous emission and to stabilize the solitons. With regard to theory, a fast SA enters the evolution equation through an instantaneous nonlinear gain, whereas SOA and slow SA exhibit genuine nonlinear dynamics, which is characterized by gain/loss saturation, finite recovery time and nonlinear dispersive effects. It can be be described reasonably by using the Agrawal-Olson model [12]. There have been several attempts to identify stable dissipative solitons in this system. Most of them have relied on solutions of complex or modified complex Ginzburg-Landau equations (GLE). One arrives at this equation when several effects, such as finite recovery time and gain/loss saturation, are disregarded or approximated (cubic-quintic GLE). These solutions have been found either by applying perturbation theories to Schrödinger solitons [13, 14] or by searching for analytic, zero-parameter dissipative soliton solutions ([7], [14]–[16]). To really identify stable dissipative solitons which were found in the experiment [9], one has to account for all relevant physical effects and to perform a careful theoretical study of the bifurcation behavior of continuous wave (c.w.) and soliton solutions. This analysis reveals that stable dissipative solitons require, for their existence, a balance between linear/nonlinear gain and loss as well as a non-instantaneous saturating nonlinearity. Basically, the bifurcation behavior of c.w. solutions and solitons are closely related, because solitons usually emerge from a c.w. solution at a critical point. Thus, theory predicts that stable solitons exist predominantly near sub-critical c.w. bifurcations ([17]–[20]) or c.w. bistability ([21]–[26]). For super-critical c.w. bifurcation, solitons also emanate super-critically [7]. They are unstable because the trivial solution of the soliton background is unstable. Prominent examples of this instability are soliton solutions of the complex or modified complex cubic Ginzburg-Landau equation [7], which describes, e.g., pulse propagation in fibers with linear gain, frequency-dependent losses and fast saturable absorption by nonlinear optical loop mirrors. Although this instability may be fairly weak, the solitons will ultimately decay because of the background instability. For certain parameter sets, fast saturable absorption can also trigger a sub-critical c.w. bifurcation [7, 14]. However, for instantaneous saturable absorption, the related soliton branch lacks the "back-bending" required for stable dissipative solitons. Thus, even in this case, the whole object (soliton plus background) eventually becomes unstable, now because of the instability of the pulse itself.

The aim of this contribution is to study, theoretically and experimentally, the existence and dynamics of dissipative solitons in an extremely versatile system with frequency-dependent losses and two competing, non-instantaneous saturating nonlinearities. In practice, the system consists of a recirculating fiber loop with SOA, slow SA and in-line filters (see Fig. 1). Beyond its relevance for investigating fundamental physical effects, this fiber loop set-up is of outstanding practical importance for optical communications. It essentially represents a long-haul transmission system with compact SOAs as an alternative to erbium-doped fiber amplifiers. These SOAs may serve as optical amplifiers in wavelength ranges where no rare-earth-doped amplifiers are available.

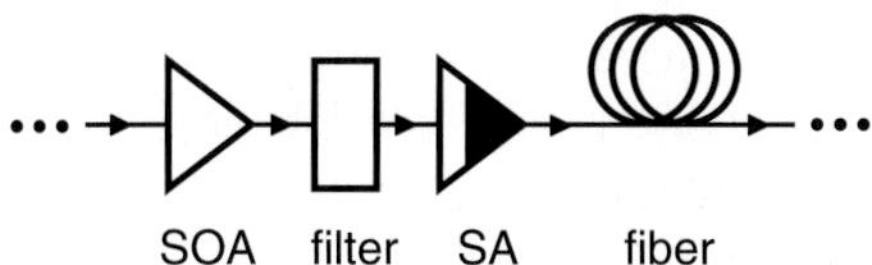

Fig. 1. Schematic of the fiber loop system

We demonstrate that stable localized solutions are defined by the interplay of saturable gain and loss, provided by SOA and SA, respectively. A pre-requisite for soliton formation is that there must be different saturation behavior, with regard to power and energy, occurring on the different time scales of gain and loss recovery. We show that these solitons bifurcate sub-critically from super-critical Ginzburg-Landau solitons, once the non-instantaneous character of the nonlinearity starts to be important. After addressing the theoretical background, we experimentally study excitation, as well as dynamics, of solitons. If there is a strong, abrupt change in the losses, then we observe that these changes trigger soliton switching or relaxation. For the first time, critical slowing down of dissipative optical solitons is experimentally verified. Furthermore, for fixed values of the losses, we predict a Hopf bifurcation of solitons, i.e., the existence of oscillating solutions. Finally, we provide experimental evidence that dissipative solitons exhibit a greatly reduced timing jitter compared to the Gordon-Haus jitter of Schrödinger solitons.

2 Mathematical Models

2.1 Lumped Model

We start by establishing a model which describes the underlying experimental set-up appropriately. To allow for quasi-infinite propagation distances, the experiments were performed in a re-circulating fiber loop. The basic elements of the loop are a standard single mode fiber of length $L_f = 25\,\mathrm{km}$, a SOA, a 3 nm bandpass filter and a SA (see Fig. 1 and, for more details, see [27]). Note that the order of these elements in the loop has a significant influence on the operation of the loop set-up, e.g., saturable absorption is only efficient after amplification of the pulses. Other components, like isolators, polarization controllers and monitoring equipment, do not influence the physical properties of the system and will not be discussed in detail. Firstly, we derive equations which describe the field or carrier dynamics of the individual elements. This establishes the lumped model. Then, we proceed in "smearing out" the effect of all elements over the fiber length (the averaged model), as is usually done in nonlinear fiber optics.

The mean wavelength, λ, of the pulses generated by a semiconductor laser may vary around the zero-dispersion point of the fiber at $\lambda = 1.306\,\mu\mathrm{m}$. The optical fiber is characterized by its group velocity dispersion ($|D_\omega| < 1\,\mathrm{ps}^2/\mathrm{km}$), effective mode area $A_{\mathrm{eff}} = 60\,\mu\mathrm{m}^2$, a Kerr nonlinearity ($n_2 = 2.5 \times 10^{-20}\,\mathrm{m}^2/\mathrm{W}$) and losses ($\alpha_f \approx 0.35\,\mathrm{dB/km}$).

The pulse evolution along the fiber is well-described by a Nonlinear Schrödinger equation with losses included

$$\left[\mathrm{i}\frac{\partial}{\partial Z} - \frac{D_\omega}{2}\frac{\partial^2}{\partial T^2} + \mathrm{i}\frac{\alpha_f}{2} + \frac{2\pi}{\lambda_0}\frac{n_2}{A_{\mathrm{eff}}}\left|U(Z,T)\right|^2\right]U(Z,T) = 0\,, \qquad (1)$$

where the rapidly-varying terms (frequency $\overline{\Omega} = 2\pi c/\lambda_0 and$ propagation constant $\overline{\beta}$ of the fiber mode) have been removed to obtain the slowly-varying envelope of the pulse $U(Z,T)$ in units of $\sqrt{\mathrm{W}}$. Losses are compensated for by an SOA exhibiting a saturation energy of $E_{\mathrm{sat}}^{\mathrm{SOA}} = 2\,\mathrm{pJ}$, a recovery time $T_{\mathrm{SOA}} = 200\mathrm{ps}$ and a Henry factor (a measure of saturation-induced refractive index changes) $h_{\mathrm{SOA}} = 5$. The optical response of the SOA may be described by a simple transmission function which neglects propagation effects and internal losses in the amplifier, and relates input $U_{\mathrm{in}}^{\mathrm{SOA}}(T)$ and output fields $U_{\mathrm{out}}^{\mathrm{SOA}}(T)$ as [12].

$$U_{\mathrm{out}}^{\mathrm{SOA}}(T) = \exp\left[(1 - ih_{\mathrm{SOA}})\,G_{\mathrm{SOA}}(T)\right]U_{\mathrm{in}}^{\mathrm{SOA}}(T) + \Gamma_{\mathrm{SOA}}(T)\,. \qquad (2)$$

During each passage through the amplifier, noise is added by using a stochastic function $\Gamma_{\mathrm{SOA}}(T)$. A rate equation defines the evolution of the corresponding gain function, G_{SOA}, of the SOA [12]:

$$\begin{aligned}\frac{\partial}{\partial T} G_{\mathrm{SOA}}(T) = &- \frac{1}{E_{\mathrm{sat}}^{\mathrm{SOA}}} \left[\left| U_{\mathrm{out}}^{\mathrm{SOA}}(T) \right|^2 - \left| U_{\mathrm{in}}^{\mathrm{SOA}}(T) \right|^2 \right] \\ &- \frac{1}{T_{\mathrm{SOA}}} \left[G_{\mathrm{SOA}}(T) - G_{\mathrm{SOA}}^0 \right] \\ = &- \frac{\exp\left[2G_{\mathrm{SOA}}(T)\right] - 1}{E_{\mathrm{sat}}^{\mathrm{SOA}}} \left| U_{\mathrm{in}}^{\mathrm{SOA}}(T) \right|^2 - \frac{\left| \Gamma_{\mathrm{SOA}}(T) \right|^2}{E_{\mathrm{sat}}^{\mathrm{SOA}}} \\ &- \frac{1}{T_{\mathrm{SOA}}} \left[G_{\mathrm{SOA}}(T) - G_{\mathrm{SOA}}^0 \right] , \end{aligned} \quad (3)$$

where G^0_{SOA} is the equilibrium value of the gain ($G^0_{\mathrm{SOA}} \approx 2.42$ in the underlying loop, corresponding to 20dB for the signal power) and signal and noise are assumed to be incoherent. The saturable absorber, which is placed behind the amplifier, is basically an SOA but with ion implantation at the output facet. Hence, its optical response and internal dynamics can be described by the same equations, but using different device parameters. In particular, here the equilibrium gain is negative ($G^0_{\mathrm{SA}} < 0$), accounting for a net absorption ($G^0_{\mathrm{SA}} \approx -1.73$ in the loop, corresponding to -15dB for the signal power). Because of the low carrier density and the defects, saturating the SA requires less energy $\left(E^{\mathrm{SA}}_{\mathrm{sat}} = 0.2 E^{\mathrm{SOA}}_{\mathrm{sat}}\right)$, it recovers faster ($T_{\mathrm{SA}} = 0.1 T_{\mathrm{SOA}}$) and the induced refractive index changes are smaller ($h_{\mathrm{SA}} = 2$) compared to the SOA.

Another important element of the loop is the spectral filter. Its effect can conveniently be described in the frequency domain. (Henceforth, quantities in the frequency domain will be labeled by $\widehat{\ }$). Thus we have

$$\widehat{U}^{\mathrm{filter}}_{\mathrm{out}}(\Omega) = \widehat{\tau}_{\mathrm{filter}}(\Omega)\, \widehat{U}^{\mathrm{filter}}_{\mathrm{in}}(\Omega) , \quad (4)$$

with a transmission function $\widehat{\tau}_{\mathrm{filter}}(\Omega)$. $\widehat{\tau}_{\mathrm{filter}}$ also accounts for losses originating from other elements, e.g. splices and polarization controllers. For the filters used in the experiment, $\widehat{\tau}_{\mathrm{filter}}$ has a well-defined maximum $\widehat{\tau}^{\max}_{\mathrm{filter}}$. Keeping in mind that $\widehat{U}(\Omega)$ is a slowly-varying function, the mean frequency Ω_0 can be set equal to the center frequency of the filter. Hence, the spectrum is shifted so that the transmission function has its maximum at $\Omega_0 = 0$. Taking advantage of (1), (3), (2), (4) and the corresponding SA equation, the dynamics along the loop can be accurately modelled. Formally, in the lumped model, the field after one round- trip is related to the input field by

$$U(L_f, T) = M_{\mathrm{SA}} * M_{\mathrm{filter}} * M_{\mathrm{SOA}} * M_{\mathrm{fiber}} * U(0, T) , \quad (5)$$

where the M's denote the transmission functions corresponding to the elements.

2.2 Averaged Model

However, a deeper theoretical understanding in terms of nonlinear dynamics is hampered by the discrete nature of the model. Therefore, we are now going to derive an averaged version of (5). The justification for this treatment is based on the fact that several round-trips are needed for the evolution of the physically relevant changes, e.g., dispersive broadening and nonlinear phase modulation. Hence, the field $U(Z,T)$ can be decomposed into a product containing a periodic part $V(Z)$ and an average amplitude $W(Z,T)$, i.e.,

$$U(Z,T) = V(Z)\, W(Z,T) \ . \tag{6}$$

(For details of this so-called "guiding center soliton" concept in systems with linear gain, see [28]). The amplitude, W, evolves slowly on propagation and is mainly controlled by the accumulated dispersive and nonlinear effects. In order to derive an evolution equation for W, we first have to define the periodic function V, which is subject to strong changes caused by losses in the system and subsequent amplification in the SOA. Absorptive losses in the fiber cause the optical power to decay exponentially, and this is reflected in the evolution of V along the fiber as follows:

$$V(Z) = \exp\left(-\frac{\alpha_f}{2} Z\right) \text{ for } 0 \le Z \le L_f \tag{7}$$

With this definition, $|W|^2$ is automatically set equal to the power measured at the loop entrance. From (5), it is evident that the SOA, filter and SA are located in this order after the fiber. Hence, we have $V_{\text{in}}^{\text{SOA}} = V(Z = L_f) = \exp\left(-\frac{\alpha_f}{2} L_f\right)$, $V_{\text{in}}^{\text{filter}} = V_{\text{out}}^{\text{SOA}}$ and $V_{\text{in}}^{\text{SA}} = V_{\text{out}}^{\text{filter}}$. We assume that V experiences the maximum transmission of the filter:

$$V_{\text{out}}^{\text{filter}} = \hat{\tau}_{\text{filter}}^{\max} V_{\text{in}}^{\text{filter}} , \tag{8}$$

and suffers small signal absorption in the SA:

$$V_{\text{out}}^{\text{SA}} = \exp\left[(1 - i h_{\text{SA}})\, G_{\text{SA}}^0\right] V_{\text{in}}^{\text{SA}} \ . \tag{9}$$

The accumulated losses are compensated for by the SOA. To restore V as a periodic function, apart from a trivial phase factor, its transmission function in the SOA has to be given by

$$V_{\text{out}}^{\text{SOA}} = \exp\left[(1 - \mathrm{i} h_{\text{SOA}})\left(\frac{\alpha_f}{2} L_f - G_{\text{SA}}^0 - \ln \hat{\tau}_{\text{filter}}^{\max}\right)\right] V_{\text{in}}^{\text{SOA}} \ . \tag{10}$$

Note that both the SA and the SOA may actually impose a slightly different gain or loss on the complete optical field U. The effect on V only accounts for some mean values. Having defined V, we are now going to derive the evolution equation for W. Based on (7), (1) is modified to [28]

$$\left[\mathrm{i}\frac{\partial}{\partial Z} - \frac{D_\omega}{2}\frac{\partial^2}{\partial T^2} + \frac{2\pi}{\lambda_0}\frac{n_2}{A_{\text{eff}}} |W(Z,T)|^2 \exp(-\alpha_f Z)\right] W(Z,T) = 0 \tag{11}$$

with a Z-dependent nonlinearity. Provided that W does not change significantly, we can average the contribution of the Kerr nonlinearity, ending up with

$$\left[\mathrm{i}\frac{\partial}{\partial Z} - \frac{D_\omega}{2}\frac{\partial^2}{\partial T^2} + \frac{2\pi}{\lambda_0}\frac{n_2}{A_{\mathrm{eff}}}\frac{1-\exp\left(-\alpha_f L_f\right)}{\alpha_f L_f}\left|W\left(Z,T\right)\right|^2\right] W\left(Z,T\right) = 0\,. \tag{12}$$

In the next step, the effect of the residual part ot the transmission function of the filter is distributed to include it in (12). Making use of the fact that all fields are attracted to the center frequency of the filter ($\Omega_0 = 0$) and that the spectral width of the pulses is much less than that of the filter, the frequency-dependent part of the transmission function in (4) can be expanded. However, note that only the modulus of τ has an extremum at $\Omega = 0$, and that its phase ϕ_τ does not. An expansion therefore yields:

$$\begin{aligned}
\widehat{W}_{\mathrm{out}}^{\mathrm{filter}}\left(\Omega\right) &= \frac{\widehat{\tau}_{\mathrm{filter}}\left(\Omega\right)}{\widehat{\tau}_{\mathrm{filter}}^{\max}}\,\widehat{W}_{\mathrm{in}}^{\mathrm{filter}}\left(\Omega\right) \\
&= \exp\left\{\ln\left|\widehat{\tau}_{\mathrm{filter}}\left(\Omega\right)\right| - \ln\left|\widehat{\tau}_{\mathrm{filter}}^{\max}\right| + \mathrm{i}\left[\phi_\tau\left(\Omega\right) - \phi_\tau\left(\Omega_0\right)\right]\right\}\,\widehat{W}_{\mathrm{in}}^{\mathrm{filter}}\left(\Omega\right) \\
&\approx \exp\left\{\frac{1}{2\left|\widehat{\tau}_{\mathrm{filter}}^{\max}\right|}\frac{\partial^2}{\partial\Omega^2}\left|\widehat{\tau}\left(\Omega\right)\right|\bigg|_{\Omega=0}\Omega^2\right\} \\
&\quad\times\exp\left\{\mathrm{i}\left[\frac{\partial}{\partial\Omega}\phi_\tau\left(\Omega\right)\bigg|_{\Omega=0}\Omega + \frac{1}{2}\frac{\partial^2}{\partial\Omega^2}\phi_\tau\left(\Omega\right)\bigg|_{\Omega=0}\Omega^2\right]\right\}\widehat{W}_{\mathrm{in}}^{\mathrm{filter}}\left(\Omega\right)\,.
\end{aligned} \tag{13}$$

Expression (13) is equivalent to the solution of the differential equation

$$\begin{aligned}
&\left\{\frac{\partial}{\partial Z} - \frac{1}{2\left|\widehat{\tau}_{\mathrm{filter}}^{\max}\right| L_f}\frac{\partial^2}{\partial\Omega^2}\left|\widehat{\tau}\left(\Omega\right)\right|\bigg|_{\Omega=0}\Omega^2\right. \\
&\left.-\frac{\mathrm{i}}{L_f}\left[\frac{\partial}{\partial\Omega}\phi_\tau\left(\Omega\right)\bigg|_{\Omega=0}\Omega + \frac{1}{2}\frac{\partial^2}{\partial\Omega^2}\phi_\tau\left(\Omega\right)\bigg|_{\Omega=0}\Omega^2\right]\right\}\widehat{W}\left(\Omega,Z\right) = 0
\end{aligned} \tag{14}$$

integrated over the length of the fiber L_f and starting from the initial condition $\widehat{W}\left(\Omega, Z=0\right) = \widehat{W}_{\mathrm{in}}^{\mathrm{filter}}\left(\Omega\right)$. Transforming back to the temporal domain, Ω and Ω^2 become first and second derivatives with respect to T, and the resulting expression can be incorporated into (12) as

$$\begin{aligned}
&\left\{\mathrm{i}\frac{\partial}{\partial Z} + \frac{\mathrm{i}}{L_f}\frac{\partial}{\partial\Omega}\phi_\tau\left(\Omega\right)\bigg|_{\Omega=0}\frac{\partial}{\partial T}\right. \\
&-\frac{1}{2}\left[D_\omega + \frac{1}{L_f}\frac{\partial^2}{\partial\Omega^2}\phi_\tau\left(\Omega\right)\bigg|_{\Omega=0}\right. \\
&\left.-\mathrm{i}\frac{1}{\left|\widehat{\tau}_{\mathrm{filter}}^{\max}\right| L_f}\frac{\partial^2}{\partial\Omega^2}\left|\widehat{\tau}\left(\Omega\right)\right|\bigg|_{\Omega=0}\right]\frac{\partial^2}{\partial T^2} \\
&\left.+\frac{2\pi}{\lambda_0}\frac{n_2}{A_{\mathrm{eff}}}\frac{1-\exp\left(-\alpha_f L_f\right)}{\alpha_f L_f}\left|W\left(Z,T\right)\right|^2\right\}W\left(Z,T\right) = 0\,.
\end{aligned} \tag{15}$$

Two of these new terms affect the field evolution only marginally. The term with the first derivative with respect to T can be incorporated into the group velocity term which was removed by introducing time in the co-propagating frame in (1). Also, the term with the second derivative of the phase of the filter function just forms a minor correction to the group velocity dispersion of the fiber. Moreover, for a Fabry-Perot cavity, mostly used as bandpass filter, this contribution vanishes completely. The only term which is physically relevant is the spectral variation of the modulus of the transmission. This introduces an imaginary group velocity dispersion or a kind of diffusion into the equations, thus adding to dissipation.

Eventually, the residual action of the SOA and of the SA on W can be included in (15) in a similar way. We make use of the exponential form of (2), while removing terms which have already been incorporated into V. Thus, we end up with the final evolution equation for the averaged field W:

$$\begin{aligned}\Bigg\{ \mathrm{i}\frac{\partial}{\partial Z} - \frac{1}{2}\left[D_\omega - \mathrm{i}\frac{1}{|\hat{\tau}^{\max}_{\text{filter}}|\,L_f}\frac{\partial^2}{\partial \Omega^2}\left|\hat{\tau}(\Omega)\right|\Big|_{\Omega=0}\right]\frac{\partial^2}{\partial T^2} \\ + \frac{2\pi}{\lambda_0}\frac{n_2}{A_{\text{eff}}}\frac{1-\exp(-\alpha_f L_f)}{\alpha_f L_f}\left|W(Z,T)\right|^2 \\ - \frac{\mathrm{i}}{L_f}(1-i\,h_{\text{SOA}})\,N_{\text{SOA}}(Z,T) + \frac{\mathrm{i}}{L_f}(1-i\,h_{\text{SA}})\,N_{\text{SA}}(Z,T) \\ - \frac{\mathrm{i}}{L_f}(1-i\,h_{\text{SOA}})\,\delta G\Bigg\} W(Z,T) = \frac{\mathrm{i}}{L_f}\Gamma_{\text{SOA}}(Z,T)\;.\end{aligned} \tag{16}$$

Here, we have assumed that the major part of the noise is generated in the amplifier and that fluctuations induced in the absorber can be neglected. In addition, the deviation of gain and loss from the equilibrium values G^0_{SOA} and G^0_{SA} are introduced, respectively, as

$$N_{\text{SOA}}\,(Z,T) = G_{\text{SOA}}\,(Z,T) - G^0_{\text{SOA}}\;, \qquad N_{\text{SA}}\,(Z,T) = G^0_{\text{SA}} - G_{\text{SA}}\,(Z,T)\;.$$

Moreover, the small signal net gain has been defined by

$$\delta G = G^0_{\text{SOA}} + G^0_{\text{SA}} - \frac{\alpha_f}{2}L_f + \ln\left(|\hat{\tau}^{\max}_{\text{filter}}|\right)\;.$$

Now the set of required equations has to be completed by defining the averaged equations for the time-dependent gain/loss in the SOA and SA, respectively. To this end, (6) is used to replace U by W and V. Using (7), the input power of the SOA is defined as

$$\left|U^{\text{SOA}}_{\text{in}}(T)\right|^2 = \left|W(Z=L_f,T)\right|^2\exp\left(-\alpha_f L_f\right)\;. \tag{17}$$

Finally, if this result is inserted into (3), and the resulting expression is generalized for arbitrary Z, we end up with the averaged evolution equation of the SOA gain:

$$
\begin{aligned}
&\frac{\partial}{\partial T} N_{\mathrm{SOA}}(Z,T) = \\
&-\frac{\exp\left[2G^0_{\mathrm{SOA}} + 2N_{\mathrm{SOA}}(Z,T)\right] - 1}{E^{\mathrm{SOA}}_{\mathrm{sat}}} \exp\left(-\alpha_f L_f\right) |W(Z,T)|^2 \\
&-\frac{|\Gamma_{\mathrm{SOA}}(Z,T)|^2}{E^{\mathrm{SOA}}_{\mathrm{sat}}} - \frac{N_{\mathrm{SOA}}(Z,T)}{T_{\mathrm{SOA}}} .
\end{aligned}
\tag{18}
$$

Using (8), (10) and (7), the field at the start of the SA is given by

$$
\begin{aligned}
\left|U^{\mathrm{SA}}_{\mathrm{in}}(T)\right|^2 &= |W(Z = L_f, T)|^2 \left|V^{\mathrm{SA}}_{\mathrm{in}}\right|^2 \\
&= |W(Z = L_f, T)|^2 \exp\left(-2G^0_{\mathrm{SA}}\right) ,
\end{aligned}
\tag{19}
$$

while the averaged evolution equation of the saturable absorber reads:

$$
\begin{aligned}
&\frac{\partial}{\partial T} N_{\mathrm{SA}}(Z,T) = \\
&\frac{\exp\left[2G^0_{\mathrm{SA}} - 2N_{\mathrm{SA}}(Z,T)\right] - 1}{E^{\mathrm{SA}}_{\mathrm{sat}}} \exp\left(-2G^0_{\mathrm{SA}}\right) |W(Z,T)|^2 \\
&-\frac{1}{T_{\mathrm{SA}}} N_{\mathrm{SA}}(Z,T) ,
\end{aligned}
\tag{20}
$$

As a final step, (17), (18) and (20) are normalized to obtain a scaled system of dimensionless evolution equations. Since the simulations have shown that pulse formation is primarily determined by the interplay of SOA, SA and filter, a convenient normalization is based on the parameters of these elements. The propagation distance Z, the time T and the field envelope W are normalized with respect to the loop length L_f, the recovery time T_{SA} of the SA and the square root of the saturation power of the SA $\sqrt{P^{\mathrm{SA}}_{\mathrm{sat}}} = \sqrt{E^{\mathrm{SA}}_{\mathrm{sat}}/T_{\mathrm{SA}}}$, respectively. Thus, we end up with the basic system of equations describing pulse dynamics in the averaged model:

$$
\begin{aligned}
&\Bigg[\mathrm{i}\frac{\partial}{\partial z} - \left(\frac{D}{2} + \mathrm{i}g\right)\frac{\partial^2}{\partial t^2} - \mathrm{i}\delta G + \chi_{\mathrm{Kerr}}|u|^2 \\
&\qquad - (\mathrm{i} + h_{\mathrm{SOA}}) N_{\mathrm{SOA}} + (\mathrm{i} + h_{\mathrm{SA}}) N_{\mathrm{SA}}\Bigg] u = \gamma \\
&\frac{\partial}{\partial t} N_{\mathrm{SOA}} = -\frac{1}{E_{\mathrm{Q}}}\left[\exp\left(2G^0_{\mathrm{SOA}} + 2N_{\mathrm{SOA}}\right) - 1\right] \exp\left(-\alpha_f L_f\right) |u|^2 \\
&\qquad - \frac{N_{\mathrm{SOA}}}{\tau_{\mathrm{Q}}} - \frac{|\gamma|^2}{E_{\mathrm{Q}}} \\
&\frac{\partial}{\partial t} N_{\mathrm{SA}} = -\left[1 - \exp\left(2G^0_{\mathrm{SA}} - 2N_{\mathrm{SA}}\right)\right] \exp\left(-2G^0_{\mathrm{SA}}\right) |u|^2 - N_{\mathrm{SA}} ,
\end{aligned}
\tag{21}
$$

where we have introduced the scaled dimensionless quantities

$$
z = \frac{Z}{L_f} , \quad t = \frac{T}{T_{\mathrm{SA}}} , \quad u = \frac{W}{\sqrt{P^{\mathrm{SA}}_{\mathrm{sat}}}} \exp\left(-\mathrm{i}h_{\mathrm{SOA}}\delta G \frac{Z}{L_f}\right) ,
$$

$$E_{\mathrm{Q}} = \frac{E_{\mathrm{sat}}^{\mathrm{SOA}}}{E_{\mathrm{sat}}^{\mathrm{SA}}} , \qquad \tau_{\mathrm{Q}} = \frac{T_{\mathrm{SOA}}}{T_{\mathrm{SA}}} , \qquad D = \frac{L_f D_\omega}{T_{\mathrm{SA}}^2} ,$$

$$\chi_{\mathrm{Kerr}} = \frac{2\pi n_2 E_{\mathrm{sat}}^{\mathrm{SA}} \left[1 - \exp\left(-\alpha_f L_f\right)\right]}{\lambda_0 A_{\mathrm{eff}} T_{\mathrm{SA}} \alpha_f L_f} , \tag{22}$$

$$g = -\frac{1}{2\left|\widehat{\tau}_{\mathrm{filter}}^{\mathrm{max}}\right| T_{\mathrm{SA}}^2} \left. \frac{\partial^2}{\partial \Omega^2} \left|\widehat{\tau}\left(\Omega\right)\right| \right|_{\Omega=0} .$$

The coupled system (21) describes all relevant physical effects, i.e. fiber dispersion (D) and instantaneous fiber nonlinearity ($\chi_{\mathrm{Kerr}} |u|^2$). The filter introduces an effective diffusion (g). The resonant nonlinearities of the SA and SOA are saturable (saturation energies 1 and E_{Q}, respectively) and non-instantaneous (response times 1 and τ_{Q}, respectively), and they evoke additional refractive index changes ($-h_{\mathrm{SOA}} N_{\mathrm{SOA}} + h_{\mathrm{SA}} N_{\mathrm{SA}}$). The small signal net gain (δG) is the total gain or loss, which a small or linear signal experiences during the passage through the whole system. It also includes all additional losses accumulated e.g. in the filter, in splices or in polarizers. However, its modulus is always small compared with, e.g., the total losses in the fiber, $\alpha_f L_f$, the equilibrium gain in the amplifier, G_{SOA}^0, and the loss in the saturable absorber, G_{SA}^0. In the experiment, it is easily tuned by varying the loop losses with an acousto-optic modulator. Hence, in all further considerations, we will only vary the small signal net gain, δG, while we keep all other parameters (e.g. G_{SA}^0, G_{SOA}^0 and $\alpha_f L_f$) constant. In what follows, it will serve as the bifurcation parameter in our analysis.

3 Solutions and Their Stability

3.1 Continuous Wave Solution

As is usual in characterizing a nonlinear system, one starts by studying the c.w. solutions of system (21) and their bifurcation behavior. To allow for stationary solutions, we neglect noise, while setting $\gamma = 0$ in (21). The *ansatz* for the optical field is:

$$u = u_{\mathrm{cw}} \exp\left(\mathrm{i}\beta z - \mathrm{i}\omega t\right) , \tag{23}$$

where β is the deviation from the linear propagation constant $\overline{\beta}$ and ω is the frequency shift from the filter resonance $\overline{\Omega}$. Since the system tends to adopt the frequency with minimum loss, we restrict ourselves to $\omega = 0$. The trivial solution of (21) is

$$u_{\mathrm{cw}} = 0 , \qquad N_{\mathrm{SOA}} = 0 , \qquad N_{\mathrm{SA}} = 0 . \tag{24}$$

Linear stability analysis reveals that this trivial solution destabilizes at a critical point ($\delta G = 0$), where a non-trivial c.w. solution bifurcates supercritically. Thus, the trivial solution is stable below ($\delta G < 0$) and unstable beyond ($\delta G > 0$) the critical point (see Fig. 2a). The non-trivial homogeneously stable solution represents an equilibrium between saturated gain and loss, i.e., $N_{\rm SA} - \delta G - N_{\rm SOA} = 0$, but it is modulationally unstable, except for very narrow filters. Due to this super-critical bifurcation and the modulational instability, one may guess that bright solitons with unstable backgrounds could exist for $\delta G > 0$, as is the case for the well-studied soliton solutions of the cubic complex Ginzburg-Landau equation [7].

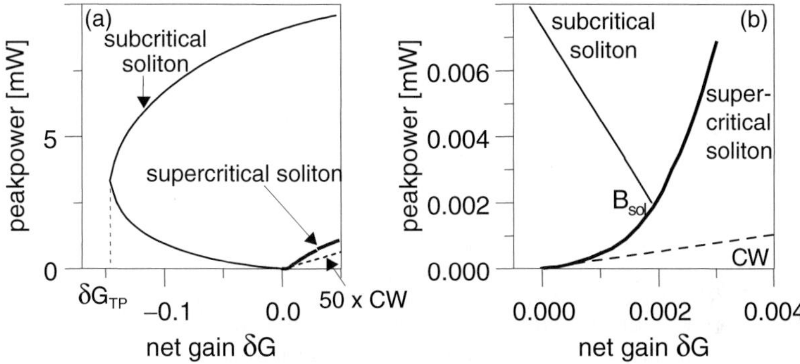

Fig. 2. (**a**) Numerically-determined bifurcation diagram, (**b**) close-up view of the bifurcation points

3.2 Soliton Solutions – Limiting Cases

Before commencing a numerically search for soliton solutions of (21), we analytically study two limiting cases, viz. the long and the short pulse limits, where the pulse length is compared with the SA/SOA relaxation times.

Long Pulses

To investigate the dynamics of long pulses ($T_{\rm sol} \gg \max(1, \tau_{\rm Q})$, i.e., cases where the pulse length exceeds about 100 ps), close to the critical point, ($\delta G = 0$), we use a multiple scale approach up to third-order in ε^3 ($\varepsilon \ll 1$), with the following scaling

$$u = \varepsilon u_1 + \varepsilon^2 u_2 + \cdots \,, \qquad \frac{\partial}{\partial t} = \varepsilon \frac{\partial}{\partial T} + \cdots \,, \qquad \frac{\partial}{\partial z} = \varepsilon^2 \frac{\partial}{\partial Z} + \cdots \,, \tag{25}$$

$$N_{\rm SOA/SA} = \varepsilon^2 N^{(2)}_{\rm SOA/SA} + \cdots \,,$$

where the bifurcation parameter δG varies as ε^2. In doing so, we end up with the Cubic Complex Ginzburg Landau equation (CGLE):

$$\left(\mathrm{i}\frac{\partial}{\partial Z}-\frac{D}{2}\frac{\partial^2}{\partial T^2}+\chi\left|u_1\right|^2\right)u_1=\mathrm{i}\left(\delta G+g\frac{\partial^2}{\partial T^2}+\nu\left|u_1\right|^2\right)u_1\,. \tag{26}$$

Here, the real and imaginary nonlinear coefficients depend on small signal gain/losses, Henry factors and saturation power as follows:

$$\begin{aligned}\chi=\chi_{\mathrm{Kerr}}+h_{\mathrm{SOA}}\left[\exp\left(2G^0_{\mathrm{SOA}}\right)-1\right]\frac{\exp\left(-\alpha_f L_f\right)}{P_Q}\\ -h_{\mathrm{SA}}\left[\exp\left(-2G^0_{\mathrm{SA}}\right)-1\right]\end{aligned} \tag{27}$$

$$\nu=\exp\left(-\alpha_f L_f\right)\left[\exp\left(2G^0_{\mathrm{SOA}}\right)-1\right]\frac{1}{P_Q}-\left[\exp\left(-2G^0_{\mathrm{SA}}\right)-1\right] \tag{28}$$

Note that the sign of the nonlinear gain ($\sim\nu$) is defined by the SOA/SA parameters. In particular, the ratio of the saturation powers,

$$P_Q=\frac{E_{\mathrm{Q}}}{\tau_{\mathrm{Q}}}=\frac{P^{\mathrm{SOA}}_{\mathrm{sat}}}{P^{\mathrm{SA}}_{\mathrm{sat}}}\,,$$

rather than that of the energies, enters the equation. Equation (26) reproduces the bifurcating c.w. solution near the critical point:

$$\left|u_{\mathrm{cw}}\right|^2\approx-\frac{\delta G}{\nu}\,,\quad \beta=\chi\left|u_{\mathrm{cw}}\right|^2\,. \tag{29}$$

On inserting the value of ν, (29) can be explicitly written as

$$\left|u_{\mathrm{cw}}\right|^2\approx\frac{\delta G}{\left[\exp\left(-2G^0_{\mathrm{SA}}\right)-1\right]-\exp\left(-\alpha_f L_f\right)\left[\exp\left(2G^0{}_{\mathrm{SOA}}\right)-1\right]/P_Q}\,,$$

$$\beta=\chi\left|u_{\mathrm{cw}}\right|^2\,.$$

Thus, if

$$P_Q<\frac{\exp\left(-\alpha_f L_f\right)\left[\exp\left(2G^0_{\mathrm{SOA}}\right)-1\right]}{\left[\exp\left(2G^0_{\mathrm{SA}}\right)-1\right]/P_Q-\left[1-\exp\left(2G^0_{\mathrm{SA}}\right)\right]} \tag{30}$$

then the bifurcation is super-critical, because only $\delta G>0$ leads to a positive $\left|u_{\mathrm{cw}}\right|^2$. This holds for dominant gain saturation (P_Q small and inequality (30) satisfied), which is just the case for the loop setup studied here ($P_Q=0.5$, $G^0_{\mathrm{SOA}}=2.42$, $G^0_{\mathrm{SA}}=1.73$); see Fig. 2a. For the reverse situation, i.e., dominant loss saturation, the c.w. bifurcation would be sub-critical. Hence, in the area of net loss ($\delta G<0$), an unstable non-zero c.w. solution co-exists with the stable trivial state. Equation (26) also has analytic soliton solutions. These are unstable, because either the background or the pulse itself destabilizes the soliton, depending on the system's parameters (see [7] and the references therein). The instability can be easily explained. Together with the non-trivial c.w. solution, the soliton bifurcates, with an infinitely large width and infinitesimally small amplitude, from the critical point ($\delta G = 0$). Thus the nonlinear response is quasi-instantaneous and the center of the soliton behaves exactly like the non-trivial c.w. solution. For a sub-critical bifurcation, the zero background is stable but the pulse itself is on the unstable back-bent branch of the non-zero c.w. solution. In the framework of the cubic complex Ginzburg Landau equation (26), this branch cannot turn because this would require an additional higher-order nonlinearity than the cubic one, e.g. a quintic term [7]. In the complete model (21), corresponding terms are induced by the saturation of the SOA or SA. For the super-critical soliton bifurcation, the pulse is stable but the surrounding background lacks stability. This situation, which is relevant to our experiments, is displayed in detail in Fig. 2b.

Short Pulses

By contrast, for short pulses, $[T_{\mathrm{sol}} \ll \min(1, \tau_Q)]$, i.e., pulse lengths less than a few ps, the nonlinearity acts as an integrating nonlinearity, recovery effects can be disregarded, and saturation energies play the major role for the system dynamics. Since the saturation energy of the SA is much smaller than that of the SOA, the situation is reversed, i.e., the SA is now saturated more strongly than the SOA. For our analytical study, we use a model introduced in [29]. Furthermore, we restrict ourselves to the primary physical effects that are responsible for soliton formation in our system, i.e., we neglect fiber dispersion, because the carrier wavelength is close to the zero-dispersion point, and we ignore all nonlinear refractive index changes. For weak gain and absorption saturation, the following real-valued dynamical equation is obtained:

$$\left\{ \frac{\partial}{\partial z} - g\frac{\partial^2}{\partial t^2} - \delta G + K(E_Q)\int_{-\infty}^{t} u^2(t')\mathrm{d}t' + L(E_Q)\left[\int_{-\infty}^{t} u^2(t')\mathrm{d}t'\right]^2 \right\} u = 0\,. \tag{31}$$

Here the coefficients $K(E_Q)$ and $L(E_Q)$ depend essentially on the saturation energies and losses. Equation (31) can be solved analytically by using the *ansatz*

$$u = \frac{A}{\cosh[B(t - vz)]}\,.$$

The amplitude A is a solution of a quadratic equation, provided that absorption saturation is dominant (i.e. $E_Q > 1$, $E_Q = 5$ in the loop setup), and this indicates that the bifurcation is sub-critical. Thus, the important conclusion to be drawn is that solitons in the loop setup with saturable SA/SOAs exhibit a transition in bifurcation behavior which occurs when the pulse length is changed. This feature is triggered by the differing relaxation times of the SA and SOA, which, in turn, cause the differences in the ratios of saturation energies and powers. Thus, the aim is to seek system (1) exact solutions that bifurcate sub-critically.

3.3 Numerical Soliton Solutions

The two limiting cases [(26) and (31)] are linked by the soliton solution of the complete system (21). Numerical studies reveal that solitons bifurcate super-critically from the critical point $\delta G > 0$ and that they are identical to cubic Ginzburg-Landau solitons of (26) (see Figs. 2 a,b). Due to the long pulse duration $[T_{\text{sol}} \gg \max(1, \tau_{\text{Q}})]$, the ratio of the saturation powers $P_Q < 1$ is the crucial parameter.

Increasing control parameter δG decreases the soliton width, and the non-instantaneous character of the nonlinearity comes into play. However, the soliton is still situated on the super-critical branch, because gain saturation prevails, and although pulses become shorter, they remain longer than the SA recovery time ($T_{\text{sol}} > 1$) and their power keeps growing monotonically with increasing net gain.

However, there is a second critical point (B_{sol} in Fig. 2a) on this super-critical soliton branch. This occurs where absorption and gain saturation balance and a new back-bent soliton branch emerges. These solitons are short and can thus bleach the SA. They may even exist in parameter domains without non-trivial c.w. solutions ($\delta G < 0$). In this back-bent part, these solitons are unstable, but at $\delta G = \delta G_{\text{TP}}$, this new branch exhibits a turning point. Beyond this turning point, and provided that $\delta G < 0$, the soliton is stable on a stable zero background.

4 Soliton Experiments

We have experimentally verified our theoretical predictions (see Fig. 3). Figure 3a shows the experimentally-determined branches of the sub-critical soliton. The measured auto-correlation function shows that the soliton shape is practically invariant (see Fig. 3b). To approximately determine the unstable soliton branch, we adopted a method similar to that used for c.w. signals in [30]. We made use of the fact that, when not too close to the bifurcation point, stable and unstable solitons differ mainly in their energies. An acousto-optic modulator was inserted into the loop to decrease the net gain abruptly for one loop round-trip. For a certain net gain after the resultant step-like decrease of the soliton energy, the soliton either recovered or decayed; the unstable soliton branch roughly marks the transition between these two possibilities (see Fig. 3c). This branch was determined as an average taken from a full train of solitons that filled the loop. Since that state is unstable, approximately one-half of the solitons decayed and one-half recovered to form stable solitons. Thus, if the unstable soliton was matched by the energy reduction, the final averaged power reached ~50% of the initial value.

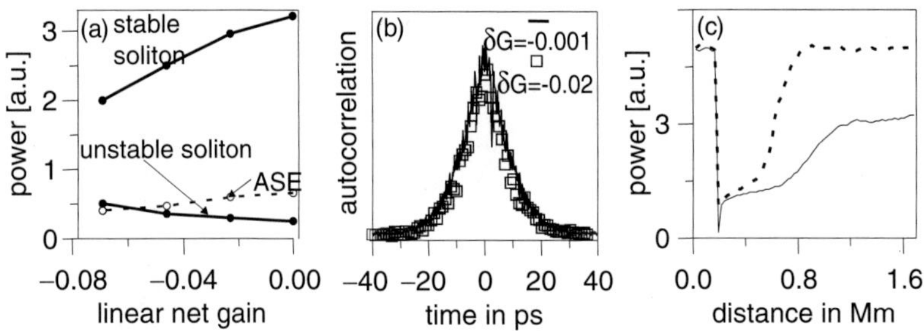

Fig. 3. Experimental observation of solitons: (**a**) bifurcation diagram, (**b**) autocorrelation function for two values of the net gain, (**c**) relaxation of perturbed pulse trains

Due to the fact that the SOA constantly inserts noise into the loop, a trivial c.w. solution can be regarded only as an approximate solution. In fact, below the non-trivial c.w. threshold ($\delta G = 0$), there is some amplified spontaneous emission (ASE) because of the joint action of noise injection and recovery of this perturbation to the trivial state. The corresponding equilibrium value depends on the distance to the bifurcation point $\delta G = 0$. Therefore, the averaged power of ASE increases for increasing net gain (see Fig. 3a), and a weak stable, but noisy, background, surrounds every soliton.

With the present experimental set-up, CGLE-like solitons could not be identified experimentally. Beyond the bifurcation of the sub-critical soliton $\delta G > B_{\text{sol}}$, the super-critical soliton itself is unstable. Only with a smaller net gain, $0 < \delta G < B_{\text{sol}}$, did we succeed in exciting a Ginzburg-Landau-like soliton. However, it disappears owing to growing ASE on the unstable zero background after just a few round trips.

5 Soliton Dynamics

5.1 Critical Slowing Down

We experimentally investigated the dynamic behavior of the stable soliton excited at δG_1 ($\delta G_{\text{TP}} < \delta G_1 < 0$) by reducing the bifurcation parameter to δG_2. In contrast to previous investigations, we keep this value unchanged for a long propagation distance (see Fig. 4). Provided that δG_2 is still inside the soliton hysteresis loop ($\delta G_{\text{TP}} < \delta G_2 < 0$), rapid relaxation towards another stable soliton can be observed (see Fig. 4a). Finally, the acousto-optic modulator is switched off and the soliton recovers to its original state. If $\delta G_2 < \delta G_{\text{TP}}$, the soliton decays to the stable trivial background (see Fig. 4b). One can also achieve this change in the bifurcation parameter δG optically by launching counter-propagating control pulses, either into the SOA (net gain reduction) or into the SA (net gain increase). In the co-propagating con-

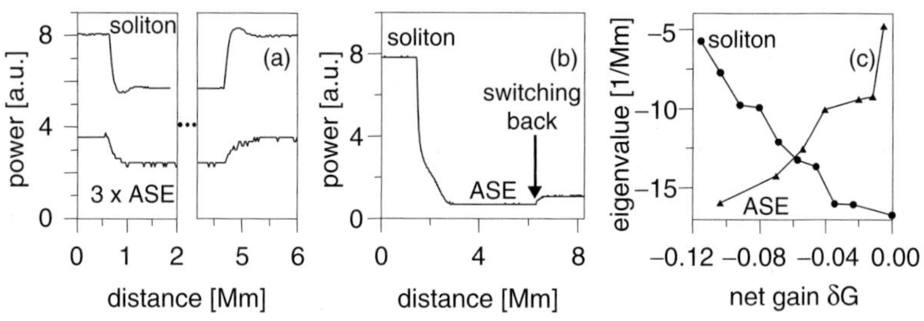

Fig. 4. Experimental results showing (**a**) soliton relaxation, (**b**) soliton decay and (**c**) dependence of relaxation parameters on net gain

figuration, either orthogonally-polarized or wavelength-shifted pulses can be used. Hence, fast all-optical logical operations can be performed.

To further explore the soliton dynamic response, we study relaxation processes. In particular, we experimentally demonstrate the critical slowing down of solitons. This had previously been investigated only for c.w. signals (see [[31] and references therein]). A weak perturbation was added to disturb the stable soliton, as had been done for unstable branch measurements. Linear stability analysis predicts that such a perturbation decays exponentially, $\sim \exp(\lambda_{\max} z)$, where $\lambda_{\max}$ is the maximum eigenvalue of (21) linearized about the soliton solution. This relaxation law is valid if other eigenvalues are far away from $\lambda_{\max}$, as in our experimental situation. For the stable soliton, the corresponding eigenvalue is negative, but at the turning point $\delta G = \delta G_{\mathrm{TP}}$, it approaches zero. Therefore, the relaxation process becomes extremely slow. We determined $\lambda_{\max}$ experimentally for various values of the net gain by fitting an exponential function ($\sim \exp(\lambda_{\max} z)$) to the evolution of the perturbed soliton. Fig. 4c (points) shows the measured critical slowing down of solitons, i.e., $|\lambda_{\max}|$ decreases with the net gain (filled circles). The critical point for the trivial c.w. solution (background) is $\delta G = 0$. Therefore, the relaxation processes of ASE become increasingly slower as the control parameter comes closer to $\delta G = 0$ (see Fig. 4c, triangles).

5.2 Hopf Instabilities of Solitons

We have shown in Sect. 3 that turning and bifurcation points on the c.w. hysteresis curve represent zero eigenvalues of the corresponding linear operator which defines the dynamics around the dissipative soliton. However, the stability behavior changes even if only the real part of the eigenvalue changes sign. These complex-valued, or Hopf, instabilities are related to two complex conjugate eigenvalues. Initially, the corresponding perturbations grow in an oscillatory manner, and, eventually they either reach a stable orbit or destroy the soliton. In fact, for a certain parameter set, our simulations predict a Hopf bifurcation on the upper branch of the hysteresis curve close to the turning point. Approaching the Hopf bifurcation point from the stable side,

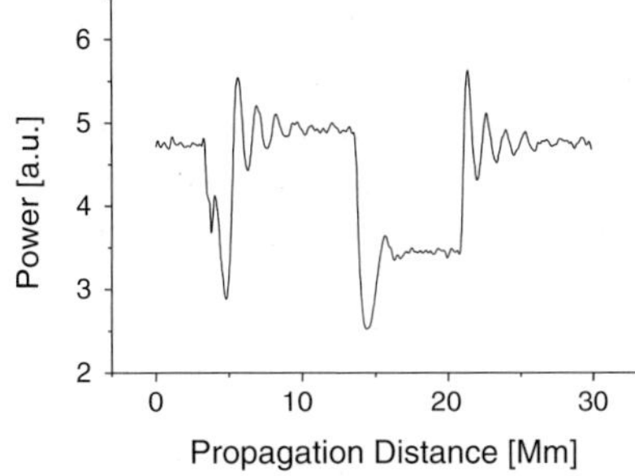

Fig. 5. Relaxation oscillations of a dissipative soliton close to a Hopf-instability (averaged power of the signal during its propagation through the system – the net gain δG was varied periodically to perturb the pulse)

i.e., increasing $|\delta G|$, we find that perturbations die out, but in an oscillatory manner. This behavior can be experimentally verified. In the corresponding control parameter range, a perturbed soliton still recovers, while exhibiting damped oscillations around its stationary state (see Fig. 5). On crossing the bifurcation point, the oscillating instability grows quickly and the soliton decays. Numerical simulations predict a small range of stable solutions, which nonetheless oscillate. Until now, this fact could not be confirmed experimentally. The oscillating solutions are probably too fragile to persist in the presence of noise.

Numerical simulations can elucidate the physical mechanism behind the oscillations. To this end, we set both Henry factors to zero, neglecting saturation-induced refractive effects. This results in the suppression of any oscillatory instability. Obviously, oscillations are caused by the interplay of the carrier-induced frequency shift and dispersive losses introduced by the filter. Strong pulses become red-shifted towards the edge of the filter. Subsequently, the pulse looses energy because of growing dispersive losses. As a consequence, the carrier-induced frequency shift is reduced and the pulse can return to the filter center wavelength. Subsequent excess amplification starts the cycle over again.

5.3 Jitter Reduction

In the previous sections, we have addressed some peculiarities of dissipative solitons. Their robustness, with respect to varying initial conditions and to accumulated noise on propagation, makes them ideal optical bits in communications. Furthermore, dissipative solitons have another advantage when compared with their conservative counterparts. Conservative (Schrödinger) solitons suffer from the so-called "Gordon-Haus jitter" [32]. This is a consequence of amplifier-induced noise. In particular, for conservative solitons, frequency noise translates into velocity noise, i.e., jitter. This jitter grows with the cube of the distance and limits the maximum transmission capacity. However, dissipative solitons are completely defined by system parameters, so their frequencies and hence velocities, are fixed. If noise is added to the soliton, it quickly returns to its original state, and thus also to its original center frequency. Hence, any noise induces only a transient temporal shift, but no permanent change of velocity. The actual variation of the pulse position $\Delta T(z)$ is the sum of all noise-induced shifts $\delta\tau(z')$ experienced at previous positions z':

$$\Delta T(z) = \int_0^z dz' \, \delta\tau(z') \, . \tag{32}$$

Since noise-induced shifts are uni-directional ($\langle \delta\tau(z') \rangle = 0$), the mean value of the pulse position remains at its original value ($\langle \Delta T(z) \rangle = 0$). Noise generated on a stable zero background is quickly damped (usually within a few round trips). Consequently, noise-induced shifts from different positions in the fiber loop are not correlated and do not depend on the actual propagation distance, i.e.,

$$\langle \delta\tau(z)\delta\tau(z') \rangle = \langle \delta\tau^2 \rangle \, \delta(z - z') \, . \tag{33}$$

Using (33) and (32), we can calculate the r.m.s. evolution of the pulse position:

$$\langle \Delta T(z)^2 \rangle = \int_0^z dz' \int_0^z dz'' \, \langle \delta\tau(z') \, \delta\tau(z'') \rangle = z \langle \delta\tau^2 \rangle \, . \tag{34}$$

Thus, the timing jitter increases linearily with propagation distance, i.e., the pulse is subject to a type of diffusion.

In the experiment, we find that the timing jitter consists of two rather distinct parts. One originates from the soliton formation and the other one accumulates during propagation (see Fig. 6). Despite the relatively high noise level in our system, the growth of the timing jitter is almost negligible, but it does follow the above predictions. It increases with the propagation distance as:

$$\frac{\partial}{\partial z} \langle \Delta T(z)^2 \rangle \approx 2.25 * 10^{-4} \, \frac{\mathrm{ps}^2}{\mathrm{km}} \, . \tag{35}$$

We found it to be almost independent of the pulse and device parameters. In particular, an increase in the bit-rate had no significant effect on its growth during propagation (compare the two traces in Fig. 6). In contrast, the part of the jitter which is induced during the formation of the dissipative soliton is not only larger, but also depends critically on the bit-rate (see Fig. 6) and wavelength (see Fig. 7). Since the initial pulses are rather distinct from the ultimate dissipative solitons, the pulse in its early stage lacks the particular properties of a dissipative soliton. Before the delicate balance of gain and loss and, in particular, the self-frequency stabilization, have taken effect, the pulse is subject to noise and interaction-induced spectral shifts just as occurs

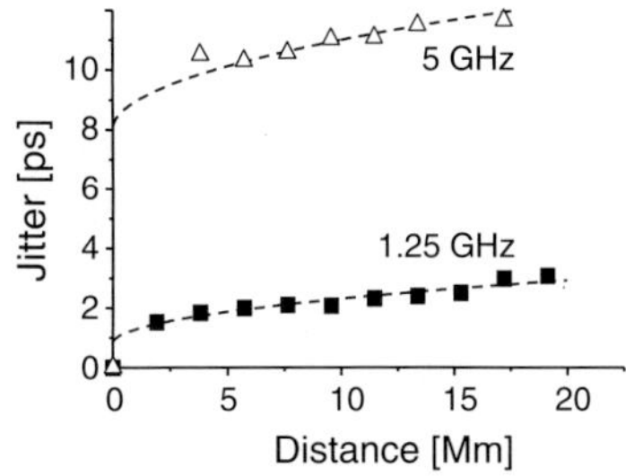

Fig. 6. Experimental results showing the evolution of the mean displacement of the signal pulse for two representative adjustments. The *dashed* lines are fitted, assuming a square-root dependence after the formation of the auto-soliton (*ca.* 1500 km)

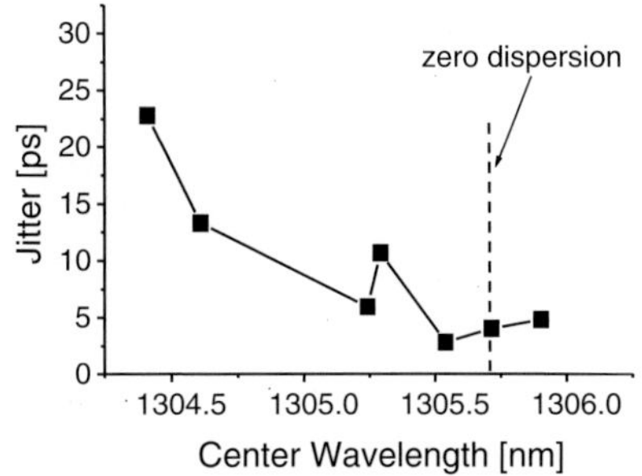

Fig. 7. Dependence of the timing jitter $\sqrt{\langle \Delta T(z)^2 \rangle}$ at 10 Mm on the operation wavelength (filter centre)

for a Schrödinger soliton. In fact, the jitter accumulated during pulse formation depends on the actual group velocity dispersion (see Fig. 7). Hence, the higher the frequency-induced velocity shifts, the higher is the resulting initial jitter. One strategy to reduce this jitter consists in choosing more appropriate initial conditions. However, it should be stressed that propagation-induced timing jitter of dissipative solitons is almost negligible for realistic propagation distances. In contrast to Schrödinger solitons, timing jitter is not a limiting factor in regard to transmission capacity. Similar results have been obtained many years ago [33] for solitons with bandwidth-limited amplification in the framework of a perturbation theory for Schrödinger solitons, where the perturbation existed only in the linear terms on the r.h.s. of (26). From the point of view of dissipative solitons, these solutions can be characterized as dissipative solitons with a weakly-unstable background.

6 Conclusions

In conclusion, we have theoretically predicted and experimentally proved that stable dissipative solitons exist near a super-critical c.w. bifurcation. These solitons bifurcate sub-critically from Ginzburg–Landau-like solitons because of the non-instantaneous nonlinear response. Excitation, switching, relaxation and, in particular, critical slowing down of solitons were investigated. These solitons are ideal candidates for all-optical signal transmission and processing because their properties are entirely defined by the system parameters and do not depend on the mode of excitation. Moreover any noise (dispersive waves and amplified spontaneous emission) is effectively suppressed and propagation-induced timing jitter is reduced.

Acknowledgements

We are grateful to the Deutsche Forschungsgemeinschaft for financial support and the Heinrich-Hertz-Institut, Berlin, for providing us with the saturable absorber.

References

1. M. C. Cross and P. C. Hohenberg, Rev. Mod. Phys., **65**, 851 (1993).
2. S. Trillo and W. Torruellas (eds.), *Spatial Solitons*, Springer Series on Optical Sciences, vol. 82, (Springer, Berlin, Heidelberg, New York, 2001).
3. F. Lederer (ed.), *Feature Issue on Spatial Solitons*, IEEE J. Quantum Electron., **39**, 1 (2003).
4. G. I. Stegeman and M. Segev, Science., **286**, 1518 (1999).
5. A. Hasegawa and Y. Kodama, *Solitons in Optical Communications*, (Clarendon Press, Oxford 1995).
6. Y. A. Kivshar and G. P. Agrawal, *Optical Solitons. From Fibers to Photonic Crystals,* (Academic Press, Amsterdam, 2003).
7. N. Akhmediev and A. Ankiewicz, *Solitons: nonlinear pulses and beams*, (Chapman & Hall, London, 1997).
8. L. A. Lugiato (ed.), *Feature Issue on Cavity Solitons*, IEEE J. Quantum Electron., **39**, 193 (2003).
9. Z. Bakonyi, D. Michaelis, U. Peschel, G. Onishchukov, and F. Lederer, J. Opt. Soc. Am. B **19**, 487 (2002).
10. E. Ultanir, G. I. Stegeman, D. Michaelis, C. H. Lange, and F. Lederer, Phys. Rev. Lett. **90**, 253903 (2003).
11. N. J. Smith and N. J. Doran, Electron. Lett., **30**, 1084 (2000).
12. G. Agrawal and N. Olson, IEEE J. Quantum Electron., **QE-25**, 2297 (1989).
13. S. Wabnitz, Opt. Lett., **20**, 1979 (1995).
14. S. K. Turitsyn, Phys. Rev. E, **20**, R3125 (1996).
15. A. Mecozzi, Opt. Lett., **20**, 1616 (1995).
16. V. S. Grigoryan, Opt. Lett., **21**, 1882 (1996).

17. S. Fauve and O. Thual, Phys. Rev. Lett., **64**, 282 (1990).
18. W. van Saarloos and P. C. Hohenberg, Phys. Rev. Lett., **64**, 749 (1990).
19. S. Longhi, Phys. Scr., **56**, 611 (1997).
20. K. Staliunas, V. Sanchez-Morcillo, Opt. Commun., **139**, 306 (1997).
21. M. Tlidi, P. Mandel, R. Lefever, Phys. Rev. Lett., **73**, 640 (1994).
22. W. J. Firth and A. J. Scroggie, Phys. Rev. Lett., **76**, 1623 (1996).
23. S. Trillo, M. Haelterman, and A. Sheppard, Opt. Lett. **22**, 1514 (1997).
24. K. Staliunas and V. J. Sanchez-Morcillo, Phys. Rev. A, **57**, 1454 (1998).
25. U. Peschel, D. Michaelis, C. Etrich, and F. Lederer, Phys. Rev. E, **58**, R27435 (1998).
26. R. Gallego, M. San Miguel, and R. Toral, Phys. Rev. E, **61**, 2241 (2000).
27. Z. Bakonyi, G. Onishchukov, C. Knöll, M. Gölles, and F. Lederer, Electron. Lett., **36**, 1790 (2000).
28. A. Hasegawa and Y. Kodama, Opt. Lett., **15**, 1443 (1990).
29. N. Akhmediev, M. J. Lederer, and B. Luther-Davis, Phys. Rev. E, **57**, 3664 (1998).
30. F. Mitschke, C. Boden, W. Lange, and P. Mandel, Opt. Comm., **71**, 385 (1989).
31. J. Danckaert, G. Vitrant, R. Reinisch, and M. Georgiou, Phys. Rev. A, **48**, 2324 (1993).
32. J. P. Gordon and H. A. Haus, Opt. Lett., **11**, 665 (1986).
33. Y. Kodama and A. Hasegawa, Opt. Lett., **17**, 31 (1992).

Soliton Dynamics in Mode-Locked Lasers

S.T. Cundiff

JILA, National Institute of Standards and Technology and University of Colorado, Boulder, CO 80309-0440 USA
cundiffs@jila.colorado.edu

Mode-locked lasers generate ultra-short optical pulses, with durations ranging from hundreds of picoseconds (ps) down to a few femtoseconds (fs). The pulse circulating in the cavity of a mode-locked laser can be thought of as a dissipative soliton, where the dissipation is due to the inevitable presence of loss, which must be compensated by gain. In addition to gain and loss, the pulse experiences nonlinearity and dispersion, the key ingredients for any soliton system. These effects occur in physical elements that do not completely fill the cavity, thus there is a similarity to the dispersion management that is used in telecommunications systems.

There are a number of features that make a mode-locked laser an excellent system for experimental study of soliton dynamics. Since a pulse is emitted once per round trip, it is possible to watch the evolution of the pulse. Furthermore, an essentially infinite propagation distance can be observed. There are no variations in the parameters from round-trip to round-trip, which makes simulation easier, although it does restrict the applicability of the results to real telecommunications systems, which can display statistical variations. It is possible to examine parameter regimes that are not easily accessible otherwise, for example peak power or characteristic nonlinear or dispersion lengths. And finally, it is relatively easy to make a change of parameters for the entire propagation distance at once.

In this chapter, the basics of mode-locked lasers will be presented first. Then the experimental observation of polarization evolution in a mode-locked fiber laser, which demonstrated both the presence of polarization-locked vector solitons and instability of the fast axis, will be discussed. This will be followed by a section on soliton instabilities, dubbed "exploding solitons", in a mode-locked Ti:sapphire laser. The next section summarizes the recent advances on phase control of mode-locked lasers, which brings up interesting new issues regarding how nonlinearity impacts both the group and phase velocities inside the laser cavity.

1 Mode-Locked Laser Basics

A laser is essentially an optical oscillator and, as such, requires the two basic constituents of any oscillator, namely amplification and feedback. The

S.T. Cundiff: *Soliton Dynamics in Mode-Locked Lasers*, Lect. Notes Phys. **661**, 183–206 (2005)
www.springerlink.com

amplification is provided by stimulated emission in a gain medium. Feedback is provided by the laser "cavity", which is a set of mirrors that cause the light to reflect back on itself. One of the mirrors, called the output coupler, transmits a small fraction of the incident light to provide output [1].

The word "mode-locked" refers to the frequency domain description of how ultra-short pulses are generated by a laser. The requirement that the electromagnetic field be unchanged after one round trip in the laser means that lasing only occurs for frequencies such that the cavity length is an integer number of wavelengths. These are referred to as the longitudinal modes and occur at frequencies $\nu_j = jc/2nl$, where j is an integer that indexes the modes, c is the speed of light, n is the average index of refraction in the cavity and l is the cavity length. If multiple modes lase at the same time, then a short pulse can be formed, but only if the modes are locked in phase, i.e., the laser is mode-locked. As illustrated in Fig. 1, this actually produces a pulse train, where the time between pulses equals the cavity round trip time. It should be noted that there are others types of lasers that generate pulses, the most common being called a "q-switched" laser. However, the pulse evolution in those lasers is unrelated to soliton dynamics. Non-pulsed lasers are designated as "continuous wave" (CW).

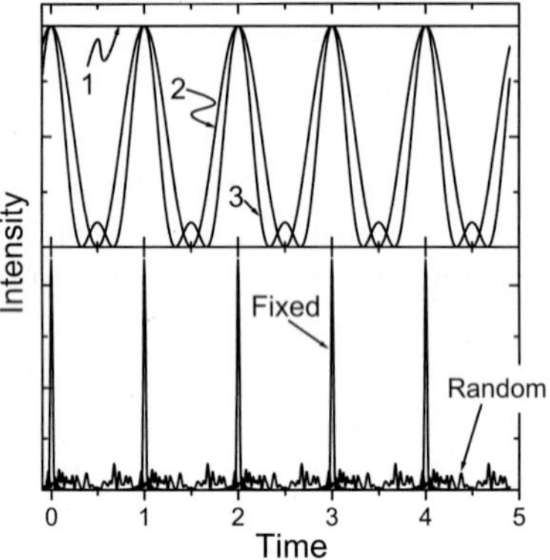

Fig. 1. Formation of a pulse train due to constructive interference among longitudinal modes. The intensities, $I(t) = |E(t)|^2$ for one, two and three longitudinal modes are shown in the *upper* panel. The *lower* panel shows 30 modes for both fixed phases, i.e., mode-locked, and random phases

Despite the fact that the term "mode-locking" comes from the frequency domain description, the dynamics are typically described in the time domain. The key element, which distinguishes mode-locked lasers from other lasers, is the presence of an element in the cavity that causes it to modelock. This can be an active element, essentially a shutter that is opened once per round trip by an external drive (active mode-locking), or a passive element, in which case the pulse itself causes it to "open" [2]. The latter case of passive mode-locking produces the shortest pulses. Passive mode-locking can be described in terms of a saturable absorber, i.e., one where the absorption saturates, so that a higher intensity is less attenuated than a lower intensity. This could be a due to a real absorber, such as dye molecules or a semiconductor, or it could be virtual. A virtual saturable absorber generally relies on the Kerr effect, which means that the index of refraction depends on intensity, $n = n_0 + n_2 I$ where n_0 is the linear index of refraction, n_2 is the Kerr coefficient and I is the intensity.

A schematic diagram showing the elements present in a mode-locked laser is shown in Fig. 2. Note that the different elements are not necessarily distinct, as shown in the figure, for example the gain medium often is also dispersive and nonlinear. Additionally, loss is not specifically shown; the output coupler causes loss and there maybe some loss in the other elements. As implied by the discrete elements, the various effects typically are not spread throughout the cavity, but rather only occur for a sub-section of it. This suggests a connection with dispersion management in telecommunications, which indeed has been applied to mode-locked lasers [3].

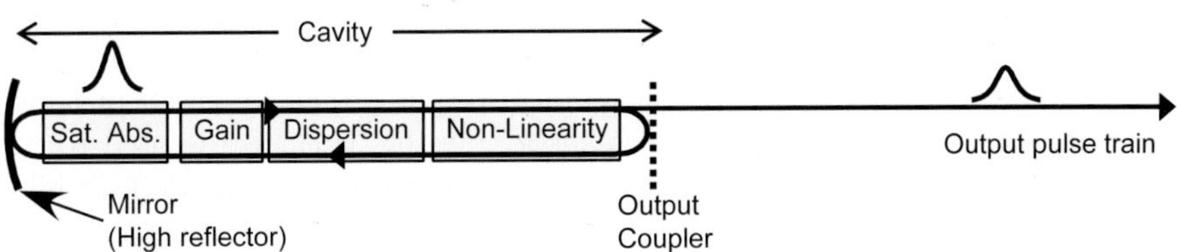

Fig. 2. Block diagram showing the essential elements of a mode-locked laser

Thus all of the physical processes associated with dissipative solitons are present in mode-locked lasers. Saturable absorption, which is also a dissipative process, is present. In the steady state, the gain medium in a laser is strongly saturated, hence gain saturation must be considered as well. In addition, birefringence, which means that the index of refraction is not a scalar but rather a tensor, may be present as well. This can cause the polarization state to evolve with propagation.

2 Soliton Polarization Evolution

Optical fiber is the most common medium for studying temporal optical solitons. This has been strongly motivated by the application to telecommunications. A key technology for optical telecommunications is the erbium-doped fiber amplifier (EDFA) [4]. The erbium dopant atoms provide gain to amplify optical signals that are attenuated due to loss in the transmission fiber. Naturally, this gain medium can also be used to construct a laser. Figure 3 shows

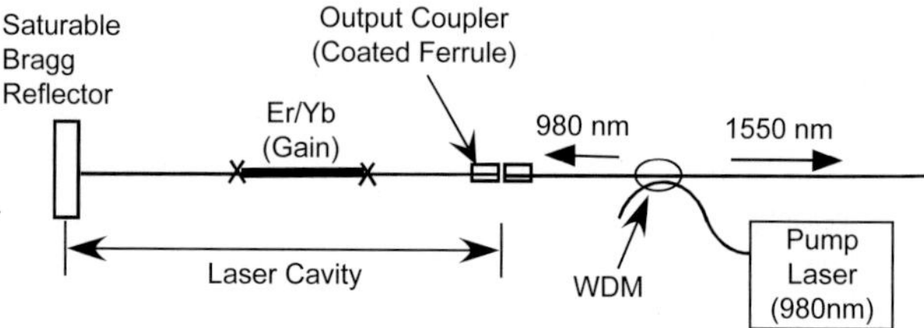

Fig. 3. Diagram of mode-locked fiber laser

a schematic of a mode-locked fiber laser that uses a saturable Bragg reflector (SBR) for mode-locking [5]. This laser uses a section of fiber co-doped with erbium (Er) and ytterbium (Yb) for gain. The gain fiber is pumped by a 980 nm diode laser, which enters the cavity through the output coupler. The wavelength division multiplexer (WDM) separates the pump light from the 1550 nm laser emission. The remainder of the cavity is made up of standard telecommunications fiber (simply called "single mode fiber" or SMF), which has anomalous dispersion at 1550 nm, whereas the gain fiber has normal dispersion. The output coupler, which is a mirror coated directly onto a cleaved fiber, serves as one end of the cavity. The other end is the SBR, which is a monolithic semiconductor structure combining a mirror and a single quantum well, which provides saturable absorption. The quantum well provides "real" saturable absorption and thus has a finite life time, as opposed to the more common fiber lasers that use nonlinear polarization to mode-lock [6].

Hasegawa and Tappert predicted the existence of temporal solitons in optical fiber in 1973 [7]. This pioneering paper ignored the fact that all "single" mode optical fiber actually supports two orthogonal polarization modes; the term "single" mode refers only to the transverse profile. If optical fiber were perfectly isotropic, the polarization modes would be completely degenerate and this treatment would be correct. In reality, manufacturing imperfections, externally applied stress, or bending lifts the degeneracy between the modes. Thus fiber supports two orthogonally-polarized modes with differing propagation constants; i.e., fiber is birefringent [8]. The experimental observation of optical solitons [9] and the subsequent explosion of work [10] proved that this omission was nevertheless justified in many circumstances.

The difference in phase velocities of the two modes causes the polarization state of a pulse to evolve as it propagates. In general, the group velocities are also unequal. The differing group velocities result in temporal pulse splitting, a phenomenon known as polarization mode dispersion and currently of great concern for non-soliton long distance optical communication systems [11]. The differing group velocities might also be expected to prevent the formation of a soliton with energy in both orthogonal polarization modes (i.e., along both principal axes of the birefringence). However solitons are remarkably robust; they do exist under these conditions and propagate as a unit [12]. To cancel the group velocity difference, the orthogonally-polarized components of the soliton shift their center frequency slightly. This phenomenon has been described theoretically using a pair of coupled non-linear Schrödinger equations [13]. Since the phase velocities are still different, the polarization state evolves with propagation.

A "vector soliton" is a multi-dimensional entity that propagates in an invariant or periodic manner in an environment that is destructive in the absence of compensating nonlinearity [14]. The situation described above can be denoted as a "group-velocity-locked vector soliton" as it has some amplitude on both principal axes and is therefore multi-dimensional.

Prior to their observation in a fiber laser [15], states that also display locking of the phase velocities, in addition to the group velocities, had only been characterized theoretically [16, 17, 18, 19, 20, 21]. These states are elliptically-polarized and they exist due to a dynamic balance between linear and non-linear birefringence. The nonlinear birefringence arises from a combination of self and cross phase-modulation. Figure 4 shows a schematic comparison between just linear birefringence [Fig. 4a] and the balance that can occur when nonlinear processes are included [Fig. 4b]. Coherent energy coupling (also called four-wave-mixing) provides a stabilizing mechanism that maintains the exact power distribution for the balance to occur. These states are designated "polarization-locked vector solitons" (PLVS) as they have power along both principal axes, and thus are multi-dimensional.

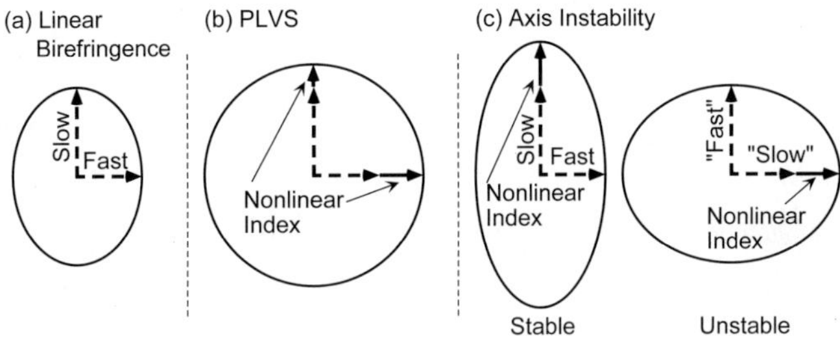

Fig. 4. Schematic of the interplay between linear and nonlinear birefringence. (**a**) Linear birefringence alone; (**b**) balance between linear and nonlinear birefringence that occurs for a PLVS; (**c**) conditions for the fast axis instability. Note that, in (**c**), the axes are labeled by their net birefringence due to both linear and nonlinear contributions. The ellipses are an aid for visualizing the relative magnitude of the indices of refraction

In addition to the PLVS, states with fixed linear polarization are also observed in the fiber laser. These occur at larger values of birefringence than the PLVS and occur due to axis instabilities [22, 23, 24]. Axis instabilities arise when the nonlinear birefringence can cause the phase velocity of the fast axis to become comparable to, or slower than, that of the slow axis [see Fig. 4c]. In this case, the fast axis becomes unstable.

The first observation showed that the polarization state in a fiber laser could spontaneously lock, although the mechanisms were unclear [25]. This triggered theoretical work to show that the PLVS, predicted for strictly conservative systems [16, 17, 18, 19, 20, 21], could exist in the non-conservative, i.e., dissipative, environment of a fiber laser [26]. Based on this theoretical work, it was possible to positively identify the polarization-locked states that were occurring [15, 27, 28].

2.1 Theoretical Background

The full vector model describing the propagation of the two components polarized along the orthogonal principal axes of a lossless fiber are given by the conservative coupled nonlinear Schrödinger equations (CNLSE) [13, 29, 30]:

$$\mathrm{i}\frac{\partial u}{\partial z} + \mathrm{i}\delta\frac{\partial u}{\partial t} + \gamma u + \frac{1}{2}\frac{\partial^2 u}{\partial t^2} + (|u|^2 + A|v|^2)u + Bv^2u^* = 0\,, \tag{1}$$

$$\mathrm{i}\frac{\partial v}{\partial z} - \mathrm{i}\delta\frac{\partial v}{\partial t} - \gamma v + \frac{1}{2}\frac{\partial^2 v}{\partial t^2} + (|v|^2 + A|u|^2)v + Bu^2v^* = 0\,, \tag{2}$$

where u and v are the component envelopes along the slow and fast axes respectively, z and t are the normalized time and distance, 2δ and 2γ are the normalized group and phase velocity differences, and A and B are the cross phase modulation (XPM) and coherent energy exchange coefficients, respectively.

These equations admit two linearly-polarized fundamental soliton solutions:

$$\begin{aligned} u(t,z) &= u_0 \mathrm{sech}(t-\delta z)\exp(\mathrm{i}\gamma z) \quad & v(t,z) &= 0 \\ u(t,z) &= 0 & v(t,z) &= v_0 \mathrm{sech}(t+\delta z)\exp(\mathrm{i}\gamma z)\,. \end{aligned} \tag{3}$$

For an isotropic and conservative medium, $B = 1 - A$ in 1 and 2. When $A = 1$ [the self phase modulation (SPM) and XPM coefficients are equal, and the coherent energy exchange vanishes] and $\delta = 0$, the CNLSE are integrable, with stationary phase-locked solutions [31]. The experimental observation of these simple hyperbolic secant shaped solutions, known as "Manakov" solitons, is difficult, since $A \neq 1$ for most low-loss materials. Through engineering of the SPM and XPM coefficients in an anisotropic waveguide ($A \sim 0.95$), spatial "Manakov" solitons have been observed [32]. Spatial solitons are governed by CNLSE analogous to 1 and 2. In isotropic media such as standard single mode fiber, $A = 2/3$ and rigorous temporal "Manakov" solitons cannot occur.

If the linear birefringence is significantly larger than the nonlinear birefringence, then the relative optical phase between components varies so rapidly that all phase-dependent phenomena effectively average to zero on propagation [13]. Thus, the coherent energy coupling and phase velocity difference terms in 1 and 2 can be ignored, resulting in the fact that coupling between the two components will only be due to incoherent cross-phase modulation. In this case, the solutions approximate "Manakov" solitons. This condition is designated "high birefringence." It occurs in SMF if the peak power is less than 1 W, or in polarization-maintaining fiber, which has very large birefringence intentionally manufactured into it. For high birefringence, the solutions to 1 & 2 correspond to two orthogonally-polarized pulses along the birefringent axes that mutually trap each other. They propagate as a non-dispersing group velocity-locked vector soliton [13, 14]. The cross-phase modulation causes the central optical frequency of one component to increase and the other to decrease. In conjunction with a frequency-dependent group velocity, these shifts equalize their group velocities [13]. In the absence of nonlinearity (i.e., cross-phase modulation), the components retain their central frequencies and travel at unequal group velocities; this causes them to split temporally. Due to the large phase velocity difference between the components, the polarization state of this vector soliton evolves rapidly with propagation. At any given point, the same polarization state applies to the entire pulse because the two components have the same amplitude and phase profiles [33]. Standard soliton communications systems operate in this high birefringence regime and utilize group velocity-locked vector solitons. Hence, these soliton systems are relatively immune to the detrimental pulse splitting effects of random and unequal group velocities (polarization mode dispersion) that often impair the performance of non-soliton communications systems [11, 34].

The designation "low birefringence" means that the linear and nonlinear birefringence are comparable (the latter depends on polarization state). In this case, the difference between group velocities can typically be ignored. The two polarization components are now coherently coupled, and the relative optical phase, phase velocity differential, and coherent energy exchange between the orthogonal polarization components must be retained. Theoretical analysis has found three lowest-order stationary solutions for the low birefringence case: two fundamental soliton solutions that are linearly-polarized along either the fast or slow axis (3) , and a (numerical) elliptically- polarized solution [18, 19, 20]. The elliptically- polarized solution is a PLVS because it contains energy in both components and propagates without change in its polarization state. The components have a relative phase of $\pm\pi/2$, but do not necessarily have amplitude profiles of the same functional shape. Hence, the polarization state is not uniform across the pulse [18, 19, 20]. Other higher-order stationary solutions have also been found [14, 16, 19]. The elliptically-polarized solitons possess a weak oscillatory instability, in contrast to solitons that are linearly-polarized along the fast axis, which are unstable [23].

For a PLVS to survive propagation with a constant polarization state, the phase velocities of the two components must be identical. As shown in Fig. 4b, a nonlinear index difference is created by an unequal distribution of energy between the two axes. The resulting difference (or nonlinear birefringence) in the nonlinear index compensates the linear birefringence exactly. This means that the phase velocities of the two axes are identical.

The magnitude of the nonlinear birefringence depends on the difference between the intensities of the components, i.e., the ellipticity of the polarization state, since the relative phases must be $\pm\pi/2$. Thus, the ellipticity of the polarization state depends directly on the linear birefringence. Within the approximation of equal component profiles, this can be expressed as

$$|V|^2 - |U|^2 = \frac{\gamma}{g}, \tag{4}$$

where U and V are the time integrated amplitudes of u and v, $g \approx q(1-A)$, and q is a soliton parameter that is inversely proportional to the soliton period and proportional to the square of the pulse energy [17, 18]. Equation 4 also shows that $|V|^2 \geq |U|^2$ with the component along the fast axis possessing greater intensity. The magnitude of the nonlinear birefringence of a pulse is limited by its energy and width (implicitly here through normalization) such that a PLVS cannot exist for linear birefringence $\gamma > (|V|^2 + |U|^2)g$. Equivalently, this limit occurs as $|U|^2 \to 0$ and the polarization state approaches linear polarization along the fast axis.

For large values of linear birefringence, a soliton that is linearly-polarized along the fast axis (second line in 3) becomes unstable. Instability of the fast axis due to nonlinearity was first described for CW propagation [22]. It also occurs for solitons [23], and has been carefully addressed using soliton perturbation theory [18]. Because the fast axis is unstable, a pulse that is initially polarized along it will evolve away from it towards the slow axis, which is a stable point, and typically will undergo oscillations around it. Because the fiber laser is non-conservative, the oscillations are damped, resulting in solitons that are linearly-polarized along the slow axis. The role played by nonlinearity provides an important distinction between the fast axis instability and the PLVS. In absence of nonlinearity, the PLVS states are not stable and do not exist, i.e., nonlinearity creates a new state. In contrast, nonlinearity destroys the stability of the fast axis, with the result that there is only a single stable state (the slow axis).

2.2 Experiment

Measurements have been performed on several implementations of the fiber laser with essentially identical results; the details vary from laser to laser (mainly the fiber lengths). Typical output optical spectra are shown in Fig. 5. The clean sech^2 spectrum shows that soliton-like pulse shaping dominates the

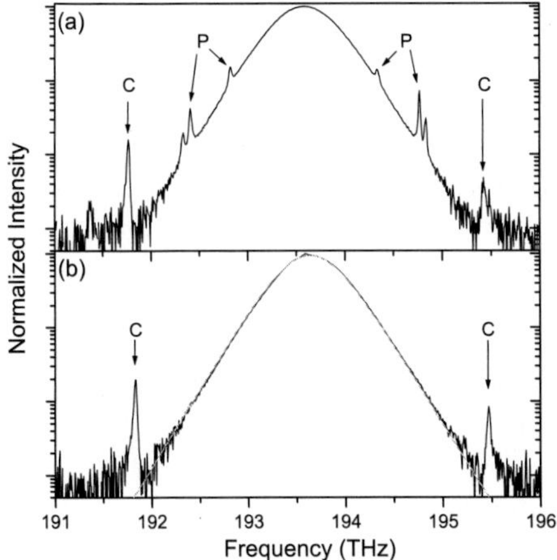

Fig. 5. Optical spectrum of the fiber laser output for (**a**) non-polarization-locked and (**b**) polarization-locked operation. The side-bands marked "C" are due to the periodic perturbation by the cavity, while those marked "P" are due to periodic perturbation of the polarization. In (**b**), the *gray* line shows a fit to a hyperbolic secant spectrum

characteristics of the pulse. The pulse width can be adjusted by changing the pulse energy.

The soliton period, assuming a linearly-polarized pulse, is given by [29]

$$z_0 = \frac{\tau_0^2}{2|\beta_2|} \tag{5}$$

where τ_0 is the pulse width and β_2 is the GVD. For the shortest pulses (350 fs), $z_0 = 3.5\,\mathrm{m}$, i.e., approximately the round-trip length of the cavity. For the longest pulses, $z_0 = 20\,\mathrm{m}$. Periodic perturbations to a soliton cause side-bands in the optical spectrum, due to phase- matching of the radiation shed by the soliton as it adjusts after being perturbed [35, 36]. The side-bands marked by "C" in Fig. 5 are due to the periodic perturbation by the cavity (the side-bands designated by "P" will be discussed later). Their relatively small magnitude (<0.1% of total energy) shows that the soliton is not strongly perturbed as it circulates in the cavity.

The linear birefringence in the cavity is controlled by wrapping a portion of the cavity fiber around two 5.5 cm diameter disks (known as fiber polarization controller paddles) [37]. Each disk has three wraps, which for standard single mode fiber, provides approximately $\pi/2$ total linear retardation at 1550 nm with the fast axis in the bend plane of the fiber; i.e., they act like quarter wave-plates. The azimuthal angles, θ_1 and θ_2, of these paddles are specified relative to a common arbitrary reference plane (i.e., the table). The remainder of the fiber constituting the laser cavity is mechanically secured so that the magnitude and principal axes of its birefringence (due to bends, strain, splices, etc.) is constant. The magnitude of the residual birefringence (see below) can be determined and is found to be less than $\pi/4$; it is typically around $\pi/8$. The total cavity birefringence is dominated by the paddles and by adjusting θ_1 and θ_2, the retardance can be varied from 0 to slightly greater than 3 rad. Approximately 60% of the cavity fiber is contained in the paddles and these lengths are considerably shorter than the soliton period. Therefore, characterizing the total cavity retardance by its net value is justified.

The entire measurement apparatus, including the fiber laser, is shown in Fig. 6. It can determine if the pulses are evolving (and how rapidly) as they circulate in the cavity. For conditions where the polarization does not evolve, the azimuthal orientation of the principal axes in the cavity and the complete output polarization state can be determined. The measurement of the rate of polarization evolution is equivalent to measuring the cavity retardance.

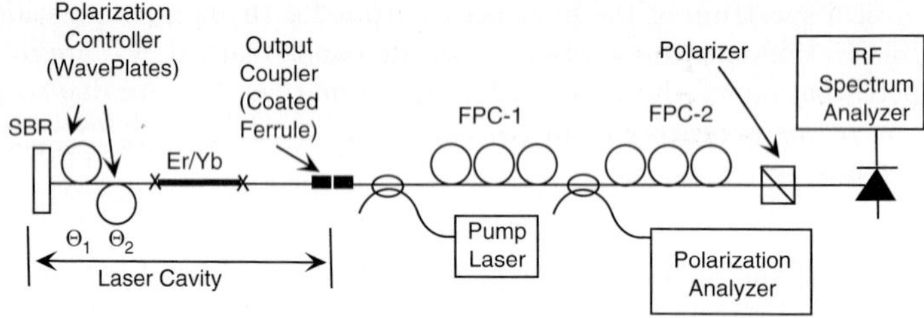

Fig. 6. Diagram of experiment used to observe polarization evolution in a mode-locked fiber laser

As the pulse circulates in the cavity, its polarization state evolves under the influence of the net (linear plus nonlinear) birefringence in the cavity. The polarization state is effectively sampled each time the pulse reflects off the output coupler. To measure the evolution of the polarization, the output is passed through a linear polarizer and the transmitted intensity is measured with a fast, polarization- insensitive photodiode. The linear polarizer maps the evolution of the polarization into amplitude modulation. The amplitude modulation is detected by measuring the radio-frequency (RF) spectrum of the signal emitted by the photodiode, and observing the frequencies of the resulting side-bands.

The RF spectrum consists of a series of spikes at integer multiples of the cavity repetition frequency. Each spike has upper and lower side-bands at relative frequency $\Delta = \gamma c/n\pi$, where n is the average index of refraction in the cavity. The product $\Delta\tau_c$, where τ_c is the round-trip time for the cavity, is the number of round trips required for the pulse to undergo a full polarization evolution and return to its original state, i.e., experience a total of 2π retardance. Hence the magnitude of the total round-trip cavity birefringence is $\beta = 2\pi\Delta\tau_c = 2l_c\gamma$ (in radians). From this ,we can see that measuring Δ is equivalent to measuring either β or γ. For a cavity birefringence of larger than π, Δ is aliased to below $1/(2\tau_c)$, because the polarization is only sampled once per round trip. If the polarizer is aligned along either principal axis, the amplitude modulation is not present. Hence, the orientation of the cavity axes can be determined by rotating the polarizer until the side-bands vanish. However, this measurement does not determine which axis is fast and which is slow. See [38] for a derivation of the RF spectrum.

In circumstances where the polarization is not evolving, it is useful to have a complete characterization of the polarization state. This is obtained by using a commercial polarization state analyzer.

To be useful, the measurement of both the cavity axes and the complete polarization state need to be made on the light at the output coupler. Any birefringence in the fiber intervening between the output coupler and the measurement apparatus will change the polarization state. It is therefore necessary to use fiber polarization controllers to compensate the extra-cavity birefringence (FPC-1 and FPC-2 in Fig. 6) [27].

2.3 Results

The fundamental measurement is to determine the polarization evolution frequency, Δ, as a function of the angles θ_1 and θ_2 of the polarization controller paddles inside the laser cavity. Typical results are shown in Fig. 7.

As mentioned above, the measurement of Δ is equivalent to measuring the round trip retardance of the laser cavity. The overall structure of $\Delta(\theta_1, \theta_2)$ is well reproduced if the total cavity birefringence is calculated using a Jones vector formulation [25]. The retardance of the paddles and the

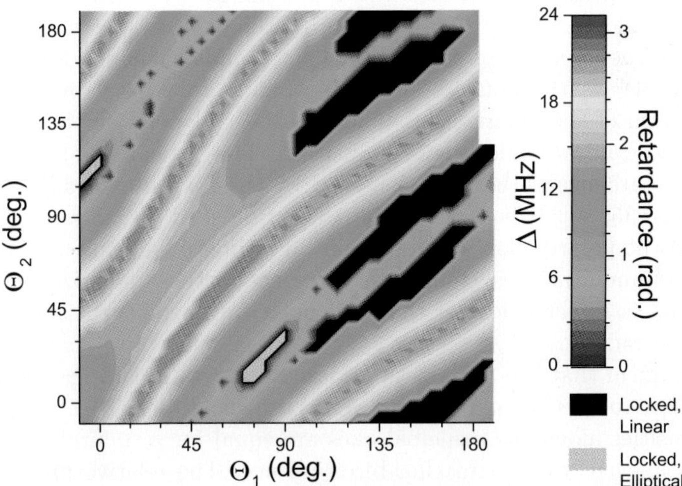

Fig. 7. The polarization evolution frequency, Δ, as a function of the azimuthal angles of the two intra-cavity paddles, θ_1 and θ_2

residual birefringence (retardance and axis orientation) used in the calculation are adjusted to obtain agreement with the data. The paddle retardance is found to be slightly under the design value of a quarter wave. The data shown in Fig. 7 are taken with the laser mode-locked; the overall structure of $\Delta(\theta_1, \theta_2)$ is identical if the SBR is replaced with a high-reflector so that the laser runs CW. This demonstrates that nonlinear birefringence is weak.

The side- bands in the RF spectrum disappear for certain values of intra-cavity birefringence. The lack of side-bands means that the polarization evolution has ceased. This occurs when the polarization is locked and is not evolving as the pulse circulates in the cavity. These regions of locked polarization are indicated in Fig. 7. The cavity still has net non-zero retardance, which is confirmed by making measurements of Δ when the laser is operating CW. The sizes of the regions of locked polarization are found to depend on pulse energy [38]. The pulse energy-dependent sizes and positions of the regions suggest that nonlinearity is responsible for the polarization locking. This is confirmed by the fact that locking is not observed when the laser is operating CW. As discussed above, there are two mechanisms that can lead to pulses with a fixed state of polarization: polarization-locked vector solitons and the fast axis instability. These are differentiated by their polarization; PLVS have elliptical polarization, while the axis instability results in linear polarization. Measurement of the polarization state in the locked regions shows that the region centered about zero retardancc is elliptically-polarized, while those centered at finite values of retardance are linearly-polarized. Based on these observations, the elliptically- polarized regions are tentatively assigned to the formation of PLVS and the linearly-polarized regions to the fast axis instability.

The two optical spectra shown in Fig. 5 are for cases where the polarization is (a) unlocked and (b) locked. Comparison of these spectra is interesting because of the appearance of additional side-bands in the unlocked case (denoted by "P" in the figure). These side bands are caused by the periodic perturbation of the polarization. Their positions depend on the intra-cavity birefringence in an analogous fashion to Δ. The optical side-bands are due to phase matching of the radiation shed by the soliton as it adjusts to perturbations [35, 36]. It is believed that radiation shed due to random perturbation of the polarization will represent an ultimate limit in very high-speed soliton communication systems. The presence of these polarization-induced side-bands is direct evidence that the polarization perturbations do cause solitons to radiate.

To confirm that the polarization locked state with elliptical polarization is a PLVS, further measurements were performed (see [38]). As expected, the intensities along the principal axes are equal for zero birefringence, and increase linearly with increasing birefringence. The relative phase between the two components is found to be approximately constant at either $+\pi/2$ or $-\pi/2$. Both of these exactly match our prediction for how a PLVS maintains stability. Additionally, the pulse energy dependence of the maximum retardance and relative optical bandwidths agree well with numerical simulations [28, 38]. Finally, it was observed that the locking is hysteretic with respect to the cavity birefringence, which again is consistent with the explanation of the locking mechanism.

To verify the origin of the locked regions with linear polarization, the polarization angle was measured as a function of θ_2. This agreed well with slow axis of the cavity as a function of θ_2, as expected if the fast axis is unstable. The boundaries of the linearly-locked region were also consistent with it being due to a slow axis instability.

2.4 Summary

The output of a mode-locked fiber laser can have a locked polarization state for certain settings of the intra-cavity birefringence. Extensive experiments examined the underlying physical mechanism. For very low values of birefringence, the output is elliptically-polarized and is due to the formation of polarization-locked vector solitons. For slightly larger retardance, the output is linearly-polarized and is assigned to the instability of the fast axis. Theoretical simulations provide good agreement with the experiment. Because the round trip gain and loss in the fiber laser is small, it provides a better approximation to a conservative system than that provided by a standard telecommunications long distance transmission system. It also provides and essentially infinite propagation distance.

3 Soliton Explosions

Dissipative systems display more complicated dynamics than integrable systems. These dynamics can have exquisite dependence on parameters. Examples include periodic or chaotic behavior or switching among stable states. In certain situations, simulations show that solitons undergo dramatic transients and then return to the initial conditions [39]. These transients have been dubbed "exploding solitons" [40]. During an explosion, the soliton energy and spectrum undergo dramatic changes, but return to the steady state value afterwards. The existence of such instabilities may influence the design of long-distance communications systems and ultra-fast mode-locked lasers. Such transients are detrimental to either application. System design that avoids them requires knowledge that they exist and accurate, experimentally-verified models.

Capturing a transient event, such as an exploding soliton, is challenging. As mentioned above, mode-locked lasers provide the ability to continuously monitor the evolution of a soliton as it circulates. This facilitates capturing rare events. The explosions last for approximately 10 μs, which corresponds to a few hundred round trips.

3.1 Experiment

Soliton explosions are detected by temporally recording the output spectrum of the laser and the integrated energy of the pulse. Full characterization by a technique such as frequency-resolved optical gating [41] would be preferable. However, all such techniques utilize optical nonlinearities and cannot make single shot measurements on the nano-joule pulses that are directly produced by a mode-locked oscillator. Thus, measurement of the temporally-resolved spectrum is the only technique that can provide information about the pulse dynamics on timescales comparable to the round-trip time of the laser cavity.

Figure 8 shows the experimental setup. The laser has a "stretched" cavity that is 4 times longer than typical (40 ns roundtrip time). The laser spectrum shows strong resonant side-bands from phase-matching of the dispersive waves shed as the 50 fs soliton undergoes periodic perturbation [35, 36]. These are stronger than typically observed in a Kerr-lens mode-locked (KLM) laser, due to the stretched cavity and lack of spectral filtering. The gain bandwidth is broader than the pulse spectrum and is limited by the mirror reflectivity. The use of a stretched cavity is crucial, because it means that the pulse spectrum is dominated by soliton dynamics through side-band formation, rather than through explicit spectral filtering due to a tuning element or mirror reflectivity. This is because the frequency spacing of the side-bands is inversely proportional to cavity length. The output of the laser is spectrally dispersed by a diffraction grating across an array of six detectors. The spectral dispersion is adjusted so that each channel has a spectral width of ~12 nm (full-width half- maximum). The data from all of the channels are synchronously recorded with an electronic bandwidth corresponding to averaging over approximately five successive pulses.

The spectrally-integrated intensity (i.e., total pulse energy) is also synchronously recorded on a separate channel. An optical spectrum analyzer is used to record the steady-state spectrum. The six channels are positioned on

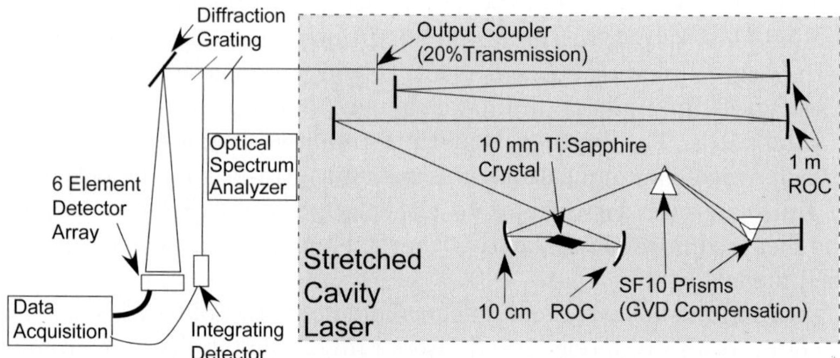

Fig. 8. Experimental set-up for observing exploding solitons. The radii of curvature (ROC) of the laser mirrors are indicated; those not designated are flat

the short wavelength side of the steady state spectrum, because the spectral transients occur in that direction. This was verified by running the optical spectrum analyzer in "peak hold" mode during a large number of explosions, capturing the highest intensity at a given wavelength. The long wavelength side of the captured spectrum was identical to the steady state spectrum. A typical steady- state spectrum and the spectral response of the individual channels are shown in Fig. 9.

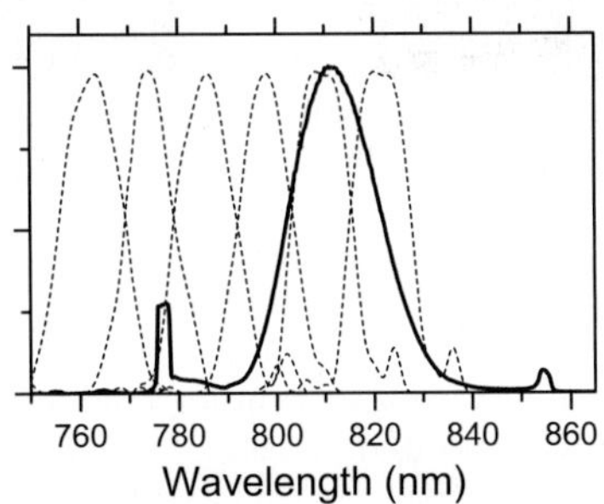

Fig. 9. Typical steady-state output spectrum of the laser and the spectral response of each of the channels (*dashed*)

3.2 Results

Data exhibiting a typical soliton explosion are shown in Fig. 10. The bulk of the integrated spectrum is concentrated around the central frequency of 812 nm. When an explosion occurs, the spectrum becomes asymmetric, abruptly blue shifts and subsequently returns to the original position. This can occur as either an isolated event, as shown in Fig. 10a, or in a burst, as shown in Fig. 10b. The intra-cavity dispersion controls which of these occurs. Small oscillations that slowly increase in amplitude precede an explosion and slowly die out after the explosion. Time-averaged auto-correlations show that the steady-state pulse is the same before and after an explosion, and a transform-limited pulse can be obtained with external dispersion compensation. This fact, together with the observation that the pulse spectrum *narrows* during the explosion, is proof that the pulse becomes longer in time, as predicted [39], because of the Fourier transform limit.

Typical measured intra-cavity dispersion, corresponding to the data in Fig. 10, is shown in Fig. 11. Measurements were performed in situ by measuring the repetition rate as a function of wavelength [42]. The change in slope and curvature show that higher-order dispersion is present and that it changes between the two cases, making it difficult to get a simple quantitative assessment of the change in dispersion.

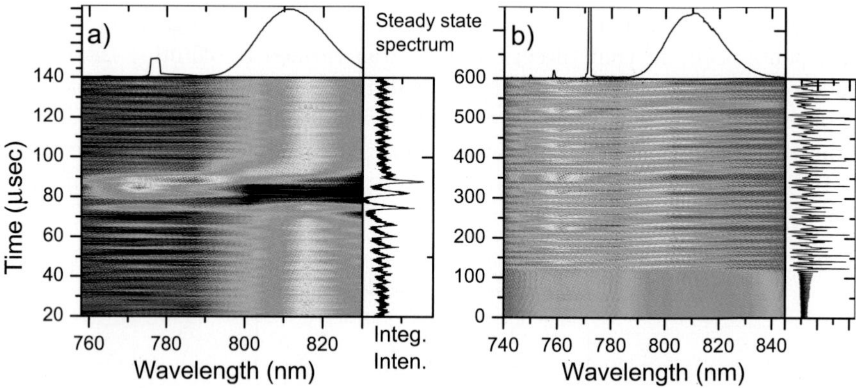

Fig. 10. Typical data showing soliton explosions. In (**a**) the dispersion is adjusted to yield solitary explosions, while in (**b**) it is adjusted to yield bursts. The resolution and positions of the spectral channels has been optimized in (**b**). The *upper* panel in each case shows the steady-state spectrum for reference. The *right* panels show the output of the spectrally-integrated detector

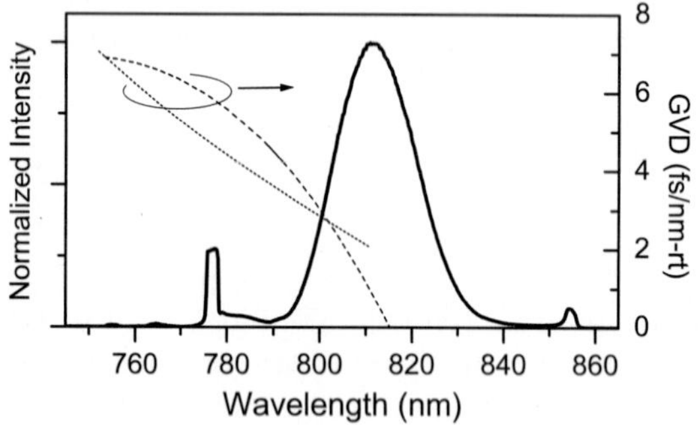

Fig. 11. Measured intra-cavity dispersion corresponding to Fig. 10(**a**) (*dotted*) and (**b**) (*dashed*). The *solid* line shows the steady-state laser spectrum for reference

The rate at which explosions occur can be controlled by changing the pump power, as this determines the steady-state soliton energy. The range of pump powers for which explosions occur depends on the intra-cavity dispersion. A histogram of the frequency of the time between spontaneously-occurring explosions (called the "laminar time" because intermittency was first studied in the context of the onset of turbulence) shows a power-law dependence, with $P(\tau) \sim \tau^{-1}$ (Fig. 12). Strong external perturbations, such as a sharp rap on the laser, can trigger an explosion if the pump power is in the range for which explosions occur. No correlations with typical vibrations in the laboratory are observed. Thus the explosions are not simply due to technical noise. Although the onset of the explosions is not our focus here,

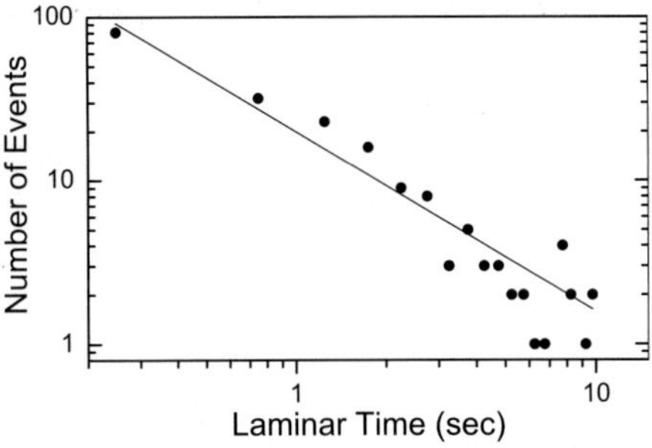

Fig. 12. Histogram of the number of events as function of the time between events ("Laminar time"). Line is a fit to a power-law with an exponent of 1.1

power-law behavior in the frequency of laminar times is a signature of "intermittency" in a system exhibiting chaotic dynamics [43].

The bursts of explosions, as seen in Fig. 10b, on average last for 0.95 ms [for the same intra-cavity dispersion as Fig. 10b]. However, a histogram of burst lengths [see Fig. 13] shows a minimum duration of around 0.4 ms and a general decrease in the number with increasing length.

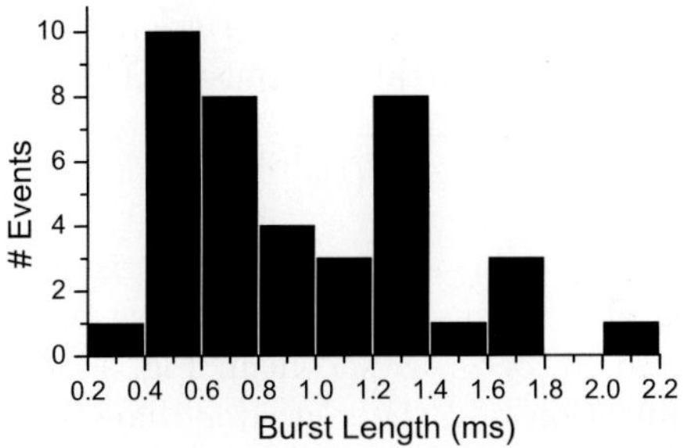

Fig. 13. Histogram of the length of bursts such as is seen in Fig. 10b

If explicit spectral filtering is imposed on the laser, explosions do not occur. Instead, strong oscillations eventually break out at pump powers lower than those characteristic of the occurrence of explosions. We attribute these oscillations to relaxation phenomena, because the damping time of relaxation oscillations increases as the pump power is lowered towards the lasing threshold [1]. The oscillations are distinct from explosions in that the large abrupt change in center wavelength does not occur; on the contrary, the center wavelength is constant, and only the width and pulse energy oscillate.

3.3 Summary

The experimentally-observed features are similar to those predicted theoretically in the continuous model [39]. However, the real system is not continuous. The discreteness of the laser must be taken into account to verify that the predictions still hold. Including the discreteness of the actual laser does not change the results qualitatively [40]. The asymmetry of the experimentally-observed explosions is reproduced in the calculations when higher-order dispersion is included. This is because higher-order dispersion breaks the symmetry between higher and lower frequencies.

In summary, the following features are observed: (1) explosions occur occasionally; (2) the rate is sensitive to pump power; (3) the laser is close to the mode-locking threshold; (4) the explosions are similar, but not identical; (5) an explosion can occur spontaneously, but can also be triggered by external perturbations; (6) there is no precursor to an explosion. Similar features are observed in the simulations.

4 Carrier-Envelope Phase

Over the last few years, one of the most exciting developments in mode-locked lasers has been the ability to stabilize the carrier-envelope phase of the output pulses [44, 45, 46]. The electric field of a pulse, $E(t)$, can be written

$$E(t) = \hat{E}(t)\mathrm{e}^{\mathrm{i}(\omega t + \phi_{CE})} \tag{6}$$

where $\hat{E}(t)$ is the "slowly" varying envelope that is superimposed on a carrier wave oscillating at a frequency ω. The phase between the carrier and envelope is designated by ϕ_{CE}, as shown in Fig. 14. Note that 6 can only be used when the duration of $\hat{E}(t)$ is greater than that of a single cycle of the carrier wave, otherwise it no longer describes a propagating solution to

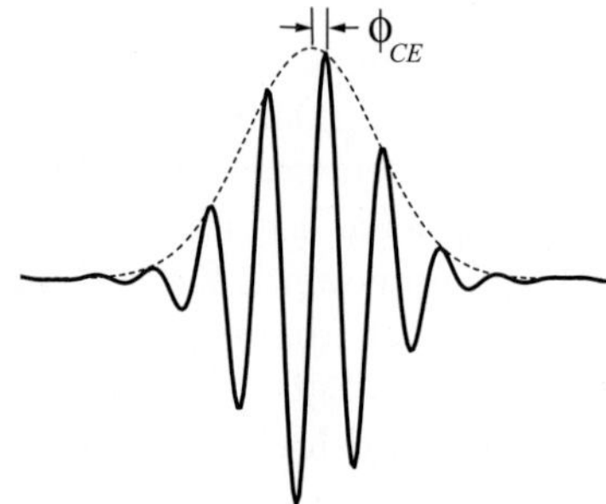

Fig. 14. The electric field of a few-cycle pulse showing the definition of the carrier-envelope phase, ϕ_{CE}. The *dashed* line shows the envelope

Maxwell's equations. For more complex waveforms, there is ambiguity in the decomposition into carrier and envelope. However, in practical situations, it is usually clear how to do it.

Normally in optics, the overall phase of the electric field is irrelevant, because measurements are always made of intensity, not of the field itself. Only relative phases, for example between beams traversing the two arms of an interferometer, matter. However, the ability to measure ϕ_{CE} overcomes this, and will allow the optical phase to be measured relative to the pulse envelope, the phase of which can in turn be measured relative to an arbitrary radio-frequency clock signal. The optical phase becomes relevant in extreme nonlinear optics, where processes occur which are sensitive to the electric field of the pulse, rather than the intensity [47]. By simply stabilizing ϕ_{CE}, the frequency spectrum can be controlled, and this has dramatically simplified optical frequency metrology [44, 48, 49] and optical atomic clocks [50, 51].

This advance makes the phase dynamics of the intra-cavity pulse a subject of strong interest. Heretofore, the phase evolution of solitons has largely been ignored, in no small part because it was not relevant to experiments. Thus the phase dynamics of solitons is now becoming a topic of interest.

4.1 Frequency Domain Stabilization of ϕ_{CE}

The difference between group and phase velocities inside the cavity of a mode-locked laser causes ϕ_{CE} to evolve on a pulse-to-pulse basis, as shown in Fig. 15a. The pulse-to-pulse change is designated $\Delta\phi_{CE}$ and is given by

$$\Delta\phi_{CE} = \frac{\omega_c v_g}{f_{rep}} \left(\frac{1}{v_g} - \frac{1}{v_p} \right) , \tag{7}$$

where ω_c is the carrier frequency, f_{rep} is the repetition rate, v_g is the intra-cavity group velocity and v_p is the intra-cavity phase velocity. Stabilization of ϕ_{CE} means controlling $\Delta\phi_{CE}$, with the simplest case being $\Delta\phi_{CE} = 0$, which means that all of the pulses have identical ϕ_{CE}, although its actual value is unknown.

The evolution of ϕ_{CE} is measured in the frequency domain. The spectrum of a pulse train is a comb of discrete lines separated by f_{rep}. If the pulses are all identical, the frequencies of the comb lines are simply integer multiples of f_{rep}. However, evolution of ϕ_{CE} causes a rigid shift of the comb by an amount

$$f_0 = \Delta\phi_{CE} f_{rep}/2\pi \, . \tag{8}$$

Thus the optical frequency of comb line n is $\nu_n = nf_{rep} + f_0$. This correspondence between time and frequency is shown in Fig. 15. (For a full derivation, see [46].) Measurement of $\Delta\phi_{CE}$ is actually implemented by measuring f_0 using a "self-referencing" technique [44, 52]. In self-referencing, comb line n is frequency-doubled and then heterodyned with comb line $2n$. Heterodyning yields the frequency difference, $2\nu_n - \nu_{2n} = 2(nf_{rep} + f_0) - (2nf_{rep} + f_0) = f_0$.

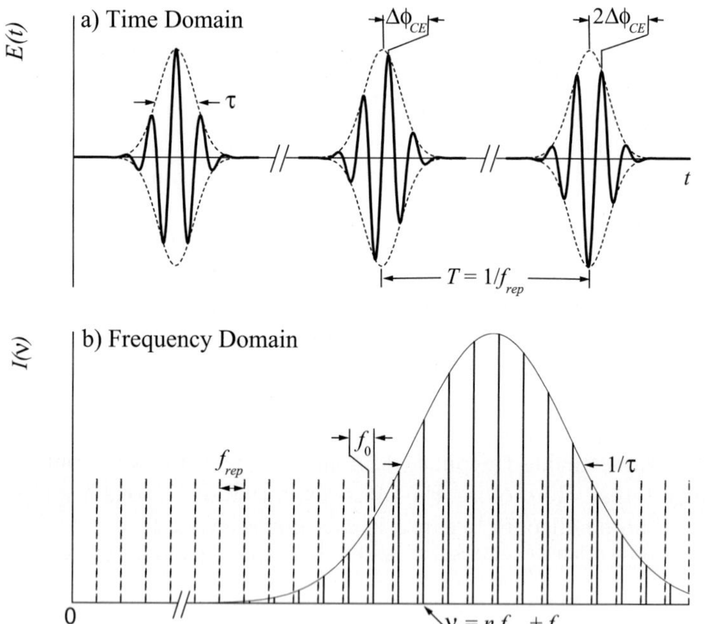

Fig. 15. Diagram showing (**a**) the time domain picture of a pulse train including evolution of ϕ_{CE} , and (**b**) the frequency spectrum showing the offset frequency f_0

4.2 Time-Domain Measurement of $\Delta\phi_{CE}$

A cross-correlator is used to perform time-domain measurement of the evolution of the carrier envelope phase [44, 53]. A cross-correlator consists of a Michelson interferometer followed by nonlinear detection (see Fig. 16). To obtain a cross-correlation between successive pulses emitted by the laser, one arm of the interferometer includes a delay that corresponds to the round trip time of the laser cavity (for the specific case shown here, it is actually two round trip times, so the correlation between pulse i and pulse $i + 2$ is actually obtained). In the absence of nonlinear detection, a field correlation is obtained; this is just the Fourier transform of the spectrum and it cannot distinguish between a short pulse and broad-band white light. The nonlinear interferogram that is obtained by using a detection scheme that includes a second-order nonlinearity clearly distinguishes between these cases and can provide information about pulse characteristics. Because air has significant dispersion for ultra-short pulses, the interferometer must be in an evacuated enclosure.

Two typical interferograms are shown in Fig. 16b. If the pulses were identical, i.e., $\Delta\phi_{CE} = 0$, then the interferograms would be perfectly symmetric. The presence of a pulse-to-pulse phase shift means the interference fringes are shifted with respect to the overall envelope, thus breaking the symmetry.

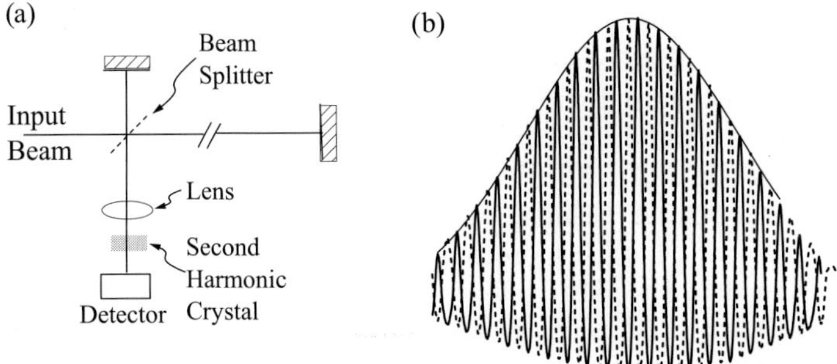

Fig. 16. (**a**) Schematic of a cross-correlator for measuring pulse-to-pulse phase shift. (**b**) Results showing stabilization of the pulse-to-pulse phase shift for two cases differing by $\sim\pi$

To aid in visualizing and extracting the phase of the fringes, the black line shows a fit to the peaks of the fringes. The two instances shown in Fig. 16b correspond to changing f_0 by $f_{rep}/2$, which clearly yields the expected π shift in the fringes. Systematically varying f_0 yields exactly the change in $\Delta\phi_{CE}$ predicted by 8.

4.3 Phase-Intensity Dynamics

These advances make it possible to study the phase dynamics of the circulating pulse in ways there have not previously been possible. The most interesting aspect of this is how phase evolution depends on the pulse intensity. From a fundamental point-of-view, this is interesting because it arises from a complex interplay between nonlinearity and dispersion, the key ingredients in any soliton system. But it is also of great practical importance because it couples intensity fluctuations into phase fluctuations. This can be detrimental, but can also be put to use by actively controlling the pump power, thereby controlling the phase evolution.

One of the early papers on carrier-envelope phase evolution by Xu et al. [53] noticed that $\Delta\phi_{CE}$ depended on the pump power, which in turn controls the pulse intensity. A simple analysis considered the pulse as a soliton, but yielded the opposite sign for $\partial\Delta\phi_{CE}/\partial I$, where I is the intensity, compared to the experiment. This analysis only included the fact that the phase velocity is intensity-dependent because of the Kerr effect. The discrepancy between theory and experiment was attributed to the fact that, in the experiment, there was a spectral shift with intensity, which, coupled with group velocity dispersion, gives a large contribution to $\partial\Delta\phi_{CE}/\partial I$ with the opposite sign from the expected nonlinear contribution. The empirical observation was sufficient to enable the use this effect for controlling $\Delta\phi_{CE}$ [54].

A subsequent analysis included the fact that the group velocity also is intensity-dependent (often called the "shock" term) [55]. This work found that the nonlinear change of the group velocity was larger than the nonlinear change in the phase velocity, which predicts the opposite sign change in $\partial\Delta\phi_{CE}/\partial I$, as compared with the previous analysis.

More recent careful experiments show that both signs $\partial\Delta\phi_{CE}/\partial I$ can actually occur [56]. Careful measurements of the spectral shift show that it behaves in a similar fashion, turning around at approximately the same intensity as does f_0. Furthermore, this work showed that $\partial\Delta\phi_{CE}/\partial I$ depends on the pulse width. Specifically, for very short pulses, with very large spectral bandwidth, $\partial\Delta\phi_{CE}/\partial I$ tends to be small. This is consistent with spectral shifts being important; for very short pulses, the spectrum is so wide that it is prevented from shifting due to implicit spectral filtering by various intra-cavity elements.

A different theoretical approach is to numerically solve Maxwell's equations. Such a direct solution avoids the approximations made in deriving the NLSE, although it is computationally more difficult and does not permit analysis. The results confirm the soliton analysis at low intensities. However, at high intensities, pulse reshaping can become appreciable and effectively change the sign of $\partial\Delta\phi_{CE}/\partial I$ [57]. This is accompanied by spectral shifts, however it occurs over much larger changes in intensity than observed in the experiments.

A complete understanding of the intra-cavity phase dynamics of a mode-locked laser has not been obtained so far. A complete analysis, for example including the dispersion map, should provide better insight.

5 Summary

Hopefully, this chapter has shown that mode-locked lasers are a very interesting "playground" for temporal solitons. Specifically, they provide a means for observing phenomena that are hard to observe in other systems. This can be due to the fact the evolution must be carefully monitored over long times, or because of the occurrence of infrequent events that are difficult to capture. This is a two-way street, since the lasers provide a test-bed for soliton dynamics, but at the same time the soliton theory helps to improve the operation of the lasers.

Acknowledgements

The author would like to thank Nail Akhmediev and Jose Soto-Crespo for stimulating discussions regarding the work in Sects. 2 and 3. In addition, contributions by Brandon Collings, Wayne Knox and Keren Bergman to the experiments presented in Sect. 2 are gratefully acknowledged. Contributions by David Jones, Tara Fortier, Jun Ye and John Hall to Sect. 4 are gratefully acknowledged.

References

1. For a textbook-level discussion of lasers, see for example A. E. Siegman, *Lasers* (University Science Books, Mill Valley, 1986) or J. T. Verdeyen, *Laser Electronics* (Prentice-Hall, Englewood Cliffs, 1995).
2. E. P. Ippen, Appl. Phys. B **58**, 159 (1994).
3. Y. Chen, F. X. Kärtner, U. Morgner, S. H. Cho, H. A. Haus, E. P. Ippen, J. G. Fujimoto, J. Opt. Soc. Am. B **16**, 1999 (1999).
4. E. Desurvire, *Erbium-Doped Fiber Amplifiers* (John Wiley & Sons, New York, 1994).
5. B. C. Collings, K. Bergman, S. T. Cundiff, S. Tsuda, J. N. Kutz, J. E. Cunningham, W. Y. Jan, M. Koch, W. H. Knox, IEEE J. Sel. Top. Quantum Electr. **3**, 1065 (1997).
6. L. E. Nelson, D. J. Jones, K. Tamura, H. A. Haus, E. P. Ippen, Appl. Phys. B **65**, 277 (1997).
7. A. Hasegawa, F. Tappert, Appl. Phys. Lett. **23**, 142 (1973).
8. I. P. Kaminow, IEEE J. Quantum Electron. **17**, 15 (1981).
9. L. F. Mollenauer, R. H. Stolen, J. P. Gordon, Phys. Rev. Lett. **45**, 1095 (1980).
10. H. A. Haus, W. S. Wong, Rev. Mod. Phys. **68**, 423 (1996).
11. C.D. Poole, J. Nagel, Polarization effects in lightwave systems. In: *Optical Fiber Telecommunications IIIA*, ed. by I. P. Kaminow, T. L. Koch (Academic, San Dieogo 1997) p. 114.
12. M. N. Islam, C. D. Poole, J. P. Gordon, Opt. Lett. **14**, 1011 (1989).
13. C. R. Menyuk, Opt. Lett. **12**, 614 (1987); IEEE J. Quantum Electron. **23**, 174 (1987); J. Opt. Soc. Am. B **5**, 392 (1988).
14. D. N. Christodoulides, R. I. Joseph, Opt. Lett. **13**, 53 (1988).
15. S. T. Cundiff, B. C. Collings, N. N. Akhmediev, J. M. Soto-Crespo, K. Bergman, W. H. Knox, Phys. Rev. Lett. **82**, 3988 (1999).
16. M. V. Tratnik, J. E. Sipe, Phys. Rev. A **38**, 2011 (1988).
17. N. Akhmediev, A. Buryak, J. M. Soto-Crespo, Opt. Commun. **112**, 278 (1994).
18. N. Akhmediev, J. M. Soto-Crespo, Phys. Rev. E **49**, 5742 (1994).
19. N. N. Akhmediev, A. V. Buryak, J. M. Soto-Crespo, D. R. Andersen, J. Opt. Soc. Am. B **12**, 434 (1995).
20. J. M. Soto-Crespo, N. Akhmediev, A. Ankiewicz, Phys. Rev. E **51**, 3547 (1995).
21. Y. Chen, J. Atai, J. Opt. Soc. Am. B **12**, 434 (1995).
22. H. G. Winful, Opt. Lett. **11**, 33 (1986).
23. K. J. Blow, N. J. Doran, D. Wood, Opt. Lett. **12**, 202 (1987).
24. Y. Barad, Y. Silberberg, Phys. Rev. Lett. **78**, 3290 (1997).
25. S. T. Cundiff, B. C. Collings, W. H. Knox, Optics Express **1**, 12 (1997).
26. N. N. Akhmediev, J. M. Soto-Crespo, S. T. Cundiff, B. C. Collings, W. H. Knox, Opt. Lett. **23**, 852 (1998).
27. B. C. Collings, S. T. Cundiff, N. N. Akhmediev, J. M. Soto-Crespo, K. Bergman, W. H. Knox, J. Opt. Soc. Am. B **17**, 354 (2000).
28. J. M. Soto-Crespo, N. N. Akhmediev, B. C. Collings, S. T. Cundiff, K. Bergman, W. H. Knox, J. Opt. Soc. Am. B **17**, 366 (2000).
29. G. P. Agrawal, *Nonlinear Fiber Optics* (Academic Press, San Diego 1995).
30. N. N. Akhmediev and A. Ankiewicz, *Solitons: Nonlinear pulses and beams*, (Chapman & Hall, London 1997).
31. S.V. Manakov, Sov. JETP **38**, 248 (1974).

32. J. U. Kang, G. I. Stegeman, J. S. Aitchison, N. Akhmediev, Phys. Rev. Lett. **76**, 3699 (1996).
33. S. G. Evangelides, L. F. Mollenauer, J. P. Gordon, N. S. Bergano, J. Lightwave Technol. **10**, 28 (1992).
34. L. F. Mollenauer, K. Smith, J. P. Gordon, C. R. Menyuk, Opt. Lett. **14**, 1219 (1989).
35. S. M. J. Kelley, Electron. Lett. **28**, 806 (1992).
36. J. P. Gordon, J. Opt. Soc. Am. B **9**, 91, (1992).
37. H. C. Lefevre, Electron. Lett. **16**, 778 (1980).
38. S. T. Cundiff, B. C. Collings, K. Bergman, CHAOS **10**, 613 (2000).
39. J. M. Soto-Crespo, N. Akhmediev, A. Ankiewicz, Phys. Rev. Lett. **85**, 2937 (2000).
40. S. T. Cundiff, J. M. Soto-Crespo N. N. Akhmediev, Phys. Rev. Lett. **88**, 073903 (2002).
41. R. Trebino, *Frequency-Resolved Optical Gating: the Measurement of Ultrashort Laser Pulses* (Kluwer, Boston 2002).
42. W. H. Knox, Opt. Lett. **17**, 514 (1992).
43. H. G. Schuster, *Deterministic Chaos*, 3rd ed. (VCH, Weinheim 1995).
44. D. J. Jones, S. A. Diddams, J. K. Ranka, A. Stentz, R. S. Windeler, J. L. Hall, S. T. Cundiff, Science **288**, 635 (2000).
45. A. Apolonski, A. Poppe, G. Tempea, C. Spielmann, T. Udem, R. Holzwarth, T. W. Hänsch, F. Krausz: Phys. Rev. Lett. **85**, 740 (2000).
46. S. T. Cundiff, J. Phys. D **35**, R43 (2002).
47. A. Baltuška, T. Udem, M. Uiberacker, M. Hentschel, E. Goulielmakis, C. Gohle, R. Holzwarth, V. S. Yakoviev, A. Scrinzi, T. W. Hänsch, and F. Krausz, Nature **421**, 611 (2003).
48. S. A. Diddams, D. J. Jones, J. Ye, T. Cundiff, J. L. Hall, J. K. Ranka, R. S. Windeler, R. Holzwarth, T. Udem, and T. W. Hänsch, Phys. Rev. Lett. **84**, 5102 (2000).
49. S. T. Cundiff, J. Ye, Rev. Mod. Phys. **75**, 325 (2003).
50. S. A. Diddams, T. Udem, J. C. Bergquist, E. A. Curtis, R. E. Drullinger, L. Hollberg, W. M. Itano, W. D. Lee, C. W. Oates, K. R. Vogel, D. J. Wineland, Science **293**, 825 (2001).
51. J. Ye, L.-S. Ma, J. L. Hall, Phys. Rev. Lett. **87**, 270801 (2001).
52. H. R. Telle, G. Steinmeyer, A. E. Dunlop, J. Stenger, D. H. Sutter, U. Keller, Appl. Phys. B **69**, 327 (1999).
53. L. Xu, C. Spielmann, A. Poppe, T. Brabec, F. Krausz, and T. W. Hänsch, Opt. Lett. **21**, 2008 (1996).
54. A. Poppe, R. Holzwarth, A. Apolonski, G. Tempea, C. Spielmann, T. W. Hänsch, F. Krausz, Appl. Phys. B **72**, 977. (2001).
55. H. A. Haus, E. P. Ippen, Opt. Lett. **26**, 1654 (2001).
56. K. W. Holman, R. J. Jones, A. Marian, S. T. Cundiff, J. Ye, Opt. Lett. **28**, 851 (2003).
57. P. Goorjian, S.T. Cundiff, Opt. Lett. **29**, 1363 (2004).

Temporal Multi-Soliton Complexes Generated by Passively Mode-Locked Lasers

J.M. Soto-Crespo[1] and Ph. Grelu[2]

[1] Instituto de Óptica, C.S.I.C., Serrano 121, 28006 Madrid, Spain
iodsc09@io.cfmac.csic.es
[2] Laboratoire de Physique de l'Université de Bourgogne, UMR 5027, B.P. 47870, 21078 Dijon, France
Philippe.Grelu@u-bourgogne.fr

Abstract. We review various experimental observations of multiple pulsing in passively mode-locked lasers and give several possible mechanisms that have been addressed in the literature. We then propose some criteria to distinguish dissipative multi-soliton complexes among them, and relate them to the theoretical literature. A particular distributed model, which includes the main laser features usually involved in the formation of multi-soliton complexes, viz. the cubic-quintic complex Ginzburg-Landau equation (CQCGLE), is detailed. We put emphasis on the attractors of soliton pairs, and show that some features predicted by the CQCGLE model are corroborated by experiments in a passively mode-locked fiber laser. However, some other experimental features of soliton pair attractors remain unexplained within the frame of a distributed model. We then develop a more realistic model which includes discreteness and periodicity, and this in turn leads to a large number of soliton pair attractors. The important influence of the dispersion regime is stressed.

1 Introduction

Passively mode-locked lasers rely on the use of nonlinearity to favor stable operation with high-peak-intensity pulses. Basically, this means that nonlinearity is used to make continuous wave operation feel more intra-cavity losses than pulsed operation. There are several schemes to achieve this. Mainly, and depending on the kind of laser system used, the experimentalist can choose among saturable absorption [1, 2], addition-pulse mode-locking (APM) [3, 4], Kerr-lens mode locking (KLM) [5, 6], and nonlinear polarization rotation [7]. In the last three principal schemes, the optical Kerr nonlinearity plays a key role. Since we know that, in conjunction with anomalous chromatic dispersion, Kerr nonlinearity has been associated with the formation of temporal solitons, we may wonder whether passively mode-locked lasers can emit solitons. This is difficult to answer, since real experimental set-ups are far more complicated than the simple passive optical fiber for which temporal optical solitons where first predicted [8, 9], and later experimentally observed [10]. In fact pulses propagating along any material suffer unavoidable losses which, if not properly compensated and after long enough propagation distance,

J.M. Soto-Crespo and Ph. Grelu: *Temporal Multi-Soliton Complexes Generated by Passively Mode-Locked Lasers*, Lect. Notes Phys. **661**, 207–239 (2005)
www.springerlink.com

would extinguish them. Apart from the balance between (Kerr) nonlinearity and chromatic dispersion, a second balance between loss and gain is necessary in order to generate stable pulses, which therefore can be called dissipative solitons.

The closest laser to a Schrödinger-soliton laser would be a fiber laser made of a loop of a single fiber featuring, at the same time, Kerr nonlinearity, anomalous dispersion, and distributed gain compensating locally for all losses. In contrast to this, a real laser has a localized output port, which means that the pulse intensity is not uniform in the cavity. However, if one places a gain section immediately before or after the laser output port, and extends the cavity by a relatively long length of passive anomalous fiber, one may then get a laser source emitting pulses which are very close to Schrödinger solitons. In this spirit, we may cite the early use of soliton-like pulse reshaping in a fiber as one passive element of a mode-locked color-center laser [11].

The development of mode-locked fiber lasers exploded in the last decade, starting with their first realizations in the early 1990s [12, 13, 14]. These tried to exploit the many advantages of a fiber laser system compared to bulk laser systems, and were boosted by the demand for compact, cheap and reliable picosecond sources in the telecom domain [15]. One key point was the development of efficient diode-laser pumped, erbium-doped fibers providing high gain per unit length in the 1520–1560 nm second telecom window [16, 17]. The total length (several tens of meters) does not make the fiber laser bulky, since the fiber may be spooled on a disk of about ten centimeters of diameter. Also, there is no misalignment issue such as the one that occurs with non-fiber laser systems. However, fiber lasers of long cavity length have a few drawbacks. The cavity may require external stabilization, since a long fiber length brings about an increased sensitivity to temperature-induced drifts of cavity length and birefringence. A long cavity also provides a lower fundamental output pulse repetition rate.

One key discovery of the early all-fiber soliton laser experiments was the ability to obtain multiple pulses in one cavity round trip [12, 14, 18]. This is easily explained by the properties of the single-mode silica telecom fiber on one hand, which provide the energy of the fundamental Schrödinger soliton, and by the efficiency of fiber lasers on the other hand. The fundamental Schrödinger soliton energy E is given by $E = 3.53|\beta_2|/\sigma/T$, where β_2 stands for the group velocity dispersion (GVD), σ for the nonlinear coefficient, and T for the full temporal width at half-maximum. As an illustration, we may consider, for a standard telecom fiber operating near 1550 nm, a GVD around $-20\,\mathrm{ps}^2/\mathrm{km}$ and a nonlinear coefficient around $1.2\,\mathrm{W}^{-1}\,\mathrm{km}^{-1}$. Dealing with 5 ps pulses, we have an energy of 10 picojoules. Considering a repetition rate of 10 MHz and one soliton in the cavity, this means that the intra-cavity average energy is only 100 microwatts. Although described as basically a three-level system, a pumped erbium-doped fiber can have a significant pump-to-signal conversion efficiency of several percent, thanks to the pump field confinement in the doped core, as well as an arbitrary long length of the amplifying fiber. With a pump power around one hundred milliwatts, and according to mode-locking settings, there can be several tens of intra-cavity pulses.

One question then arises: what are all these soliton pulses doing together in the cavity? More specifically, concerning experimental observations in fiber lasers, we shall address this issue in Sects. 2.1 and 2.2. We shall then detail multi-soliton complexes observed in fiber lasers in Sect. 3.

In the course of the large number of investigations that have been undertaken with solid-state bulk lasers to achieve sub-10-fs pulse durations [19, 20], there have also been discoveries of multi-soliton-like behavior, and we shall also address these in the present chapter. In contrast to the features of fiber lasers, we should stress that in solid-state bulk lasers, the length of the nonlinear medium is of the order of a centimeter. Efficient Kerr lens mode-locking (KLM) requires large optical field intensities, achieved at pump powers of around several watts, although recent work has demonstrated the scaling down of the pump power required for mode-locking [21]. The requirement of large pump is one of the reasons why the observation of multiple pulsing in bulk lasers has generally been limited to two or three pulses. Another feature of the ultra-short solid-state laser is the usual interplay of higher-order dispersion terms that makes a comparison between experiment and theory more difficult. Although the identification of multi-soliton complexes is thus not always possible from the experimental results, a series of papers has given convincing arguments for the existence of multi-soliton complexes in solid-state bulk lasers, and this is addressed in Sect. 2.3.

In Sects. 4 and 5, we concentrate on the main theoretical results published to date on this subject. Each one of the different theoretical models proposed provides a valuable explanation concerning several aspects of some of the experimental results mentioned. Section 4.1 addresses one of the most universal models for passively mode-locked lasers, namely the complex cubic-quintic Ginzburg-Landau equation, which provides bound soliton states that are relevant for both fiber and bulk solid state lasers. Sections 4.2 and 4.3 present more specific distributed models whose predictions have been confirmed within particular experimental set-ups. An important point is the influence of discontinuities and lumped sections that are almost unavoidable in realistic cases. This is discussed in Sect. 5, where a specific non-distributed model is detailed as a clear illustration to show new experimental phenomena that cannot be described by a distributed model. Finally Sect. 6 summarizes the chapter.

2 Experimental Evidence for Multi-Soliton Formation in Passively Mode-Locked Laser Cavities

2.1 Harmonic Mode Locking: Hopes and Deceptions (and Vice-Versa)

In the early experiments reporting multiple pulsing within fiber lasers, various behaviors were observed. The intra-cavity pulses could have erratic relative motions, stabilize themselves at more or less random relative positions, or distribute themselves regularly all along the cavity. The latter perspective, called harmonic mode-locking, seemed very attractive in achieving an increase of the output repetition rate of a fiber laser, in view of telecom applications, since experimentalists could achieve gigahertz repetition rates [22]. The tendency to spontaneously obtain harmonic mode-locking with regular soliton spacing is explained as a consequence of a weak force that is due to gain depletion and recovery time between subsequent solitons [23]. This force has a dissipative nature that arises from the laser medium. Since one soliton, when passing through the gain section, extracts energy and reduces population inversion, the next soliton needs the gain medium to recover inversion in order to "feel" the same gain. The consequence is a weak repulsive force between subsequent pulses. The weakness of this force comes from the many orders of magnitude difference in the ratio between the cavity transit time (100 ns for example) and the excited state lifetime of the erbium ion (10 ms). However, laser cavity dynamics may be sensitive to very tiny effects, unseen in single transmission experiments, since significant pulse pattern evolution may take place on a time scale of one second, which corresponds to a propagation distance of around 200, 000 km.

Unfortunately, harmonic mode-locked fiber lasers did not fully meet telecom sources requirements. The pulse-to-pulse separation suffers a large level of timing jitter, and the weak interaction described is not able to suppress it. For this reason, since 1994, many efforts have been made to achieve harmonic mode-locking with the help of an intra-cavity modulator, externally driven by a reference clock. This is called active mode-locking, which is a useful key for developing practical telecom sources, but it also removes lots of interesting cavity dynamics that we address in the present chapter. In particular, active mode-locking makes the cavity repetition rate limited by the speed of the driving electronics.

However, recently there have been very interesting alternatives. One possible scheme is to develop a short-cavity passively-mode locked fiber laser in which fewer pulses are needed to obtain telecom repetition rates [24]. Another scheme can be called "passive-active mode-locking". Basically, it consists of pulse rate stabilization using the optical output to generate an electrical signal that is applied to an internal modulator acting as a feed-back element [25, 26]. In the frame of active mode-locking, it is also possible to exploit the harmonics of phase modulated light to extend pulse repetition rate beyond the speed of electronics [27].

2.2 The Intriguing Pulse-Bunching Behavior

"Pulse-bunching" is another frequent behavior of the set of intra-cavity soliton pulses that was readily observed and reported [13]. At first, it did not attract much attention, since it was eclipsed by the perspectives of harmonic mode-locking. Pulse-bunching corresponds to the ability of several pulses to group themselves in a stable and tight packet whose duration is much smaller than the cavity round trip time. In most cases, the resolution of the bunch structure requires the recording of an optical auto-correlation trace, since the delay between subsequent solitons may be on the order of a few picoseconds, and is thus smaller than the response time of a photodiode.

Pulse bunches may comprise up to several tens of equal soliton-like pulses [28]. They may be regularly distributed to form a high-repetition rate pulse train, whose origin was rather mysterious at the beginning. Indeed, several mechanisms could be suspected, besides direct soliton-soliton interaction, from modulational instability [29], to birefringence-induced channeled spectrum formation [30] or acousto-optical interaction. The latter phenomenon comes from the electrostrictive force created by light pulses [31]. It results in a slight radial distortion of the fiber core, which in turn changes the refractive index (in the 10^{-10}–10^{-11} range) in the wake of a pulse. Hence, the following pulses feel a weak force, which can be either attractive or repulsive, depending on the time delays of the following pulses. This force can play an important role in bunches comprising solitons separated by more than 100 ps, when direct interaction through the soliton tails becomes negligible [32, 33].

As we shall see in Sect. 2.3, the experimental study of bunched soliton states in fiber lasers has regained some interest in the last few years (2001–2003), by unveiling interesting multi-soliton physics at play. Before entering into more details, we need to stress some experimental studies undertaken in the past decade with solid-state ultra-fast passively mode-locked laser systems such as Ti:sapphire and Cr:forsterite lasers.

2.3 Multi-Soliton Complexes in Solid-State Bulk Lasers

Most of the studies concerning mode-locked solid-state laser systems were aimed at achieving the shortest pulse durations, from 100 fs to below 10 fs, along with the largest pulse energies (>10 nJ intra-cavity). Ti:sapphire and Cr:forsterite lasers operate at wavelengths where chromatic dispersion is positive for the gain medium and the optical lenses. In order to achieve ultra-short pulse durations, low-loss negative dispersion elements, such as prism pairs, are thus required inside the cavity to compensate for positive dispersion and self-phase modulation. This gives what was called a "solitary laser" rather than a pure soliton laser [34]. Indeed, the pulses circulating inside the cavity experience self-phase modulation and negative dispersion separately. Self-phase modulation alone adds frequency chirping, which is approximately linear in the central portion of the pulse, but nonlinear in the pulse tails. Thus, the negative dispersion is only able to compress the central part of the pulse. The tails become distorted and eventually split off.

If we stop the explanation here, it is only bad news for the formation of soliton-like structures in these lasers. However, dissipation comes in to help, since, with the mode-locking mechanism, the pulse tails experience more loss than the central part of the pulse. There is the possibility of reaching a double balance for a soliton-like pulse, between nonlinearity and dispersion on one hand, and between gain and loss (both linear and nonlinear) on the other hand. Indeed, considering a "moderate" performance of a Ti:sapphire laser with single pulse operation featuring a duration of around 100 fs, the optical auto-correlation and the optical spectrum are found, in many cases, to be compatible with temporal profiles close to a hyperbolic secant (*sech*), the shape expected for a Schrödinger soliton. Even in some reports of the shortest pulse achievements (around 10 fs or below), use of sech^2 profiles to fit the auto-correlation traces were convincing [19, 35]. Also, experimental evidence of second-order soliton formation in a Ti:sapphire was found [36]. This occurred a few years after their discovery in the more delicate set-up of a colliding-pulse passively mode-locked (CPM) ring dye laser [37, 38].

In 1994, a new type of phenomenon was observed in a Ti:sapphire laser; it was called "pulse splitting" [39]. In contrast to higher-order soliton formation, which entails periodic pulse splitting and recombination with a temporal periodicity that does not match the cavity round trip time, the pulse splitting described by Spielmann and colleagues refers to a transient field evolution leading to the formation of two separated pulses in the cavity. A possible mechanism for pulse splitting was given in [40], involving the fourth-order dispersion. The authors assumed that the laser cavity, designed for pulses as short as a few tens of femtoseconds, operates near the zero group velocity dispersion point, and showed that, with a small third-order dispersion and a relatively large amount of fourth-order dispersion, the formation of phase-matched spectral sidebands would give rise to pulse splitting and to the formation of two stable separate pulses in the cavity. Numerical simulations showed that the two pulses had different temporal profiles. However, they converged to a fixed separation in time, associated with the formation of stable spectral modulations, in contrast to the oscillating spectra which are characteristic of higher-order solitons.

New experimental data on pulse splitting were provided in 1997 [41, 42]. The reported pulse splitting leads to the formation of a bunched pair of pulses, whose separation was stable in time, but could be varied according to cavity settings, and associated with stable spectral modulations. In [41], the separation between two 35 fs pulses was found to vary step-wise from 129 to 318 fs with an increment of around 64 fs, whereas in [42], the separation between two 85 fs pulses could be tuned continuously from 100 fs to 1 ps. However, considering the difficulty of achieving a precise measurement of third- and fourth- order dispersion, it was not possible to relate these experimental data to the mechanism proposed in [40]. Another regime of a pulse pair with separation on the nanosecond scale was attributed [42] to the particular linear cavity configuration used, where the two pulses may cross and interact nonlinearly in the gain medium.

Considering the large number of possible effects coming into play in ultrashort laser systems, there are probably numerous mechanisms leading to pulse splitting, with some of them leading to the formation of stable patterns of bound pulses. For example, the birefringence of the Ti:sapphire laser rod, associated with Fresnel reflections from the intra-cavity surfaces, was shown to be responsible for the formation of highly modulated output spectra [30]. In this case, the spectral inter-fringe is fixed by the rod length, and corresponds to a pulse pair of fixed separation in time; this contrasts with the reports of [41] and [42] featuring adjustable separation between pulses. In 1998, multi-pulse operation featuring two and three bound 50-fs pulses, separated by 120 to 400 fs, was demonstrated in a Cr:forsterite laser. A numerical analysis similar to [40] was used for this, but it showed that a large amount of third-order dispersion was not a limiting factor for pulse splitting and stabilization in this case [43].

Experimental recordings, as well as the numerical simulations, showed that these multi-pulse regimes are generally not made of equal pulses, in contrast to what was described in the case of fiber lasers. It seems that the closer the pulses stand, the more different they are. However, a subsequent observation of multiple pulsing in a Ti:sapphire laser reported the generation of up to six identical bound 35-fs pulses, with irregular separations lying in the 100 fs to 3 ps range [44].

Quite obviously, the mode-locking mechanism used in the laser set-up strongly influences the ability to achieve multiple pulsing. When using a semiconductor saturable-absorber mirror, it is possible, under high pump power, that the intra-cavity fluence energy is several times greater than the saturable absorber fluence energy. Hence, a reduced discrimination between single and multiple pulsing takes place. Upon gain bandwidth limitation, the transition from single to double pulse may be accompanied by an increase in pulse duration and a narrowing of the output spectrum, leading to a net preference of the laser system for multiple pulsing with an increase of the pump power. These multiple pulsing behaviors were reported experimentally [45, 46, 47]. References [45, 46] clearly showed the formation of 1, 2 and 3 intra-cavity solitons when the pump power was increased, or equivalently when the absolute value of the net anomalous dispersion was decreased, in agreement with the Schrödinger soliton theory. However, the two or three solitons remained separated by more than 90 ps, and could not form any bound state. This could be due to an enhanced force arising from gain depletion and recovery, and, indeed, the solitons tended to form a harmonically mode-locked state (that was probably an operation preferred by the experimentalists involved). This contrasts with the results of reference [47], where a very large range of multiple pulse behavior was experimentally reported. It featured several types of bound states, namely widely-separated as well as closely-spaced bound 100-fs solitons. Solitons with large separations (of more than 1 ps) are found to be identical, whereas closely-spaced pulses are found to be different in most cases. In this reference, an extensive theoretical study was also undertaken, on the basis of a Ginzburg-Landau model equation, and found to be compatible with the observed bound states. We shall explain the model and its predictions in more detail in Sect. 4.

3 Multi-Soliton Complexes in Various Fiber Laser Configurations

When operating with pulse durations above 100 fs, and non-zero group velocity dispersion, fiber lasers can often be successfully modeled without taking higher-order dispersion terms into account. These lasers are then a perfect choice to study multi-soliton formation in passively mode-locked cavities and to compare with theoretical predictions. In 2001, there were new results concerning bunched- and bound-soliton states in fiber lasers.

3.1 New Light on Bunched Solitons

Considering bunched states of more than two pulses, the experiments that followed were performed with very different laser cavity set-ups. However, the set-ups had in common the tailoring of the path-averaged cavity dispersion with the use of both normal and anomalous fibers in the cavity. This is quite practical for silica fibers operating around 1550 nm. When the resulting average dispersion is low compared to the local dispersion in the fibers, it may also be referred to as "dispersion management", as in the field of optical transmission lines [48, 49]. Dispersion management has led to the concept of the dispersion-managed soliton ("DM soliton") [50]. Although the transmitted pulse is never stationary inside the repeated unit, which is called a map, at each point of the map, the temporal profile of the pulse converges towards a fixed profile, which is called the DM soliton. DM solitons approach chirp-free solitons at certain locations (usually two: one in the anomalous fiber and one in the normal fiber), but otherwise have a relatively high degree of frequency chirping.

In [51], Seong et al. reported the self-adjustment of pulses, under pump power changes, in order to form a stable train of solitons with separations in the nanosecond scale. The laser was a figure-of-eight type, featuring, for mode-locking, a dispersion-imbalanced nonlinear optical loop mirror [52] whose length was more than 120 meters.

In [53], the cavity was a dispersion-managed 7-meter-long cavity operating in a slightly anomalous path-averaged dispersion regime (+3 ps/(nm km)). More details on this set-up are given in Sect. 3.3. Bunching of an increasing number of 450-fs-solitons, with equal spacings of 20.7 ps, was reported to stabilize the cavity stationary state when the pump power was increased. We give details on this experiment below.

For single pulse operation, let us first recall that, when the pump power is increased, the pulse energy tends to increase, whereas its minimum duration in the cavity tends to reduce. This general trend of mode-locked lasers is compatible with the Schrödinger soliton theory, since the $N = 1$ soliton energy should be inversely proportional to its duration. This trend is also associated with the following closely-related behaviors of the multi-pulse regime, namely, that there is a "fuzzy" energy quantization in the cavity with respect of the number of pulses, and that there are hysteresis and multi-stability with respect to the number of pulses, when a main cavity parameter is varied.

With an increase of the pump power, many fiber laser experiments also include a growth of the pulse wings. These wings are related to the formation of spectral side-bands. These side-bands represent a phase-matched process between some frequency components of the soliton spectrum and some spectral components of the dispersive waves radiated out of the perturbed soliton. Such side-bands were described by Kelly in 1992, and later have been referred to as "Kelly side-bands" [54]. We shall emphasize the importance of these side-bands in the formation of multi-soliton states in Sect. 5.

Figure 1 presents one aspect of the self-stabilization process, *viz.* stabilization through additional pulse formation. This figure presents, on a logarithmic scale, the central part of a background free auto-correlation trace. The curve (a) corresponds to a low-energy single soliton in the cavity (15 pJ at the laser output), and is compatible with a hyperbolic-secant temporal profile over 3 orders of magnitude. The pump power is increased from 70 to 90 mW, and this gives the auto-correlation (b), always for single pulse operation (25 pJ at the output), but this time featuring broader wings with a small pedestal. At such a point, the output intensity develops 10 to 20 percent of oscillations in the kHz range. The pulsed stationary regime is destabilized. As the pump power is further increased, the laser dynamics answers by creating a second soliton pulse. The two solitons have identical energy and typically travel separated by 20 ps. Quite remarkably, the central part of the auto-correlation trace (c) takes the same profile as the one obtained with one single low energy soliton, (a). The non-solitonic wings have almost vanished, whereas output intensity kHz oscillations disappear.

One other aspect of self-stabilization is the formation of soliton bunches, once several solitons co-exist in the cavity. In the experiment described here [53], bunching was leading to the most stable stationary states, allowing the

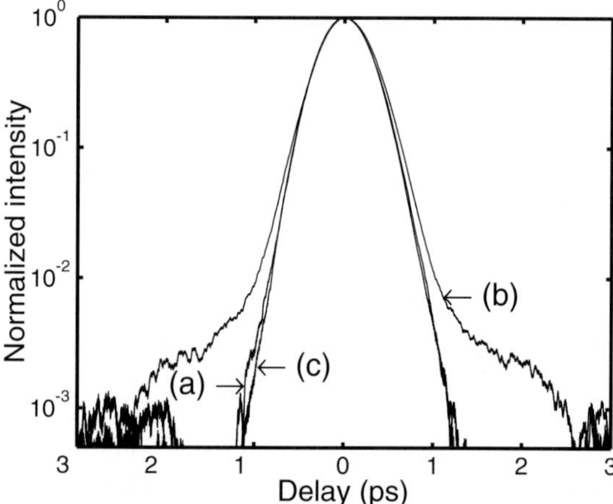

Fig. 1. Stabilization through additional soliton creation. In the above background-free auto-correlation traces, the maximum intensity has been normalized to unity. (**a**) low-energy single soliton (15 pJ) (**b**) higher-energy single soliton (25 pJ) and (c) after formation of a two-soliton bunch (31 pJ in total). From [53]

laser to remain mode-locked with the same multi-pulse characteristics for several hours, without any external stabilization of the cavity length or temperature. The build-up of a bunch of solitons is presented in Fig. 2. To add more solitons into the bunch, more pump power could be coupled, and some small adjustments of the cavity could also be made. Under these changes, Fig. 2 shows that the separation between subsequent solitons remained constant (20.7 ± 0.1 ps). As expected with identical solitons in the bunch, the overall shape of the auto-correlation function is triangular.

In this experiment, although solitons were separated by more than 40 pulse widths, the temporal cross-correlation between the first and the last solitons of the pulse train showed no evidence of broadening when compared with the central peak of the auto-correlation trace. It indicated that the level of timing jitter was extremely low, if any. Further investigations with a high resolution monochromator showed that the optical spectra were bearing high-contrast interference patterns, indicating that the different solitons had a fixed-phase relationship between themselves [55]. An example of such a channeled spectrum is displayed in Fig. 3. The fixed-phase relationship between solitons is linked with the expected properties of stationary multi-soliton states (also called bound solitons in the literature). The experimental observation of multi-soliton bound states is detailed in the following section.

3.2 Multi-Soliton Bound States

Bound soliton pairs were first reported in a 10-m long fiber cavity operating in the normal dispersion regime [56]. They were characterized by 340-fs pulses

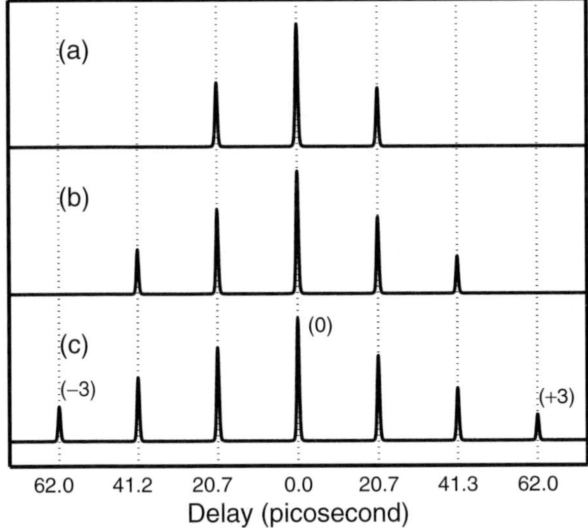

Fig. 2. Auto-correlation traces showing the buildup of a train of regularly-spaced identical solitons, comprising (**a**) two, (**b**) three, and (**c**) four solitons. The total output energy is (**a**) 34 pJ, (**b**) 61 pJ, and (**c**) 64 pJ. From [53]

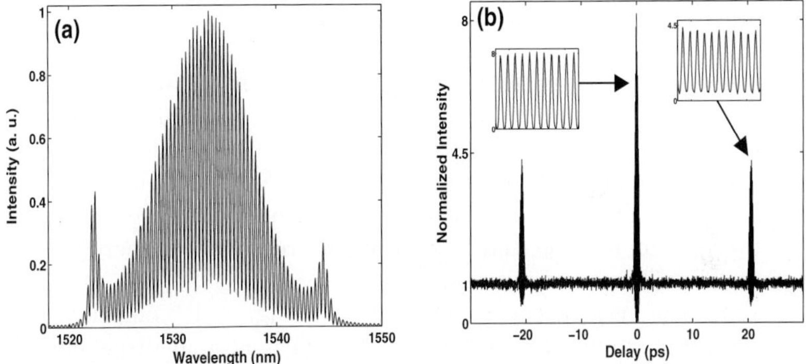

Fig. 3. Spaced soliton pair featuring a fixed phase relationship. (**a**) Channeled optical spectrum, and (**b**) corresponding interferometric auto-correlation trace. The two side peaks in (**b**) represent temporal cross-correlation between the two solitons, and bear fringes of interference due to the fixed phase relationship between them. From [55]

separated by either 1.16 ps or 2.28 ps. The most striking experimental feature was the ability to generate and stabilize several soliton pairs in the cavity, with one soliton pair acting as a new soliton unit.

The discovery of bound soliton pairs featuring a near $\pm\pi/2$ phase relationship [57], as expected from theoretical work based on the Ginzburg-Landau model (see Sect. 4.1), revived the comparison between experimental results and dissipative multi-soliton complexes, with the help of numerical simulations. The experimental set-up used is detailed in Sect. 3.3, while related experimental results are provided in Sects. 3.4 and 3.5. A numerical model of the experimental set-up is given in Sect. 5 to explain most of the experimental results.

3.3 Dispersion-Managed Fiber Ring Laser: Experimental Set-Up

In some way, dispersion management could provide to fiber lasers a situation lying between conventional fiber soliton lasers, where pulse shaping is dominated by propagation in one main length of anomalous fiber, and solid-state bulk lasers, where pulses experience propagation through completely different lumped elements. In fact, dispersion-managed fiber lasers change their emission characteristics greatly when the path-averaged dispersion and the pump power are varied [58]. The first implementation of strong dispersion management in a fiber laser [59] was motivated by the desire to achieve shorter pulses and higher energies than in past fiber laser cavities. The idea was to build a cavity made of segments of large positive- and negative-dispersion fibers, so that the pulse would experience temporal stretching in each section. The stretching factor depends firstly on the path-averaged dispersion and secondly on the pump power. For normal path-average dispersion, and large pump power, it can be as large as 10 to 50. Thus, relative to constant-dispersion fiber lasers, the peak-intensity is lower most of the time, as is the accumulated nonlinearity. This then results in the possibility of getting as much as 1 nJ intra-cavity with sub-100 fs pulses [60], whereas, when used in the slightly anomalous dispersion regime, the laser emits average energy soliton-like pulses. The latter situation is quite convenient for our purpose of getting several (but not too many) solitons of "reasonable" energy in order to study multi-soliton bound states.

The fiber ring laser set-up is illustrated in Fig. 4. It comprises, in series, a 1400-ppm erbium-doped fiber (EDF) with normal dispersion $D = -40$ (ps/nm)/km, a polarization-insensitive coupler-isolator (WDM-IS) into which 980-nm pump light is injected, an optional 10 percent output coupler, a length of SMF-28 fiber with anomalous dispersion $D = +16.5$ (ps/nm)/km, and a 50-cm long open-air section containing wave-plates and polarizing cubes. Due to nonlinear polarization evolution that takes place during propagation in the fibers [7], the transmission through the polarizer P1 is intensity-dependent, and an appropriate adjustment of the preceding wave-plates triggers the mode-locking operation. Depending on which experiment is performed, the EDF length is adjusted between 1.6 and 2.5 m, while the choice of the SMF-28 length sets the path-averaged cavity dispersion either in a slightly anomalous regime $\bar{D} < 5$ (ps/nm)/km, or in a slightly normal regime $\bar{D} > -5$ (ps/nm)/km. The field can be monitored at three different

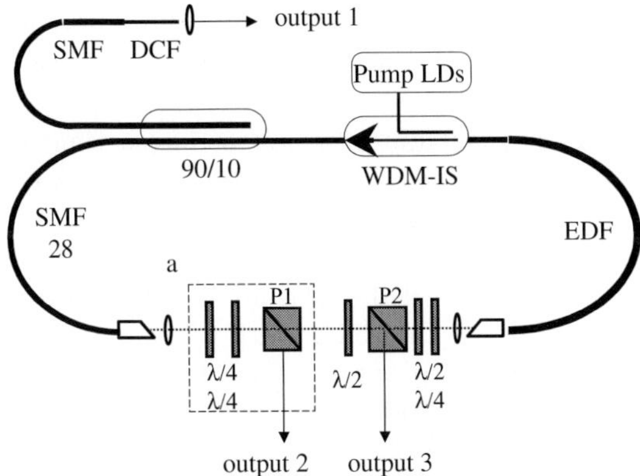

Fig. 4. Experimental set-up of the fiber ring laser

output ports, labeled 1, 2 and 3. Only output 1 is always required for mode-locking operation, while the other ports are optional. Port 2 gives a field proportional to the field re-injected in the cavity, whereas Port 3 is used in the last series of experiments (Sect. 3.5) in order to compress the chirped pulses by using an appropriate length of dispersion compensating fiber (DCF) spliced in series at this output.

Temporal characterization of the output intensity is performed using an optical auto-correlator that uses a 1-mm thick beta-barium borate (BBO) crystal in type I second harmonic generation (SHG) and that can be set in interferometric or non-interferometric mode. For spectrum analysis, we use a high-resolution monochromator (HR 640).

3.4 $\pm\pi/2$ Phase-Locked Soliton Pairs

The laser self mode-locks for a pump power above 35 mW. Operating with a slightly anomalous path-averaged dispersion of about +3 ps/nm/km, we observe low-pedestal sech-profiled pulses, whose FWHM durations vary between 400 fs and 650 fs. Multi-soliton bunched states are easily obtained when the pump power is increased above 80 mW, as described in [53, 55].

When searching for bound soliton pairs, most of the time the pairs are formed with a large separation of around 20 ps between the two solitons. It is then difficult to determine the relative phase between them. However, it is possible to act on the pair separation by adjusting the pump power. When the pump power is reduced, this separation changes, step-wise, to smaller values. Let us point out that the step-wise variation, in contrast with a continuous variation, is linked to the high stability of the operation regime in which a pulse pair remains precisely phase-locked, even in the presence of small perturbations. When the separation is less than about 15 pulse widths, and if the envelope of the spectrum (which would correspond to a single soliton spectrum) is symmetric, then the phase relationship between the two phase-locked solitons can be inferred with good precision from the position of the fringes in the pair spectrum.

Figure 5a shows the spectrum of a pair of interacting pulses. It is well-fitted by the spectrum resulting from two 610-fs FWHM sech-profiled pulses separated by 6.8 ps, and having a phase difference of $\pi/2$. Pump power is 31 mW.

The 610-fs pulse-width was deduced from the auto-correlation trace, and the time-bandwidth product of 0.32 ± 0.01 is compatible with unchirped sech-profiled pulses. Another proof of phase locking is obtained with the optical auto-correlation set in the interferometric mode. The trace is shown in Fig. 5b, and the presence of lateral peaks, as well as their equal amplitudes, confirms the pulse separation. Moreover, the fact that the lateral peaks comprise fringes, as shown in the insets, proves that there is a stable phase relationship between the two pulses. The auto-correlation traces have been normalized so that the SHG average recorded intensity is unity where the pulses do not overlap. With this normalization and when the pulses have identical amplitudes, the fringes of the lateral peaks are expected to extend between 0.5 and 4.5.

Another spectrum, recorded at a lower pump power of 29 mW, is displayed in Fig. 5c and is fitted by the spectrum of two pulses separated by 2.7 ps and having a $-\pi/2$ phase relationship. The corresponding interferometric auto-correlation function is shown in Fig. 5d. The $\pm\pi/2$ phase relationship, calculated from the previous spectra, had been predicted theoretically using a model [61] based on the complex Ginzburg-Landau equation (CGLE) (see Sect. 4.1).

It is highly remarkable that this relation holds for pulses separated by more than 10 pulse-widths, in a highly perturbed dynamical system such as a stretched-pulse fiber laser. However, this relation is obtained when the interaction between the two solitons is kept quite low. The previous condition is fulfilled either when the soliton energy is low, which is the case in the present situation, or when solitons are situated far apart. In the latter situation, an individual soliton can therefore possess a higher energy. This is typically the case for solitons separated by around 20 ps, where a phase relationship close to $\pm\pi/2$ could also be deduced from recent investigations, using the "Kelly sidebands" as frequency markers in the spectrum [62]. In otherwise general experimental conditions, the phase relationship between the two solitons is not restricted to $\pm\pi/2$, which is apparently not compatible with the predictions obtained from the CGLE model. Another experimental fact which is not supported by the CGLE model is the observation, for the same experimental parameters, of a "rosary" of possible separations in the 19–22 ps range between the solitons. All these experimental facts need further explanation, requiring a non-distributed model as a more realistic model [62]. Such a non-distributed model is addressed in Sect. 5.

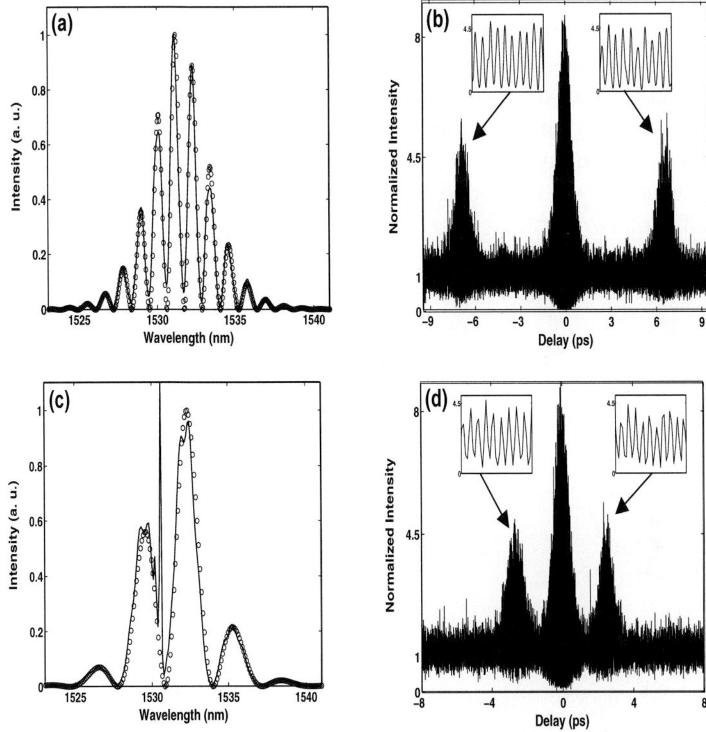

Fig. 5. Experimental observation of $\pm\pi/2$ phase-locked soliton pairs. (**a**) Optical spectrum of a bound soliton pair of two 610-fs solitons separated by 6.8 ps (in *solid line*), fitted with a phase relationship of $-\pi/2$ (in *circles*). (**b**) Corresponding interferometric auto-correlation trace. (**c**) Optical spectrum of a pair of 540-fs solitons separated by 2.7 ps (in *solid line*), fitted with a phase relationship of $+\pi/2$ (in *circles*). From [57]

3.5 Soliton Pairs, from Anomalous to Normal Average Dispersion Regime

As the previous results were obtained in the anomalous average regime, we may wonder if bound soliton pairs could form in a dispersion-managed fiber cavity operating in the normal dispersion regime. By removing an appropriate length of anomalous fiber, the average dispersion is set to $D = -2.5(\pm 0.5)$ ps/nm/km. As expected for operation in the normal regime, the required injected pump power for mode-locking is significantly higher than in the anomalous regime [58]. A pump power above 150 mW is required for the formation of single pulses, whereas the spectral width and the pulse energy have increased when compared with the anomalous regime.

Increasing the injected pump power above 250 mW, we observe the formation of pulse pairs or triplets. According to the mode-locking settings and to the pump power, several stable pulse-to-pulse separations can be observed, from less than 1 ps to more than 10 ps. We detail here the observation of a close pulse pair. Auto-correlation functions taken from outputs 2, 3 and from output 1 without DCF have the same, somewhat triangular shape (see Fig. 6a), whereas the recorded spectra all present the characteristic fringe pattern of a pulse pair.

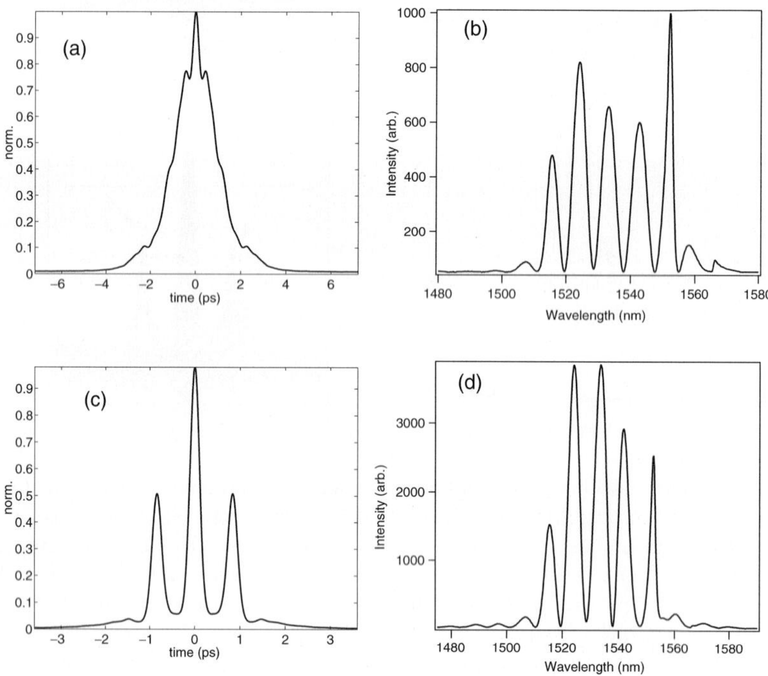

Fig. 6. Close bound soliton pair, observed in the normal path-averaged dispersion regime. (**a**) Auto-correlation trace recorded at output 3 of the experimental setup and (**b**) corresponding optical spectrum. (**c**) Compressed soliton pair at output 1, with DCF fiber, and (**b**) corresponding optical spectrum. From [85]

The channeled spectrum structure, as shown in Fig. 6b, has a high contrast which reveals precise phase-locking of the two pulses. According to the spectrum, we should find two pulses separated by around 850 fs. The spectral envelope has a width close to 35 nm, which means it could support 100 fs FWHM minimum pulse durations, assuming a Gaussian pulse shape. Thus, Fig. 6a would be the auto-correlation function of two closely-separated chirped pulses. In order to verify this, we spliced 52 cm of DCF to the end of the SMF fiber at output 1. This DCF length compensates for the dispersion accumulated by the pulses in the SMF fibers when traveling from the end of the EDF to the end of the SMF, after the 10% output coupler. Gradually reducing the DCF length, we obtained an optimum compression with a length of 36 cm, revealing the pulse pair structure. The corresponding auto-correlation trace and spectrum are displayed in Fig. 6c and d. The asymmetry of the spectrum envelope prevents a direct accurate retrieval of the phase relationship, as was performed in Sect. 3.4. However, phase retrieval could be achieved by using a more elaborate measurement set-up, such as frequency-resolved optical gating [63].

The auto-correlation trace is compatible with two pulses of the same amplitude, separated by 870 fs. The main peak of the auto-correlation trace has a 230 fs FWHM, yielding 150 fs FWHM pulses. However, we note the presence of small humps in the peak's sides, which means that the compressed pulse pair is not the superposition of two pure Gaussian pulses.

As was stated, in the case of a single laser pulse [58], the pulse shape is greatly affected by a change in the cavity path-averaged dispersion, from the anomalous to normal regime. Spectral width and pulse energy are larger in the normal regime than in the anomalous regime. In our experiment, the pulse pair has an intra-cavity energy of 0.65 nJ, which is around 10 times the intra-cavity energy of the soliton pairs reported in [57], for the anomalous regime. As the pulse propagates in the cavity, the pulse duration undergoes large changes due to the dispersion management effect. This is the so-called "stretched-pulse" operation of the fiber laser. An estimate of the stretching factor is 10 in the present experiment, whereas it is less than 3 in the anomalous regime. A more detailed comparison between soliton pairs in the anomalous regime and soliton pairs in normal regime is provided, along with the numerical simulations, in Sect. 5.2.

4 Multi-Soliton Complexes in Distributed Models

It is common practice [64] to model a laser as a distributed system if the pulse shape changes only slightly during each round trip. There are many advantages in using a distributed model, governed by a continuous equation, since it allows, to some extent, an analytical study. We can look for analytical solutions, either exactly or approximately, determine their stability through linear stability analysis, apply some perturbation theory [65], search for conserved quantities, etc. Even numerical simulations, which are often needed, are usually simpler.

4.1 The Complex Cubic-Quintic Ginzburg-Landau Equation

The cubic-quintic Ginzburg-Landau equation (CGLE) is a generic equation used in different fields for describing a number of physical phenomena. In optics, passively mode-locked lasers are often modeled using this equation as a continuous approximation to the field dynamics inside the cavity [66]. In this context, this equation has the following form: [67]:

$$\mathrm{i}\psi_z + \frac{D}{2}\,\psi_{tt} + |\psi|^2\psi + \nu|\psi|^4\psi = \mathrm{i}\delta\psi + \mathrm{i}\epsilon|\psi|^2\psi + \mathrm{i}\beta\psi_{tt} + \mathrm{i}\mu|\psi|^4\psi\,, \qquad (1)$$

where z is the normalized propagation distance along the cavity, t is the retarded time, ψ is the normalized envelope of the field, D is the group velocity dispersion coefficient, with $D = \pm 1$, depending on whether the group velocity dispersion (GVD) is anomalous or normal, respectively, δ is the linear gain (/loss) coefficient. Furthermore, $\mathrm{i}\beta\psi_{tt}$ accounts for spectral filtering or linear parabolic gain ($\beta > 0$), $\epsilon|\psi|^2\psi$ represents the nonlinear gain (which arises, e.g., from saturable absorption), while the term with μ ($\mu < 0$) represents the saturation of the nonlinear gain, and the term with ν corresponds, if negative, to the saturation of the nonlinear refractive index.

Stationary solutions of the CGLE have been found in a number of papers. For specific relations between its parameters, analytical solutions can be obtained [68, 69]. In the general case, localized solutions with exponentially decaying tails can be easily found [70] using a shooting method to solve an ordinary differential equation. Linear perturbative analysis can then be carried out to determine their stability. Alternatively, the full (1) can be solved to look for stable localized solutions [69, 71], which we shall call solitons.

We are interested here in studying the interaction between solitons. For this purpose, we assume that, from any of the above-mentioned methods, we already know the field profile of the soliton. Let us call it $\psi(t,z) = \psi_\circ(t)\exp(\mathrm{i}qz)$. Because a double balance, between dispersion and nonlinearity, and gain and losses must be satisfied, only one or a few soliton solutions exist for a given set of equation parameters. This fact implies that, during the interaction of two solitons, basically only two parameters change: their separation, ρ, and their phase difference, α [61]. Therefore, the formation of pairs of solitons can be analyzed in the corresponding 2-D space, named the interaction plane [72], where these two variables (ρ, α), are the polar coordinates describing each possible situation of the pair of solitons. Specifically, we assume that the evolution of a pair of solitons can be well-approximated by:

$$\psi(t,z) = \psi_0(t - \rho/2) + \psi_0(t + \rho/2)\exp(i\alpha)\,, \qquad (2)$$

where both ρ and α are functions of z.

Although the CGLE has no known conserved quantities, some equations can be written down for the rate of change of the energy (Q) and momentum (M) with respect to z, namely:

$$\frac{d}{dz}Q = F[\psi]\,, \qquad (3)$$

where $Q = \int_{-\infty}^{\infty} |\psi|^2\, dt$, and the functional $F[\psi]$ is given by

$$F[\psi] = 2 \int_{-\infty}^{\infty} \left[\delta|\psi|^2 + \epsilon|\psi|^4 + \mu|\psi|^6 - \beta|\psi_t|^2\right] \, dt \, . \tag{4}$$

Similarly, the momentum is $M = Im \left(\int_{-\infty}^{\infty} \psi_t^* \psi \, dt \right)$, and its rate of change is defined by

$$\frac{d}{dz} M = J[\psi] \, , \tag{5}$$

where the real functional $J[\psi]$ is given by

$$J[\psi] = 2 \, Im \int_{-\infty}^{\infty} \left[(\delta + \epsilon|\psi|^2 + \mu|\psi|^4)\psi + \beta\psi_{tt}\right] \, \psi_t^* dt \, . \tag{6}$$

For stationary pairs of solitons, the allowed values of ρ and ϕ should keep Q and M constant, i.e.:

$$F[\psi] = 0 \tag{7}$$

and

$$J[\psi] = 0 \, . \tag{8}$$

Figure 7a shows the zeros of these functionals on the interaction plane for $\delta = -0.01$, $\beta = 0.5$, $\mu = -0.05$, $\nu = 0$ and $\epsilon = 1.8$. The zeros of F and J are found in the interval $0.4 < \rho < 4$, because the separation ρ must be of the same order as, but larger than, the width of a single soliton (indicated by a dashed circle in the figure). Smaller ρ correspond to merging of the solitons while at larger ρ, the interaction between the solitons is too weak. The solid lines in Fig. 7 show the locus of points where $F[\psi] = 0$, while the dotted lines show those where $J[\psi] = 0$. It can be seen from this figure that the functional $F[\psi]$ has two zeros in the interval $-\pi/2 < \phi < \pi/2$, but only one zero in the interval $\pi/2 < \phi < 3\pi/2$. (Here we have not included the obvious zero at $\rho = \infty$).

The functional $J[\psi]$ is zero everywhere on the horizontal axis of the interaction plane. This shows that every intersection of a solid curve with the horizontal axis corresponds to a bound state of two solitons. There are three examples of this type of bound state in Fig. 7a – they are labeled $S_i, i = 1, 2, 3$. These solutions have 0 or π phase difference between the component solitons. In addition, the functional $J[\psi]$ has zeros along two almost circular curves. The intersections of the outer circle with the solid curve (points F_1 and F_2) correspond to the new bound states where the phase difference between the solitons is close to $\pi/2$. These predictions depend on the validity of the assumption that (2) is a good approximation. Furthermore, the above analysis does not provide any information about the stability of the predicted bound states of two solitons. At this stage, one needs to solve the CGLE numerically, taking, as the initial condition, the superposition of two stable solitons with

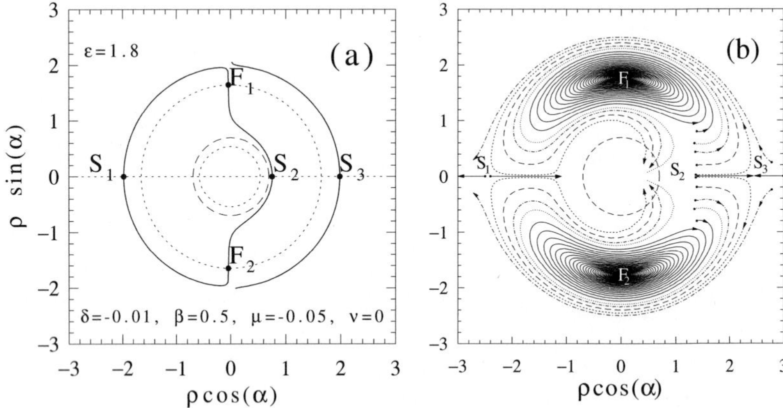

Fig. 7. (**a**) Zeros of $F[\psi_0, \rho, \phi]$ (*solid lines*) and $J[\psi_0, \rho, \phi]$ (*dotted lines*) on the interaction plane for the parameters shown in the figure. The intersection points (*bold points*) correspond to bound states of two solitons. (**b**) Trajectories showing the evolution of two-soliton solutions. The five singular points in (**b**) correspond to the five bound states depicted in (**a**). Trajectories converging to the center describe the merging of two solitons. From [61]

arbitrary separation (ρ) and phase difference (α) between them. Figure 7b shows the results of such simulations. It indicates that, for the given set of parameters, there are five singular points. Within the accuracy of the method, these coincide with the solutions which are found above using the balance equations. The differences, which are very minor, have their origin in the fact that the exact bound solutions differ slightly from those which are approximated by (2). Three of these singular points (S_1, S_2 and S_3) are saddles, with the phase difference between the solitons being zero or π. Clearly, these are unstable bound states of two solitons, and indeed they were unstable for all the sets of parameters which we used in our numerical simulations. In addition, there are two symmetrically-located stable foci (F_1 and F_2) which correspond to stable bound states of two solitons in quadrature, i.e. the phase difference between them is plus or minus $\pi/2$. These are the bound states with asymmetric phase profiles predicted above.

Changing the parameters in the (1) may convert stable foci into centers (or elliptic points) and further into unstable foci. They can even disappear [61]. On the other hand, we cannot rule out the possibility that, in the frame of the CGLE, pairs of solitons with phase differences other than $\pi/2$ could be stable for some other parameters. Thus far, at least, they have not been found.

As a consequence of the existence of two-soliton solutions, three- and more soliton solutions also exist. [73]

4.2 Swift-Hohenberg Equation

The spectral filtering in the CGLE model is limited to the second-order term and can only describe a spectral response with a single maximum. In the experiment, the gain spectrum is usually wide and may have several maxima. It is clear that, within the scope of ultra-short pulse generation, higher-order filtering terms may be necessary, both to make the model more realistic and to describe more involved pulse generation effects.

The addition of a fourth-order derivative term, $i\gamma\Psi_{tttt}$, into the CGLE transforms it into the quintic-complex Swift-Hohenberg equation (CSHE).γ is a complex coefficient that can be expressed as $\gamma = \gamma_1 + \mathrm{i}\gamma_2$, where γ_1 is a higher-order dispersion coefficient and γ_2 takes into account the more complicated spectral response mentioned above.

Ignoring the fourth-order order dispersion, the CSH equation with the quintic nonlinear term can be written in the form:

$$\mathrm{i}\psi_z + \frac{1}{2}\psi_{tt} + |\psi|^2\psi + \nu|\psi|^4\psi = \mathrm{i}\delta\psi + \mathrm{i}\epsilon|\psi|^2\psi + \mathrm{i}\mu|\psi|^4\psi + \mathrm{i}\beta\psi_{tt} + \mathrm{i}\gamma_2\psi_{tttt}\,. \tag{9}$$

Therefore, the last term in (9) is the additional one with respect to the CGLE (1). In principle, the double peak structure of the spectral response could facilitate the generation of double-pulse solutions, as their spectra could match the spectral response of the system. In particular, this could apply with double pulse solutions featuring a π-phase difference, where the optical spectra which comprise two equal maxima could match the spectral response of the system. This indeed occurs in the numerical simulations.

Figure 8a, c, shows two stable pairs of solitons obtained for this equation for the following parameters: $\delta = -0.5$, $\beta = 0.3$, $\nu = 0$, $\mu = -0.1$,$\gamma_2 = 0.05$. Phase and intensity profiles are plotted in the same part of the figure, while their corresponding spectra are plotted in parts (b) and (d). Apart from the usual pair of solitons with $\pi/2$ phase difference, the double-pulse solution with π phase difference also appears to be stable for some values of the parameters. The 2 types can even co-exist for the same parameters. Further, there are two sorts of double-pulse solutions with approximately π phase difference between the pulses; they differ in their pulse separations. However, they appear at different values of ϵ . Pulsating solutions formed by two pulses whose phase difference oscillates around π have been also reported with this model [74].

4.3 Non-Local Model

Sometimes, the response of the mechanism responsible for passive mode-locking in the laser is not instantaneous. The use of a semiconductor saturable absorber mirror [75] with a relatively slow response time and even slower

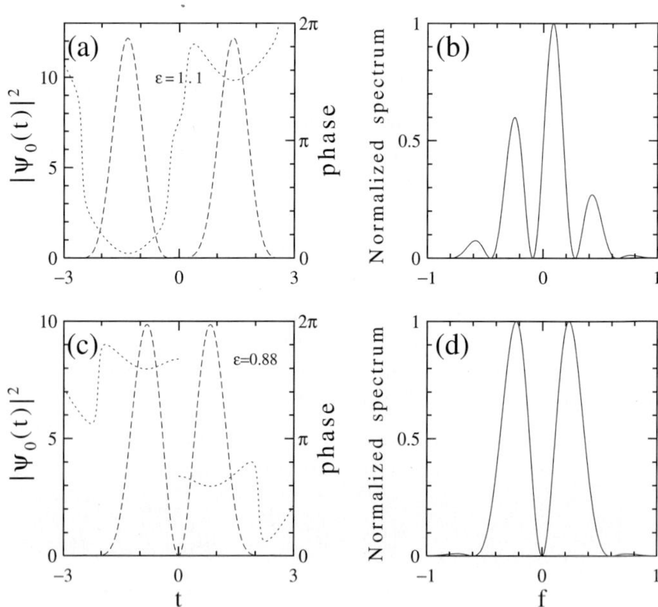

Fig. 8. Intensity (*dashed line*), phase (*dotted line*) profile and corresponding spectra (*continuous line*) for two different pairs of solitons. (**a**) (and (**b**)) with $\pi/2$ phase difference between the pulses and (**c**) (and (**d**)) with π in the CSHE model. From [74]

recovery time has been suggested for a passively mode-locked soliton laser [76]. Experimental verification of this possibility has been reported in [77].

Assuming, as before, that in each round-trip the pulse shape hardly changes, the dynamics can be described by a modified CGLE, namely: [76, 78]:

$$\mathrm{i}\psi_z + \frac{D}{2}\psi_{tt} + |\psi|^2\psi = \mathrm{i}[g(Q) - \delta_s(|\psi|^2)]\psi + \mathrm{i}\beta\psi_{tt} \, , \tag{10}$$

where $g(Q)$ is the cavity gain and $\delta_s(|\psi|^2)$ represents the losses in the cavity and in the slow saturable absorber. The remaining variables and parameters have the same meaning as in the previous cases.

The gain term $g(Q)$ in (10) describes a typical solid state laser gain medium with a recovery time much longer than the round-trip time of the cavity. Therefore $g(Q)$, which accounts for gain depletion, depends on the total pulse energy in the following way:

$$g(Q) = \frac{g_0}{1 + \frac{Q}{E_L}} \, . \tag{11}$$

Here E_L is the saturation energy and g_0 is the small signal gain. The value of $g(Q)$ decreases as the energy increases, so that only a limited number of pulses can exist inside the cavity.

The loss modulation in the saturable absorber can be described by the following rate equation [76]:

$$\frac{\partial \delta_s}{\partial t} = -\frac{\delta_s - \delta_0}{T_1} - \frac{|\psi|^2}{E_A}\,\delta_s\,, \tag{12}$$

where T_1 is the recovery time of the saturable absorber, δ_0 is the loss introduced by the absorber in the absence of pulses, and E_A is the saturation energy of the absorber.

Soliton solutions of (10) exist for a certain range of the parameters. For the pulse to be stable, the gain g must depend on the energy Q as in (11). The Q-dependence of g serves as a feedback mechanism which stabilizes the pulse for a given value of E_L. When the pump power is high enough, two or more pulses may be generated sequentially. Because the total gain/loss is influenced by the pulses, two co-existing pulses inside the cavity influence each other, and the distance between them can be fixed.

The maximum number of pulses per round trip is determined by the gain parameters E_L and g_0 for otherwise fixed parameters. In the regime where two pulses are possible, they can propagate independently if they are far apart from each other. On the other hand, if the pulses get close to each other, they can form a bound state. The distance and phase difference between the pulses are determined by complicated dynamics inside the cavity, and especially by the balance between gain and loss [61]. The new feature of bound states here is that, due to loss relaxation, the pulse shapes (and their amplitudes) also depend on the relative distance between them. This, in turn, influences the overall balance between gain and loss. There are two types of bound states for the same set of parameters. One is asymmetric, with unequal pulse amplitudes (see Fig. 9a), and the other is symmetric, with equal pulse amplitudes (see Fig. 9b). We denote the former pair as A-type and the latter as B-type.

Solitons in A-type pairs are bound due to the balance of gain and loss at a certain distance ρ_A between the pulses. The pulse amplitudes also vary with the relative distance because the gain/loss difference for the second pulse is different from that for the first. Then, the overall balance is determined by the distance. The phases of the two solitons are independent. Due to the unequal amplitudes, the rates at which their phases change are different, and then the phase difference between the solitons increases monotonically. Trajectories on the interaction plane [61] corresponding to these pairs are almost circles around a certain point which is close to the origin but does not coincide with it (see Fig. 9c). The shift of the center of the orbit from the origin indicates that the overall balance also depends on the relative phase difference between the two pulses in the bound state – pulses tend to approach when they are in-phase and separate when they are out-of-phase. This periodic change from attraction to repulsion, and vice versa, results in zero average force between the pulses at a certain pulse separation. There is

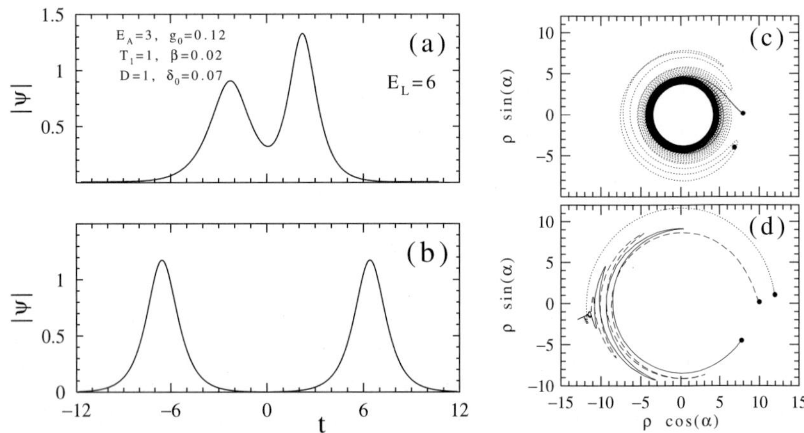

Fig. 9. Stable soliton pairs of (**a**) type A and (**b**) type B. The parameters used in the simulation are shown in the figure. Trajectories showing the evolution of (**c**) A-type and (**d**) B-type two-soliton solutions on the interaction plane. The *solid* circle shows the starting point. From [79]

a limit cycle to which all nearby trajectories converge. As a result, the bound state A is stable.

The two solitons in a B type pair are located at a distance (ρ_B) much larger than the recovery time T_1. Hence, the interaction is small and weakly influences the amplitudes of the two pulses. Nevertheless, the two solitons interact through their tails (the field amplitude between the pulses is $\approx 10^{-2}$). Although the interaction is small, its presence is indicated by the fact that the phase difference between the pulses in this case is fixed and close to π. Due to the asymmetry of the mutual interaction, the phase difference is slightly less than π. In this case, the equilibrium separation between the pulses is due to the balance between the repulsion of solitons with opposite phases and their attraction through the gain/loss mechanism. The fixed point in this case appears to be the stable focus. Trajectories which start from distances around the equilibrium converge to this singular point (see Fig. 9d). The convergence shows that the pair B is also stable.

Triplets also exist in two types [79]. Either two pulses are combined into a pair of type A and the third one is at a distance ρ_B from them, or each separation is equal to ρ_B. Four or more soliton solutions are also possible. The multi-pulse generation described above is in qualitative agreement with experimental observations (see [47]).

5 Non-Distributed Model

In a real laser, the different effects occur in different parts of the cavity. It usually happens that, in each round-trip, the solution experiences such large changes that distributed models turn out to be inappropriate. And even if this were not the case, they would still fail to explain any phenomena arising from the periodicity of the effects taking place at each round-trip inside the laser cavity. A first attempt to include the periodicity in the model was made by using the CGLE with periodically-changing equation parameters [80, 81].

In this section, we shall describe a model of the fiber ring laser in which the most recent experimental results on multi-soliton complexes have been observed [57]. The model includes all the main physical effects taking place in the fiber laser, and at the same time contains a reduced number of parameters. A more accurate model [55] has also being used, but the results obtained are practically the same, although the number of parameters needed is higher.

The model is shown in Fig. 10. It consists of an Erbium-doped fiber (EDF) taken as isotropic, a quarter wavelength plate, a birefringent single-mode fiber (SMF) and a polarizer. Linear losses are incorporated between these elements at points a), b), c) and d). The wave propagation in the EDF is described by the NLS equation with an additional gain term featuring saturation and limited bandwidth:

$$\mathrm{i}U_z + \frac{D}{2}U_{tt} + \Gamma|U|^2U = \mathrm{i}g(Q)U + \mathrm{i}\beta U_{tt} \,, \tag{13}$$

where $D = \beta_2^{Er}/\beta_2^{SMF}$ is the dimensionless dispersion parameter, where $\beta_2^{Er,SMF}$ are the dispersion coefficients related to the two pieces of fiber, respectively. As the EDF operates in the normal dispersion regime, and the SMF in the anomalous regime, D is thus negative. $\Gamma = A_{eff}^{SMF}/A_{eff}^{Er}$ (A_{eff} being the effective area in each type of fiber) and β is the strength of the spectral filtering due to the gain limited bandwidth. The function $g(Q)$ represents the gain in the Er-doped fiber: $g(Q) = g_0(z)/(1 + Q/E_L)$, where g_0

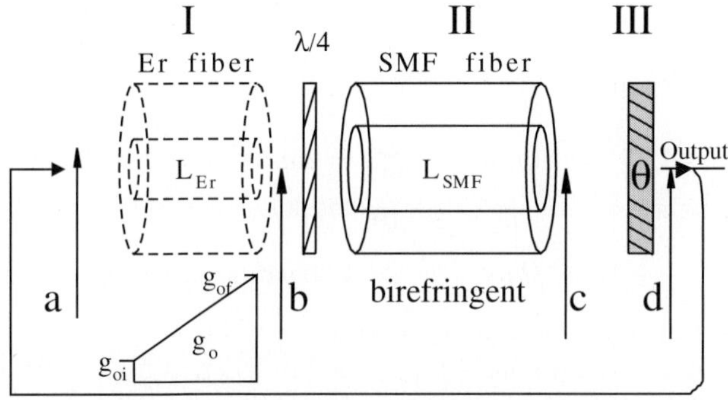

Fig. 10. Scheme for the numerical model

is the small signal gain and E_L is the saturation energy. The small signal gain generally depends on the position z in the fiber, because of pump depletion. However, when the fiber length and the pump power are appropriately chosen, this dependence is small. Thus, in the frame of a model with a minimal number of parameters, the z dependence of the small signal gain can be left out. We have checked that, using a first-order approximation such as: $g_0(z) = g_{0i} + (g_{0f} - g_{0i})z/L_{Er}$, where L_{Er} is the length of the EDF, the main physical results are not changed. For the purpose of model simplification, we adopt linear polarization all along the EDF. Nonlinear birefringence, required for pulse formation in this laser, acts during propagation in the passive SMF fiber, described by the two following coupled NLS equations:

$$i\phi_z + \gamma\phi + \frac{1}{2}\phi_{tt} + |\phi|^2\phi + \frac{2}{3}|\psi|^2\phi + \frac{1}{3}\psi^2\phi^* = 0$$

$$i\psi_z - \gamma\psi + \frac{1}{2}\psi_{tt} + |\psi|^2\psi + \frac{2}{3}|\psi|^2\phi + \frac{1}{3}\phi^2\psi^* = 0 \,, \tag{14}$$

where ϕ and ψ are the normalized orthogonal components of the optical field defined at the fiber input by $(\phi, \psi) = (1, i)U/\sqrt{2}$, and γ is the half-difference between the propagation constants. The polarizer III mixes the two components into a single component $U = \phi \cos(\theta) + \psi \sin(\theta)$. The latter is injected into the EDF. Observing the pulse at a certain point of the cavity in each round-trip, we can see its convergence to some stationary profile.

The relation between the dimensionless magnitudes: $z, t, U(\psi, \phi)$ used above and the real ones: $Z, T, E(E_x, E_y)$ is the following:

$$z = Z/Z_o$$

$$Z_o = \frac{T_o^2}{|\beta_2^{SMF}|}$$

$$t = (T - \frac{Z}{v_g})/T_o$$

$$U = E\sqrt{RT_o^2/|\beta_2^{SMF}|}\,, \quad R = \frac{2\pi n_2}{\lambda A_{eff}^{SMF}}$$

Typical values are:

$$A_{eff}^{SMF} = 80\mu m^2 \,, \; A_{eff}^{Er} = 26\mu m^2 \implies \Gamma \approx 3 \,.$$

$$n_2 = 2.5\,10^{-20} m^2/W \,.$$

$$\beta_2^{SMF} = -21\,ps^2/km \;\; (\beta_2^{Er} = +50\,ps^2/km) \implies D = -2.3 \,.$$

Taking $Z_o = 1\,\mathrm{m}$ fixes our time unit to be $T_0 = 145\,\mathrm{fs}$.

The part of the cavity consisting of the birefringent fiber and the polarizer acts as a fast saturable absorber with a transmission coefficient that depends on the instantaneous intensity of the optical field. The laser starts to generate single pulses for a large range of values of the parameters. On increasing the gain, two or more pulses appear. The interaction between them, that gives rise to the formation of multi-soliton complexes, depends strongly on whether the path-averaged group dispersion is anomalous or normal.

5.1 Anomalous Dispersion Regime

Equations (13) and (14) are solved, taking an arbitrary localized input as the initial condition. Fixing the cavity parameters, and properly choosing the angle θ of the polarizer, we observe a solution which converges to a stationary state consisting of one or several solitons. When the two-pulse solution appears, its relative phase difference (α) and separation (ρ) may converge to fixed values.

In order to find more two-soliton stationary states with different values of the phase difference and separation between the solitons, one starts simulations with two single solitons of the above equations with arbitrary relative phase and separation. The single soliton solutions are obtained for the same cavity parameters, but with the gain half of that required to observe two soliton solutions. We let them evolve until a stationary solution is reached. These simulations are repeated with many initial values of (ρ_o, α_o). Each stationary state has a basin of attraction, and the two-pulse solution converges to it. Figure 11 shows several trajectories in the (ρ, α) plane corresponding to different pairs of pulses initially located at arbitrary separations and phase differences. Depending on the initial conditions, the solution evolves, converging to some state with fixed separation and phase difference between the solitons. The phase difference for each of these states is close, but not equal, to $\pi/2$, getting closer and closer to $\pi/2$ as the separation ρ increases. Let us point out that, as a matter of symmetry, the same number of attractors are found around the phase difference of $-\pi/2$.

One of the main consequences of the periodic evolution of the soliton shape is the radiation of small amplitude waves, at fixed frequencies, located symmetrically at each side of the soliton spectrum in the form of side-bands. Spectral side-bands, related to periodic perturbation of a soliton, were described in the original paper by Kelly [54]. More detailed theory based on perturbation theory of solitons [82] was presented later in [83]. The radiation serves as a "bath" for solitons in the cavity, and can relate the separation and phase differences between the solitons. The interaction between the solitons will be stronger when the solitons are located closer to each other. At the same time, the interaction between the solitons will have periodic components because of the specific wavelength of the radiation fields. This periodicity, in the case of fiber links, has been noted in [84]. Clearly, it should exist in the case of multiple pulse generation by lasers.

As a result of the interference between two single solitons, the spectrum of a two-soliton solution shows a fringe pattern whose frequency is the inverse of the separation between the two pulses. The envelope of these spectra always

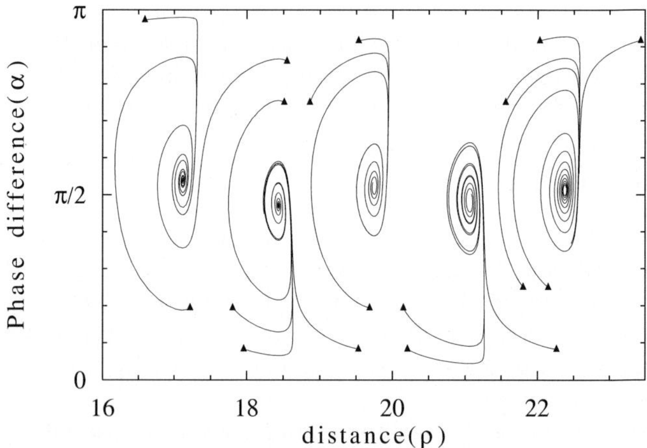

Fig. 11. Trajectories of pulse pairs in the anomalous dispersion regime. The initial value is marked with a small triangle The parameter values are: $L_{SMF} = 4.8$, $L_{EDF} = 1.6$, $D = -2.3$, $\Gamma = 3$, $\gamma = 0.09$, $E_L = 0.9$, $g_{oi} = g_{of} = 1.5$, $\beta = 0.05$, $\theta = 70°$

coincides numerically with the single pulse spectrum. Each pair of solitons has one fringe of its spectrum coinciding with a Kelly side-band [62]. This observation proves that radiation related to the Kelly side-bands plays a major role in soliton coupling. It is remarkable that the phase difference between the solitons is fixed at $\pi/2$. This makes the solution asymmetric, so that only one side of the spectrum is attached to the side-band. Taking into account the above asymmetry, the discrete time separation ρ_N between the centers of the two solitons can be easily calculated; it is found to be $\rho_N = (2N-1)/(4\,\delta\nu_1)$, where $\delta\nu_1$ is the frequency displacement of the lowest order Kelly side-band. The latter can be estimated using (5) of [54], or taken directly from the experimental data. Let us point out that, for small separations, the stronger interaction between the two solitons makes the phase difference depart significantly from the $\pi/2$ value.

5.2 Normal Dispersion Regime

If the length of the SMF fiber is made shorter and shorter, the average dispersion gradually changes from being anomalous and increasingly becomes normal. However, one easily finds stable soliton solutions, but these are for higher values of the gain. This is related to the formation of more and more energetic pulses. As a consequence, the energetic pulse will experience greater changes in each cavity round-trip. Therefore, when the formation of pairs of pulses occurs, their interaction will be much stronger than in the previous case. Figure 12 explicitly shows a comparison between two stable pairs of pulses evolving periodically. The figure shows the evolution, during one round-trip, of a stationary pair of solitons for both cases: (a) normal and (b) anomalous path-averaged dispersion regime. The pulse profiles in the Er fiber are plotted as thicker lines than those in the SMF fiber. The parameters used in both simulations are: (a) $L_{SMF} = 3.6$, $L_{EDF} = 1.88$, $D = -2.3$, $\Gamma = 3$, $\gamma = 0.2$, $E_L = 8$, $g_{oi} = g_{of} = 1.5$, $\beta = 0.08$, $\theta = 142°$. (b) $L_{SMF} = 4.8$, $L_{EDF} = 1.6$, $D = -2.3$, $\Gamma = 3$, $\gamma = 0.09$, $E_L = 0.9$, $g_{oi} = g_{of} = 1.5$, $\beta = 0.05$, $\theta = 70°$.

As Fig. 12 shows, the interaction is completely different in each regime. In the normal average dispersion regime, there is a strong overlap between pulses in some parts of the cavity, so that the accumulated interaction is much stronger than in the anomalous average dispersion regime. Also, in marked contrast to the anomalous dispersion regime (see Fig. 11, where the soliton pair spirals towards a $\pi/2$ phase-locked state, here in the normal dispersion regime, either it converges much faster to a state with a π phase difference, or the solitons repel each other forever. This is shown in Fig. 13, which corresponds to Fig. 11, but now is for normal path-averaged dispersion. The discrete number of possible separations, almost equally-spaced, observed in the anomalous regime, disappears for the normal case. Clearly, now the interaction through radiative waves generated by the periodic perturbation of the soliton pair does not play a significant role.

For other values of the equation parameters, one sees that sometimes several separations, though far fewer than in the anomalous case, between

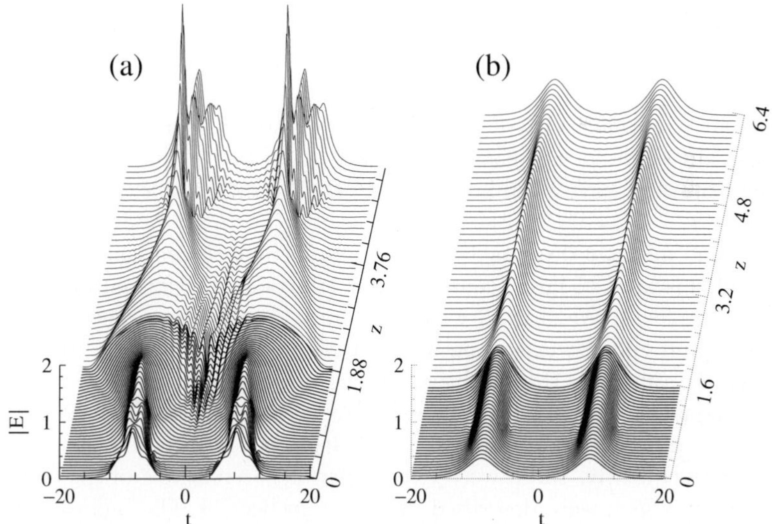

Fig. 12. Pulse profiles along the dispersion-managed cavity in (**a**) normal and (**b**) anomalous path-averaged dispersion regimes. The propagation in the Er fiber is indicated by a *thicker* line. From [85]

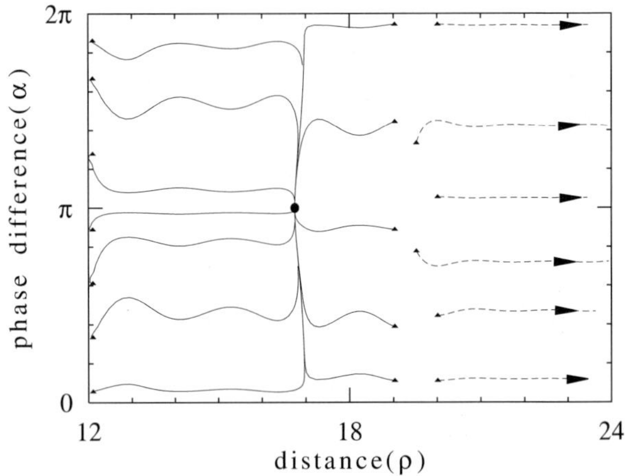

Fig. 13. Trajectories of pulse pairs. The *solid* circle denotes the stable focus, while the *solid* triangles indicate the initial values for the simulations. The equation parameters for these simulations are the same as in Fig. 12a. From [85]

solitons of the pair are possible for the same set of parameters, and that the pair can also be in-phase.

Pairs of pulses in lasers operating in the normal dispersion regime have also been reported recently in [86]

6 Conclusions

We have extensively reviewed the experimental observation of multiple pulses in passively mode-locked lasers, and have described a number of possible mechanisms that have been addressed in the literature to explain their formation.

Further, we have described the most important theoretical models used to model this kind of laser, centering our efforts on the formation of multi-soliton complexes. We started with distributed models governed by a single equation, and ended with a non-distributed model. Among the former, the most widely-studied is the one based on the cubic-quintic complex Ginzburg-Landau equation. It predicted the existence of phase-locked pairs of solitons with phase differences of 0, π and $\pm\pi/2$ degrees. Only these last ones were stable. Subsequently, there were experimental observations of $\pm\pi/2$ phase-locked pairs which were described at length in Sect. 3. The inclusion of additional terms in the master equation, as in the case of the CSH equation, revealed that pairs of solitons with a π phase-difference could become stable. For completeness, we briefly described laser systems where the response of the nonlinear saturation mechanism leading to the passive mode-locking is not instantaneous.

As distributed models fail to take account of the periodic changes that occur in the cavity, we introduced a model, which is thought to match a particular experimental set-up, with the minimal possible number of parameters in order to make it numerically tractable. It explained the discrete number of allowed inter-pulse distances observed in the experiment, and the great differences between the multi-soliton complexes observed for normal and anomalous averaged group dispersions. The agreement with the experiment, even quantitatively, was excellent.

References

1. A. J. DeMaria, D. A. Stetser and H. Heynau, Appl. Phys. Lett. **8**, 174 (1966).
2. E. M. Garmire and A. Yariv, IEEE J. Quant. Elect. **QE-3**, 222 (1967).
3. J. Goodberlet, J. Wang, J. G. Fujimoto and P. A. Schulz, Opt. Lett. **14**, 1125 (1989).
4. E. P. Ippen, H. A. Haus, and L. Y. Liu, J. Opt. Soc. Am. **B 6**, 1736 (1989).
5. D. E. Spence, J. M. Evans, W. E. Sleat and W. Sibbett, Opt. Lett. **16**, 1762 (1991).
6. D. Huang, M. Ulman, L. H. Acioli, H. A. Haus and J. G. Fujimoto, Opt. Lett. **17**, 511 (1992).
7. V. J. Matsas, T. P. Newson, D. J. Richardson and D. N. Payne, Electron. Lett. **28**, 1391 (1992).
8. V. E. Zakharov and A. B. Shabat, Sov. Phys. JETP **34**, 62 (1972).
9. A. Hasegawa and F. Tappert, Appl. Phys. Lett. **23**, 142 (1973).
10. L. F. Mollenauer, R. H. Stolen, and J. P. Gordon, Phys. Rev. Lett. **45**, 1095 (1980).
11. L. F. Mollenauer and R. H. Stolen, Opt. Lett. **9**, 13 (1984).
12. Irl N. Duling III, Opt. Lett. **16**, 539 (1991).
13. D. J. Richardson, R. I. Laming, D. N. Payne, M. W. Phillips, and V. J. Matsas, Electron. Lett. **27**, 730 (1991).
14. M. Nakazawa, E. Yoshida, and Y. Kimura, Appl. Phys. Lett. **59**, 2073 (1991).
15. *Rare-earth doped fiber lasers and amplifiers*, Editor M. J. F. Digonnet, second edition, Marcel Dekker, New York (2001).
16. R. J. Mears, L. Reekie, S. B. Poole, and D. N. Payne, Electron. Lett. **22**, 159 (1986).
17. E. Desurvire, J. R. Simpson, and P. C. Becker, Opt. Lett. **12**, 888 (1987).
18. D. J. Richardson, R. I. Laming, D. N. Payne, V. J. Matsas and M. W. Phillips, Electron. Lett. **27**, 1451 (1991).
19. M. T. Asaki, C. Huang, D. Garvey, J. Zhou, H. C. Kapteyn and M. M. Murnane, Opt. Lett. **18**, 977 (1993).
20. A. Stingl, M. Lenzner, Ch. Spielmann, F. Krausz and R. Szipcs, Opt. Lett. **20**, 602 (1995).
21. A. M. Kowalevicz Jr., T. R. Schlibli, F. X. Kärtner and J. G. Fujimoto, Opt. Lett. **27**, 2037 (2002).
22. A. B. Grudinin, D. J. Richardson and D. N. Payne, Electron. Lett. **29**, 1860 (1993).

23. J. N. Kutz, B. C. Collings, K. Bergman, and W. H. Knox, IEEE J. Quantum Electron. **34**, 1749 (1998).
24. B. C. Collings, K. Bergman, and W. H. Knox, Opt. Lett. **23**, 123 (1998).
25. C. X. Yu, H. A. Haus, E. P. Ippen, W. S. Wong and A. Sysoliatin, Opt. Lett. **25**, 1418 (2000).
26. N. H. Bonadeo, W. H. Knox, J. M. Roth and K. Bergman, Opt. Lett. **25**, 1421 (2000).
27. K. S. Abedin, N. Onodera and M. Hyodo, Opt. Lett. **24**, 1564 (1999).
28. M. J Guy, D. U. Noske and J. R. Taylor, Opt. Lett. **18**, 1447 (1993).
29. M. Nakazawa, K. Suzuki and H. A. Haus, Phys. Rev. **A 38**, 5193 (1988).
30. X. Zhu, M. Pich, G. D. Goodnob, R. J. D. Miller, Opt. Commun. **145**, 123 (1998).
31. E. M. Dianov, A. V. Luchnikov, A. N. Pilipetskii and A. N. Starodumov, Opt. Lett. **15**, 314 (1990).
32. A. N. Pilipetskii, E. A. Golovchenko and C. R. Menyuk, Opt. Lett. **20**, 907 (1995).
33. A. B. Grudinin and S. Gray, J. Opt. Soc. Am. **B 14**, 144 (1997).
34. T. Brabec, Ch. Spielmann and F. Krausz, Opt. Lett. **16**, 1961 (1991).
35. I. D. Jung, F. X. Kärtner, N. Matuschek, D. H. Sutter, F. Morier-Genoud, G. Zhang, U. Keller, V. Sheuer, M. Tilsch and T. Tschudi, Opt. Lett. **22**, 1009 (1997).
36. T. Tsang, Opt. Lett. **18**, 293 (1993).
37. F. Salin, P. Grangier, G. Roger and A. Brun, Phys. Rev. Lett. **56**, 1132 (1986).
38. F. W. Wise, I. A. Walmsley and C. L. Tang, Opt. Lett. **13**, 129 (1988).
39. Ch. Spiclmann, P. F. Curley, T. Brabec and F. Krausz, IEEE J. of Quantum Electron. **30**, 1100 (1994).
40. J. Herrmann, V. P. Kalosha and M. Miller, Opt. Lett. **22**, 236 (1997).
41. C. Wang, W. Zhang, K. F. Lee and K. M. Koo, Opt. Commun. **137**, 89 (1997).
42. M. Lai, J. Nicholson and W. Rudolph, Opt. Commun. **142**, 45 (1997).
43. B. Chassagne, G. Jonusauskas, J. Oberl and C. Rullire, Opt. Commun. **150**, 355 (1998).
44. H. Kitano and S. Kinoshita, Opt. Commun. **157**, 128 (1998).
45. J. Aus der Au, D. Kopf, F. Morier-Genoud, M. Moser and U. Keller, Opt. Lett. **22**, 307 (1997).
46. B. C. Collings, K. Bergman and W. H. Knox Opt. Lett. **22**, 1098 (1997).
47. M. J. Lederer, B. Luther-Davies, H. H. Tan, C. Jagadish, N. Akhmediev and J. M. Soto-Crespo, J. Opt. Soc. Am. B, **16**, 895 (1999).
48. M. Nakazawa and H. Kubota, Electron. Lett. **31**, 216 (1995).
49. W. Forysiak, J. F. L. Devaney, N. J. Smith and N. J. Doran, Opt. Lett. **22**, 600 (1997).
50. M. Nakazawa, H. Kubota, A. Sahara and K. Tamura, IEEE Phot. Technol. Lett. **8**, 1088 (1996).
51. N. H. Seong, D. Y. Kim and S. K. Oh, Electron. Lett. **37**, 157 (2001).
52. W. S. Wong, S. Namiki, M. Margalit, H. A. Haus and E. P. Ippen, Opt. Lett. **22**, 1150 (1997).
53. F. Gutty, Ph. Grelu, N. Huot, G. Vienne and G. Millot, Electron. Lett. **37**, 745 (2001).
54. S. M. J. Kelly, Electron. Lett. **28**, 806 (1992).
55. Ph. Grelu, F. Belhache, F. Gutty and J. M. Soto-Crespo, J. Opt. Soc. Am. **B 20**, 863 (2003).

56. D. Y. Tang, W. S. Man, H. Y. Tam and P. D. Drummond, Phys. Rev. A, **64**, 033814 (2001).
57. Ph. Grelu, F. Belhache, F. Gutty and J. M. Soto-Crespo, Opt. Lett. **27**, 966 (2002).
58. K. Tamura, L. E. Nelson, H. A. Haus and E. P. Ippen, Appl. Phys. Lett. **64**, 149 (1994).
59. K. Tamura, E. P. Ippen, H. A. Haus and L. E. Nelson, Opt. Lett. **18**, 1080 (1993).
60. K. Tamura and M. Nakazawa, Appl. Phys. Lett. **67**, 3691 (1995).
61. N. Akhmediev, A. Ankiewicz and J. M. Soto-Crespo, Phys. Rev. Lett., **79**, 4047 (1997).
62. J. M. Soto-Crespo, N. Akhmediev, Ph. Grelu and F. Belhache, Opt. Lett. **28**, 1757 (2003).
63. R. Trebino, K. W. DeLong, D. N. Fittinghoff, J. N. Sweetser, M. A. Krambugel, B. A. Richman and D. J. Kane, Rev. Sci. Instrum. **68**, 3277 (1997).
64. C.-J. Chen, P. K. A. Wai, and C. R. Menyuk, Opt. Lett. **19**, 198 (1994).
65. B. Malomed, Phys. Rev. A **44**, 6954 (1991).
66. J. D. Moores, Opt. Commun. **96**, 65 (1993).
67. N. Akhmediev and A. Ankiewicz, *Solitons: Nonlinear Pulses and Beams*, (Chapman & Hall, London, 1997).
68. N. Akhmediev, V. V. Afanasjev and J. M. Soto-Crespo, Phys. Rev E **53**, 1190 (1996).
69. J. M. Soto-Crespo, N. N. Akhmediev, V. V. Afanasjev and S. Wabnitz, Phys. Rev. E **55** 4783 (1997).
70. J. M. Soto-Crespo, Nail Akhmediev and Kin S. Chiang Phys. Lett. A **291**, 115 (2001).
71. J. M. Soto-Crespo, N. Akhmediev, and V. V. Afanasjev, J. Opt. Soc. Am. B **13**, 1439 (1996).
72. V. V. Afanasjev and N. Akhmediev, Phys. Rev. E **53**, 6471 (1996).
73. N. Akhmediev, A. Ankiewicz and J. M. Soto-Crespo, J. Opt. Soc. Am. B **15**, 515 (1998).
74. J. M. Soto-Crespo and N. Akhmediev, Phys. Rev. E. **66**, 066610 (2002).
75. M. J. Lederer, B. Luther-Davies, H. H. Tan and C. Jagadish, Appl. Phys. Lett., **70**, 3428 (1997).
76. F. X. Kärtner and U. Keller, Opt. Lett. **20**, 16 (1995).
77. I. D. Jung, F. X. Kärtner, L. R. Brovelli, M. Kamp and U. Keller, Opt. Lett. **20**, 1892 (1995).
78. H. A. Haus, J. Appl. Phys., **46**, 3049 (1975).
79. J. M. Soto-Crespo and N. Akhmediev, J. Opt. Soc. Am. B, **16**, 674 (1999).
80. N. Akhmediev, F. Zen, and P. Chu, Opt. Comm. **201**, 217 (2002).
81. S. T. Cundiff, J. M. Soto-Crespo and N. Akhmediev, Phys. Rev. Lett. **88**, 073903 (2002).
82. J. P. Gordon, J. Opt. Soc. Am. B **9**, 91 (1992).
83. J. N. Elgin and S. M. J. Kelly, Opt. Lett. **18**, 787 (1993).
84. L. Socci and M. Romagnoli, J. Opt. Soc. Am. B **16**, 12 (1999).
85. Ph. Grelu, J. Béal and J. M. Soto-Crespo, Opt. Express, **11**, 2238 (2003).
86. A. Hideur, B. Ortaç, T. Chartier, M. Brunel, H. Leblond and F. Sánchez, Opt. Comm. **225**, 71 (2003).

Mode-Locking of Fiber Lasers via Nonlinear Mode-Coupling

J.N. Kutz

Department of Applied Mathematics, University of Washington, Seattle, WA 98195-2420, USA
kutz@amath.washington.edu

Abstract. We consider mode-locking in optical fiber lasers where the pulse propagation is strongly influenced by a nonlinear mode-coupling interaction. This nonlinear coupling in conjunction with the cubic Kerr nonlinearity and chromatic dispersion leads to the formation of attracting soliton states, i.e. dissipative solitons. A review of each model is given with an evaluation of its merit and mode-locking features.

1 Introduction

In an optical fiber laser cavity, the underlying wave behavior is governed by the the nonlinear Schrödinger equation (NLS). The NLS, which is derived from a high-frequency asymptotic expansion of Maxwell's equations, characterizes the physical effects arising from chromatic dispersion and a weak Kerr nonlinearity [1]:

$$\mathrm{i}\frac{\partial U}{\partial Z} + \frac{1}{2}\frac{\partial^2 U}{\partial T^2} + |U|^2 U = 0 \,. \tag{1}$$

Here U is the electric field envelope, Z represents the propagation distance in the fiber, and T is the time in the rest frame of a propagating pulse. The NLS has been investigated extensively in this context with particular emphasis given to the robust and stable soliton solutions which result from a fundamental balance between the linear dispersion and the cubic nonlinearity [2].

Although many other effects are present in the fiber, such as linear attenuation, third-order dispersion, polarization dynamics, higher-order nonlinear effects, or saturable absorption, it is the combined effects of the cubic nonlinearity and temporal dispersion which eventually dominate the stability and behavior of the resulting soliton pulse train. Of course, in the context of this book the term soliton is used in a much broader sense than the strict mathematical definition of a solution of a completely integrable nonlinear partial differential equation. Rather, it refers to any pulse-like structure which is robust and persists under a large class of perturbations. This includes stable localized solutions of dissipative systems which are clearly not integrable or Hamiltonian.

J.N. Kutz: *Mode-Locking of Fiber Lasers via Nonlinear Mode-Coupling*,
Lect. Notes Phys. **661**, 241–265 (2005)
www.springerlink.com

From a physical standpoint, the successful operation of a mode-locked laser [3, 4] is achieved by utilizing an *intensity discrimination* perturbation in the laser cavity in conjunction with bandwidth limited gain [5, 6]. In its simplest form, intensity discrimination preferentially attenuates weaker intensity portions of individual pulses. This attenuation is compensated by the gain medium (e.g. Erbium-doped fiber) which acts to preserve the total cavity energy. Thus pulse shaping occurs since the peak of a pulse experiences a higher net gain per round trip than its lower intensity wings. This time-domain narrowing (compression) of a propagating pulse is limited, however, by the bandwidth of the gain medium (typically $\approx$20–40 nm [3, 4]). This mode-locking mechanism, which is only a small perturbation to the NLS (1), has been successfully achieved in a wide variety of experimental configurations including: a fiber ring laser with a linear polarizer [7, 8, 9, 10], the figure-eight laser with nonlinear interferometry [11, 12, 13], a linear-cavity configuration with a semi-conductor saturable absorber [14, 15, 16], and a laser cavity with an acousto-optic modulator [17, 18].

The canonical model which qualitatively describes both the energy saturation and the pulse stabilization process in a solid-state laser is the *master mode-locking* equation [5, 6]. The master mode-locking model is a perturbation from the NLS model which was proposed to capture two key effects: the equilibration of the pulse energy, and the consequent balance of nonlinearity and dispersion in forming stable pulses. Because it achieves both these goals, the master mode-locking equation is widely considered to be the most successful model developed for a phenomenological description of mode-locking. However, the master mode-locking description is limited to systems where spatial [19] and higher order [20] dispersion effects are negligible. Further, it does not include mode-coupling, polarization dynamics, active modulation, or a quantitatively meaningful saturable absorption component. It is by construction, a qualitative model. Section 2 of this manuscript reviews the various qualitative behaviors and features in a mode-locked laser cavity described by the master mode-locking model.

Quantitative modeling of mode-locked laser cavities requires improved model descriptions which account for the additional physical effects allowing for the mode-locking to occur. Thus individual mode-locked lasers must be modeled independently. Like master mode-locking, these models are perturbations from the NLS equation (1) and rely on the underlying hyperbolic-secant pulse solutions for ideal mode-locking. However, the models can vary drastically from one another as they account for significantly different mechanisms for intensity discrimination in the laser cavity. Section 3 of this work considers specific models for which the underlying mode-locking dynamics is a result of nonlinear mode-coupling which leads to intensity-discrimination.

2 Master Mode-Locking

Mode-locked solitons in optical fiber lasers result from a balance between the nonlinearity and dispersion given by the NLS (1). Achieving the soliton state, however, requires a form of intensity-discrimination. Although we will focus on the intensity-discrimination which arises in nonlinear mode-coupling, we first consider the pre-dominant analytic description. In the following section, the standard mode-locking model is explored which incorporates a phenomenological intensity discrimination term that leads to the stable development of mode-locked pulse trains. Thus we begin with the master mode-locking model [5, 6], which is the generic and canonical description. The analysis of this model serves as a pre-cursor to the consideration of more complicated models.

The key elements in deriving the master mode-locking model is a physical system which exhibits nonlinearity, dispersion, bandwidth limited gain, energy saturation, and intensity dependent attenuation. These elements are crucial for self-starting and stabilization of constant-energy pulses. Combining these various physical effects results in the master mode-locking model [5, 6]:

$$\mathrm{i}\frac{\partial U}{\partial Z} + \frac{1}{2}\frac{\partial^2 U}{\partial T^2} + (1-\mathrm{i}\beta)|U|^2 U + \mathrm{i}\gamma U - \mathrm{i}g(Z)\left(1+\tau\frac{\partial^2}{\partial T^2}\right)U = 0\,. \quad (2)$$

where

$$g(Z) = \frac{2g_0}{1+\|U\|^2/e_0}\,, \quad (3)$$

and U represents the electric field envelope normalized by the peak field power $|U_0|^2$. Here the variable T represents the physical time normalized by $T_0/1.76$ where T_0 is the full width at half maximum (FWHM) of the pulse (e.g. $T_0 = 200\,\mathrm{fs}$). The variable Z is scaled on the dispersion length $Z_0 = (2\pi c)/(\lambda_0^2 \bar{D})(T_0/1.76)^2$ corresponding to the average dispersion. This then gives the peak field power $|U_0|^2 = \lambda_0 A_{\mathrm{eff}}/(4\pi n_2 Z_0)$. Note that $\|U\|^2 = \int_{-\infty}^{\infty}|U|^2 dT$. Further, for all models discussed $n_2 = 2.6\times10^{-16}\,\mathrm{cm^2/W}$ is the nonlinear coefficient in the fiber, $A_{\mathrm{eff}} = 60\,\mu\mathrm{m}^2$ is the effective cross-sectional area of the fiber, $\Gamma = 0.2\,\mathrm{dB/km}$ is the fiber loss, and $\lambda_0 = 1.55\,\mu\mathrm{m}$ and c are the free-space wavelength and speed of light respectively.

The remaining parameters can vary largely from one experiment to another. In terms of the scaling variables, this gives $e_0 = E_0/|U_0|^2$, $g_0 = G_0 Z_0$, $\gamma = \Gamma Z_0$, $\beta = B|U_0|^2 Z_0$, and $\tau = (1/\Omega^2)(1.76/T_0)^2$. The parameter E_0 is the saturated gain in the cavity, G_0 is the gain, and the parameter B (in 1/Watt/meter) is the intensity discrimination that controls the strength of the phenomenological model of the mode-locking element. For a gain bandwidth which can vary from $\Delta\lambda = 20$–$40\,\mathrm{nm}$, we find that $\Omega = (2\pi c/\lambda_0^2)\Delta\lambda$ so that τ varies from $\approx 0.04-0.16$ for a pulse width scaling of $T_0 = 200\,\mathrm{fs}$. Note that τ is the parameter which largely controls the resulting pulse width of the stabilized mode-locked pulse stream since it characterizes the spectral gain bandwidth of the mode-locking process.

We consider pulse solutions to the master mode-locking model (2). By assuming that the constants g_0, γ, and β are small, equation (2) can be analyzed using perturbation techniques [21, 22, 23]. The leading order NLS description has the well known soliton solution

$$U_0(Z,T) = \eta \operatorname{sech}(\eta T) \exp(\mathrm{i}\eta^2 Z/2) \tag{4}$$

where $\eta > 0$. A perturbation analysis reveals that this soliton persists for small g_0, γ, and β provided

$$4\sqrt{2}\beta\eta^3 + 2(\beta - \tau g_0)e_0\eta^2 - 6\sqrt{2}\gamma\eta + 3e_0(2g_0 - \gamma) = 0\,. \tag{5}$$

The primary effect of the perturbation on the soliton is an additional phase chirp. In fact, the leading order mode-locked pulse solution for (2) is given by [24]

$$U(Z,T) = \eta \operatorname{sech}(\eta T)^{1+\mathrm{i}A} \exp(\mathrm{i}\eta^2 Z/2) \tag{6}$$

with

$$A = \frac{2\tau g_0 e_0 \eta^2 - 2\sqrt{2}\gamma\eta + e_0(2g_0 - \gamma)}{\eta^2(e_0 + 2\sqrt{2}\eta)}\,. \tag{7}$$

If all constants aside from β are fixed such that $\gamma < 2g_0$, then there is a number $\beta^* > 0$ that depends upon all the fixed parameters so that the following is true: For $\beta \leq 0$, there is a unique soliton. For $0 < \beta < \beta^*$, there are precisely two solitons; one of these has large energy. Lastly, for $\beta > \beta^*$, there are no solitons since the nonlinear gain is too large to sustain a pulse with finite energy [22, 23].

The dynamics and stability of the mode-locking behavior is illustrated with numerical simulations of the governing equation (2). The numerical procedure employed for this model and all others considered here uses a fourth-order Runge-Kutta method in Z and a filtered pseudo-spectral method in T. In the calculations and simulations that follow, we scale all parameters on the FWHM pulse width $T_0 = 200\,\mathrm{fs}$ and assume that the master mode-locking parameters are $e_0 = 4.0$, $g_0 = 0.1$, $\tau = 0.05$ and $\gamma = 0.1$. The remaining free parameter β is the bifurcation parameter which determines the stability of the pulse solutions. These values give $\beta_* = 0.0025$ (radiation-mode instability) and $\beta^* = 0.0087$ (saddle-node instability) [22, 23]. Thus, there are four distinct regions of behavior.

To illustrate the possible mode-locking behaviors and the various stability regimes, we first consider the stable mode-locking regime for which $\beta_* = 0.0025 < \beta < 0.0087 = \beta^*$. Figure 1a depicts the evolution to a steady-state pulse solution given an arbitrary initial condition and $\beta = 0.0085$. Over approximately $Z \approx 400$ units, the pulse modifies its height, width, and energy to reach an equilibrium pulse solution described by (6). This stable evolution is the ideal behavior of the master mode-locking model and has been achieved in successful mode-locked lasers. The instabilities are also easily captured computationally. For $\beta = -0.0125 < 0$, the unstable radiation

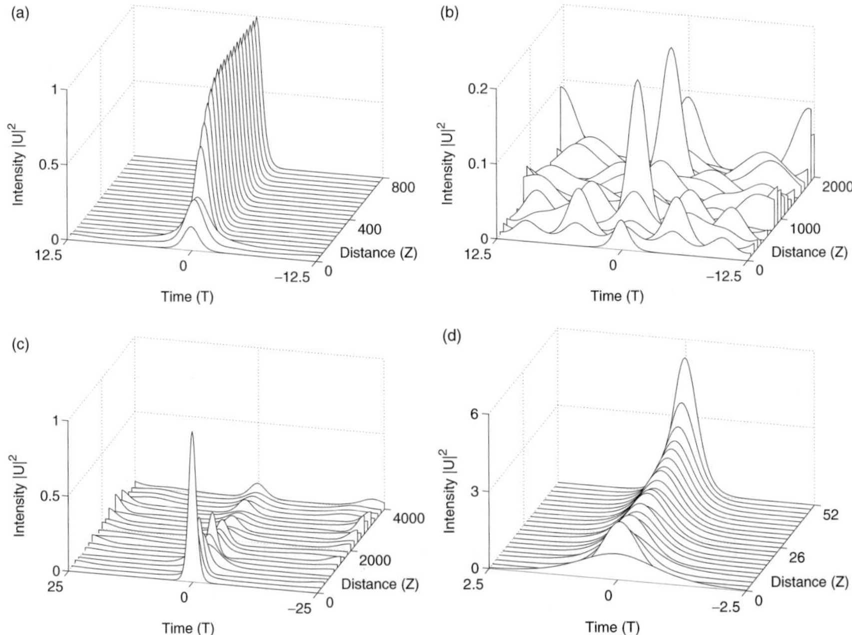

Fig. 1. Mode-locking dynamics in the master mode-locking model. (**a**) Stable mode-locking evolution for $\beta = 0.0085$ which quickly settles to the steady-state pulse solution. (**b**) Unstable evolution for $\beta = -0.0125$ which quickly evolves into a non-localized weak-turbulence regime which persist over time. (**c**) Unstable evolution for $\beta = 0.00125$ which quickly evolves into a nonlocalized quasi-periodic solution. Although the cavity energy is controlled by the time-dependent gain of the master mode-locking equation, the quasi-periodic oscillations persist. (**d**) Unstable evolution for $\beta = 0.00875$ where the pulse quickly blows-up. In this case, the cavity energy is not controlled by the time-dependent gain of the master mode-locking equation

modes together with nonlinear damping lead to the instability illustrated in Fig. 1b [22, 23]. In this case, the initial localized pulse energy is quickly transfered into quasi-periodic behavior. So although the pulse energy is kept in control by the gain model, the pulse continues to oscillate as depicted by the continued fluctuations in Fig. 1b. The oscillatory dynamics persist as $Z \to \infty$. This behavior is characteristic of simulations with $\beta < 0$. For $0 < \beta < 0.0025$, the pulse is destabilized by radiation modes [22, 23]. Since the radiation modes are, strictly speaking, only present in the infinite domain, the computational method with periodic boundary conditions gives rise to the instability depicted in Fig. 1c. In this case, the amplitude continues to fluctuate and a steady-state is never achieved. The energy which accumulates at the boundary is a manifestation of a finite computational domain. Regardless, the pulse destabilizes and although the cavity energy reaches equilibrium, oscillations persist as $Z \to \infty$. Finally, for values of $\beta > 0.0087$, the nonlinear gain is too high to be overcome by the linear loss mechanisms present in the master mode-locking equations [22] which are due to linear attenuation and diffusion. The solution quickly grows and blow-up occurs in a self-similar fashion. The onset of blow-up is shown in Fig. 1d for $\beta = 0.00875$. Note that by $Z = 52$, the pulse intensity has grown to five times its original value. Just after $Z = 56$, the pulse solution undergoes self-similar collapse, and it is no longer computationally feasible to resolve the evolution. Thus the three regions of unstable behavior are markedly different (see Figs. 1b–d) and present significant limitations to the use of the master mode-locking model. Specifically, only a narrow window of parameter space, i.e. $\beta \in [0.0025, 0.0087]$, allows for the behavior which is experimentally sought.

Since the master mode-locking equation is the canonical model describing both the energy saturation and the pulse stabilization process in a solid-state laser in the absence of spatial and higher order effects, regions of stable operation can be predicted within an analytic framework [22, 23]. The successful operation of the laser is achieved by having a sufficient amount of intensity discrimination. Otherwise, the laser is subject to a radiation mode instability. Likewise, if the nonlinear gain is too high, the linear attenuation terms are unable to prevent the pulse from blowing up, suggesting the breakdown of the master mode-locking model. Although it is a valuable tool for analysis, the master mode-locking model fails to capture many physically relevant behaviors exhibited in practical mode-locked systems. Modifications to this model are extensive, with the simple addition of a quintic absorption term being the standard modification since it greatly extends the range of stable solutions which exist for $\beta > 0$. Indeed, the addition of a saturating quintic term is a standard trick for regularizing Ginzburg-Landau type equations [25] including the master modelocking model. The range of mode-locking behavior (attracting states) is also greatly increased with a wide variety of stationary solutions, breathers, and so-called exploding solitons exhibited [26, 27, 28, 29, 30].

3 Mode-Locking via Nonlinear Mode-Coupling

The master mode-locking model gives a simple phenomenological description of the intensity discrimination element present in a laser cavity. However, it has only a narrow window in parameter space for supporting stable solutions and fails to give quantitative agreement with experiment. Unlike the master mode-locking model presented in the previous section, the models presented in this section aim to be of a quantitative nature. Thus an effort is made to try and include all the relevant physical effects present in a given mode-locked laser cavity configuration. As will be demonstrated by the various models, there are many ways to achieve an intensity dependent discrimination required for mode-locking using nonlinear mode-coupling. And by modeling the system from a quantitative perspective, a realistic model can be obtained for a given laser cavity configuration. This allows for a characterization of the model which is beyond the generic qualitative approach. In particular, the breakdown of the mode-locking behavior and the optimal laser performance can be investigated.

The first step in these models is to consider the quantitative effects of the pulse propagation in the absence of a mode-locking mechanism. Thus we begin by once again considering the pulse propagation in the optical fiber (solid state configuration) in the absence of a mode-locking term (e.g. $\beta = 0$ in (2)). Thus the governing evolution is given by

$$\mathrm{i}\frac{\partial U}{\partial Z} + \frac{1}{2}\frac{\partial^2 U}{\partial T^2} + |U|^2 U + \mathrm{i}\gamma U - \mathrm{i}g(Z)\left(1 + \tau\frac{\partial^2}{\partial T^2}\right)U = 0\,. \tag{8}$$

where U is the normalized electric field envelope scaled as in (2). Note that this model differs from the experimental set up for which different fiber segments are utilized with differing dispersion values and effective cross-sectional areas. Although corrections to the average dispersion can be included, numerical results suggest that they can be neglected with only a minimal amount of error incurred. The gain in the fiber is incorporated through the dimensionless parameter $g = g(Z)$ just as in (3).

In addition to (8), an accurate model is required for the mode-locking mechanism. Certainly every parameter used in (8) has a direct correlation to an experimentally measurable physical value. Thus it only remains to provide a quantitatively accurate model of the mode-locking mechanism to achieve an accurate and physically realizable model for the mode-locked laser cavity. In the subsections that follow, various models for the mode-locking mechanism will be considered and shown to generate stable mode-locked pulse trains.

3.1 Mode-Locking with a Long Period Fiber Grating

Fiber gratings have long been used for photonic applications. For instance, a long-period fiber grating (LPFG) can be used as a band-rejection filter for all-optical signal processing [31] or as a mode-locking element in a laser cavity. The concept of the LPFG as a mode-locking mechanism arises from the intensity dependent mode-coupling which occurs between co-propagating core and cladding modes [32]. Nonlinear mode-coupling was first investigated by Jensen [33] for co-propagating core modes in the context of a dual-core, nonlinear directional coupler. This nonlinear dual-core coupler was later proposed by Winful and Walton [34] as a mechanism for generating stable mode-locked pulses [34, 35]. The underlying concept is as follows: a resonant and linear mode-coupling interaction transfers energy periodically between core and cladding modes [36]. Nonlinearity, however, can be used to detune the resonant interaction by shifting the propagation constant of each mode via the self-phase and cross-phase modulation introduced by the nonlinearity [32, 33]. Thus the LPFG can be manufactured so that a resonant linear interaction occurs between the core mode and a given cladding mode. Then the low-intensity parts of a pulse which propagate through the grating are efficiently coupled to the cladding and attenuated. In contrast, the peak of the pulse is detuned from the resonant coupling (i.e. the transmission minimum of the LPFG is shifted to longer wavelengths) and is transmitted more efficiently. Thus, the wings of a pulse are attenuated slightly more than its peak giving the necessary pulse shaping required to achieve stable mode-locked operation after many round trips of the fiber laser cavity.

The numerical model presented here is based on the fiber laser configuration of Fig. 2. The ideal laser cavity consists of a segment of optical fiber (≈1 m), an output coupler (≈2%), a reflector, and a short section of the LPFG (≈5–10 cm). This simple passive device is sufficient of producing stable mode-locked pulse trains.

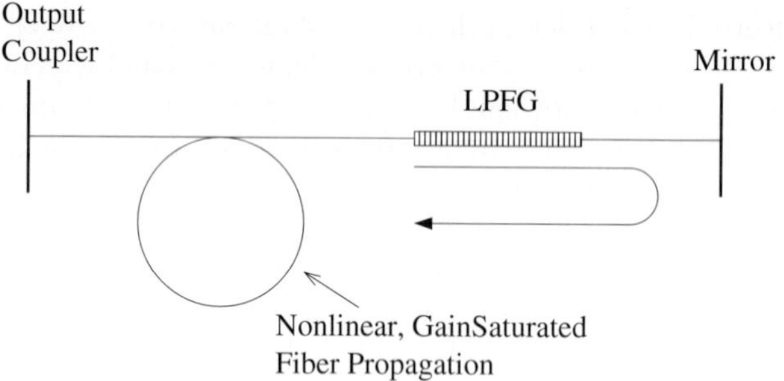

Fig. 2. Experimental setup

The asymptotic derivation of the nonlinear coupled mode equations for the co-propagating core and cladding modes in a LPFG are carried out in Kutz et al. [37]. This derivation assumes quasi-monochromatic waves along with the paraxial and rotating-wave approximations. These asymptotic assumptions are justified for the LPFGs under consideration.

To induce coupling between co-propagating modes, the grating period is adjusted so that a resonant interaction occurs between the core mode and a given cladding mode. Each cladding mode will have a particular coupling strength with the core mode. Thus the index of refraction of the grating is given by

$$n(x, y, z) = n_0(x, y)\,[1 + \Delta n_{uv} \cos(\gamma Z)] \tag{9}$$

where $n_0(x, y)$ accounts for the transverse index profile of the grating and Δn_{uv} measures the ultraviolet induced index of refraction change in the grating in the longitudinal direction. Since the grating is periodic, we assume there is a dominant Fourier mode component which measures the period of the grating. Thus the parameter γ measures the grating period which is crucial for the co-propagating coupling.

As derived previously [37], the coupling between co-propagating core (i) and cladding (j) modes is given by

$$
\begin{aligned}
&\mathrm{i}\frac{\partial U}{\partial Z} + (c_{ii}|U|^2 + 2c_{ij}|V|^2)U + d_{ij}V - \frac{\Delta}{2}U = 0 \\
&\mathrm{i}\frac{\partial V}{\partial Z} + (2c_{ij}|U|^2 + c_{jj}|V|^2)V + d_{ij}U + \frac{\Delta}{2}V = 0
\end{aligned}
\tag{10}
$$

where the normalized nonlinear and linear coupling coefficients are given by

$$
c_{ij} = \frac{4\pi^2 n_2}{\lambda A_{\text{eff}}} \frac{(|\phi_{i0}|^2, |\phi_{j0}|^2)_r}{(|\phi_{10}|^2, |\phi_{10}|^2)_r} Z_0 |E_0|^2 \tag{11}
$$

$$
d_{ij} = \frac{\Delta n_{\text{UV}}\pi}{2\lambda}(V(r)\phi_{i0}, \phi_{j0})_r \, Z_0 \, . \tag{12}
$$

and the parameter

$$
\Delta = [\gamma - (\beta_i - \beta_j)/2k] \; Z_0 \tag{13}
$$

measures the detuning from resonant coupling ($\Delta = 0$). It will be assumed in the simulations which follow that resonant coupling between co-propagating modes can be achieved. The nonlinearity acts to shift the resonance for high intensity pulses so that they are transmitted through the grating with a minimal amount coupling to the cladding.

The linear coupling strength, which is proportional to the ultraviolet induced index changes, is given by the parameter d_{ij}. The parameters c_{ii} and c_{jj} give the strength of the self-phase modulation acting on the core and cladding modes respectively. Similarly, the parameter c_{ij} measures the strength of the cross-phase modulation between the core and cladding. At sufficiently high intensities, the self-phase and cross-phase modulation terms act to shift the transmission minimum of the grating to longer wavelengths by inducing an intensity dependent phase-shift.

The pulse shaping provided by the long-period fiber grating is demonstrated by simulating (10) for various pulse shapes and intensities. This gives a clear picture of the dynamics of both the core and cladding modes and their respective energy content. For illustrative purposes, we assume the grating length to be $Z = 0.7$ which corresponds to a grating of length 11.76 cm for the scalings given for (8). Further, we take the normalized linear and nonlinear coupling values to be $d_{14} = 2$, $c_{11} = c_{44} = 0.3$ and $c_{14} = 0.5$. The propagation of an initial Gaussian core mode pulse with an amplitude of $a = 4$ is demonstrated in Fig. 3. Note that after passing through the grating, the initial pulse is narrowed considerably while energy is exchanged to the cladding mode. Since the peak of the pulse is detuned from resonance due to its intensity, it does not couple energy to the cladding. The output pulse for the cladding and core modes is shown in Fig. 4. This shows the pulse narrowing process in greater detail.

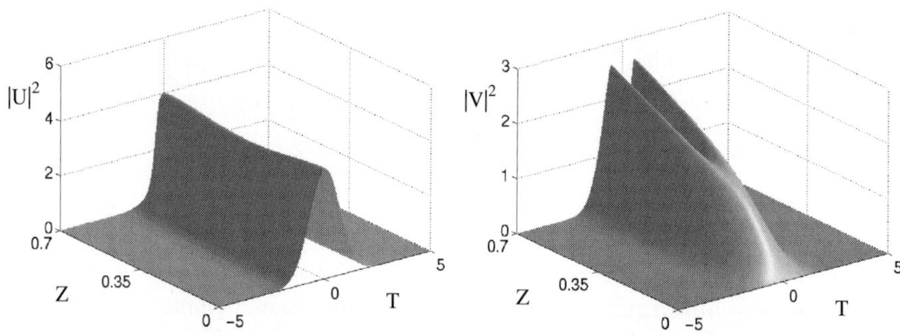

Fig. 3. Evolution of an initial Gaussian core mode pulse over a grating of length $Z = 0.7$ with a computational domain of $T = 10$, and coupling coefficients $d_{14} = 2$, $c_{11} = c_{44} = 0.3$, and $c_{14} = 0.5$

The pulse shaping which occurs due to the interaction of the core and cladding modes in the LPFG can lead to stable mode-locking in the laser cavity. Figure 5 demonstrates this stable mode-locked operation. In this case, we have taken a very short LPFG of length $Z = 0.1$ with linear and nonlinear coupling coefficients $d_{14} = 12$, $c_{11} = c_{44} = 0.2$ and $c_{14} = 20$. Additionally, the gain parameter is taken to be $g_0 = 1$. In this case, the cavity mode-locks the pulse after roughly fifteen roundtrips of the cavity. The corresponding cladding field is also depicted. Note that most of the energy transfer from core to cladding occurs on the wings, thus giving the necessary pulse shapping for

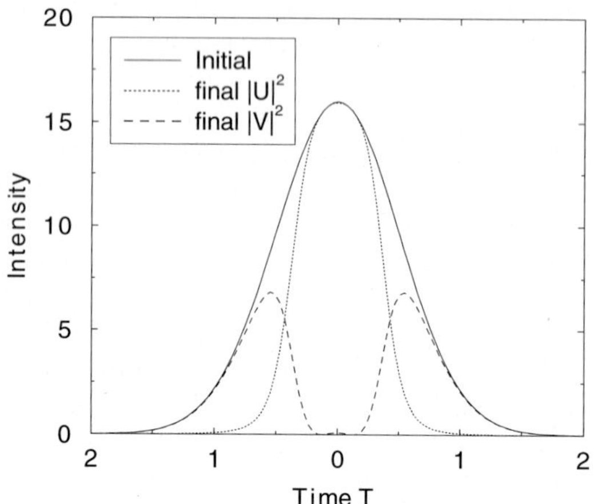

Fig. 4. Input and output pulse profiles through a grating given an initial Gaussian core mode pulse profile progated over a grating of length $Z = 0.7$ with a computational domain of $T = 10$ (magnified here to $T \in [-2, 2]$, and coupling coefficients $d_{14} = 2$, $c_{11} = c_{44} = 0.3$, and $c_{14} = 0.5$

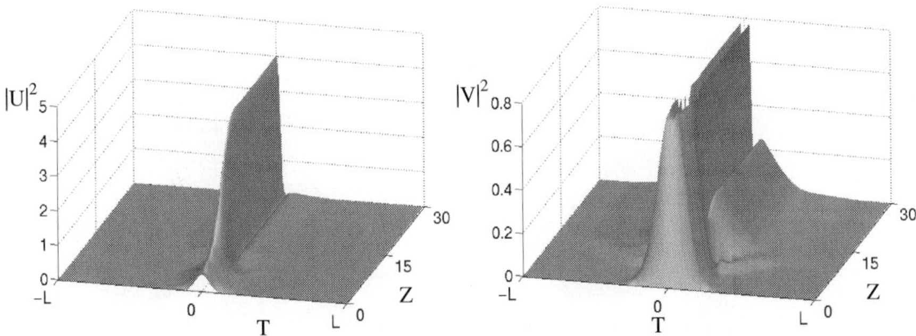

Fig. 5. Stable mode-locking pulse dynamics for a laser cavity with $Z = 0.1$ and coupling coefficients $d_{14} = 12$, $c_{11} = c_{44} = 0.2$, and $c_{14} = 20$. The gain parameter in this case is $g_0 = 1$

mode-locking. The cladding field is then almost completely attenuated once butt-coupling back to the fiber cavity occurs.

Other behaviors also occur in the cavity. In particular, the gain parameter g_0 acts as a bifurcation parameter for the pulse dynamics. Figure 6 depicts the evolution of the pulse with the gain parameter being $g_0 = 0.5, 1.0, 2.0$ and 3.0. If the cavity gain is too low ($g_0 = 0.5$) the pulse slowly dies in the cavity and there is no mode-locking. For larger values of gain ($g_0 = 1.0$) the pulse shaping allows for the stabilization of a mode-locked pulse. Higher values of gain generate periodic mode-locked trains or chaotic pulse trains, neither of which are desirable for application purposes.

3.2 Mode-Locking with Waveguide Arrays

Waveguide arrays have recently been of great interest due to their applications in all-optical signal processing and switching. The particular interest here is in the basic waveguide coupling phenomena which occurs through evanescent mode coupling [36]. The linear mode coupling in conjunction with nonlinear self- and cross-phase modulation effects has been proposed [39] due to a wide variety of interesting soliton like effects and dynamics [39, 40, 41, 42, 43]. The particular application of mode-locking considered here relies on the intensity dependent mode-coupling aspects much as with the long-period fiber grating laser of the previous subsection (see Fig. 7 for the mode-locked laser configuration).

The equations which govern the nearest neighbor coupling of the waveguides is given by [39, 40, 41, 42, 43]

$$\mathrm{i}\frac{dA_n}{dz} + (A_{n-1} + A_{n+1}) + \gamma|A_n|^2 A_n = 0 \tag{14}$$

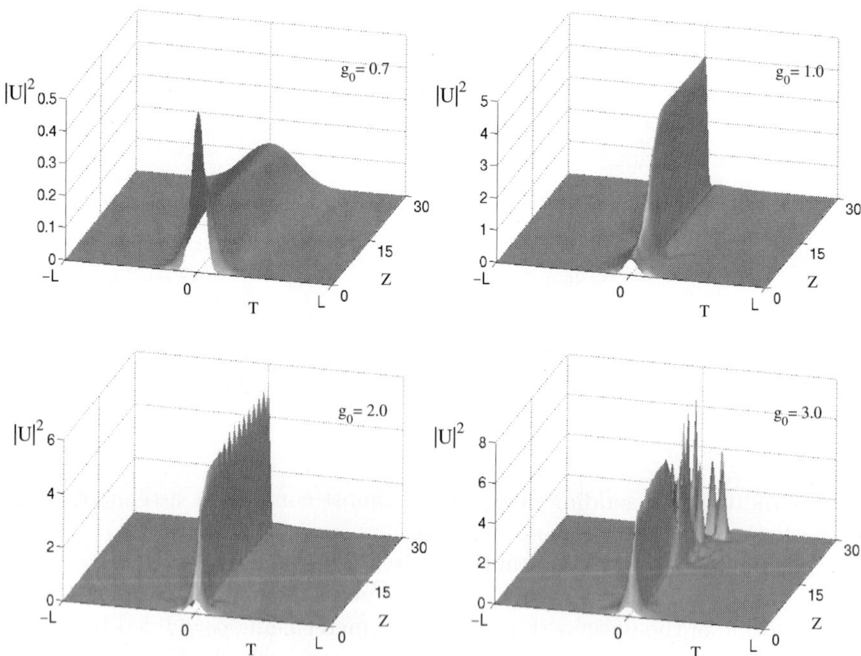

Fig. 6. Pulse dynamics for a laser cavity with $Z = 0.1$ and coupling coefficients $d_{14} = 12$, $c_{11} = c_{44} = 0.2$, and $c_{14} = 20$. The gain parameter in this case varies from $g_0 = 0.5, 1.0, 2.0$ and 3.0. Note that close to the ideal cavity gain ($g_0 \approx 1$) the cavity stabilizes and mode-locks a pulse

where A_n represents the electric field amplitude in the nth waveguide and we have rescaled the equations so that the coupling coefficient is unity. It has been assumed that the waveguide array is periodically spaced and lossless. The waveguide spacing is fixed so that nearest neighbor linear coupling dominates the interaction between waveguides. Further, the waveguide material is assumed to have a cubic Kerr nonlinearity and a sufficient number of waveguides so that no boundary effects occur in the dynamics.

A typical representation of the grating array is shown in Fig. 7. The array is inserted into a optical fiber ring cavity as the mode-locking element responsible for generating intensity-dependent discrimination in a propagating temporal pulse. A simple simulation of (14) serves to illustrate the dynamics of the waveguide array. By butt-coupling the array to the waveguide array, electromagnetic radiation initially enters only into a single waveguide of the array. We denote this initial waveguide as A_0. The initial energy in A_0 will couple through nearest neighbor interactions to the neighboring waveguides. These will in turn couple to their neighbors. Figure 8 demonstrates the effective spatial diffraction of energy in the waveguide array for two different initial conditions and for $\gamma = 1$. In the first (left), the initial field $A_0 = 1$

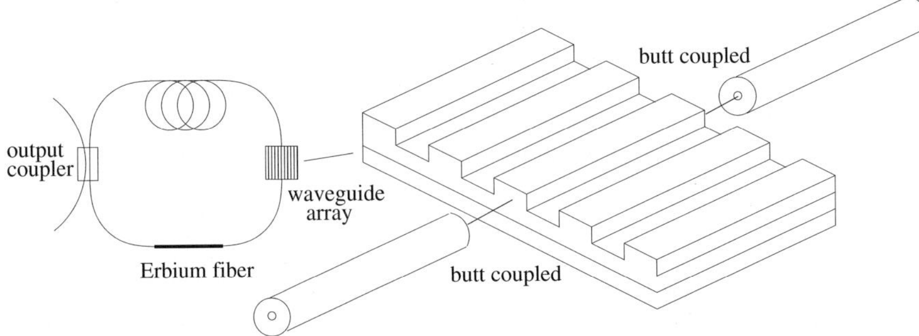

Fig. 7. Configuration of a mode-locked laser cavity based upon a waveguide array element for intensity discrimation. The waveguide array is a set of periodically spaced waveguides which couple evanescently to nearest neighbors. Only the energy which remains in the original launch waveguide is of interest in the mode-locking, i.e. the energy in the neighboring waveguides is lost after butt-coupling back into the laser cavity

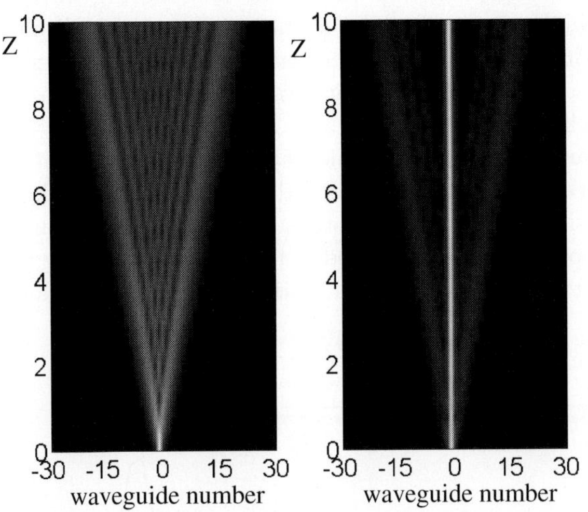

Fig. 8. Mode-coupling dynamics in the waveguide array as given by (14) with $\gamma = 1$. For both figures, energy is launched into the center waveguide with amplitude $A_0 = 1$ (*left*) and $A_0 = 3$ (*right*). For the lower intensities, the energy is quickly coupled to the neighboring waveguides and diffracted throughout the structure. For higher intensities, the energy remains highly localized in the launch waveguide with only a small portion escaping to the neighboring waveguides

so that the nonlinearity has little effect on the linear coupling dynamics. For higher initial intensities (right), i.e. $A_0 = 3$, the nonlinearity provides self-focusing which helps to confine the initial energy to the launch waveguide. Only a small amount of energy is coupled to the neighboring waveguides in this case.

A simple understanding of this localization comes from thinking of (14) as a finite-difference discretization of the NLS equation (1), i.e. the second-order finite-difference discretization of the second derivative couples to nearest neighbors. Thus the effective diffraction equation in the continuum limit is a *spatial* NLS in the waveguide array [39]. For weak nonlinearity, diffractive spreading occurs so that the energy propagates away from the center waveguide. However, when diffraction and self-phase modulation are balanced, a steady-state soliton profile in the spatial direction is achieved and energy remains in the launch waveguide.

As with the long-period fiber grating, the temporal pulse shaping aspects of the waveguide array are what determine the suitability of the array for mode-locking. To illustrate the analog of Fig. 4 of the LPFG, a temporal initial condition is launched into the waveguide array subject to (14). Figure 9 illustrates the pulse shaping which occurs for a train of two noisy initial Gaussian pulses. The light solid lines are the initial data while the heavy lines represent the output of the center waveguide. Note that the low amplitude pulse is strongly coupled out to the neighboring waveguides and lost, whereas the high amplitude pulse is subject to strong self-phase modulation and remains localized in the center waveguide. Note further that the higher intensity pulse is substantially narrowed in the time-domain as is expected of a mode-locking mechanism.

In summary, the basic principle of mode-locking operation relies on the same ideas as of the LPFG mode-locked laser. The waveguide array via mode-coupling provides the necessary intensity discrimination which allows stable mode-locked pulses to be supported in the ring cavity of Fig. 7. The temporal pulse shaping exhibited in Fig. 9 leads to similar mode-locking results to Figs. 5 and 6.

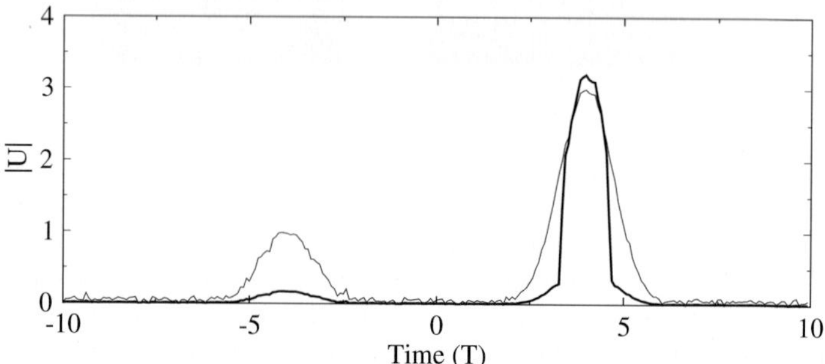

Fig. 9. Temporal input (*light lines*) and output (*heavy lines*) in the center waveguide A_0 after propagation through the waveguide array of Fig. 8. Note the required pulse shaping which occurs for higher intensities versus lower intensities. Here the electric field input and output is denoted by U since it is coupled back into the laser cavity dynamics (8)

3.3 Dual Core Fiber Laser

Yet another example of mode-locking induced by mode-coupling arises from dual-core optical fibers. In principle, the mode-locking mechanism is identical to that of the LPFG and waveguide array, i.e. high intensity portions of a pulse are almost completely transmitted in one core of the fiber whereas low intensity portions are coupled to the second core and lost. A typical cavity configuration is illustrated in Fig. 10. Of importance to the laser is to engineer the dual core fiber spacing and length so that in the *linear* regime electromagnetic energy is completely coupled from one core to its neighbor. This coupling, as in the waveguide array, is an evanescent mode-coupling. Thus low intensity portions of a pulse will be effectively attenuated and the appropriate intensity discrimation is established. Note that for longer fibers, energy from the second core will begin coupling back to the original core mode.

The normalized equations which govern the nearest neighbor coupling of the waveguides is given by [33, 36]

$$\begin{aligned} & \mathrm{i}\frac{dA_1}{dz} + A_2 + \gamma|A_1|^2 A_1 = 0 \\ & \mathrm{i}\frac{dA_2}{dz} + A_1 + \gamma|A_2|^2 A_2 = 0 \end{aligned} \tag{15}$$

where A_n $(n = 1, 2)$ represents the electric field amplitude in the nth core and we have rescaled the equations so that the coupling coefficient is unity. It has been assumed that the dual core fiber is lossless.

A simulation of (15) serves to illustrate the dynamics of the dual-core coupling. Initially, electromagnetic radiation initially enters only into a single core A_1. The initial energy in A_1 will couple through evanescent wave

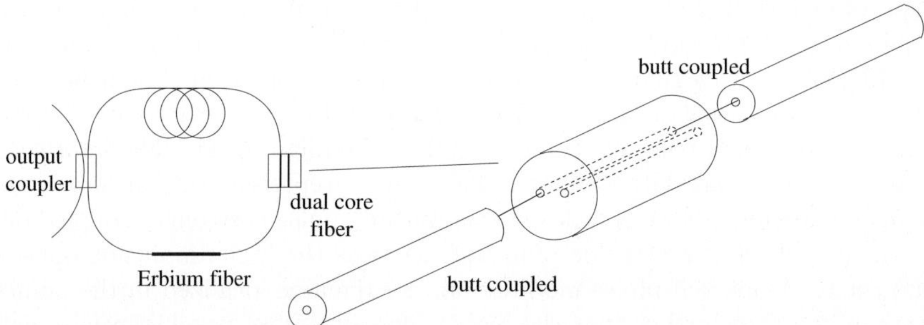

Fig. 10. Configuration of a mode-locked laser cavity based upon a dual core fiber for intensity discrimation. The dual core fiber couples via evanescence from core to core. Only the energy which remains in the original launch core is of interest in the mode-locking, i.e. the energy in the second core is lost since it is not coupled back into the laser cavity

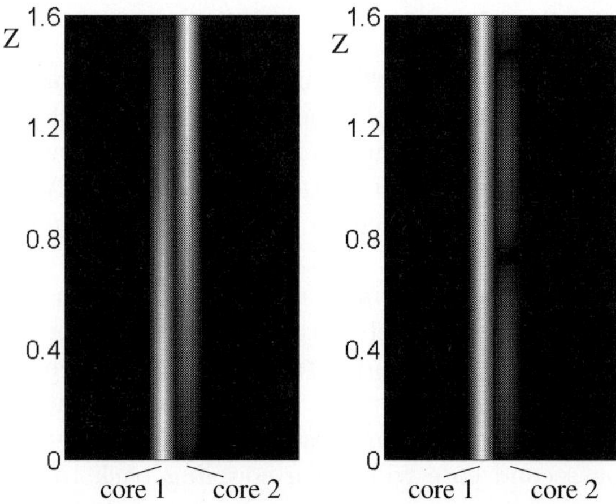

Fig. 11. Mode-coupling dynamics in the dual core fiber as given by (15) with $\gamma = 1$. For both figures, energy is launched into the left core with amplitude $A_1 = 1$ (*left*) and $A_1 = 3$ (*right*). For the lower intensities, the energy is coupled evanescently to the second core. For higher intensities, the energy remains highly localized in the launch waveguide with only a small portion escaping to the second core

interaction to the neighboring core mode A_2. Figure 11 demonstrates the effective spatial exchange of energy in the dual-core fiber for two different initial conditions and for $\gamma = 1$. In the first (left), the initial field $A_1 = 1$ so that the nonlinearity has little effect on the linear coupling dynamics. For higher initial intensities (right), i.e. $A_1 = 3$, the nonlinearity provides self-focusing which helps to confine the initial energy to the launch core-mode. Only a small amount of energy is coupled to the neighboring core-mode in this case.

As with the LPFG and waveguide array lasers, the temporal pulse shaping aspects of the dual-core coupling are what determine the suitability of the dual-core fiber for mode-locking. To illustrate the analog of Figs. 4 and 9 of the LPFG and waveguide array, a temporal initial condition is launched into the waveguide array subject to (15). Figure 12 illustrates the pulse shaping which occurs for a train of two noisy initial Gaussian pulses. The light solid lines are the initial data while the heavy lines represent the output of the launch waveguide. Note that the low amplitude pulse is strongly coupled out to the neighboring waveguides and lost, whereas the high amplitude pulse is subject to strong self-phase modulation and remains localized in the launch waveguide. Note further that the higher intensity pulse is substantially narrowed in the time-domain as is expected of a mode-locking mechanism.

In summary, the basic principle of mode-locking operation relies on the same ideas as of the LPFG and waveguide array mode-locked lasers. The dual-core array via mode-coupling provides the necessary intensity discrimination which allows stable mode-locked pulses to be supported in the ring

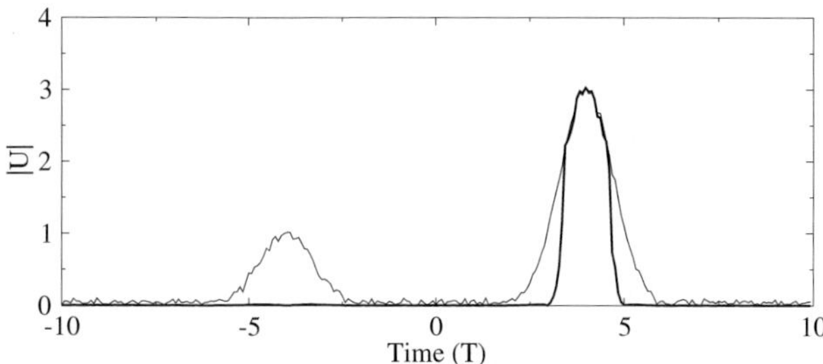

Fig. 12. Temporal input (*light lines*) and output (*heavy lines*) in the launch core A_1 after propagation through the dual core fiber of Fig. 10. Note the required pulse shaping which occurs for higher intensities versus lower intensities. Here the electric field input and output is denoted by U since it is coupled back into the laser cavity dynamics (8)

cavity of Fig. 10. Similar results to Figs. 5 and 6 can be obtained. Although the model is theoretically capable of supporting mode-locked pulse operation, practical issues limit the usefulness of the dual-core mode-locked laser. Most importantly is the fact that the fiber length required to achieve complete coupling of *linear* energy from one core to the next is on the order of several meters or more. However, over this length, the dual-core fiber is highly sensitive to temperature and core-to-core spacing perturbations. These perturbations greatly effect the mode-coupling dynamics and pose practical limitations on generating mode-locked pulses from the cavity. However, with the advance of material technologies such as chalcogenide glass [46, 47], the core-to-core spacing in the dual core fiber can be decreased in order to shorten the dual-core fiber length significantly. And with the increased nonlinearity provided by the chalcogenide glass, the self-phase modulation remains sufficiently strong to provide the necessary pulse shaping over very short distances of dual-core fiber. Thus the key components of a physically realizable dual-core fiber based mode-locked laser can be achieved.

3.4 Mode-Locking with an Acousto-Optic Modulator

In contrast to the passive mode-locking models represented by the previous two subsections, this subsection considers the laser dynamics driven by active mode-locking techniques. In this case, nonlinear mode-coupling effects generate the phenomena known as *Q-switching*. In fact, in the following analysis, Q-switching can be thought of as a nonlinear two-mode interaction between a symmetric and anti-symmetric pair of unstable eigenmodes. Their nonlinear beating interaction drives the observed Q-switching phenomena.

The pulse evolution in the actively mode-locked laser system is governed by the equation [17, 18, 44]:

$$\mathrm{i}\frac{\partial Q}{\partial Z}+\frac{1}{2}\frac{\partial^2 Q}{\partial T^2}+|Q|^2Q-\mathrm{i}g(Z)\left(1+\tau\frac{\partial^2}{\partial T^2}\right)Q+\mathrm{i}M(\Gamma-\mathrm{cn}^2(\omega T,k))Q=0\,. \quad (16)$$

where Q is again the normalized electric field envelope scaled as in (8). The variable Q is used instead of U to distinguish this as an active mode-locking model. The new parameters M, Γ, and ω measure the normalized strength and frequency of the active modulation element which is responsible for generating the mode-locked pulse stream. The gain in the fiber is incorporated through the dimensionless parameter $g = g(Z)$ just as in (3).

The active modulation model of mode-locking is fundamentally different from the previous subsections in that instead of a passive, nonlinear response generating pulse shaping, here it is the linear, time-dependent, periodic forcing $\mathrm{cn}^2(\omega T, k)$ which leads to stabilized pulses. This periodic forcing generates peaks and troughs in the gain as a function of the time T. Thus cavity energy will accumulate at the local peaks of gain whereas cavity energy will be attenuated at the local troughs in the gain. This preferential localization of cavity energy gives the necessary pulse shaping required to form and stabilize localized pulses in a periodic wavetrain.

To proceed analytically, we have generalized the usual periodic forcing given by $\cos^2 \omega T$ [17, 18] to the Jacobi elliptic [45] cosine function $\mathrm{cn}^2(\omega T, k)$ [44]. Here $0 \le k \le 1$ is the elliptic modulus. In the limit $k = 0$, the modulation reduces to a purely sinusoidal forcing as given previously [17, 18]. For values of $k < 0.9$ the potential is virtually indistinguishable from the sinusoidal modulation. However, as $k \to 1$, the forcing becomes a series of well separated hyperbolic secant shaped modulations. This modulation is introduced not only for its mathematical generality and ease, but also because exact analytic solutions to (16) with a constant gain parameter $g(Z) = g =$constant can be found. In particular, two families of solutions exist. The first takes the form (setting $\omega = 1$ without loss of generality):

$$Q(Z,T) = k\,\mathrm{cn}(T,k)\exp[-\mathrm{i}(1/2 - k^2)Z]\,, \quad (17)$$

where $M = 2k^2 g\tau$ and $\Gamma = (2k^2 + 1/\tau - 1)/2k^2$.
This solution branch represents a periodic train of pulses where adjacent pulses are separated by a node. This nodal separation is critical to the stability of the pulse trains, i.e. adjacent pulses need to be out-of-phase in order to be stabilized. Alternatively, an in-phase solution is found of the form

$$Q(Z,T) = \mathrm{dn}(T,k)\exp[-\mathrm{i}(k^2/2 - 1)Z]\,, \quad (18)$$

where $M = 2k^2 g\tau$ and $\Gamma = k^2(2\tau(1 - 1/k^2) + 1/k^2 + \tau(2 - k^2)/k^2)/2$. This solution branch represents a periodic train of pulses where adjacent pulses are not separated by a node and solutions go unstable.

The behavior of both the out-of-phase and in-phase pulse train evolution is depicted in Figs. 13 and 14. Figure 13 illustrates the stabilizing influence of the active modulation whereas Fig. 14 demonstrates the unstable evolution of both an out-of-phase and in-phase initial solution. The stabilized pulse trains of Fig. 13 settle quickly to the exact solution given by (17). For both these simulations, the elliptic modulus is $k = 0.999$ so that the localized solutions are nearly hyperbolic secant pulses which are well separated. For Fig. 14, the out-of-phase solution has significant overlap with neighboring pulses since $k = 0.75$. This nearest neighbor interaction results in instability of the uniform pulse train. The in-phase solutions are always unstable [49].

The exact solutions (17) and (18) of the governing equation (16) can be analyzed with a standard linear stability analysis. The underlying idea of linear stability is to perturb around an exact solution and derive an eigenvalue problem which predicts the asymptotic behavior of the perturbation, whether it be growth, decay, or oscillatory behavior [48]. This is one of the few analytic methods available for understanding the stability dynamics near exact solutions. By considering the out-of-phase (17) and in-phase (18) solutions under perturbation, the long-term behavior within the laser cavity can be better characterized.

Beginning with the $\mathrm{cn}(T, k)$ solution (17), the eigenvalues of the linearization are calculated as a function of the elliptic modulus k. Dependence of the stability on the elliptic modulus has been observed previously in simulations [44]. Unlike the continuous spectrum, the discrete spectral content exhibits instability. Specifically, there are between two and four eigenvalues in the right-half (unstable regime) of the complex plane. The specific location and number is determined by the value of the elliptic modulus k [49]. For low k values, the four unstable modes are present which are composed of two double roots. At a value of $k \approx 0.4$, one of the double roots separates with one eigenvalue increasing and the second decreasing. At a value of $k \approx 0.6$, the

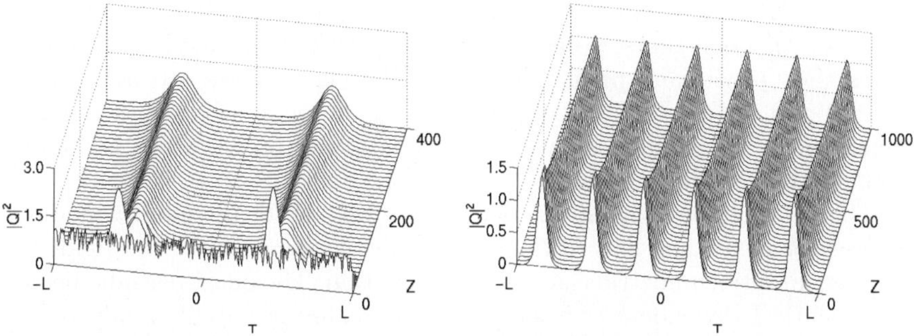

Fig. 13. (**a**) Stable evolution of the mode-locking pulse train starting near the exact solution given by (17). Here the initial amplitude is perturbed from the exact solution by 0.2. Three periods are considered with the parameters used being $k = 0.999$, $g = 0.3$ and $\tau = 0.1$. (**b**) Stable evolution of the mode-locking pulse train starting with initial noise. One period is considered with the parameters used being $k = 0.999$, $g = 0.3$ and $\tau = 0.1$. Note the formation of two pulses from this noise realization

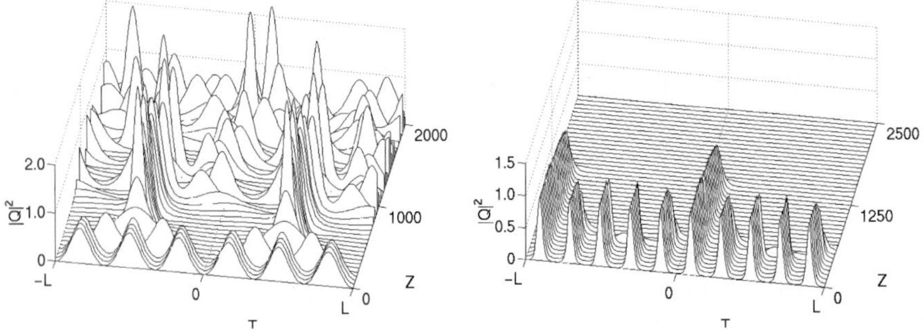

Fig. 14. (**a**) Unstable evolution of the mode-locking pulse train starting near the exact solution given by (17). One period is considered with the parameters used being $k = 0.75$, $g = 0.3$ and $\tau = 0.1$. In contrast to Fig. 13, the reduced separation of the neighboring pulses causes the pulses to destabilize and evolve to a chaotic pulse train. (**b**) Evolution of the mode-locking pulse train starting near the exact solution given by (18) with $k = 0.999$. Here the we begin with the exact solution and show that it destabilizes and decays to zero. Thus the in-phase solution is unstable

second double root diverges in a similar fashion. Thus at a value of $k \approx 0.65$, only two unstable modes remain. The largest unstable growth is achieved at $k \approx 0.82$. Beyond this, the unstable growth rates quickly decrease to zero. In the limit $k \to 1$, the growth rate is identically zero so that neutral stability is established (See Fig. 13).

The eigenmode structure of the two unstable modes which persist beyond $k \approx 0.6$ suggests the source of the Q-switching instability. These modes persist for $k \in [0, 1)$ and clearly show a symmetric/anti-symmetric structure. Specifically, the shape of the eigenvectors resemble Jacobi elliptic functions of the $\mathrm{cn}(T, k)$ and $\mathrm{dn}(T, k)$ type. And since there is a difference in their phase-rotation constants, a beating occurs between these modes which generates the Q-switching phenomena [49]. The eigenvectors are not exactly $\mathrm{cn}(T, k)$ and $\mathrm{dn}(T, k)$ functions. Rather, they are qualitatively described by them. The true eigenmodes have a nontrivial phase profile which is difficult to characterize analytically.

The Q-switching dynamics is depicted in Fig. 15 for the elliptic modulus values of $k = 0.5$ and $k = 0.75$. Both of these cases exhibit the predicted symmetric/anti-symmetric eigenmode interaction. The quasi-periodic behavior is clearly k dependent as is suggested by the linear stability analysis [49]. Thus Q-switching can be understood as a consequence of the quasi-periodic behavior. Specifically, every round trip the quasi-periodic electric field oscillations traverse an output coupler which launches the mode-locked pulses down a optical fiber waveguide. If, for instance, the output coupler is at every $Z = 1$, then the intensity profile $|Q(Z, T)|^2$ is transmitted into the fiber at $Z = n$ where $n = 1, 2, 3, \cdots$. By retrieving these slices of the intensity profile at the various times $Z = n$ and concactenating them together, the

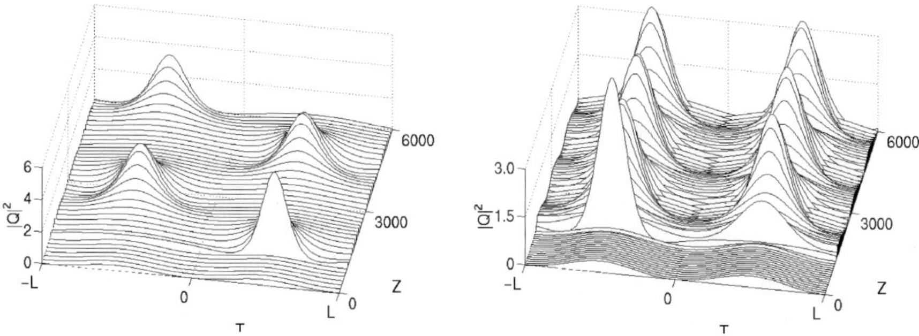

Fig. 15. Q-switching evolution of the mode-locking pulse train starting near the exact solution given by (17). (**a**) One period is considered with the parameters used being $k = 0.5$, $g = 0.3$ and $\tau = 0.1$. (**b**) One period is considered with the parameters used being $k = 0.75$, $g = 0.3$ and $\tau = 0.1$

mode-locked pulse train is constructed. Assuming for convenience that the output coupler is at $Z = n$, then the mode-locked pulse streams for the k values of Fig. 15 are given in Fig. 16. This shows the dependence of the Q-switching amplitude and frequency on the elliptic modulus k as predicted from the approximate model in the previous subsection. This behavior is the essense of Q-switching, i.e. a slow modulation of the periodic wave train. By controlling the elliptic modulus, the maximum Q-switching power level can be maximized or minimized depending upon the specific engineering application.

To make connection with nonlinear coupled mode theory, an approximation to the mode coupling dynamics is constructed. In a linear system, the interaction could be easily characterized by superposition. The presence of self-phase modulation in (16) gives a cubic term which does not allow for the application of superposition theory. Regardless, an attempt can be made at characterizing the Q-switching using the superposition principle. Thus a solution of the form

$$U(Z,T) = A(Z)\mathrm{cn}(T,k) + B(Z)\mathrm{dn}(T,k) \tag{19}$$

is considered where the amplitudes $A = A(Z)$ and $B = B(Z)$ depend upon the propagation distance. Inserting (19) into (16) and combining the rotating phase approximation with $\tau \ll 1$ gives

$$\begin{aligned} &\mathrm{i}\frac{dA}{dZ} + \frac{2k^2-1}{2}A + 2(1-k^2)|B|^2 A = 0 \\ &\mathrm{i}\frac{dB}{dZ} + \frac{2-k^2}{2}B + 2(1-1/k^2)|A|^2 B = 0\,, \end{aligned} \tag{20}$$

whose solutions are

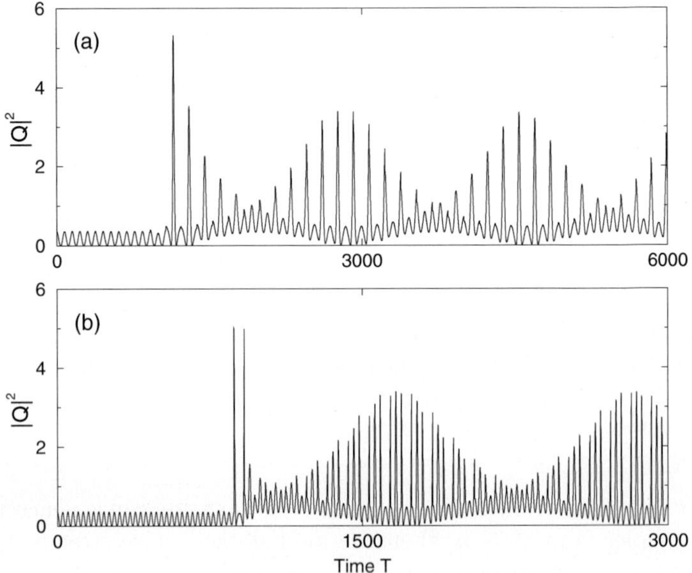

Fig. 16. Resulting pulse train dynamics for Fig. 15 in (**a**) and (**b**) respectively. Note the distinctive Q-switching pattern which results from the quasi-periodic cavity behavior

$$\begin{aligned} A(Z) &= A_0 \exp\left[\mathrm{i}((k^2 - 1/2) + 2(1 - k^2)B_0^2)Z\right] \\ B(Z) &= B_0 \exp\left[\mathrm{i}((1 - k^2/2) + 2(1 - 1/k^2)A_0^2)Z\right] \ . \end{aligned} \tag{21}$$

The real constants A_0 and B_0 are determined from consistency conditions under the rotating-wave approximation. This gives $A_0 = \pm k/\sqrt{3}$ and $B_0 = \pm 1/\sqrt{3}$. Of course, this is only an approximate solution which in no way captures the correct quantitative behavior. However, it does do well in capturing the qualitative behavior observed numerically. The difficulty in making quantitative comparison is due to the marginal validity of the rotating-wave approximation. Regardless, this analytic approach captures the simple two-mode nonlinear interaction which can lead to the experimentally and computationally observed Q-switching phenomena.

4 Conclusions and Discussion

Considerable interest and research in the past few years has focused on the use of Erbium-doped fibers in actively and passively mode-locked fiber lasers operating in both the normal and anomalous dispersion regimes. Many mode-locking schemes have been successfully demonstrated. A generic feature of each mode-locked laser is the intensity-dependent loss which is achieved by the mode-locking mechanism. The additional loss imparted upon low-intensity parts of the pulse, whether it be dispersive radiation or the wings of a pulse, gives the necessary pulse shaping required to achieve stable mode-locking operation, i.e. a dissipative soliton becomes an attracting state of the system. These compact sources of optical pulses near wavelengths of 1.55 microns are key enabling technologies for high speed fiber optic communication systems and interconnection networks.

The mode-locking models considered here are driven by a nonlinear mode-coupling interaction. For mode-locking with a long-period fiber grating, waveguide array, or dual-core coupler, the nonlinear mode-coupling resonantly couples low-intensity light to a second mode which is attenuated. High-intensity light is detuned from resonance and largely unattenuated during propagation. This intensity discrimination is suitable for generating stable dissipate solitons for a large range of parameter space. In contrast, it is nonlinear mode-coupling which drives the observed quasi-periodic instability in an active mode-locked laser. Here, the nonlinear beating between symmetric and anti-symmetric modes gives the characteristic Q-switching behavior, i.e. the slow modulation of a propagating pulse train. In all these models, comparison with experiment can be made since they are quantitative in nature. This is in contrast to the master mode-locking dynamics which is a qualitative description. Because the of the quantitative aspects of the models, the design and operating regimes for realistic lasers can be fully explored.

Acknowledgements

The author would especially like to thank Karen Intrachat, Todd Kapitula, Jennifer O'Neil, Joshua Proctor, and Bjorn Sandstede for significant collaborative efforts pertaining to the various models described in this work.

References

1. A. Hasegawa, F. Tappert, Appl. Phys. Lett. **23**, 142 (1973).
2. H. A. Haus, W. S. Wong, Rev. Mod. Phys., **68**, 423 (1996).
3. H. A. Haus, IEEE J. Sel. Top. Quant. Electr., **6**, 1173 (2000).
4. I. N. Duling, M. L. Dennis, *Compact sources of ultrashort pulses.* (Cambridge University Press, Cambridge, 1995).
5. H. A. Haus, J. G. Fujimoto, E. P. Ippen, J. Opt. Soc. Am. B **8**, 2068 (1991).
6. H. A. Haus, J. G. Fujimoto, E. P. Ippen, IEEE J. Quant. Elec., **28**, 2086 (1992).
7. K. Tamura, H. A. Haus, E. P. Ippen, Electr. Lett., **28**, 2226 (1992).
8. H. A. Haus, E. P. Ippen, K. Tamura, IEEE J. Quant. Elec., **30**, 200 (1994).
9. M. E. Fermann, M. J. Andrejco, Y. Silberberg, M. L. Stock, Opt. Lett., **29**, 447 (1993).
10. D. Y. Tang, W. S. Man, H. Y. Tam, Opt. Comm., **165**, 189 (1999).

11. I. N. Duling, Electr. Lett., **27**, 544 (1991).
12. D. J. Richardson, R. I. Laming, D. N. Payne, V. J. Matsas, M. W. Phillips, Electr. Lett. **27**, 542 (1991).
13. M. L. Dennis, I. N. Duling, Elec. Lett. **28**, 1894 (1992).
14. F. X. Kärtner, U. Keller, Opt. Lett., **20**, 16 (1995).
15. B. Collings, S. Tsuda, S. Cundiff, J. N. Kutz, M. Koch, W. Knox, K. Bergman, IEEE J. Selec. Top. Quant. Elec., **3**, 1065 (1997).
16. S. Tsuda, W. H. Knox, E. A. DeSouza, W. J. Jan, J. E. Cunningham, Opt. Lett., **20**, 1406 (1995).
17. F. X. Kärtner, D. Kopf, U. Keller, J. Opt. Soc. Am. B **12**, 486 (1994).
18. H. A. Haus. IEEE J. Quant. Elec. **11**, 323 (1975).
19. I. P. Christov, V. Stoev, M. Murname, H. Kapteyn, Opt. Lett., **21**, 1493 (1996).
20. I. P. Christov, M. Murname, H. C. Kapteyn, J. P. Zhou, C. P. Huang, Opt. Lett. **19**, 1465 (1994).
21. C. J. Chen, P. K. A. Wai, C. R. Menyuk, Opt. Lett. **17**, 417 (1992).
22. T. Kapitula, J. N. Kutz, B. Sandstede, J. Opt. Soc. Am. B **19**, 740 (2002).
23. T. Kapitula, J. N. Kutz, B. Sandstede, Indiana J. of Math. (to appear) (2004).
24. N. R. Pereira, L. Stenflo, Phys. Fluids., **20**, 1733 (1977).
25. T. Kapitula, Physica D **116**, 95 (1998).
26. M. Romagnoli, S. Wabnitz, P. Franco, M. Midrio, L. Bossalini, and F. Fontana, J. Opt. Soc. Am. B **12**, 938 (1995).
27. M. Romagnoli, S. Wabnitz, P. Franco, M. Midrio, F. Fontana, and G. E. Town, J. Opt. Soc. Am. B **12**, 72 (1995).
28. N. Akhmediev, J. M. Soto-Crespo, and G. Town, Phys. Rev. E **63**, 056602 (2001).
29. J. M. Soto-Crespo and N. Akhmediev, Phys. Rev. E **66**, 066610 (2002).
30. J. M. Soto-Crespo, N. Akhmediev, K. S. Chiang, Phys. Lett. A **291**, 115 (2001).
31. A. M. Vengsarkar, P. J. Lemaire, J. B. Judkins, V. Bhatia, T. Erdogan, J. E. Sipe, J. Light. Tech., **14**, 58 (1996).
32. B. J. Eggleton, R. E. Slusher, J. B. Judkins, J. B. Stark, A. M. Vengsarkar, Opt. Lett., **22**, 883 (1997).
33. S. M. Jensen, IEEE J. Quant. Elec. **QE18**, 1580 (1982).
34. H. G. Winful, D. T. Walton, Opt. Lett., **17**, 1688 (1992).
35. Y. Oh, S. L. Doty, J. W. Haus, R. L. Fork, J. Opt. Soc. Amer. B **12**, 2502 (1995).
36. D. Marcuse, *Theory of Dielectric Optical Waveguides*, (Academic, New York, 1974).
37. Kutz J. N., Eggleton B. J., Stark J. B., Slusher R. E. (1997) Nonlinear pulse propagation in long-period fiber gratings: theory and experiment. IEEE J Selec Top in Quan Elec 3:1232–12451.
38. K. Intrachat, J. N. Kutz, IEEE J. Quant. Elec., **39**, 1572 (2003).
39. D. N. Christodoulides, R. I. Joseph, Opt. Lett. **13**, 794 (1988).
40. H. S. Eisenberg, Y. Silberberg, R. Morandotti, A. R. Boyd, J. S. Aitchison, Phys. Rev. Lett., **81**, 3383 (1998).
41. A. C. Aceves, C. De Angelis, T. Peschel, R. Muschall, F. Lederer, S. Trillo, S. Wabnitz, Phys. Rev. E **53**, 1172 (1996).
42. H. S. Eisenberg, R. Morandotti, Y. Silberberg, J. M. Arnold, G. Pennelli, J. S. Aitchison, J. Opt. Soc. Am. B **19**, 2938 (2002).
43. U. Peschel, R. Morandotti, J. M. Arnold, J. S. Aitchison, H. S. Eisenberg, Y. Silberberg, T. Pertsch, F. Lederer, J. Opt. Soc. Am. B **19**, 2637 (2002).

44. J. J. O'Neil, J. N. Kutz, B. Sandstede, IEEE J. Quant. Electr., **38**, 1412 (2002).
45. M. Abramowitz, I. A. Stegun, *Handbook of Mathematical Functions*, (National Bureau of Standards, Washington DC, 1964).
46. M. Asobe, T. Kanamori, K. Kubodera, IEEE J. Quantum Electron., **29**, 2325 (1993).
47. M. Asobe, T. Ohara, I. Yokohoma, T. Kaino, Electron. Lett., **32**, 1611 (1996).
48. P. G. Drazin, *Nonlinear Systems*, (Cambridge University Press, Cambridge, 1992).
49. J. Proctor, J. N. Kutz, (submitted to IEEE J. Quant. Elec., 2004).

Dissipative Solitons in Reaction-Diffusion Systems

H.-G. Purwins, H.U. Bödeker, and A.W. Liehr

Institut für Angewandte Physik, Corrensstraße 2/4, D-48149 Münster, Germany
purwins@nwz.uni-muenster.de
boedeker@nwz.uni-muenster.de
obi@uni-muenster.de

1 Introduction

A major goal of natural science is to understand the formation of spatially-extended patterns in all kinds of physical, chemical, biological and other systems. In many cases, it is advantageous to interpret the overall pattern under consideration in terms of a superposition of certain spatially well-localized elementary patterns that we may refer to as "particles". In the simplest case, all these particles are of the same kind and the complex behavior of the extended pattern can be described in terms of simple individual properties of the particles and their interaction. A clear illustrative example for this approach is the concept of atoms. In this case, the elementary pattern or particle is the atom and the complex spatially-extended pattern is, e.g., the crystal.

From a theoretical point of view, pattern forming systems are described by field equations with infinitely many degrees of freedom. However, a powerful technique for describing their temporal evolution is to use a "particle approach". In this approach, well-localized solutions of the field equation are viewed as particles. The dynamic behavior and the interaction of these particles are described by ordinary differential equations, using center-of-mass co-ordinates and possibly some other variables. The decisive advantage of such an approach is that the underlying field equations, with infinitely many degrees of freedom, can be reduced to order-parameter equations with a finite and possibly small number of degrees of freedom, without losing the important information. An extremely powerful and far-reaching application is the notion of atoms.

We recall that macroscopic physical systems can be separated into two classes, according to their long-time behavior. One class approaches thermodynamic equilibrium, resulting in a vanishing exchange of energy with the surroundings. The second class is characterized by external driving "forces" which lead to a finite energy transfer to the system, and, correspondingly, to a finite dissipation in the long run.

For the first class of systems, general techniques to find physical solutions have been developed. Systems in thermodynamic equilibrium can be described by a thermodynamic potential, of which one has to find the absolute

H.-G. Purwins, H.U. Bödeker, and A.W. Liehr: *Dissipative Solitons in Reaction-Diffusion Systems*, Lect. Notes Phys. **661**, 267–308 (2005)
www.springerlink.com

extremum. Well-known examples for such systems are atoms and molecules forming crystals, defects and precipitations in solid materials, magnetic and electric dipoles leading to domain structures in condensed matter and islands on solid surfaces.

The class of dissipative systems is much less well-understood than the class of thermodynamic equilibrium systems. As a rule, no concepts like the thermodynamic potential exist. Nevertheless, dissipative systems represent an extremely promising area of research as they exhibit an overwhelming diversity of spatially-extended self-organized patterns. Typical representatives can be found in physics, chemistry, geology, biology and even sociology. Examples for patterns of this kind are wind-driven waves of fluids, electrical field-driven lightning, spiral patterns in chemical reactions, periodic sedimentation, nerve pulses and accumulations of amoebae. In numerous cases, it is advantageous to define localized solitary structures that, in many respects, behave like individual objects and that are generated or annihilated as a whole. Some examples are shown in Fig. 1.

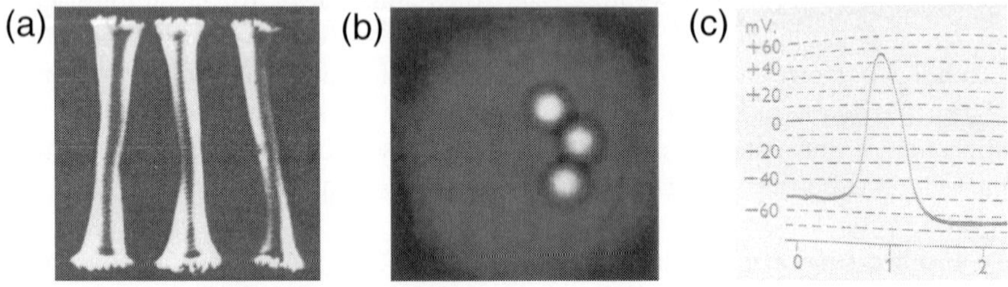

Fig. 1. Examples of localized structures in dissipative systems: (**a**) current filaments in a semiconductor that form electrical current density spots in the plane vertical to the axis of the filaments [18], (**b**) light spots in laser-driven nonlinear sodium vapor [17], (**c**) electrical potential of a propagating nerve pulse, as a function of time [19]

In our chapter, we will focus on a special class of dissipative systems, namely on reaction-diffusion systems. This class has its origin in chemistry, but today representatives can be found in many branches of natural sciences. Although different patterns and localized structures in reaction-diffusion systems have been known for a long time, there was a lack of understanding of the underlying principles. The first major breakthrough was achieved in 1952 when Turing, aiming to understand the principles of morphogenesis, published his pioneering work about pattern formation and morphogenesis in reaction-diffusion systems [16]. Since that time, modern science has struggled to get a deeper insight into different processes of pattern formation in various fields. Important milestones are given by the works of Hodgkin and Huxley on models of nerve membranes [19] and the reduction of these models by Fitz-Hugh and Nagumo to describe travelling pulses [13, 25]. We also mention the work of Fife on two-component reaction-diffusion equations [14], that on the Brusselator and Oregonator [12, 15] and the Barkley model [20], all three being used for the description of pattern formation in chemical systems. In addition, we mention various other works related to biological problems (see e.g. [24, 26]). Finally, we want to refer the reader to the treatment of auto-solitons by Kerner and Osipov [22]. Parallel to the theoretical works, there were investigations on experimental systems in which the formation of different patterns was observable under controlled conditions. Well-known examples of such systems are those with chemical reactions in gels [27, 29], reactions on surfaces [10], and also systems with charge carrier generation and annihilation, such as those known from semiconductor physics [5, 6, 8, 9] or gas discharge systems [3, 4, 7, 11, 23]. In many reaction-diffusion systems, both localized and spatially-extended structures can be found. In this chapter, we will concentrate on localized solitary structures, which we will refer to as *dissipative solitons* (DSs) [7, 28].

The chapter is organized as follows: Sect. 2 deals with basic mechanisms of pattern formation in reaction-diffusion systems. Here, the principle of local activation and lateral inhibition in the presence of diffusion plays an important role. This principle is first illustrated in Subsect. 2.1. We will then argue in more detail how a homogeneous state can be destabilized by diffusion, leading to spatially-extended patterns (Subsect. 2.2). Then, we will explain under which conditions localized solutions, in the form of DSs, can be stabilized (Subsect. 2.3). A one-dimensional experimental realization of a reaction-diffusion system by an electrical network is described in Subsect. 2.4. To stabilize several moving DSs in more than one spatial dimension, extensions of the system have to be made (Subsect. 2.5). In this way, interesting phenomena, like the formation of molecules of DSs or scattering of DSs, become possible as solutions of a reaction-diffusion equation in more than one spatial dimension. Numerical solutions for the extended equations supporting the former statement are presented in Sect. 3. The goal of Sect. 4 is to investigate selected problems on an analytical level. We will treat the onset of propagation of DSs due to a symmetry-breaking drift bifurcation (Subsect. 4.1) and derive order-parameter equations describing the dynamics in terms of ordinary differential equations near the drift bifurcation point (Subsect. 4.2). As an experimental example of a spatially two-dimensional system that carries DSs and that can be related to reaction-diffusion systems, we will consider the dynamics of current filaments in a planar gas-discharge system (Sect. 5). In Subsect. 5.1, we present the experimental set-up and its qualitative modelling by reaction-diffusion equations. In Subsect. 5.2, we report on experimental results and their evaluation using new statistical data analysis tools. The article closes with a summary and an outlook in Sect. 6.

2 Mechanism of Pattern Formation in Reaction-Diffusion Systems

In the following section, we want to clarify the nature of reaction-diffusion systems and the kinds of mechanisms acting to produce self-organized patterns. In particular, we will concentrate on the principle of local activation and lateral inhibition.

We define reaction-diffusion systems as systems that are described by the following parabolic partial differential equations of the general form

$$\dot{\boldsymbol{U}}(\boldsymbol{x},t) = \underline{\boldsymbol{D}}\Delta\boldsymbol{U}(\boldsymbol{x},t) + \boldsymbol{R}(\boldsymbol{U}(\boldsymbol{x},t)) \; . \tag{1}$$

Here, $\boldsymbol{x} \in \mathbb{R}^m$, $t \in \mathbb{R}$, $\boldsymbol{U} \in C^2(\mathbb{R}^m \times \mathbb{R} \mapsto \mathbb{R}^n)$ and $\boldsymbol{R} \in C(\mathbb{R}^n \mapsto \mathbb{R}^n)$. The name for this type of equation originally comes from chemistry, where $\boldsymbol{U}$ usually describes the concentration of some reagent. On one hand, $\boldsymbol{U}$ may also change in time, due to diffusion, and this local effect is expressed through the first summation on the right hand side. Here, $\underline{\boldsymbol{D}}$ is a diagonal matrix containing a diffusion constant for each component. Each diffusion constant can be interpreted as the square of the diffusion length, l_{U_i}, divided by the corresponding collision time, t_{U_i}. On the other hand, the concentrations may change due to local reactions, and this is expressed by the reaction function $\boldsymbol{R}$. Apart from an interpretation in the context of chemical systems, the equation is also suitable for describing many other systems from various branches of physics and other sciences. In particular, in electrical systems, the Laplacian can come into play via the Poisson equation.

Reaction-diffusion equations belong to the class of dissipative systems. As partial differential equations, they have infinitely many degrees of freedom, and a mathematical proof of the dissipative nature is difficult, as it touches upon very basic physical questions. At this point, we content ourselves with stating that diffusion can be considered as a dissipative process.

In the following, we will assume that the system (1) has at least one stationary, spatially homogeneous solution $\boldsymbol{U}_0$, obeying the relation

$$\underline{\boldsymbol{D}}\Delta\boldsymbol{U}_0(\boldsymbol{x}) + \boldsymbol{R}(\boldsymbol{U}_0(\boldsymbol{x})) = \boldsymbol{0} \; . \tag{2}$$

We now decompose an arbitrary solution of the system as

$$\boldsymbol{U} = \boldsymbol{U}_0 + \tilde{\boldsymbol{U}} \; . \tag{3}$$

In the following considerations, we will use (2) and (3) to expand the right-hand-side of (1) around $\boldsymbol{U}_0$, using Fréchet derivatives. This yields the equation

$$\dot{\tilde{\boldsymbol{U}}}(\boldsymbol{x},t) = \underline{\boldsymbol{D}}\Delta\tilde{\boldsymbol{U}}(\boldsymbol{x},t) + \boldsymbol{L}\tilde{\boldsymbol{U}} + \boldsymbol{N}(\tilde{\boldsymbol{U}}) \tag{4}$$

for the evolution of the deviation $\tilde{\boldsymbol{U}}$, where $\boldsymbol{L} = \boldsymbol{L}(\boldsymbol{U}_0)$ corresponds to the linear and $\boldsymbol{N} = \boldsymbol{N}(\boldsymbol{U}_0)$ corresponds to the nonlinear part of the Taylor expansion of the reaction term around $\boldsymbol{U}_0$.

2.1 Diffusion in the Presence of Local Activation and Inhibition

For reasons of simplicity, we will first restrict ourselves to the case $n = 2$, $m = 1$. In this case, one may rewrite (1) as

$$\begin{pmatrix} \dot{u} \\ \dot{v} \end{pmatrix} = \begin{pmatrix} D_u & 0 \\ 0 & D_v \end{pmatrix} \begin{pmatrix} \Delta u \\ \Delta v \end{pmatrix} + \begin{pmatrix} F(u,v) \\ G(u,v) \end{pmatrix} . \tag{5}$$

Consequently, the linear part of (4) can be expressed as

$$\begin{pmatrix} \dot{\tilde{u}} \\ \dot{\tilde{v}} \end{pmatrix} = \begin{pmatrix} D_u & 0 \\ 0 & D_v \end{pmatrix} \begin{pmatrix} \Delta \tilde{u} \\ \Delta \tilde{v} \end{pmatrix} + \begin{pmatrix} F_u & F_v \\ G_u & G_v \end{pmatrix}\Bigg|_{\boldsymbol{U}_0} \begin{pmatrix} \tilde{u} \\ \tilde{v} \end{pmatrix} . \tag{6}$$

From (6), one can see that the constants in the linearized reaction function can be interpreted as inverse relaxation time constants of the reaction function, which describe, in the vicinity of $\boldsymbol{U}_0$, how fast u and v, respectively, change in their dependence on the actual values of these components.

Physically, the second term on the right-hand-side of (6) describes locally activating and inhibiting processes; in other words, the presence of a component u or v can stimulate or dampen the evolution of u and v according to the positive or negative sign of the entries in the matrix. If we write out the second term on the right-hand-side of (6), the component in the evolution equation will act as an activator when the corresponding coefficients have a positive sign and as an inhibitor when it is negative. In the case of the presence of some activator, the linearized system may go to infinity, provided there is no controlling inhibitor or that the action of the inhibitor is not efficient enough. However, in the long run in real systems, due to nonlinearities, an activator will always be controlled by some inhibitor or the activator itself will be switched to an inhibitor for a large deviation of u or v from $\boldsymbol{U}_0$. As a consequence, the concentrations u and v remain finite.

Let us now include in our discussion the effect of diffusion. Here, one should have in mind that diffusion has a tendency to distribute matter in space. The influence of diffusion is represented by the first term on the right-hand-side of (5). Let us imagine a situation where the homogeneous stationary state $\boldsymbol{U}_0$ of the system (5) is a stable stationary solution for a given set of diffusion parameters D_u and D_v. The stability of the stationary state implies that any small perturbation of this state, its evolution being described by (6) in the vicinity of $\boldsymbol{U}_0$, will decay in the course of time. In particular, this is true for local perturbations. We now want to choose the diffusion constants so that the diffusion of the inhibitor that we assume, for example, to be v, is considerably larger than that of the activator, which we assume to be u. After a local perturbation of the same size, both components, for example in the form of a local positive deviation of u and v from the homogeneous stationary distribution $\boldsymbol{U}_0$, u and v start to diffuse in the vicinity of the site of the perturbation. In the course of time, there will be a decrease of u and v due to diffusion. However, since the diffusion of the activator u is weaker than that of the inhibitor v, there will be a deficiency of the inhibiting component v, and consequently the local control of the activator by the inhibitor may be less efficient. As a result of this effect, there is, locally, the possibility that the activator grows in an uncontrolled manner until it runs into saturation because of inevitable nonlinearities. We conclude that, due to the interplay of diffusion with local effects of activation and inhibition, diffusion may destabilize the homogeneous stationary state $\boldsymbol{U}_0$, thereby supporting spatially inhomogeneous patterns. This is the essential idea of Turing's pioneering investigation of diffusion-driven spontaneous patterns in reaction-diffusion systems [16]. In this work, Turing demonstrated that a reaction-diffusion system in the form of (1) is able to create spatially inhomogeneous patterns due to diffusion if the following conditions are fulfilled:

1. Constant global deviations from the stationary solution (i.e. $(\tilde{u}(x), \tilde{v}(x))^{\mathrm{T}} = (c_1, c_2) \in \mathbb{R}^2$ for all x) decay in the course of time, meaning that effects caused by the reaction function alone do not lead to a destabilization.
2. Some local deviation from the stationary state, influencing its neighborhood by the effect of diffusion, in co-action with the reaction function leads to the creation of a spatially-extended pattern.

To clarify Turing's idea and to find types of reaction-diffusion systems in which diffusion is responsible for the stabilization of spatially inhomogeneous patterns, we will carry out a Fourier-transformation of the deviation $\tilde{\boldsymbol{U}}$ from the homogeneous stationary solution. In this way, we decompose the deviation into modes $\tilde{\boldsymbol{U}}_k \mathrm{e}^{ikx}$, $k \in \mathbb{R}$ which are eigenfunctions of the Laplace operator, thereby obtaining an infinite number of ordinary differential equations instead of one partial differential equation. Starting from the example of system (5), we insert the Fourier *ansatz* into the linearized (6) and project the resulting equation onto the linearly-independent exponential functions, obtaining

$$\dot{\tilde{\boldsymbol{U}}}_k = \boldsymbol{L}\tilde{\boldsymbol{U}}_k - k^2 \underline{\boldsymbol{D}}\tilde{\boldsymbol{U}}_k \,. \tag{7}$$

The homogeneous stationary solution $\boldsymbol{U}_0$ is linearly stable against disturbances with wave number k if, and only if, the conditions

$$\mathrm{Tr}(\boldsymbol{L} - k^2 \underline{\boldsymbol{D}}) = \mathrm{Tr}\boldsymbol{L} - k^2 \mathrm{Tr}\underline{\boldsymbol{D}} < \mathrm{Tr}\boldsymbol{L} < 0 \tag{8}$$

and

$$\mathrm{Det}(\boldsymbol{L} - k^2 \underline{\boldsymbol{D}}) = \mathrm{Det}\boldsymbol{L} - (D_u G_v + D_v F_u)k^2 + D_u D_v k^4 > 0 \tag{9}$$

are fulfilled (compare e.g. [14, 39]). In particular, spatially homogeneous deviations decay in the course of time only if the mode $\tilde{\boldsymbol{U}}_0$ is stable. If we first take a look at condition (8), we see that if a mode with $k > 0$ is unstable (i.e. $\mathrm{Tr}\boldsymbol{L} - k^2\mathrm{Tr}\underline{\boldsymbol{D}} > 0$), the mode with $k = 0$ is always unstable as well ($\mathrm{Tr}\boldsymbol{L} > 0$). From this, we conclude that a destabilization by a violation of condition (8) contradicts Turing's first claim. The same consideration can be made for condition (9), and we can draw the same conclusion for the trace criterion (8) if the factor in front of the quadratic term is smaller than zero. To avoid this, we have to demand

$$D_u G_v + D_v F_u > 0 \; . \tag{10}$$

Only if condition (10) is fulfilled, can a destabilization of the stable stationary solution by a disturbance with a finite wave number occur due to diffusion (compare Turing's second claim).

We now ask the question: what kind of matrix $\boldsymbol{L}$ in the second term of (6) can fulfill the conditions (8), (9) and (10)? It turns out that there are six different equivalence classes of matrices that can be characterized by the signs of the coefficients of the matrix, tabulated in Table 1.

Table 1. List of different equivalent classes of matrices of the linearized reaction term of (6), differentiated with respect to the signs of the entries. Only in case IV is the spatially uncoupled system stable and at the same time a destabilization of the homogeneous stable stationary state of the full system is possible due to diffusion

I	$\begin{pmatrix} + & + \\ + & + \end{pmatrix}$	$\begin{pmatrix} + & - \\ - & + \end{pmatrix}$			cond. (8) violated
II	$\begin{pmatrix} + & + \\ + & - \end{pmatrix}$	$\begin{pmatrix} + & - \\ - & - \end{pmatrix}$	$\begin{pmatrix} - & + \\ + & + \end{pmatrix}$	$\begin{pmatrix} - & - \\ - & + \end{pmatrix}$	cond. (9) violated
III	$\begin{pmatrix} + & - \\ + & + \end{pmatrix}$	$\begin{pmatrix} + & + \\ - & + \end{pmatrix}$			cond. (8) violated
IV	$\begin{pmatrix} + & - \\ + & - \end{pmatrix}$	$\begin{pmatrix} + & + \\ - & - \end{pmatrix}$	$\begin{pmatrix} - & + \\ - & + \end{pmatrix}$	$\begin{pmatrix} - & - \\ + & + \end{pmatrix}$	suitable
V	$\begin{pmatrix} - & + \\ + & - \end{pmatrix}$	$\begin{pmatrix} - & - \\ - & - \end{pmatrix}$			cond. (10) violated
VI	$\begin{pmatrix} - & - \\ + & - \end{pmatrix}$	$\begin{pmatrix} - & + \\ - & - \end{pmatrix}$			cond. (10) violated

From all these classes, only class IV fulfills conditions (8), (9) and (10). The first representative of the equivalence class IV provides the name of the class, which we refer to as "activator-inhibitor systems" (sometimes also called "winner-loser systems"). Neglecting nonlinear contributions, an increase in the activator concentration u results in the growth of both components, whereas growth of the inhibitor component v causes the opposite process. In particular, u is auto-catalytic, while v is auto-inhibitoric.

2.2 Turing Patterns

We want to analyze how patterns can develop in activator-inhibitor systems. To this end, we consider a concrete example, choosing the reaction function

$$\boldsymbol{R}(u,v) = \begin{pmatrix} f(u) - \kappa_3 v \\ \frac{1}{\tau}(u - v) \end{pmatrix} \tag{11}$$

with $f(u) = \lambda u - u^3 + \kappa_1$, $\tau, \lambda, \kappa_3 \in \mathbb{R}^+, \kappa_1 \in \mathbb{R}$. In the vicinity of $(u,v)^{\mathrm{T}} = \mathbf{0}$, we may refer to u and v as activator and inhibitor, respectively. (See also shaded areas in Fig. 2 discussed below). Although a nonlinearity is now only present in the activator equation and only three of the four relaxation time constants of the reaction function can be varied (by λ, κ_3 and τ), the example covers many important properties of more general cases of reaction-diffusion systems. To simplify the notation for the following considerations, we introduce $d_u = D_u$ and $d_v = \tau D_v$, yielding

$$\begin{aligned} \dot{u} &= d_u \Delta u + \lambda u - u^3 - \kappa_3 v + \kappa_1 \\ \tau \dot{v} &= d_v \Delta v \quad + u \quad - v \,. \end{aligned} \tag{12}$$

The nullclines of the system are given by $\boldsymbol{R}(u,v) = \mathbf{0}$. Depending on λ, κ_1 and κ_3, one, two or three stationary homogeneous solutions exist. We will now focus on the last case (see Fig. 2).

For the cubic curve depicted as a solid line, the three intersection points of the two nullclines correspond to the stationary homogeneous solution, which we will label $\boldsymbol{U}_-$, $\boldsymbol{U}_0$ and $\boldsymbol{U}_+$. A detailed stability analysis of the cases

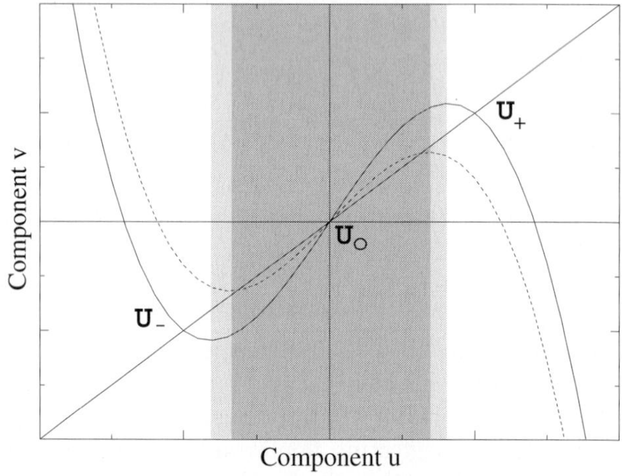

Fig. 2. Nullclines for the system (12) in the case of three intersection points $\mathbf{U}_-$, $\mathbf{U}_0$ and $\mathbf{U}_+$ for $\kappa_1 = 0$, $\kappa_3 = 1$ and different values of λ. The regimes of self-activation, corresponding to a positive gradient of the cubic curve, are marked with *gray shaded* areas

shown in Fig. 2 reveals that homogeneous stationary solutions resulting from the cubic polynomial, drawn as a solid line, do not correspond to the class IV of Table 1. The same is true for U_0 resulting from the dashed line. In contrast to this, $\boldsymbol{U}_+$ and $\boldsymbol{U}_-$, resulting from the latter, can be destabilized by diffusion in Turing's sense given above.

In what follows, we want to analyze in detail the stability of homogeneous stationary states resulting from the nullclines of Fig. 2 for the situation defined by the class IV of Table 1. In particular, we want to find the neutral stability curves of the system that are defined as the curves that mark the violation of the stability conditions (8) and (9), depending on the wave number k of the disturbance. For our specific example and the destabilization of the state $\boldsymbol{U}_-$, the curves take the form

$$f'(u_-) = \lambda - 3u_-^2 = \frac{1}{\tau} + \left(d_u + \frac{1}{\tau}d_v\right)k^2 =: U_{\mathrm{n}}^H(k) \tag{13}$$

and

$$f'(u_-) = \lambda - 3u_-^2 = \frac{\kappa_3}{1 + d_v k^2} + d_u k^2 =: U_{\mathrm{n}}^T(k) \,, \tag{14}$$

which are depicted in the right half of Fig. 3. Let us assume that the parameters of the system are chosen in such a way that, for all possible values of k, the conditions (13) and (14) hold. One may now change κ_1, thereby shifting the value of u_- (compare with Fig. 2). This shift of u_- causes a change of $f'(u_-)$ (see Fig. 3), which may lead to a violation of the stability conditions for one or several modes. If the stability condition (13) is violated first, the mode to become unstable must be $\tilde{\boldsymbol{U}}_0$, due to the parabolic shape of the neutral curve $U_{\mathrm{n}}^H(k)$.

The result of the destabilization is a *Hopf bifurcation*. After the initial growth of the critical mode, further temporal evolution is determined by the nonlinear part of the reaction function, and concrete statements demand an elaborate nonlinear stability analysis. Generally, two situations are encountered – the system can exhibit spatially homogeneous periodic oscillations (corresponding to a supercritical bifurcation) or the system can be driven far away from the homogeneous solution $\boldsymbol{U}_-$ (corresponding to a subcritical bifurcation). Apart from the Hopf bifurcation, a *Turing bifurcation* can be encountered if condition (13) is violated first. Here, a mode $\tilde{\boldsymbol{U}}_k$ with a finite wave number k will grow. If, once again, the nonlinearities are taken into account, one can encounter stationary spatial oscillations of the system around $\boldsymbol{U}_-$ in the supercritical case, or again the system can be driven far away from the stationary solution in the subcritical case. For the first possibility, the resulting pattern is called a *Turing pattern*. As one can see from the neutral curve $U_{\mathrm{n}}^T(k)$ for the Turing destabilization, wave numbers of medium size are most susceptible to destabilization (compare also condition (9) and the resulting demand (10)). The physical reason for this phenomenon is that perturbations of large wave numbers are smoothed by diffusion, which is especially strong when large gradients are present, while perturbations with small

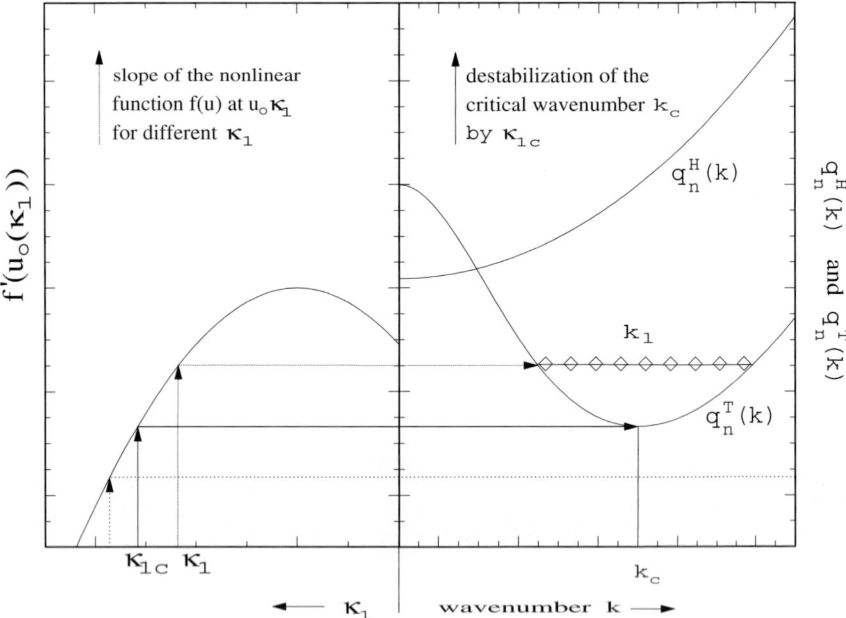

Fig. 3. Neutral stability curves and destabilization against critical wave numbers for the reaction-diffusion system (11) and $d_v > d_u$

wave numbers are not affected by diffusion, which is necessary for destabilization in systems of activator-inhibitor type (see above). Usually, the size of physical systems is limited, so that one can assume a domain Ω of finite size $\|\Omega\|$ with no-flux boundary conditions. In this case, possible modes in the system must fulfill the condition $k_l = \frac{l\pi}{\|\Omega\|}$, $l \in \mathbb{N}$. This means that, apart from varying the parameters in (13) and (14), a destabilization can also be caused by increasing the system length, which is closer to Turing's original idea of morphogenesis.

2.3 Localized Solutions

After having obtained some insight into the formation of spatially-extended patterns, one may pose the question as to whether localized solutions for systems of the type of (1) are also possible. To this end, we will again consider our specific system (12). First of all, it seems reasonable to assume that well-localized patterns in an unbounded space can only be stable when there is a stable stationary homogeneous solution that serves as a kind of "background" state for the localized structure. Therefore localized structures cannot be stable for parameter values for which a Turing destabilization is possible. Nevertheless, the mechanism of local activation and lateral inhibition is suitable for creating stable localized solutions if the parameters of the system are chosen appropriately, as we will explain in the following.

For the stabilization of a localized structure, several possibilities exist, and, of these, we will discuss the most intuitive one. To this end, for the general form (6) of the system of the linearization of (12), we assume $D_u \ll D_v$ and $|F_u|, |F_v| \ll |G_u|, |G_v|$. This means: (a) that the component u diffuses more slowly than the component v, and (b) that the relaxation times in the evolution equation of u are much larger than those in the evolution equation of v, meaning that v can follow u almost immediately.

We now want to consider a local perturbation of the homogeneous stable stationary solution of (12) (Fig. 4a) which results from the nullcline diagram Fig. 2 and which is the intersection point $\boldsymbol{U}_-$ of the straight line with the solid cubic curve. Thereby, we assume that the amplitude of the locally- perturbed values of u and v approximately reach the values of the unstable homogeneous stationary solution $\boldsymbol{U}_0 = (u_0, v_0)$ (corresponding to the intersection point in Fig. 2). We recall that, locally, near (u_0, v_0), the system may be of activator-inhibitor type (see Sect. 2.1 and 2.2). Now, in the course of time, the imposed local perturbation will broaden and qualitatively take the shape of a Gaussian distribution due to diffusion. Also, as $D_v \gg D_u$, then after some time, the width of the inhibitor distribution will become larger than the width of the activator distribution (see Fig. 4b). In addition, the amplitude of the inhibitor distribution decreases faster than the amplitude of the activator distribution, leading to a reduced control of the activator by the inhibitor in the center of the structure. Keeping in mind that, locally in the vicinity of (u_0, v_0), we are in the activator-inhibitor regime and that there is a deficiency of inhibitor in the center of the perturbation, the net result of this process is an auto-catalytic increase of u in the vicinity of the center. In the long run, the system may lose the capability of self-activation close to the center of the perturbation and will remain close to u_+. Further away from the center, the dominant inhibitor brings the system into a state where no self-activation is possible, although the state $\boldsymbol{U}_-$ is not reached completely, due to diffusion.

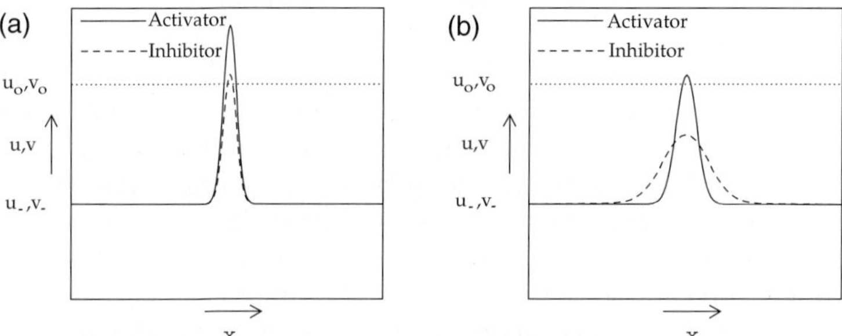

Fig. 4. Schematic presentation of a local perturbation of a homogeneous stable stationary solution of (12) (**a**) and its evolution (**b**)

In conclusion, the mechanism presented is suitable for generating well-defined stationary localized structures. Due to some special properties of the solutions that are discussed below in detail, we want to refer to these self-organized structures as *dissipative solitons* (DSs) [7, 28]. In the Russian literature, these objects are also called *auto-solitons* [22]. As indicated above, other ways exist to stabilize DSs, e.g. by choosing parameters close to the Turing bifurcation point in such a way that only one stable state, $\boldsymbol{U}_-$, exists. However, the basic mechanism is similar in all cases.

After finding localized solutions, the problem of their stability shall be discussed in more detail. To this end, one could linearize the system around the DS solution, but an exact treatment would lead to an extended amount of calculation, as one encounters linear systems with non-constant coefficients. For our concrete example, the calculations can be simplified if the cubic function $f(u)$ is approximated by a piecewise linear function [40, 41]. We will not deal with this topic in more detail, but state that for the system (12), solutions in the form of DSs have been found numerically, and, in some special cases, even analytically so that the stability could be confirmed.

If one thinks about localized solution in different systems, one might remember that some cases exist in which localized solutions may propagate, e.g. as electric pulses on nerve tracts. This rises the question if also travelling DSs in the reaction-diffusion system under consideration are possible. The question can be answered with yes if the parameters are changed appropriately. As above, we will develop a mechanism to illustrate the basic principles for propagating DSs.

Obviously, a solution which is symmetric with respect to its center will not propagate. Consequently, the symmetry of the solution has to be broken to allow for propagation. To this end, we add an anti-symmetric perturbation $g(x)$ to the activator component, which shall be proportional to the spatial derivative of the stationary activator distribution $u_{\rm s}(x)$, i.e. $g(x) = u_{{\rm s},x}(x)$ (Fig. 5a). In this way, the left hand slope of the activator distribution is raised, while the right hand slope is lowered. Therefore, the disturbance approximately corresponds to a shift of the activator distribution to the left with respect to the inhibitor distribution. If we claim $|F_u|, |F_v| \ll |G_u|, |G_v|$ as above, the inhibitor can adapt fast to the disturbance and we end up with a stable stationary DS that has undergone a shift to the left. We now choose $|F_u|, |F_v| \gg |G_u|, |G_v|$. The inhibitor is now slow and cannot follow the activator distribution immediately. We remember that in Fig. 5a, in the vicinity of $u = u_0$ on the slopes of the DS, we operate in the range where the system locally acts as activator-inhibitor system (class IV in Table 1). After adding the perturbation, the excess of activator u on the left hand slope of the DS leads to an auto-catalytic increase of u while due to the deficiency of u on the right hand side, u is decreased. At the same time, diffusion causes a shift of the left hand slope in the x-direction in Fig. 5a. This diffusion combined with further activation of u on the left side of the original DS can only be

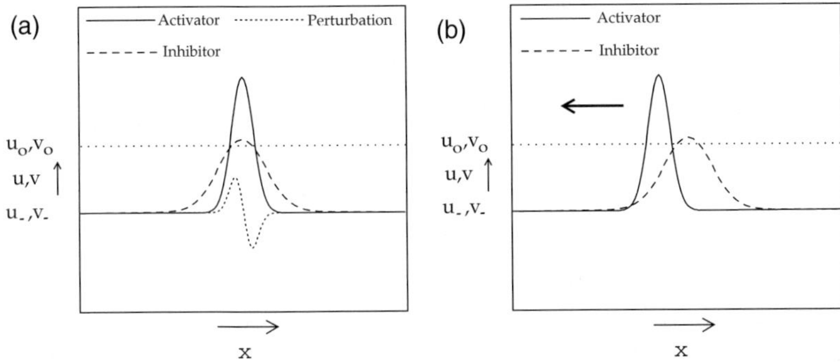

Fig. 5. Mechanism for the propagation of DSs: symmetric distributions of u and v and an anti-symmetric perturbation (**a**), result of the superposition of the symmetric distributions of u and v with the anti-symmetric perturbation (**b**)

controlled by v with a time lag. However, the control of the activator by v on the right side is no problem because there is enough surplus of inhibitor v. As a result, the perturbation leads to a continuous shift of the slopes of the original DS to the left while the inhibitor catches up with u with a time lag due to its slow reaction rates. This dynamical interplay of u and v is a mechanism leading to the propagation of DSs with constant speed.

Further investigations demonstrate that it is even possible to find solutions that may be interpreted as bound stationary and travelling states of DSs. This fact should become clear when considering that the mechanisms stabilizing the DSs works locally so that distant DSs practically do not affect each other.

Before coming to the interesting question of what may happen when two counter-propagating DSs collide, we investigate localized solutions of the reaction-diffusion system in more than one spatial dimension. It turns out that also in two or more dimensions, stable localized stationary solutions with rotational symmetry exist, their mechanism of stabilization being the same as for their one-dimensional counterparts. However, stable propagating structures cannot be observed. The reason for this initially astonishing fact is that when the symmetry of the solution is broken as described above, a stabilization has to occur in two or more spatial dimensions. A stabilization in the direction of propagation is still possible using the mechanism described above, but no stabilizing inhibitor is present to take care of perturbations perpendicular to the direction of propagation.

To overcome the described stability problem of travelling DSs in higher-dimensional space an extension of the considered two-component reaction diffusion equation of the type (12) is necessary to take care of a fast control of the spread of activator perpendicular to the direction of propagation. This can be done in a most simple manner by incorporating into the reaction function $F(u, v)$ in (12) an integral term such that the new reaction function reads as:

$$F(u,v) = f(u) - v + \kappa_1 - \frac{\kappa_2}{\|\Omega\|} \int_{\Omega} u \, \mathrm{d}x \tag{15}$$

We note that the integral term acts as a negative feedback term thus acting in an inhibitoric manner. One may also say that by the help of the integral term the systems exhibits self-control by adjusting an effective parameter of the system may be considered to yield an effective value $\kappa_{1,\mathrm{eff}} = \kappa_1 - \frac{\kappa_2}{\|\Omega\|} \int_{\Omega} u \, \mathrm{d}x$.

2.4 Experimental Realization of Dissipative Solitons: Electrical Networks

One might wonder if a reaction-diffusion system of the discussed type can be realized experimentally, especially with an approximately cubic nonlinearity of the form (11). It turns out that this is actually the case if one uses a spatially discrete system in the form of an electrical network [42]. The basic idea is that the spatially extended system is divided into N cells which are coupled to each other by identical linear resistors. The corresponding electrical circuit is shown in Fig. 6. The only nonlinearity is a resistor S with S-shaped current-voltage characteristic $S(I)$. Every cell is connected to an external voltage source U_0, protected by a series resistor R_0. One can derive differential equations for the time evolution of the voltage and the current for the i-th cell inside the network by using Kirchhoffs laws:

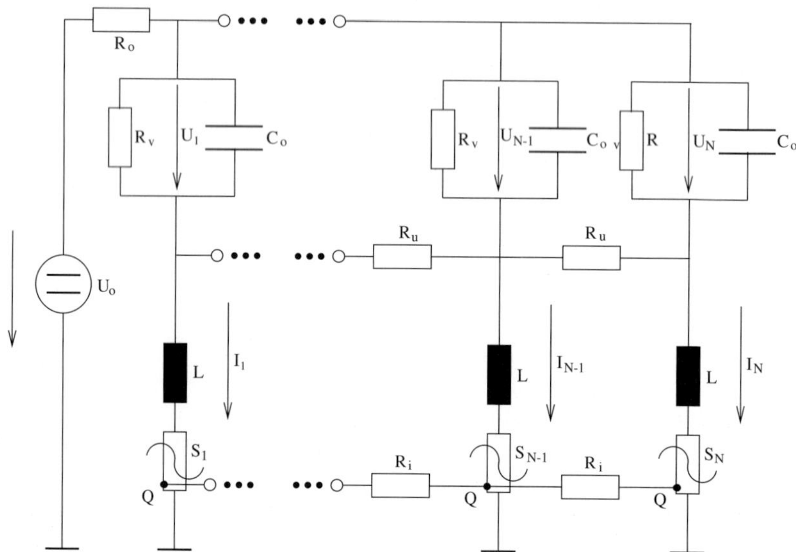

Fig. 6. Electrical equivalent circuit for the discrete realization of the two-component reaction-diffusion system (12) in one spatial dimension

$$C_0\dot{U}_i - \frac{1}{R_u}(U_{i+1} - 2U_i + U_{i-1}) + \frac{U_i}{R_v} = I_i \tag{16}$$

$$L\dot{I}_i + U_i + S(I_i) - \frac{\gamma}{R_i}(I_{i+1} - 2I_i + I_{i-1}) = U_0 - R_0\sum_i I_i \tag{17}$$

For the cells at the boundary, one finds

$$C_0\dot{U}_{1,N} - \frac{1}{R_u}(U_{2,N-1} - U_{1,N}) + \frac{U_{1,N}}{R_v} = I_{1,N} \tag{18}$$

$$L\dot{I}_{1,N} + U_{1,N} + S(I_{1,N}) - \frac{\gamma}{R_i}(I_{2,N-1} - I_{1,N}) = U_0 - R_0\sum_i I_i \tag{19}$$

Here we have to read the first index of U and I for the cell at the left hand side of the network and the second index of U and I for the cell at the right hand side of the network. We note that the nonlinear resistor has been realized by an electronic circuit containing linear electronic components and two transistors effectively forming a four layer structure that has the typical behavior of a thyristor. The resulting current-voltage characteristic $S(I)$ of the nonlinear resistor is depicted in Fig. 7a. For small and large values of the current, it exhibits a positive differential resistivity, whereas for intermediate values, it is negative. The resistor has the peculiarity that the current voltage characteristic can be shifted in the voltage proportional to the electrical current that is applied to the point Q in Fig. 6. In this way the realization of the fourth term on the left hand side of (17) is possible and γ can be considered as coupling strength. For a detailed description of this technical matter we refer the reader to [42].

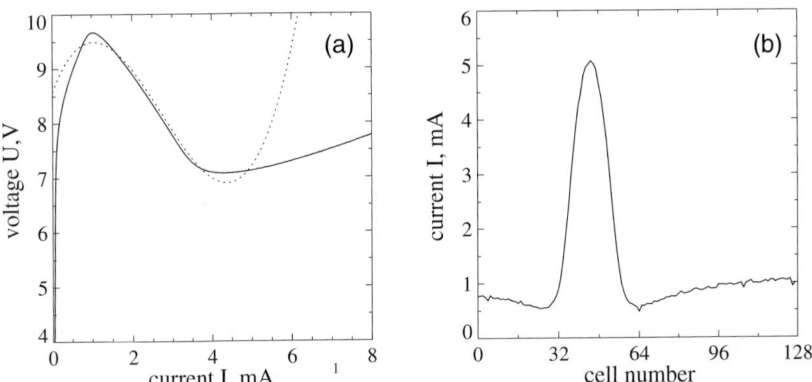

Fig. 7. (**a**) Experimentally realized current-voltage characteristic $S(I)$ and corresponding cubic fit with $I^* = 2.708\,\text{mA}$, $U^* = 8.199\,\text{V}$, $\chi = 1.168\,\frac{\text{V}}{\text{mA}}$, $\varphi = 0.1406\,\frac{\text{V}}{\text{mA}^3}$. (**b**) Stationary DS observed on the electrical network with 128 cells. Parameters: $U_0 = 15.06\,\text{V}$, $R_0 = 20\,\Omega$, $R_v = 2.4\,\text{k}\Omega$, $R_u = 3\,\Omega$, $R_i = 1\,\text{k}\Omega$, $C_0 = 0\,\text{F}$, $L = 33\,\text{mH}$

Before we come to DSs that can be observed on the electrical network, we want to establish a connection of the ordinary differential equations (16)–(19) to the the reaction-diffusion equations (12). In its relevant part, in particular the branch with negative differential resistivity, the current-voltage characteristic can be approximated by a cubic function:

$$S(I) = U^* - \chi(I - I^*) + \varphi(I - I^*)^3 \,. \tag{20}$$

To obtain an optimal fit to the data, we choose

$$\chi = -\min\left(S'(I)\right) \,, \qquad I^* = \frac{1}{2}(I_{\min} + I_{\max}) \,,$$
$$\varphi = \frac{\chi}{3(I_{\max} - I^*)^2} \,, \qquad U^* = S(I^*) \,,$$

where $I_{\max}$ and $I_{\min}$ denote the local maximum and minimum of the current-voltage characteristic. It is now reasonable to renormalize (16) and (17) and the boundary equations (18) and (19) by introducing

$$\begin{aligned}
u_i &= \sqrt{\frac{\varphi}{R_v}}(I_i - I^*) & v_i &= \sqrt{\frac{\varphi}{R_v^3}}(U_i - R_v I^*) \\
\tau &= \frac{R_v^2 C_0}{L} & t' &= \frac{R_v}{L} t \\
\kappa_2 &= \frac{N R_0}{R_v} & \kappa_1 &= \sqrt{\frac{\varphi}{R_v^3}}(U_0 - U^* - (R_v + N R_0) I^*) \\
\lambda &= \frac{\chi}{R_v} & f(u_i) &= \sqrt{\frac{\varphi}{R_v^3}}\left(U^* - S\left(\sqrt{\frac{\varphi}{R_v}} u_i + I^*\right)\right) \\
d_u' &= \frac{\gamma}{R_u R_i} & d_v' &= \frac{R_v}{R_u} .
\end{aligned}$$

In this way, the equations (16) and (17) can be transformed into

$$\begin{aligned}
\dot{u}_i &= d_u'(u_{i+1} - 2u_i + u_{i-1}) + f(u_i) - v_i + \kappa_1 - \frac{\kappa_2}{\|\Omega\|}\sum_i u_i \\
\tau \dot{v}_i &= d_v'(v_{i+1} - 2v_i + v_{i-1}) + u_i - v_i \,.
\end{aligned} \tag{21}$$

with $\|\Omega\| = N$. This equation is already quite similar to the system (12) with a global feedback term.

In order to relate the electrical network in Fig. 6 to some continuous physical system described by a reaction-diffusion equation of the kind (12) we imagine the continuous system depicted in Fig. 8 consisting of a linear layer L in parallel with a nonlinear layer N. The extension d of the system in y-direction is assumed to be small with respect to the typical length on which the voltage or the current may change. In x-direction we divide the system

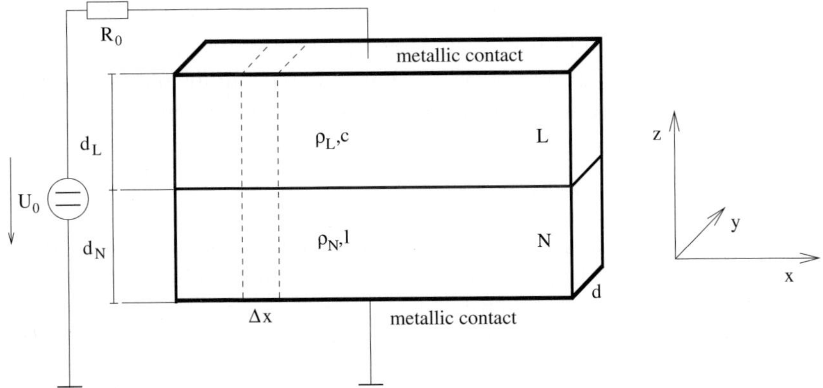

Fig. 8. Continuous system for which the circuit depicted in Fig. 6 can be considered as an equivalent circuit. L: Material with linear current-density voltage characteristic and specific capacity c. N: Nonlinear material with S-shaped current-density-voltage characteristic and specific inductivity l

into cells of width Δx. Each individual cell of Fig. 8 shall correspond to a single cell of Fig. 6. For the layer L, we assume constant specific resistivity ρ_L and a constant specific capacity c_L. In addition, current transport in x- and z-direction shall be possible. For a single cell of the nonlinear layer we assume the nonlinear voltage current characteristic depicted in Fig. 7a and a constant specific inductivity l. However we neglect lateral coupling due to lateral voltage drop and corresponding drift current. This corresponds to choosing the coupling constant of the electrical network to be $\gamma = 0$. Using the relations

$$R_v = \frac{\rho_L d_L}{d\Delta x} \qquad R_u = \frac{\rho_L \Delta x}{d_l d} \qquad d_v = d_v'(\Delta x)^2$$

$$C = c\Delta x \qquad L = \frac{l}{\Delta x} \qquad \|\Omega\| = N\Delta x$$

and making the transitions

$$u_i \to u(x) \qquad \frac{\kappa_2}{N}\sum_i u_i \to \frac{\kappa_2}{\Omega}\int\limits_\Omega u(x,t)\,\mathrm{d}x$$

$$v_i \to v(x) \qquad \frac{v_{i+1} - 2v_i + v_{i-1}}{(\Delta x)^2} \to \Delta v(x)$$

we finally come to the system

$$\dot{u} = d_u \frac{\partial^2}{\partial x^2} u + \lambda u - u^3 - \kappa_3 v + \kappa_1 - \frac{\kappa_2}{\|\Omega\|}\int\limits_\Omega u\,\mathrm{d}x$$

$$\tau\dot{v} = d_v \frac{\partial^2}{\partial x^2} v + u - v \tag{22}$$

with $u = u(x, t')$ and $v = v(x, t')$, where in the final end we have added in the first equation the term $d_u \Delta u$. This current diffusion as the result of the assumption that in the layer N, charge carrier diffusion is of relative importance for current transport in x- but not in z-direction.

2.5 The Three-Component System

Equation (12) together with a global feedback term has been investigated in many circumstances [43, 44, 45, 46]. One of the most striking features of the equation is the existence of well-localized solitary stationary and travelling solutions, which we refer to as DSs. As explained in Sect. 2.3, their mechanism of stabilization is largely due to the principle of local activation and lateral inhibition that is based on the interplay of two components, of which one is referred to as activator (at least in a certain region of the parameter space) and the other one is referred to as inhibitor (everywhere in the parameter space). Parallel to the works on (12), a lot of research has been conducted on similar two-component equations like the Brusselator and the Oregonator model and other equations [12, 15, 47], and in special cases, even systems with more than two components have been treated [48].

As it was described in Sect. 2.3, single standing and moving DSs are stable solutions of reaction-diffusion equations of the activator-inhibitor type with a cubic nonlinearity and a global feedback term like in (15) in more than one dimension. Naturally, people were interested if also multiple DSs could be stabilized, opening up the way to complex and fascinating processes like collision and scattering. It has turned out that using a global feedback term, the stabilization of stationary, equally shaped structures in two or more dimensions is unproblematic. Unfortunately, this does not hold if at least one DS is supposed to propagate. The problem in this case is that the global feedback term puts DSs in competition, and if they are not equally shaped, usually one DS can "blow up" in favor of another which is "eaten up", so that finally only one object will survive [49, 51]. Without stabilizing several propagating structures, interesting interaction processes like scattering cannot be observed, and so one has to develop a new mechanism of stabilization which works independently of the spatial dimension.

To overcome the mentioned problems, a further extension of the two-component reaction-diffusion system seems necessary. The basic idea for the extension is the introduction of a mechanism stabilizing each DS individually and locally instead of using a global mechanism. This can be achieved by generating additional inhibition by a third component that is generated in turn by the activator. Therefore we add a second inhibiting component w with a small time constant θ and a large diffusion-related coefficient d_w to the reaction-diffusion system (12) which quickly follows the activator distribution and surrounds it entirely (in contrast to the slow inhibitor which is shifted with respect to the activator during propagation processes, compare Fig. 5)

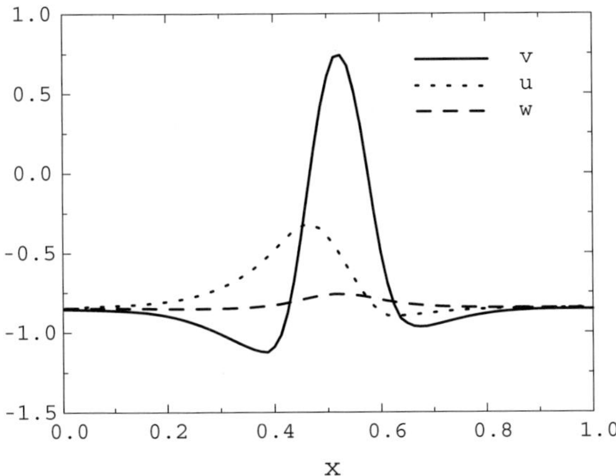

Fig. 9. Local stabilization of a two-dimensional propagating DS using a fast second inhibiting component w (numerical simulation of (23) in two dimensions). The plot shows an intersection through the center of the DS, whose direction of motion is to the right. Parameters: $\tau = 48$, $\theta = 1$, $d_u = 10^{-3}$, $d_v = 1.25 \cdot 10^{-3}$, $d_w = 0.064$, $\lambda = 2$, $\kappa_1 = -6.92$, $\kappa_2 = 0$, $\kappa_3 = 8.5$, $\kappa_4 = 1$ and no-flux boundary conditions

[49]. In addition, extending the system to arbitrary spatial dimensions the resulting three-component reaction-diffusion system takes the form:

$$\begin{aligned} \dot{u} &= d_u \Delta u + f(u) - \kappa_3 v - \kappa_4 w + \kappa_1 - \frac{\kappa_2}{||\Omega||} \int_\Omega u \, \mathrm{d}\Omega \, , \\ \tau \dot{v} &= d_v \Delta v + u - v \, , \\ \theta \dot{w} &= d_w \Delta w + u - w \, . \end{aligned} \tag{23}$$

A typical stable DS propagating in the two-dimensional space is depicted in Fig. 9.

3 Numerical Investigations of the Three-Component System

In two or more dimensions, it is generally not easy to make analytical statements, and one often has to consider numerical calculations (compare [52]). Before treating selected problems analytically, we will have a look on different simulations showing typical phenomena, starting with single DSs before passing to more complex phenomena that involve several DSs like scattering, the formation of soliton molecules, generation and annihilation [49, 53, 54].

When searching for stationary solutions corresponding to single DSs, it turns out that two qualitatively different prototypes of solutions can be found (Fig. 10), depending on the chosen parameters. In the first case, the solitons

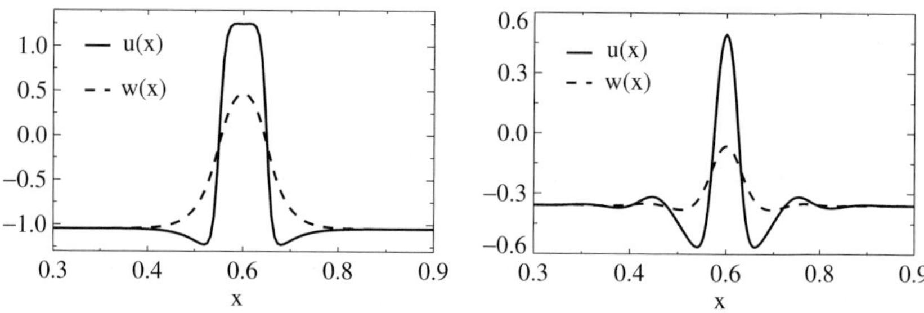

Fig. 10. (**a**) Non-oscillatory and (**b**) oscillatory decay of one-dimensional stationary DSs towards the ground state (numerical result). Parameters: a. $d_u = 0.8 \cdot 10^{-4}$, $d_v = 0$, $d_w = 10^{-3}$, $\lambda = 3$, $\kappa_1 = -0.1$, $\kappa_2 = 0$, $\kappa_3 = 1$ and $\kappa_4 = 1$, b. $d_u = 0.5 \cdot 10^{-4}$, $d_v = 0$, $d_w = 9.64 \cdot 10^{-3}$, $\lambda = 1.71$, $\kappa_1 = -0.15$, $\kappa_2 = 0$, $\kappa_3 = 1$, $\kappa_4 = 1$ and no-flux boundary conditions

decay in space non-oscillatorily towards the ground state. In contrast to this, the decay in the second case is oscillatory. As a rule of thumb one can note that close to the point of the Turing destabilization, DSs exhibiting oscillatory tails are observed, while for greater distances a non-oscillatory decay is observed.

If we now turn to interaction phenomena between DSs, one finds that in most cases, the solitons do not interact by merging their cores, but interact by their tails, so that the individual structure if the involved DSs remains preserved to a large extend. The shape of the tails therefore plays an important role for the mechanism of interaction. As rather many different processes can be observed, we give a brief classification in Table 2 before treating selected cases in more detail.

As the table indicates, the interaction between DSs possessing non-oscillatorily decaying tails is purely repulsive and can only be overcome when the velocity of the solitons is high, resulting in processes where the number of solitons changes. For more details on solitons featuring a purely repulsive

Table 2. Classification of different interaction phenomena observed in numerical simulations of the three-component system (23)

Dynamical Behavior	Decay of Tails:	
	Non-Oscillatory	Oscillatory
stationary DSs	maximal possible distance	locking on neighboring tails
slowly moving DSs	scattering	scattering and formation of molecules
fast moving DSs	annihilation by merging and generation by transient states	annihilation by extinction and generation by “replication”

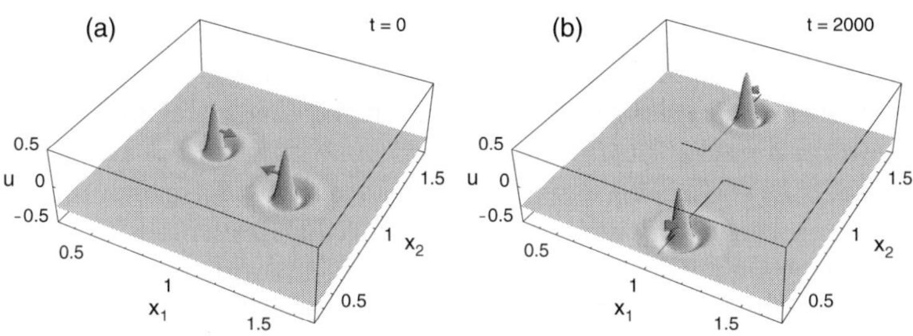

Fig. 11. Numerical solution for (23) in $\mathbb{R}^2$ showing the activator distribution for a typical scattering process of two DSs and corresponding trajectories of its centers. Parameters: $\tau = 3.345$, $\theta = 0$, $d_u = 1.1 \cdot 10^{-4}$, $d_v = 0$, $d_w = 9.64 \cdot 10^{-4}$, $\lambda = 1.01$, $\kappa_1 = -0.1$, $\kappa_2 = 0$, $\kappa_3 = 0.3$, $\kappa_4 = 1$ and no-flux boundary conditions

interaction, see [51] and [35] for a theoretical treatment. In the present chapter, we will focus on the more interesting case of DSs with oscillatory tails, but we will come back to the problem in a more general approach (Sect. 4.2).

In the following, we will discuss the three cases in the right column of Table 2. If the distance between the DSs is very large compared to their own size, their interaction is negligible. This changes when the DSs come close to each other. Stationary DSs have the tendency to arrange themselves on local maxima of the oscillatory tails of neighboring solitons in a stable way [42, 52], which under certain conditions may lead to the formation of stable crystalline structures [36]. In contrast to that, local minima of the tails are unstable positions of DSs with respect to each other [37].

Slowly moving DSs can either scatter (Fig. 11) or form travelling or rotating bound states (Fig. 12). The "lock-in" mechanism is the same as for stationary DSs, indicating that depending on the distance of the solitons, regions of attraction and repulsion must exist.

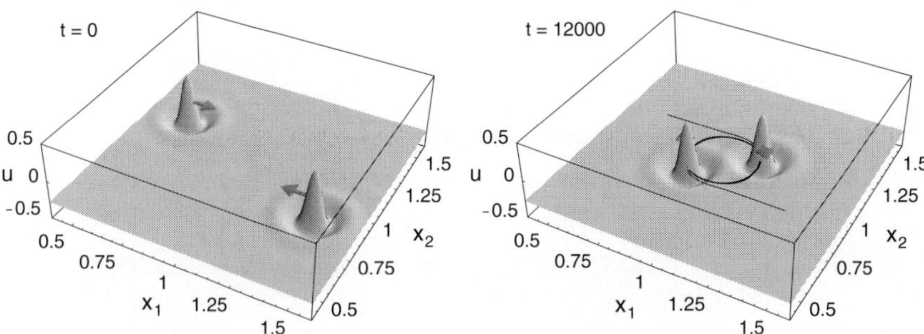

Fig. 12. Numerical solution for (23) in $\mathbb{R}^2$ where two DSs approach and "lock in" on each others tails, resulting in the formation of a stable molecule. Parameters like in Fig. 3, except for $\tau = 3.35$

A close look reveals that the shape of stationary and moving DSs is rather similar and changes in a rather insignificant way during the interaction processes. We should keep this in mind for the analytical investigations in the next section.

We now turn to the last case, i.e. fast moving DSs, for which the observed interaction processes differ from the cases already presented in a significant way. We first consider a head-on collision between two DSs, during which the structures rapidly come to rest. This goes along with the observation that the offset between the activator and the slow inhibitor distribution, being the reason of the finite speed of the DSs, vanishes. As the inhibitor amplitude of a rapidly propagating soliton is much larger than in the stationary equilibrium case, the activator distribution is strongly diminished, eventually causing the DSs to annihilate (Fig. 13).

Astonishingly enough, collisions may also lead to generation processes in a so-called "replication scenario" (Fig. 14). Here, the oscillating tails superimpose, so that locally a critical threshold is reached above which a new DS can be ignited. A high velocity is needed in this case as for the ignition, the

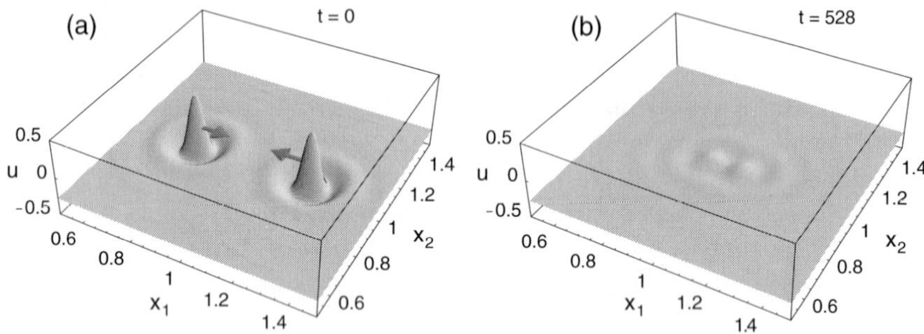

Fig. 13. Numerical solution for (23) in $\mathbb{R}^2$ demonstrating the annihilation of two fast DSs in a head-on collision. Parameters like in Fig. 3, except for $\tau = 3.59$

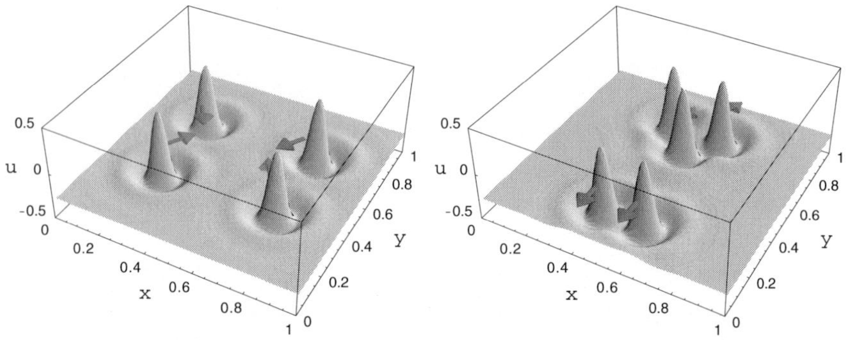

Fig. 14. Numerical solution for (23) in $\mathbb{R}^2$ demonstrating the generation of a new filament through "replication". Parameters like in Fig. 3, except for $\tau = 3.47$

cores of the solitons must come rather close [53]. One might ask the question if the parameters of the system can be chosen such that the threshold for the ignition is very low, e.g. very close to the point of the Turing destabilization, so that also slow DSs may ignite new solitons. Concerning this point, one should keep in mind that a low threshold may lead to a destabilization of the whole active area as more and more DSs are ignited like in a chain reaction [38].

4 Analytical Investigations

We now turn to some analytical investigations concerning the three-component system (23). In a first step, we will concentrate on an interesting aspect concerning the dynamics of single DSs. As it was mentioned, these structures can move intrinsically or stay at rest, depending on the system parameters. Consequently, a bifurcation between these states may take place, which we want to refer to as travelling or drift bifurcation. As mentioned above, the difference between the shape of stationary and propagating DSs is not large if the velocity of the propagation is not too high, a fact that will be exploited in the following section.

4.1 The Drift Bifurcation

As it can be shown numerically, for a certain set of parameters the system (23) has a stationary stable solution $\boldsymbol{U}_0$, possessing rotational symmetry with respect to the center of the activator and inhibitor distributions. We want to analyze how the stability can be expressed mathematically, starting from the more general form (1) of the field equation. As $\boldsymbol{U}_0$ is a stationary solution, it must fulfill relation (2). As done in Chap. 2, we rewrite an arbitrary solution as

$$\boldsymbol{U} = \boldsymbol{U}_0 + \tilde{\boldsymbol{U}} \, , \tag{24}$$

keeping in mind that $\boldsymbol{U}_0$ depends on $\boldsymbol{x}$, which makes the stability analysis more difficult. The reaction term can now be evaluated in the vicinity of $\boldsymbol{U}_0$, similar to (4), one finds

$$\begin{aligned} \dot{\tilde{\boldsymbol{U}}} &= \underline{\boldsymbol{D}}\Delta\tilde{\boldsymbol{U}} + \boldsymbol{L}\tilde{\boldsymbol{U}} + \boldsymbol{N}(\tilde{\boldsymbol{U}}) \\ &= \underline{\boldsymbol{L}}\tilde{\boldsymbol{U}} + \boldsymbol{N}(\tilde{\boldsymbol{U}}) \, . \end{aligned} \tag{25}$$

The stability of the solution demands that small deviations $\tilde{\boldsymbol{U}}$ from the stationary solution $\boldsymbol{U}_0$ decay in the course of time. In the following we want to assume that the operator $\underline{\boldsymbol{L}}$ has a spectrum, in which the information about the stability of the localized solution is contained. We will denote the eigenvalues of $\underline{\boldsymbol{L}}$ by λ_i and the corresponding eigenmodes by $\boldsymbol{\varphi}_i$:

$$\underline{\boldsymbol{L}}\boldsymbol{\varphi}_i = \lambda_i \boldsymbol{\varphi}_i \ , \tag{26}$$

the index can be discrete or continuous. For the solution $\boldsymbol{U}_0$ to be stable, all eigenmodes must have a negative or zero real part. The modes with a vanishing real part are the neutral modes, characterized by $\lambda_0 = 0$, they exist due to the translational invariance of the basic equations (23). This can be seen from the consideration that if $\boldsymbol{U}_0(\boldsymbol{x})$ is a solution of the system, so is $\boldsymbol{U}_0(\boldsymbol{x} - \boldsymbol{p})$ for arbitrary $\boldsymbol{p} \in \mathbb{R}^m$, especially for infinitesimal shifts where we obtain

$$\boldsymbol{U}_0(\boldsymbol{x} - \epsilon \boldsymbol{e}_\xi) = \boldsymbol{U}_0(\boldsymbol{x}) - \epsilon \partial_\xi \boldsymbol{U}_0(\boldsymbol{x}) \tag{27}$$

with infinitesimally small ϵ and $\xi = x_1, \ldots, x_m$. From this consideration one may see that the m spatial derivatives $\partial_\xi \boldsymbol{U}_0$ of the stationary solution $\boldsymbol{U}_0$ with respect to the spatial directions $\xi = x_1, \ldots, x_m$ form a linearly independent set of neutral eigenmodes of $\underline{\boldsymbol{L}}$, in our context they are also called *Goldstone modes* $\boldsymbol{G}_\xi$. If the system is disturbed by fluctuations, changes from the original shape will decay, nevertheless the fluctuations may contain components of one or several Goldstone modes, causing the solution to be shifted in an erratic manner.

As we have seen, we can describe the situation as follows: the existence of neutral modes of translation allow external fluctuations to displace the original solution, but this "driven" propagation does not correspond to the intrinsic propagation that we are interested in. To make more concrete statements about a possible drift-bifurcation, we have to make some restrictions to the linear operator $\underline{\boldsymbol{L}}$: below the drift bifurcation point, all eigenmodes $\boldsymbol{\varphi}_i$ shall form a full set, and in addition, all modes with large real parts shall be part of the discrete spectrum. In particular, the neutral eigenmodes shall not be degenerate. If the parameters of the system are changed, the eigenvalues and the corresponding eigenmodes may change, but m linearly independent Goldstone modes still must exist. A bifurcation occurs if the eigenvalue λ_c of an initially stable mode $\boldsymbol{\varphi}_c$ crosses the imaginary axis. In the following, we will restrict ourselves to the case $\mathrm{Im}(\lambda_c) = 0$ and a mode $\boldsymbol{\varphi}_c$ that is asymmetric in space. Depending on the properties of $\underline{\boldsymbol{L}}$, different types of bifurcations can be expected. With regard to our special system (23), we will analyze the frequently encountered case that in the bifurcation point, the mode growing unstable (i.e. $\lambda_c \to 0$) becomes a linear combination of the Goldstone modes. We now face a degeneration (a Jordan block is formed in the matrix representation of $\underline{\boldsymbol{L}}$), and the set of eigenmodes is no longer a full set. At this point, the generalized eigenmodes $\boldsymbol{P}_\xi$ corresponding to the Goldstone modes have to be considered to complete the set of modes with an additional mode. We will call the additional mode *propagator mode*, it must be a linear combination of the generalized eigenmodes corresponding to the Goldstone modes. The generalized eigenmodes fulfill the relation [54]

$$\underline{\boldsymbol{L}}\boldsymbol{P}_\xi = \boldsymbol{G}_\xi \ . \tag{28}$$

To explicitly determine the bifurcation point for the presented situation, we consider the neutral modes of the adjoint operator $\underline{\boldsymbol{L}}^\dagger$ of $\underline{\boldsymbol{L}}$,

$$\underline{\boldsymbol{L}}^\dagger \boldsymbol{G}_\xi^{\underline{\boldsymbol{L}}^\dagger} = \boldsymbol{0} \, . \tag{29}$$

The neutral modes of the adjoint operator are not the adjoints of the Goldstone modes. To avoid confusion, we well call them *complementary Goldstone modes*. Projecting these modes on the Goldstone modes yields

$$\left\langle \boldsymbol{G}_\xi^{\underline{\boldsymbol{L}}^\dagger} | \boldsymbol{G}_\xi \right\rangle = \left\langle \boldsymbol{G}_\xi^{\underline{\boldsymbol{L}}^\dagger} | \underline{\boldsymbol{L}} \boldsymbol{P}_\xi \right\rangle = \left\langle \underline{\boldsymbol{L}}^\dagger \boldsymbol{G}_\xi^{\underline{\boldsymbol{L}}^\dagger} | \boldsymbol{P}_\xi \right\rangle = 0 \, . \tag{30}$$

Equation (30) can be considered as a condition for the occurrence of the degeneration at the bifurcation point.

What we have achieved up to this point is that we have found a general expression that allows us to determine the dependency of the drift bifurcation point on the system parameters, provided that the made restrictions hold and explicit expressions for $\boldsymbol{G}_\xi$ and $\boldsymbol{G}_\xi^{\underline{\boldsymbol{L}}^\dagger}$ are known. Considering practical problems, one may encounter a problem at this point as an analytical calculation of $\underline{\boldsymbol{L}}^\dagger$ and the corresponding neutral modes may be difficult. Therefore, we now turn to the concrete system (23). To ease the calculations, we choose $\kappa_2 = 0$, i.e. we neglect the integral term, as the numerical simulations in the last section have shown that in the three-component system, DSs exist even without an integral term. It can be proven in the given case, the operator $\underline{\boldsymbol{L}}$ fulfills all mentioned requirements. Furthermore, we face the lucky situation that $\underline{\boldsymbol{L}}$ can be decomposed into the product of a diagonal matrix $\underline{\boldsymbol{M}}$ and a self-adjoint operator $\mathcal{S}$ [54]:

$$\underline{\boldsymbol{L}} = \underline{\boldsymbol{M}} \cdot \mathcal{S} = \begin{pmatrix} 1 & 0 & 0 \\ 0 & \frac{-1}{\kappa_3 \tau} & 0 \\ 0 & 0 & \frac{-1}{\kappa_4 \theta} \end{pmatrix} \cdot \begin{pmatrix} D_u \triangle + \lambda - 3\bar{u}^2 & -\kappa_3 & -\kappa_4 \\ -\kappa_3 & -\kappa_3 D_v \triangle + \kappa_3 & 0 \\ -\kappa_4 & 0 & -\kappa_4 D_w \triangle + \kappa_4 \end{pmatrix} . \tag{31}$$

The time constants now appear only in the first matrix. From (31) one finds that if $\boldsymbol{\varphi}_i$ is an eigenmode of $\underline{\boldsymbol{L}}$ for the eigenvalue λ_i, then the corresponding complementary mode (for the same eigenvalue) is given by

$$\boldsymbol{\varphi}_i^{\underline{\boldsymbol{L}}^\dagger} = \underline{\boldsymbol{M}}^{-1} \boldsymbol{\varphi}_i \, , \tag{32}$$

so that is is easy to calculate the complementary Goldstone modes. In detail we find

$$\boldsymbol{G}_\xi = \begin{pmatrix} \bar{u}_\xi \\ \bar{v}_\xi \\ \bar{w}_\xi \end{pmatrix} \quad \boldsymbol{G}_\xi^{\underline{\boldsymbol{L}}^\dagger} = \begin{pmatrix} \bar{u}_\xi \\ -\kappa_3 \tau \bar{v}_\xi \\ -\kappa_4 \theta \bar{w}_\xi \end{pmatrix} . \tag{33}$$

Inserting the relevant expressions into (30) results in

$$\tau_c = \frac{\langle \bar{u}_x^2 \rangle - \kappa_4 \theta \langle \bar{w}_x^2 \rangle}{\kappa_3 \langle \bar{v}_x^2 \rangle} \, , \tag{34}$$

the brackets denote integration over the considered domain. Equation (34) describes the bifurcation value of the time constant τ as a function of the parameters θ, κ_3 and κ_4, provided that no other bifurcation has destabilized the stationary DS before. The result can be interpreted in the following way: when the time constant τ is small, the slower inhibitor v can easily follow the activator, smoothing deviations from the rotational symmetry of the distributions. When the inhibitor becomes too slow ($\tau \geq \tau_c$), this is no longer possible. Slight disturbances break the symmetry of the structure, causing a shift of the inhibitor with respect to the activator distribution as described above. The mode responsible for the shift is the propagator mode. Of course, it is also possible to solve the relation (34) for another parameter which is varied while keeping τ fixed or even to vary several parameters to reach the bifurcation point. Nevertheless, an illustrative interpretation then would be more difficult [55, 56].

We are now interested in the equilibrium velocity of the DS close to the bifurcation point. For reasons of simplicity, in the following we will consider the case $D_v = 0$, for which (34) reduces to

$$\tau_c = \frac{1}{\kappa_3} - \theta \frac{\kappa_4}{\kappa_3} \frac{\langle \bar{w}_x^2 \rangle}{\langle \bar{u}_x^2 \rangle} \tag{35}$$

as $\bar{u} = \bar{v}$ in this case. To achieve our goal, we transform (23) into a frame system moving with the velocity c. Without loss of generality, we will consider a motion in the direction $\xi = x_1$. One then finds

$$\begin{aligned} u_t &= c u_{x_1} + d_u \Delta u + \lambda u - u^3 - \kappa_3 v - \kappa_4 w + \kappa_1 \,, \\ v_t &= c v_{x_1} + (u - v)/\tau \,, \\ w_t &= c w_{x_1} + (d_w \Delta w + u - w)/\theta \,. \end{aligned} \tag{36}$$

To find a stationary solution $\hat{\boldsymbol{U}}$, the left hand side is put to zero. The resulting system is projected onto the vector $\boldsymbol{G}_{x_1}^{\boldsymbol{L}^\dagger} = (\hat{u}_{x_1}, -\kappa_3 \tau \hat{v}_{x_1}, -\kappa_4 \theta \hat{w}_{x_1})^{\mathrm{T}}$, yielding the relation

$$c(\langle \hat{u}_{x_1}^2 \rangle - \kappa_3 \tau \langle \hat{v}_{x_1}^2 \rangle - \kappa_4 \theta \langle \hat{w}_{x_1}^2 \rangle) = 0 \,. \tag{37}$$

Making use of the stationary form of the second equation of (36), some terms can be eliminated and we obtain the result

$$c\tau^2 \langle \hat{v}_{x_1 x_1}^2 \rangle \left[c^2 - \frac{\kappa_3}{\tau^2} \frac{\langle \hat{v}_{x_1}^2 \rangle}{\langle \hat{v}_{x_1 x_1}^2 \rangle} \left(\tau - \left(\frac{1}{\kappa_3} - \theta \frac{\kappa_4}{\kappa_3} \frac{\langle \hat{w}_{x_1}^2 \rangle}{\langle \hat{v}_{x_1}^2 \rangle} \right) \right) \right] = 0 \,. \tag{38}$$

The expression is somewhat complicated, but can be approximated close to the the drift bifurcation point τ_c by

$$c\tau_c^2 \langle \bar{u}_{x_1 x_1}^2 \rangle \left[c^2 - \frac{\kappa_3}{\tau_c^2} \frac{\langle \bar{u}_{x_1}^2 \rangle}{\langle \bar{u}_{x_1 x!}^2 \rangle} \left(\tau - \tau_c \right) \right) \right] = 0 \,. \tag{39}$$

We note that below the bifurcation point, only the trivial solution $c = 0$ exists. Above the bifurcation point, this solution becomes unstable, and two new solutions appear in the course of a supercritical bifurcation, with

$$c^2 = \frac{\kappa_3}{\tau_c^2} \frac{\langle \bar{u}_{x_1}^2 \rangle}{\langle \bar{u}_{x_1 x_1}^2 \rangle} \left(\tau - \tau_c \right) . \tag{40}$$

This bifurcation is typical for symmetry-breaking bifurcations and can be observed in various systems [54, 57, 58, 59].

4.2 Reduction of the Soliton Dynamics in a "Particle Approach"

In the preceding section, the central ansatz was to use perturbation theory close to the drift bifurcation point. The perturbation was decomposed into modes of the linearized operator $\underline{\boldsymbol{L}}$. Of the infinite number of modes, only the critical modes with $\mathrm{Re}(\lambda_i) \geq 0$ were important for the dynamics. In addition, we found that with the assumptions made for the operator $\underline{\boldsymbol{L}}$, the drift bifurcation is of supercritical nature. This means that the above statement stays approximately correct in the vicinity of the bifurcation point.

We now use the drift bifurcation as a starting point for the derivation of order parameter equations, describing the time evolution of the relevant modes on the center manifold. To deal with nonlinearities, we well have to use a perturbation ansatz as described further below. The relevant modes close to the point of the drift bifurcation are the Goldstone modes and the corresponding generalized eigenmodes, respectively the propagator mode. As long as no other modes get critical, the dynamics can be understood if the time evolution of this relevant modes is known. We will first consider the behavior of a single DS, rewriting (24) in the form

$$\boldsymbol{U}(\boldsymbol{x}, t) = \boldsymbol{U}_0(\boldsymbol{x} - \boldsymbol{p}(t)) + \tilde{\boldsymbol{U}}(\boldsymbol{x} - \boldsymbol{p}(t), t) . \tag{41}$$

We have now explicitly introduced the position $\boldsymbol{p}(t)$ of the DS, taking care that the deviation $\tilde{\boldsymbol{U}}$ stays small (the original distributions is shifted towards the new distribution) and does not contain any contributions of the Goldstone modes. The ansatz can be inserted into (1), yielding

$$\begin{aligned}
&\dot{\boldsymbol{U}}(\boldsymbol{x}, t) \\
&= -\dot{\boldsymbol{p}} \cdot \nabla_{\boldsymbol{p}} [\boldsymbol{U}_0(\boldsymbol{x} - \boldsymbol{p}) + \tilde{\boldsymbol{U}}(\boldsymbol{x} - \boldsymbol{p}, t)] + \frac{\partial}{\partial t} \tilde{\boldsymbol{U}}(\boldsymbol{x} - \boldsymbol{p}, t) \\
&= \underline{\boldsymbol{D}} \Delta [\boldsymbol{U}_0(\boldsymbol{x} - \boldsymbol{p}) + \tilde{\boldsymbol{U}}(\boldsymbol{x} - \boldsymbol{p}, t)] + \boldsymbol{R}(\boldsymbol{U}_0(\boldsymbol{x} - \boldsymbol{p}) + \tilde{\boldsymbol{U}}(\boldsymbol{x} - \boldsymbol{p}, t)) .
\end{aligned} \tag{42}$$

As $\boldsymbol{U}_0(\boldsymbol{x})$ is a stationary solution, one finds (compare also (4))

$$\begin{aligned}
&-\dot{\boldsymbol{p}} \cdot \nabla_{\boldsymbol{p}} [\boldsymbol{U}_0(\boldsymbol{x} - \boldsymbol{p}) + \tilde{\boldsymbol{U}}(\boldsymbol{x} - \boldsymbol{p}, t)] + \frac{\partial}{\partial t} \tilde{\boldsymbol{U}}(\boldsymbol{x} - \boldsymbol{p}, t) \\
&= \underline{\boldsymbol{D}} \Delta \tilde{\boldsymbol{U}}(\boldsymbol{x} - \boldsymbol{p}, t) + \boldsymbol{L} \tilde{\boldsymbol{U}}(\boldsymbol{x} - \boldsymbol{p}, t) + \boldsymbol{N}(\tilde{\boldsymbol{U}}(\boldsymbol{x} - \boldsymbol{p}, t)) \\
&= \underline{\boldsymbol{L}} \tilde{\boldsymbol{U}}(\boldsymbol{x} - \boldsymbol{p}, t) + \boldsymbol{N}(\tilde{\boldsymbol{U}}(\boldsymbol{x} - \boldsymbol{p}, t)) .
\end{aligned} \tag{43}$$

In the first line of (43), one finds the product of the time derivative of the position and the spatial derivatives of the stationary solution, i.e. the Goldstone modes. Before being able to extract the relevant information from this equation, we have to rewrite the deviation as an expansion in eigenmodes of $\underline{\boldsymbol{L}}$:

$$\tilde{\boldsymbol{U}} = \sum_{j_{nc}} \lambda_j(t)\boldsymbol{\varphi}_j - \sum_{\xi} \alpha_\xi(t)\boldsymbol{P}_\xi \,. \tag{44}$$

In this expression, the eigenmodes are divided into non-critical modes and critical modes, the latter are the generalized Goldstone modes forming the propagator mode. The amplitude vector $\boldsymbol{\alpha}(t) = \{\alpha_\xi(t)\}$ belonging to the generalized Goldstone modes can even be interpreted in an intuitive way: it corresponds to the shift between the activator and the slow inhibitor distribution. We now use a projection to obtain expressions for the time derivatives of the expansion coefficients $\boldsymbol{p}(t)$ and $\boldsymbol{\alpha}(t)$. Suitable modes for this purpose are the complementary Goldstone modes $\boldsymbol{G}_\xi^{\underline{\boldsymbol{L}}^\dagger}$ and their generalized eigenmodes $\boldsymbol{P}_\xi^{\underline{\boldsymbol{L}}^\dagger}$ (compare also [51, 59]). We face the advantage that the critical modes are orthogonal to the non-critical modes, but nevertheless, the non-linear term $\boldsymbol{N}(\tilde{\boldsymbol{U}})$ can couple critical and non-critical modes. Here, a careful analysis of the relevant coupled terms is necessary [59]. The basic idea for the treatment is that the dynamics of the critical modes occurs on fast and slow time scales whereas the dynamics of the non-critical modes occurs only on fast time scales, so that after a short relaxation time, the dynamics relaxes onto the central manifold and the non-critical modes are slaved by the critical modes, an idea that was originally introduced by Haken [57]. One may therefore use a perturbation ansatz in the following way: a parameter $\epsilon = \frac{|\alpha|}{L} \ll 1$ is introduced, where L is the half width of one DS. The choice reflects the fact that the shape of a propagating structure differs only slightly from the stationary shape as the displacement of the slow inhibitor distribution with respect to the activator distribution is small compared to the extension of the DS. Now, the dynamics is considered on different time scales $T_1 = \epsilon^1 t, \ldots, T^n = \epsilon^n t$. We introduce

$$\boldsymbol{p} = \boldsymbol{p}(T_1, T_2, T_3) \,, \tag{45}$$

$$\boldsymbol{\alpha} = \epsilon\boldsymbol{\alpha}(T_1, T_2) \,. \tag{46}$$

The first equation expresses that the main reason for a change of the solution is a shift of the solution ($\boldsymbol{p}(t) \sim O(1)$), making the center $\boldsymbol{p}(t)$ of the solution an order parameter whose dynamics may occur on fast, slow and very slow time scales, of which the latter dominates the dynamics on the center manifold. The ansatz (46) reflects that the deformation of the unperturbated solution due to the appearance of the propagator mode is already less significant than the deformation by the shift ($\boldsymbol{\alpha}(t) \sim O(\epsilon)$) and exists on fast and slow timescales. Consequently, the whole derivation (44) from stationary solution can be written in the following way:

$$\begin{aligned}\tilde{\boldsymbol{U}} &= \epsilon^2 \sum_{j_{nc},\,\mathrm{s}} \lambda_j(T_1)\boldsymbol{\varphi}_j + \epsilon^3 \sum_{k_{nc},\,\mathrm{ns}} \lambda_k \boldsymbol{\varphi}_k - \epsilon \sum_{\xi} \alpha_\xi(T_1, T_2)\boldsymbol{P}_\xi \\ &= \epsilon^2 \boldsymbol{r}_{\mathrm{s}} + \epsilon^3 \boldsymbol{r}_{\mathrm{ns}} - \epsilon \boldsymbol{r}_\alpha \,. \end{aligned} \tag{47}$$

Noncritical modes which still have some significance for the dynamics ($\boldsymbol{r}_{\mathrm{f}} \sim O(\epsilon^2)$, index s) decay fast so that the dynamics takes place on the center manifold given by $\boldsymbol{p}$ and $\boldsymbol{\alpha}$, the time scale of the insignificant modes ($\boldsymbol{r}_{\mathrm{ns}} \sim O(\epsilon^3)$, index ns) is of no interest for the further calculation. The ansatz can now be inserted into the original equation and the low orders in ϵ can be evaluated by projecting onto the complementary critical modes.

As in the last section, at this point we leave the more general considerations and turn to the concrete system (23). The actual execution of the calculation of perturbations is rather extensive (for details see [59]). To keep the results simple, we will treat the special case $D_v = 0$, $\theta = 0$ and $\kappa_2 = 0$. One then arrives at the system of equations

$$\begin{aligned}\dot{\boldsymbol{p}} &= \kappa_3 \boldsymbol{\alpha}, \\ \dot{\boldsymbol{\alpha}} &= \kappa_3^2 \left(\tau - \frac{1}{\kappa_3}\right) \boldsymbol{\alpha} - \kappa_3 \underbrace{\frac{\langle (\partial_{x_1 x_1} u_0)^2 \rangle}{\langle (\partial_{x_1} u_0)^2 \rangle}}_{=:Q} |\boldsymbol{\alpha}|^2 \boldsymbol{\alpha} \,. \end{aligned} \tag{48}$$

Before discussing the physical meaning of the system (48), we want to make an extension to two interacting DSs. To this end, we assume that the distance between the solitons is so large that the shape of each solitons essentially is preserved (in contrast to the generation and annihilation processes presented in Sect. 2.5). The ansatz now has the form

$$\boldsymbol{U}(\boldsymbol{x}, t) = \boldsymbol{U}_{0,1}(\boldsymbol{x} - \boldsymbol{p_1}) + \tilde{\boldsymbol{U}}_1(\boldsymbol{x} - \boldsymbol{p_1}, t) + \boldsymbol{U}_{0,2}(\boldsymbol{x} - \boldsymbol{p_2}) + \tilde{\boldsymbol{U}}_2(\boldsymbol{x} - \boldsymbol{p_2}, t) \,. \tag{49}$$

with time scales as above. The further treatment is analogous to that used for a single DS, but a special treatment of the nonlinear term is required (for details compare also [51]). We decompose the nonlinearity as follows:

$$\begin{aligned}&\boldsymbol{N}(\boldsymbol{U}) = \boldsymbol{N}(\boldsymbol{U}_{0,1}) + \boldsymbol{N}(\boldsymbol{U}_{0,2}) \\ &+ \nabla_{\boldsymbol{U}_{0,1}} \boldsymbol{N}(\boldsymbol{U}_{0,1})(\boldsymbol{U}_{0,2} + \epsilon^2(\boldsymbol{r}_{\mathrm{s},1} + \boldsymbol{r}_{\mathrm{s},2}) + \epsilon^3(\boldsymbol{r}_{\mathrm{ns},1} + \boldsymbol{r}_{\mathrm{ns},2}) - \epsilon(\boldsymbol{r}_{\alpha,1} + \boldsymbol{r}_{\alpha,2})) \\ &+ \nabla_{\boldsymbol{U}_{0,2}} \boldsymbol{N}(\boldsymbol{U}_{0,2})(\boldsymbol{U}_{0,1} + \epsilon^2(\boldsymbol{r}_{\mathrm{s},1} + \boldsymbol{r}_{\mathrm{s},2}) + \epsilon^3(\boldsymbol{r}_{\mathrm{ns},1} + \boldsymbol{r}_{\mathrm{ns},2}) - \epsilon(\boldsymbol{r}_{\alpha,1} + \boldsymbol{r}_{\alpha,2}))\end{aligned} \tag{50}$$

taking into account terms up to third order in ϵ. The resulting equations are

$$\begin{aligned}\dot{\boldsymbol{p}}_i &= \kappa_3 \boldsymbol{\alpha}_i + F(d_{ij})(\boldsymbol{p}_i - \boldsymbol{p}_j) \,, \\ \dot{\boldsymbol{\alpha}}_i &= \kappa_3^2 \left(\tau - \frac{1}{\kappa_3}\right) \boldsymbol{\alpha}_i - \kappa_3 Q |\boldsymbol{\alpha}_i|^2 \boldsymbol{\alpha}_i + F(d_{ij}) \frac{\boldsymbol{p}_i - \boldsymbol{p}_j}{|\boldsymbol{p}_i - \boldsymbol{p}_j|}\end{aligned} \tag{51}$$

with $i, j = 1, 2$, $i \neq j$. We now have a set of four ordinary differential equations, extended by a distance-dependent interaction term. For the interaction term, one finds

$$F(d_{ij}) = \frac{\langle f'(u_{0,i}) u_{0,j} \nabla u_{0,i} \rangle}{\langle (\partial_x u_{0,i})^2 \rangle} \cdot \frac{\boldsymbol{p}_i - \boldsymbol{p}_j}{|\boldsymbol{p}_i - \boldsymbol{p}_j|} \,. \tag{52}$$

In this expression only the activator component u of the DSs appears, as in the concrete system (23) with the simplifications $D_v = 0$, $\theta = 0$, the only relevant part of the dynamics occurs in the activator equation. Equation (52) makes it possible to calculate the interaction function from the stationary distributions. A generalization to an arbitrary number of interacting solitons is possible, as a result, (51) is still valid if the restriction $i, j = 1, 2$ in (51) is dropped and sum convention is used.

The systems (48) and (51) offer an intuitive way to understand how the propagation physically takes place: as stated above, the vector $\boldsymbol{\alpha}$ can be identified with the shift between the activator and the slow inhibitor distribution, caused by the appearance of the propagator mode. If this shift increases for one soliton, the velocity increases (first of (48)). The growth of the shift is auto-catalytic for small values, but an infinite growth is limited by non-linear terms (second of (48)), corresponding to the propagation mechanism developed in Sect. 2.3. When inserting the first into the second equation, one arrives at

$$\ddot{\boldsymbol{p}} = \kappa_3^2 \left(\tau - \frac{1}{\kappa_3} \right) \dot{\boldsymbol{p}} - \frac{Q}{\kappa_3} |\dot{\boldsymbol{p}}|^2 \dot{\boldsymbol{p}} \,. \tag{53}$$

This Newton-like type of equation may be interpreted as a dynamic equation of a particle with unit mass and a velocity-dependent friction, but one should be careful seeing a direct correspondence: classical particles have an inertia due to their mass, whereas for moving DSs, one rather has a "virtual inertia", resulting from an inner degree of freedom (namely, the shift between activator and inhibitor distribution). Note that the equilibrium velocity given by (53) corresponds to (40) in the considered case $D_v = 0$, $\theta = 0$. In the case of multiple DSs, generalized interaction terms are added to both equations, meaning that both Goldstone modes and generalized eigenmodes are affected by neighboring DSs. It is also possible to convert the system (51) into one differential equation of second order by a comparison of magnitude of the terms [61]. The calculation then yields

$$\ddot{\boldsymbol{p}}_i = \kappa_3^2 \left(\tau - \frac{1}{\kappa_3} \right) \dot{\boldsymbol{p}}_i - \frac{Q}{\kappa_3} |\dot{\boldsymbol{p}}_i|^2 \dot{\boldsymbol{p}}_i + \kappa_3 F(d_{ij}) \frac{\boldsymbol{p}_i - \boldsymbol{p}_j}{|\boldsymbol{p}_i - \boldsymbol{p}_j|} \,, \tag{54}$$

showing that the influence of neighboring DSs on the Goldstone modes is much smaller than on the generalized eigenmodes. Similar to (53), there is a formal analogy to the dynamic equation of a classical particle, this time we observe motion in a potential generated by neighboring particles.

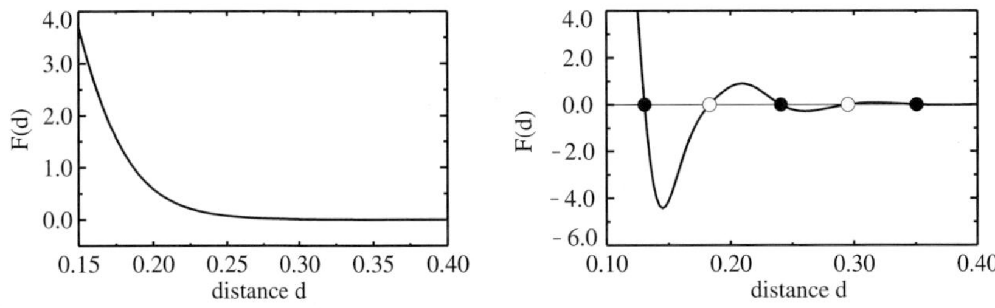

Fig. 15. Interaction function for the distributions depicted in Fig. 10a and b, calculated according to (52)

Before finishing this part of the investigation, we will have a look at typical interaction functions which can be calculated according to (52). Figure 15 shows the result for the distributions depicted in Fig. 10. For non-oscillatorily decaying DSs, the interaction is purely repulsive (Fig. 15a), whereas for DSs with oscillatorily decaying tails, the sign of the interaction function changes with the distance between the solitons, confirming the numerical observations made in Sect. 2.5. Consequently, for the latter case one finds stable fixed points of the interaction function, corresponding to stable binding distances for which stable molecules and clusters of several DSs can form. An examination of this fact in a simulation of the three-component system (23) confirms that the binding distances are the same as for the full reaction-diffusion system.

The reduction of the dynamics of the system by the derivation of order parameter equations brings many advantages. First of all, a numerical simulation of the dynamics of several thousands of DSs in the full system (23) is possible only on very large computers, and the simulation time is limited. In contrast, the numerical solving of thousand ordinary differential equations can be done by a usual personal computer and for much longer simulation times. For all results on the reduction of the dynamics presented in the last section, one should bear in mind that a perturbation ansatz using two sorts of relevant modes was made. The derived formulas are valid as long as no other modes get critical, which usually holds in the vicinity of the drift bifurcation point (see the comparison of a simulation of DSs in the full system with the reduced dynamics in Fig. 16). If other non-critical modes come into play, e.g. in generation processes, the conditions for the derivation are violated.

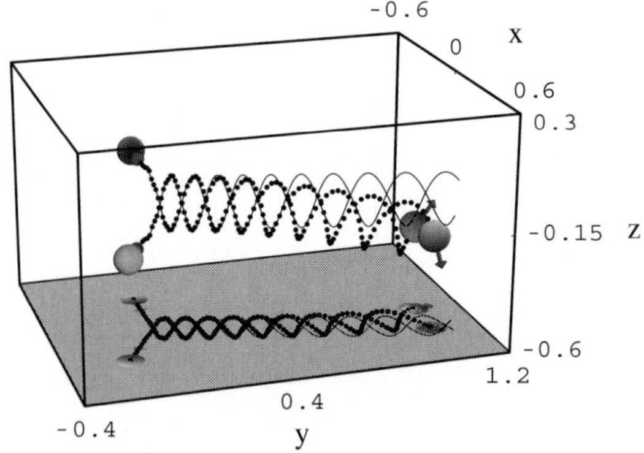

Fig. 16. Dynamics of two interacting DSs (the balls mark equi-surfaces of the activator distribution) in three spatial dimensions, calculated according to a simulation of the full three-component system (23) (*black dots*) and according to the corresponding reduced equations (51) (*solid lines*). The difference of the trajectories mainly results from numerical errors in the simulation of the full system. Parameters: $\tau = 3.36$, $\theta = 0$, $d_u = 1.1 \cdot 10^{-4}$, $d_v = 0$, $d_w = 9.64 \cdot 10^{-4}$, $\lambda = 1.01$, $\kappa_1 = -0.1$, $\kappa_2 = 0$, $\kappa_3 = 0.3$, $\kappa_4 = 1$ and no-flux boundary conditions

5 Planar Gas-Discharge Systems as Reaction-Diffusion Systems

5.1 Description of the System and Relation to Reaction-Diffusion Systems

In this chapter, we want to focus onto a special class of systems, namely planar DC gas-discharge systems with high-ohmic barriers. Such systems can be considered as reaction systems because in a local approximation, the generation and annihilation of charge carriers can formally be described in terms of reaction functions. In addition, the diffusion of charge carriers is a mayor mechanism for charge transport.

A schematic representation of an experimental realization of a planar gas-discharge system with one high-ohmic layer is depicted in Fig. 17. A thin gas layer is contacted by two electrodes, of which the left electrode consists of high-ohmic semiconductor material with a high-ohmic current-voltage characteristic, covered by a semitransparent gold layer to establish an homogeneous contact. The second electrode is a glass disc covered with a layer of indium tin oxide (ITO), which is conductive and transparent to visible light. The resistivity of the high ohmic semiconductor layer can be controlled by illumination, taking advantage of the internal photo effect. If the externally applied DC voltage is high enough, a discharge can be ignited in the discharge cell, which is accompanied by light emission from the discharge gap. The density of the radiation emitted from the discharge gap is proportional

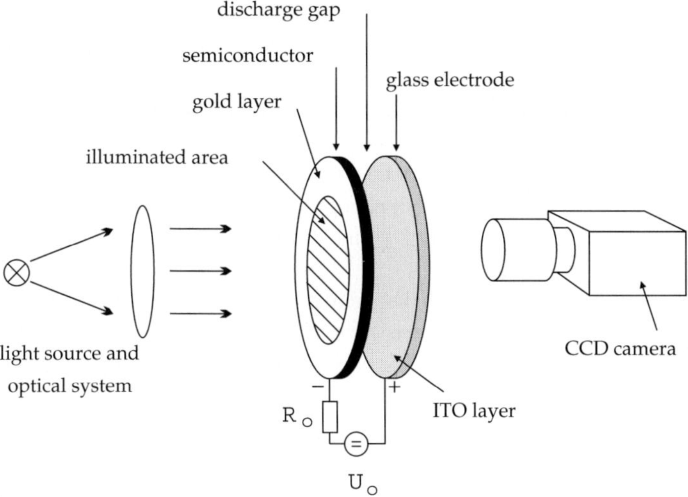

Fig. 17. Hybrid semiconductor gas-discharge system as two-dimensional experimental realization of the three-component reaction-diffusion system (23)

to the average current density in the discharge gap [60] and can be recorded through the transparent ITO electrode using a CCD camera.

Principally, one would expect a modelling of the experimental systems in terms of a drift-diffusion approximation of the transport equations for electrons and other charge carriers as it is well-known from long discharge tubes with metallic electrodes [62, 63, 64, 65]. However, so far there is no even qualitative description of pattern formation in planar DC gas-discharge systems in terms of such an approach. In particular this holds for the description of the formation and the dynamics of DSs in such systems. Due to this, one may choose a very simplified model based on electrical equivalent circuit considerations that are related to Fig. 6, Fig. 8 and the corresponding equations.

The modelling of the gas-discharge by the mentioned electrical equivalent circuit takes place in the following steps:

- Starting from Fig. 8, we interpret the linear layer as semiconductor layer.
- The nonlinear layer is supposed to be the gas layer.
- The lateral coupling in the gas is supposed to describe the diffusion of charge carriers.

Under these conditions, we end up with a system of equations identical to (22). The variables u and v can be identified as the normalized current density and the voltage drop over the semiconductor. The parameters κ_1 and κ_2 are related to the supply voltage and the series resistor. d_u formally is related to a "voltage diffusion" and d_v is related to charge carrier diffusion. The function $f(u)$ reflects the nonlinear relation between the voltage and the current density in the gas layer. The interpretation of Fig. 8 as a model for a planar DC gas-discharge system with a high-ohmic barrier is done by a straight forward generalization in two spatial dimensions.

In a second step, one has to take into account that charges on the interface between the semiconductor and the gas layer play an important role for the dynamics of the system due to the large resistivity of the semiconductor layer. This gives rise to a third component as written down in (23). Here, w is the normalized surface charge density and θ and d_w are proper time and diffusion-related constants.

In spite of the many difficulties related to the described model, it is amazing that many effects of pattern formation that have been experimentally seen in DC gas-discharge systems can be observed in the three component system (23) and vice versa as we will demonstrate in the next paragraph.

5.2 Dissipative Solitons in Gas-Discharge Systems

For a certain range of parameters, a homogeneous discharge over the whole cell can be observed as a stable state. If the parameters are changed appropriately, the homogeneous state destabilizes, and spatially inhomogeneous patterns in the current density distribution appear in the discharge gap. Among these are stripes [60, 66, 67], target patterns, spirals [68] and, in particular, bright spots, corresponding to solitary localized current filaments [60, 69, 71]. These filaments can appear as individual objects or in large clusters [71] and, like their counterparts in the three-component model systems, exhibit solitary properties like propagation [72], scattering, formation of molecules [73] as well as generation and annihilation [11].

Figure 18 shows examples of the current density distribution for two current filaments recorded for different system parameters. To eliminate the influence of noise, the profile was averaged over many frames. The correspondence between the experimental results of Fig. 18 and the results depicted in Fig. 10 obtained from solving (23) is striking.

We will now investigate the dynamics of single, non-interacting filaments. In the experiment, propagating filaments can be observed. As they preserve their identity during the propagation, is is reasonable to describe their motion in terms of the center of the corresponding current density distribution. In Fig. 19a, the trajectory for the filament depicted in Fig. 18a is shown. On a homogeneously prepared area, one would expect the structure to move on a straight line (compare also (53)), but this does not seem to be the case for the presented situation. As a lot of care has been taken to assure the spatial homogeneity of the system, the most probable cause for the observed behavior are fluctuations present in the system.

Naturally, one may pose the question whether the propagation of the filaments is due to noise or if also a deterministic component is present in the dynamics. In what follows we want to describe a method how the deterministic part of the velocity can be determined in the presence of strong

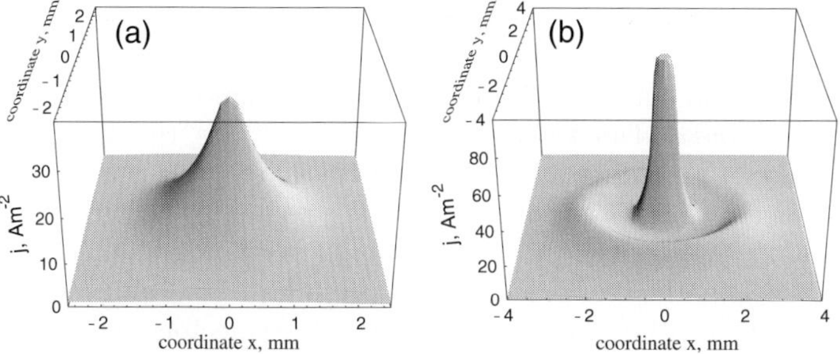

Fig. 18. Surface plots of the experimentally determined current density distributions for single current filaments, averaged over 4000 recorded frames. Parameter: (**a**) global voltage $U_0 = 2740\,\mathrm{V}$, semiconductor resistivity $\rho_{SC} = 4.95 \cdot 10^6\,\Omega\,\mathrm{cm}$, global resistor $R_0 = 20\,\mathrm{M}\Omega$, pressure $p = 280\,\mathrm{mbar}$, temperature $T_{\mathrm{SC}} = 100\,\mathrm{K}$, height of semiconductor wafer $a_{\mathrm{SC}} = 1\,\mathrm{mm}$, width of gas layer d $= 250\,\mu\mathrm{m}$, with global current $I = 46\,\mu\mathrm{A}$, (**b**) $U_0 = 3900\,\mathrm{V}$, $\rho_{SC} = 3.05\,\mathrm{M}\Omega\,\mathrm{cm}$, $R_0 = 4.4\,\mathrm{M}\Omega$, $p = 279\,\mathrm{hPa}$, $T = 100\,\mathrm{K}$, $d = 500\,\mu\mathrm{m}$, with $I = 200\,\mu\mathrm{A}$

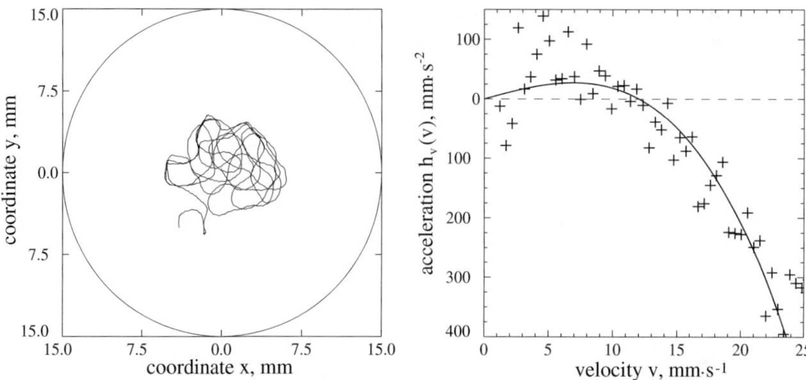

Fig. 19. Dynamics of the filament depicted in Fig. 18a: (**a**) part of the recorded trajectory, (**b**) deterministic part of the acceleration as result of the stochastic data analysis

fluctuations. The key to the separation of the deterministic and the stochastic part of the dynamics is to assume the following *Langevin equation* (i.e. an ordinary differential equation with stochastic contributions) to describe the time evolution of the velocity of the center of the filament:

$$\begin{aligned} \boldsymbol{v}(t) &= \boldsymbol{h}(\boldsymbol{v}(t)) + \underline{\boldsymbol{R}}(\boldsymbol{v}(t))\boldsymbol{\Gamma}(t) \\ &= h_v(v(t))\frac{\boldsymbol{v}(t)}{v(t)} + R(v(t))\boldsymbol{\Gamma}(t) \,. \end{aligned} \tag{55}$$

The first term on the right hand side describes the deterministic part of the motion, in analogy to (53). The second term is the product of a matrix of noise amplitudes and a normalized vector of Gaussian-distributed noise forces with vanishing mean. Due to symmetry considerations, the dimensionality of the problem can be reduced, see the second line of (55). A detailed description of the derivation of the equation and the technique applied to determine the function $h_v(v)$ is given in [72], for a brief version see also [74]. Here, we only want to show the main result, depicted in Fig. 19b, which shows the deterministic part $h_v(v)$ of the acceleration for the trajectory shown in Fig. 19a. The experimental results (crosses) can be well fitted to the cubic polynomial from the theoretical derivation (53), demonstrating that the filament propagates with finite intrinsic velocity, superimposed by fluctuations. Apart from the case presented in Fig. 19, a second case exists where the acceleration is proportional to the velocity with a negative proportionality constant. This case corresponds to a filament that would stay at rest if no noise was present, analogous to a *Brownian particle*. The filaments in the first case are therefore also referred to as *active Brownian particles* [75].

The one-to-one correspondence of the theoretical predictions of the three-component system and the experimental findings motivates a search for a transition between the two states in the experiment, i.e. a drift bifurcation. Using the specific resistivity ρ_{SC} of the semiconductor as control parameter, it is indeed possible to find the searched phenomenon (Fig. 20). If one compares this result with the derivation of the reaction-diffusion system using an electrical network, one can see that an increase of the resistivity in the upper layer (corresponding to ρ_{SC}), leads to an increase of the time constant τ which is the control parameter for the drift bifurcation in the theoretical

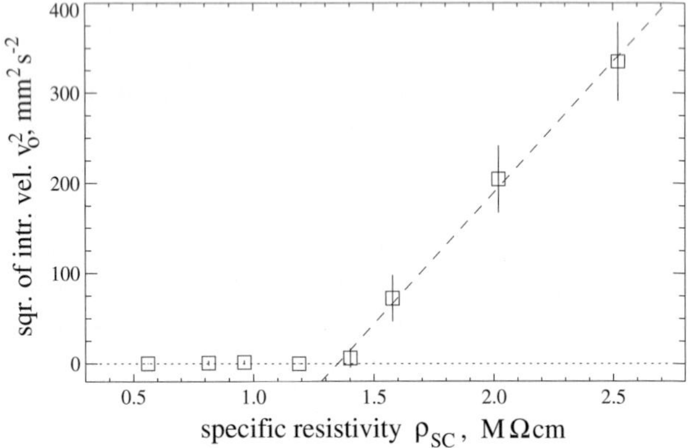

Fig. 20. Square of the intrinsic deterministic equilibrium velocity of experimentally observed DSs as a function of the specific resistivity of the high-ohmic layer determined by stochastic data analysis. Parameters: $U_0 = 3700\,\mathrm{V}$, $R_0 = 10\,\mathrm{M}\Omega$, $p = 286\,\mathrm{mbar}$, $d = 750\,\mu\mathrm{m}$, all other parameters as in Fig. 19a

model. Unfortunately, a quantitative comparison of the bifurcation point is not possible as the model system is only qualitative, but the scaling law of the velocity in dependance of the control parameter agrees with the theoretical predictions.

In the context of reducing the three-component system (23) to order parameter equations, it was possible to determine quantitatively an interaction function from the stationary distributions of two DSs. Using stochastic data analysis, it is possible to obtain an interaction function from experimental data [73]. The key approach is to extend the Langevin equation for each DS with a generalized interaction term, resulting in an equation of the form

$$\boldsymbol{v}_i(t) = h_v(v_i(t))\frac{\boldsymbol{v}_i(t)}{v_i(t)} + F(d_{ij})\frac{\boldsymbol{p}_i - \boldsymbol{p}_j}{|\boldsymbol{p}_i - \boldsymbol{p}_j|} + R(v_i(t))\boldsymbol{\Gamma}_i(t) \,. \tag{56}$$

Note the formal analogy of this expression to (54). The interaction function $F(d)$ can be determined from experimental data if only two filaments are on the active area. Fig. 21 shows a typical result of such an interaction function, again, the crosses are the experimentally determined values. As the interaction function exhibits the same oscillatory nature as the interaction function depicted in Fig. 15b, we may come to the conclusion that the qualitative validity of the three-component model (23) for the description of planar DC gas-discharge systems is strongly supported. The theoretical interaction function can be fitted using the power law

$$F(d) = \frac{-f_1}{\sqrt{d}} \exp(-f_2 d) \cos(f_3 d - f_4) \,, \tag{57}$$

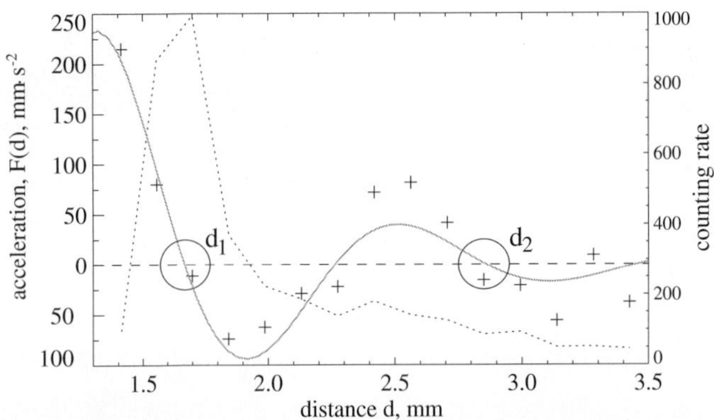

Fig. 21. Acceleration $F(d)$ determined from experimental data points (crosses), the *solid grey* line is a fit according to (57), with $f_1 = 1386.6\,\mathrm{mm}^{\frac{3}{2}} \cdot \mathrm{s}^{-2}$, $f_2 = 1.2167\,\mathrm{mm}^{-1}$, $f_3 = 5.27957\,\mathrm{mm}^{-1}$ and $f_4 = 8.47037$. The values d_1 and d_2 mark points where other filaments may "lock in". Parameters: $U_0 = 4600\,\mathrm{V}$, $\rho_{SC} = 3.5\,\mathrm{M}\Omega\,\mathrm{cm}$, $R_0 = 4.4\,\mathrm{M}\Omega$, $p = 283\,\mathrm{mbar}$, $T = 100\,\mathrm{K}$, $d = 500\,\mu\mathrm{m}$, $t_{\mathrm{exp}} = 0.02\,\mathrm{s}$, with $I = 235\,\mu\mathrm{A}$

where the f_i are fit parameters. The fit function was chosen as it reflects the decay of the first Bessel function for large values of d. It turns out that the same fit function can be used to fit the experimental data, see the solid line in Fig. 21. Furthermore, the interaction law in the presented example can be connected to oscillatorily decaying tails of the filaments. In particular, the theoretical investigation of (23) shows that the stable zeros of the interaction function correspond to stable distances of two interacting DSs with respect to each other. Having determined the interaction function experimentally, one may predict stable distances where the formation of soliton molecules may take place and compare them to directly observed distances in pairs of DSs. It turns out that these distances agree very well at least with the distance d_1 of Fig. 21, where the acceleration due to the interaction is rather strong. Again this is a very good support for the validity of the model equations (23).

Up to this point, only examples were presented in which the filaments preserve their shape and number. However, also generation and annihilation processes of filaments can be observed experimentally, two examples for this kind of phenomena are presented in Fig. 22. Both processes show noticeable similarities to the processes depicted in Sect. 3. In particular, one may note that in the generation process, the third filament is generated at a position such that all filaments lie on a equilateral triangle. This corresponds to the described replication scenario.

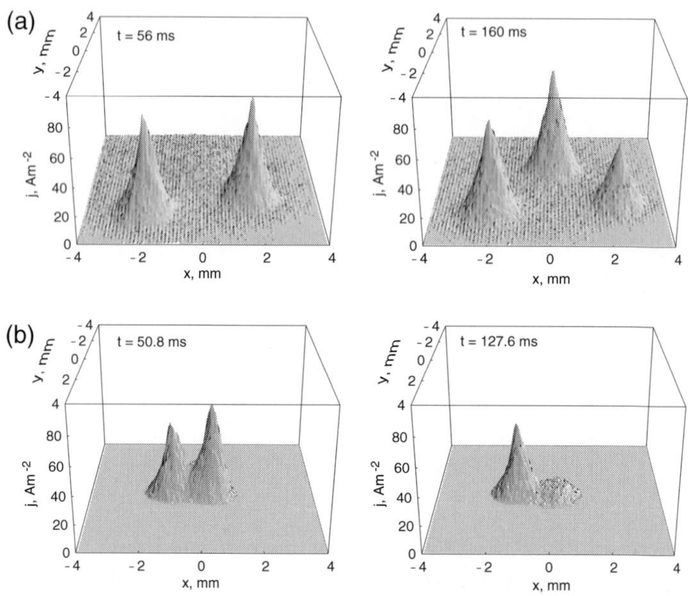

Fig. 22. (**a**) Experimentally observed generation and (**b**). annihilation of a filament, recorded with an exposure time $t_{\mathrm{exp}} = 2 \cdot 10^{-4}\,\mathrm{s}$. Parameters: $U_0 = 3800\,\mathrm{V}$, $\rho_{SC} = 4.14 \cdot 10^6\,\Omega\,\mathrm{cm}$, $R_0 = 20\,\mathrm{M}\Omega$, $p = 290\,\mathrm{hPa}$, $T = 100\,\mathrm{K}$, $d = 500\,\mu\mathrm{m}$

6 Conclusion

Reaction-diffusion systems are suitable to describe a rather large class of systems from the areas of chemistry, physics, geology and even biology. They exhibit various spatially inhomogeneous patterns. Among these patterns, one finds both spatially extended and localized solutions if the form of DSs. In particular, DSs appear for such a wide range of parameters that one may call them a generic type of pattern. DSs are very robust and exhibit fascinating properties in their dynamics and their interaction behavior. Furthermore, they can be considered as constituents of patterns of higher complexity.

Two- and three-component systems of the activator-inhibitor type allow for a paradigmatic description of basic mechanisms for the formation and the dynamics of DSs. Exemplary experimental investigations allow for a verification of many theoretical predictions made for reaction-diffusion systems. In addition, DSs are found also in various other branches of physics as in optics [2, 30, 31], hydrodynamics [32], granular media [33, 34] and semiconductor physics [5, 9]. Although the underlying physical laws in all these systems are rather different, similar behavior of the DSs is found, indicating a certain universality of the observed phenomena. The derivation of order parameter equations for single and multiple DSs in the reduced dynamics, describing the time evolution of the critical modes in terms of ordinary differential equations, is a step to separate from the particular properties of the individual system and to understand more general aspects.

The described investigations give rise to the hope that one day it will be possible to manipulate DSs in a way that they can be used in areas like material preparation, data transmission, image recognition and in many other applications. In addition, nature uses DSs since biological beings came into existence, so that it will be a future challenge to understand the underlying principles also in the field of biology.

Acknowledgement

We greatly acknowledge the support of this work by the Deutsche Forschungsgemeinschaft and by the High Performance Computing Center in Stuttgart. Furthermore, we want to thank Prof. Dr. R. Friedrich for fruitful discussions.

References

1. J. Abshagen, A. Schulz and G. Pfister, *The Taylor-Couette flow: A paradigmatic system for instabilities, pattern formation, and routes to chaos*, Lecture Notes in Physics, Eds.: S. Parisi, S. Müller and W. Zimmermann (Springer, Berlin, 1996).
2. T. Ackemann and W. Lange, Applied Physics B, **72**, 21 (2001).

3. H.-G. Purwins, Yu. A. Astrov, I. Brauer and M. Bode, *Localized Patterns in Planar Gas-Discharge Systems*, RIMS Project 2000: Reaction-Diffusion Systems: Theory and Application. Interfaces, Pulses and Waves in Nonlinear Dissipative Systems, 28–31 August 2000, Research Institute for Mathematical Sciences, Kyoto University, Kyoto, Japan, pp. 1–8, (2001).
4. H.-G. Purwins, Yu. Astrov and I. Brauer, Self-Organized Quasi Particles and Other Patterns in Planar Gas-Discharge Systems, The 5th Experimental Chaos Conference. Orlando, Florida, USA 28 June – 1 July 1999, Eds.: M. Ding, W. L. Ditto, L. M. Pecora and M. L. Spano, (World Scientific, Singapore, 2001) pp. 3–13.
5. E. Schöll, *Nonlinear Spatio-Temporal Dynamics and Chaos in Semiconductors*, Cambridge Nonlinear Science Series, V. 10, (Cambridge University Press, Cambridge, 2001).
6. E. Schöll, F. J. Niedernostheide, J. Parisi, W. Prettl and H.-G. Purwins, *Formation of Spatio-Temporal Structures in Semiconductors*, In: Evolution of Spontaneous Structures in Dissipative Continuous Systems, Eds.: F. H. Busse and S. C. Müller, pp. 446–494, 1998.
7. M. Bode and H.-G. Purwins, Physica D **86**, 53 (1995).
8. V. V. Bel'kov, J. Hirschinger, V. Novák, F. J. Niedernostheide, S. D. Ganichev and W. Prettl, Nature, **397**, 398 (1999).
9. K. Aoki, *Nonlinear Dynamics and Chaos in Semiconductors*, (Institute of Physics Publishing, Bristol and Philadelphia, 2001).
10. H. H. Rotermund, S. Jakubith, A. von Oertzen and G. Ertl, Physical Review Letters, **66**, 3083 (1991).
11. Yu. A. Astrov and H.-G. Purwins, Physics Letters A, **283**, 349 (2001).
12. R. J. Field and R. M. Noyes, Journal of Chemical Physics **60**, 1877 (1974).
13. R. FitzHugh, Biophysical Journal **1**, 445 (1961).
14. P. C. Fife, *Mathematical Aspects of Reacting and Diffusing Systems*, Lecture Notes in Biomathematics, V. 28, (Springer, Berlin, 1979).
15. I. Prigogine and R. Lefever, Journal of Chemical Physics **48**, 1695 (1968).
16. A. M. Turing, Philosophical Transactions of the Royal Society of London Series B – Biological Sciences **237**, (1952).
17. B. Schäpers, M. Feldmann, T. Ackemann and W. Lange, Physical Review Letters, **85**, 748 (2000).
18. K. M. Mayer, J. Parisi and R. P. Huebener, Zeitschrift für Physik B – Condensed Matter **71**, 171 (1988).
19. A. L. Hodgkin and A. F. Huxley, Journal of Physiology **117**, 500 (1952).
20. D. Barkley, Physica D, **49**, 61 (1991).
21. B. S. Kerner and V. V. Osipov, Uspekhi Fizicheskikh Nauk **157**, 201 (1989).
22. B. S. Kerner and V. V. Osipov, *Autosolitons. A New Approach to Problems of Self-Organization and Turbulence*, Fundamental Theories of Physics, V. 61, (Kluwer Acad. Publ., Dordrecht, 1994).
23. F.-J. Niedernostheide, M. Arps, R. Dohmen, H. Willebrand and H.-G. Purwins, Physica Status Solidi B **172**, 249 (1992).
24. H. Meinhardt, *Models of Biological Pattern Formation*, (Academic Press, London, 1982).
25. J. Nagumo, S. Arimoto and S. Yoshizawa, Proceedings of the institute of radio engineers **50**, 2061 (1962).
26. J. D. Murray, *Mathematical Biology*, (Springer, Berlin, 1993).

27. Q. Ouyang and H. L. Swinney, Nature, **352**, 610 (1991).
28. C. I. Christov and M. G. Velarde, Physica D, **86**, 323 (1995).
29. J. J. Tyson, *The Belousov-Zhabotinsky Reaction*, Lecture Notes in Biomathematics, Volume 10, (Springer, Berlin, 1976).
30. M. A. Vorontsov and W. B. Miller, *Self-Organization in Optical Systems and Application in Information Technology*, 2-nd edition, (Springer, Berlin, 1998).
31. F. T. Arecchi, S. Boccaletti and P. Ramazza, Physics Reports, **328**, 1,
32. O. Lioubashevski, H. Arbell, and J. Fineberg, Physical Review Letters, **76**, 3959 (1996).
33. P. B. Umbanhowar, F. Melo, H. L. Swinney, Nature, **382**, 793 (1996).
34. C. Crawford and H. Riecke, Physica D, **129**, 83 (1999).
35. T. Ohta, Physica D, **151**, 61 (2001).
36. Yu. A. Astrov, Physical Review E, **67**, 035203 (2003).
37. C. P. Schenk, P. Schütz, M. Bode and H.-G. Purwins, Physical Review E, **57**, 6480 (1998).
38. P. Coulett, C. Riera and C. Tresser, Physical Review Letters, **84**, 3069 (2000).
39. J. Smoller, *Shock Waves and Reaction Diffusion Equations*, Second Edition, (Springer, New York, 1994).
40. S. Koga and Y. Kuramoto, Progress of Theoretical Physics, **63**, 106 (1980).
41. T. Ohta, M. Mimura and R. Kobayashi, Physica D, **34**, 115 (1989).
42. R. Schmeling, *Experimentelle und numberische Untersuchungen von Strukturen in einem Reaktions-Diffusions-System anhand eines elektrischen Netzwerkes*, Institut für Angewandte Physik, Westfälische Wilhelms-Universität Münster, Dissertation, 1994.
43. Mimura, M. and Nagayama, M., Methods and Applications of Analysis oder Tohoko Mathematical Publications, **8**, 239 (1997).
44. Mimura, M. and Nagayama, M. and Ikeda, H. and Ikeda, T., Hiroshima Mathematical Journal, **30**, 221 (1999).
45. R. J. Field and M. Burger, *Oscillations and traveling waves in chemical systems*, (Wiley, New York, 1985).
46. Y. Nishiura and D. Ueyama, Physica D, **130**, 73 (1999).
47. A. Gierer and H. Meinhardt, Kibernetik, **12**, 30 (1972).
48. G. Bordiougov and H. Engel, Physical Review Letters, **90**, 148302 (2003).
49. C. P. Schenk, M. Or-Guil, M. Bode and H.-G. Purwins, Physical Review Letters, **78**, 3781 (1997).
50. M. Bode, A. W. Liehr, C. P. Schenk and H.-G. Purwins, Erratum to [Physica D **161**, 45 (2002)], Physica D, **165**, 127 (2002).
51. M. Bode, A. W. Liehr, C. P. Schenk and H.-G. Purwins, Physica D, **161**, 45 (2002).
52. C. P. Schenk, *Numberische und analytische Untersuchung solitärer Strukturen in zwei- und dreikomponentigen Reaktions-Diffusions-Systemen*, Institut für Angewandte Physik, Westfälische Wilhelms-Universität Münster, Dissertation, 1999.
53. A. W. Liehr, A. S. Moskalenko, M. C. Röttger, J. Berkemeier and H.-G. Purwins, Replication of Dissipative Solitons by Many-Particle Interaction, In: *High Performance Computing in Science and Engineering '02. Transactions of the High Performance Computing Center Stuttgart (HLRS) 2002*, Eds.: E. Krause and W. Jäger, pp. 48–61 (Springer, Berlin, 2003).
54. A. S. Moskalenko, A. W. Liehr and H.-G. Purwins, Europhysics Letters, **63**, 361 (2003).

55. S. V. Gurevich, H. U. Bödeker, A. S. Moskalenko, A. W. Liehr and H.-G. Purwins, Drift Bifurcation of Dissipative Solitons due to a Change of Shape, *Proceedings of the International Conference of Physics and Control St. Petersburg*, pp. 601–606, (IEEE, 2003).
56. S. V. Gurevich, H. U. Bödeker, A. S. Moskalenko, A. W. Liehr and H.-G. Purwins, Physica D, **199**, 115 (2004).
57. H. Haken, *Synergetics, An Introduction*, 3rd edition, Springer Series Synergetics, (Springer, Berlin, Heidelberg, New York, 1983).
58. R. Friedrich, Zeitschrift für Physik B – Condensed Matter, **90**, 373 (1993).
59. M. Bode, Physica D, **106**, 270 (1997).
60. E. Ammelt, Yu. A. Astrov and H.-G. Purwins, Physical Review E, **58**, 7109 (1998).
61. A. Moskalenko, *Dynamische gebundene Zustände und Drift-Rotations-Dynamik von dissipativen Solitonen*, Institut für Angewandte Physik, Westfälische Wilhelms-Universität Münster, 2002.
62. Y. P. Raizer, *Gas Discharge Physics*, 2-nd edition, (Springer, Berlin, 1997).
63. M. S. Benilov, Physical Review E, **48**, 5901 (1992).
64. M. S. Benilov, Physical Review A, **45**, 506 (1993).
65. Y. B. Golubovskii, I. A. Porokhova, A. Dinklage and C. Wilke, Journal of Physics D – Applied Physics, **33**, 517 (2000).
66. Yu. Astrov, E. Ammelt, S. Teperick and H.-G. Purwins, Physics Letters A, **211**, 184 (1996).
67. Yu. A. Astrov, Yu. A. Logvin, Physical Review Letters, **79**, 2983 (1997).
68. E. L. Gurevich, A. S. Moskalenko, A. L. Zanin, Yu. A. Astrov and H.-G. Purwins, Physics Letters A, **307**, 299 (2003).
69. I. Brauer, M. Bode, E. Ammelt and H.-G. Purwins, Physical Review Letters, **84**, 4104 (2000).
70. C. Strümpel, Yu. A. Astrov and H.-G. Purwins, Physical Review E, **65**, 066210 (2002).
71. C. Strümpel, H.-G. Purwins and Yu. A. Astrov, Physical Review E, **63**, 026409 (2001).
72. H. U. Bödeker, M. C. Röttger, A. W. Liehr, T. D. Frank, R. Friedrich and H.-G. Purwins, Physical Review E, **67**, 056220 (2003).
73. H. U. Bödeker, A. W. Liehr, M. C. Röttger, T. D. Frank, R. Friedrich and H.-G. Purwins, New Journal of Physics, **6**, 62 (2004).
74. A. W. Liehr, H. U. Bödeker, M. C. Röttger, T. D. Frank, R. Friedrich and H.-G. Purwins, New Journal of Physics, **5**, 89.1 (2003).
75. U. Erdmann, W. Ebeling, L. Schimansky-Geier and F. Schweitzer, European Physical Journal B, **15**, 105 (2000).

Discrete Ginzburg-Landau Solitons

N.K. Efremidis and D.N. Christodoulides

School of Optics, Center for Research and Education for Laser and Optics
University of Central Florida, 4000 Central Florida Blvd., Orlando, FL., U.S.A.
nefrem@mail.ucf.edu
demetri@creol.ucf.edu

Abstract. In this chapter, we present a review of recent results concerning dissipative lattices of the Ginzburg-Landau type. Firstly, we study effects such as complex discrete diffraction, as well as Bloch oscillations, arising from the linear properties of such systems. Subsequently, using a generic cubic-quintic nonlinearity, we identify self-localized dissipative discrete soliton solutions, and study their characteristics.

1 Introduction

The complex Ginzburg-Landau (GL) equation is known to play a ubiquitous role in science. This equation is encountered in several diverse branches of physics, for example in superconductivity and superfluidity, non-equilibrium fluid dynamics and chemical systems, nonlinear optics and Bose-Einstein condensates [1, 2, 3, 4, 5, 6]. In general, the very nature of the GL system is such that it can provide rich behavior, ranging from chaos and pattern formation to self-localized solutions or solitons. In the latter regime, GL dissipative solitons (or auto-solitons) are possible as a result of the interplay between linear/nonlinear gain, nonlinearity and complex dispersion [7, 8, 9, 10]. Over the years, the soliton solutions of the Ginzburg-Landau equation and their underlying dynamics have been the subject of intense investigation. Such pulse-like soliton states were first identified within the context of the cubic GL model [7, 8], and subsequently in the generalized quintic regime [4, 9, 10], and, typically, they represent chirped coherent structures (or one-dimensional defects) that are obtained through heteroclinic trajectories in the phase space of the stationary GL equation. Several types of solitary wave solutions of the continuous GL equation have been identified over the last few years. These include flat-top solutions in one and two dimensions [11, 12], erupting (exploding) and creeping solitons [13, 12] and spiraling solitons carrying topological charge [14, 15].

Quite recently, the behavior of nonlinear discrete optical systems has received considerable attention [16, 17]. Because of discreteness, the properties of this class of systems are artificially altered, and, as a result, their nonlinear behavior is known to exhibit features that are otherwise impossible in the bulk/continuous regime. In optics [16], nonlinear arrays of nearest-neighbor-coupled waveguides have provided a fertile ground where such discrete

N.K. Efremidis and D.N. Christodoulides: *Discrete Ginzburg-Landau Solitons*,
Lect. Notes Phys. **661**, 309–325 (2005)
www.springerlink.com

nonlinear interactions of the discrete nonlinear Schrödinger-type can be experimentally observed and investigated [18]. In the presence of loss or gain, these lattices become dissipative, and can be described by discrete Ginzburg-Landau (DGL) type models. Such DGL lattices are quite often used to describe a number of physical systems, such as Taylor and frustrated vortices in hydrodynamics [19] and laser arrays and semiconductor optical amplifiers in optics [20, 21]. In these latter studies, the DGL model has been predominantly used in connection with spatio-temporal chaos, instabilities and turbulence [22]. However, discrete soliton (DS) solutions of the DGL model have not been identified until very recently [23]. In addition, discrete solitons have been identified in dissipative lattices of the Ablowitz-Ladik type [24], as well as in optical cavities [25].

In this chapter, we present a review concerning the linear and nonlinear properties of Ginzburg-Landau discrete chains [23, 26]. More specifically, we first analyze the complex dispersion properties within the Brillouin zone. As a result of discreteness, the dispersion/diffraction behavior of a GL lattice differs substantially from that encountered in conservative arrays [16]. In addition, we study the effect of a linear refractive index modulation in a low-power system of active waveguides (such as laser arrays or semiconductor optical amplifier lattices), and study Bloch oscillations of the complex type. In optics, Bloch oscillations have been predicted [27] and experimentally demonstrated in waveguide arrays [28, 29]. One would naturally expect that the presence of gain and loss would prevent the existence of periodic Bloch oscillations. However, as it is shown, Bloch oscillations are still possible, even though gain and loss mechanisms allow energy exchange in this system [26], and in spite of the fact that, in general, the total power will increase or decrease exponentially. For a specific choice of the parameters, the total gain compensates for the loss, and this results, on average, in a constant power envelope. This is in contrast to regular lattices with zero refractive index modulation, where either the in-phase or the π-out-of-phase evanescently-coupled modes (corresponding to the base and the edge of the first Brillouin zone) are preferentially amplified. In the nonlinear regime, we study the properties of discrete soliton solutions [23]. In general, two new types of DGL solitons can exist under the same conditions. These solutions are located either at the base or at the edge of the Brillouin zone, and they bifurcate at different values of the linear gain. Due to discreteness, the system exhibits novel features that have no counterparts whatsoever in either the continuous GL limit or in other conservative discrete models [16, 18] [as in discrete nonlinear Schrödinger (DNLS) chains]. These include, for example, on-site and intra-site bright DGL solitons that can both be stable, as well as new bifurcation types that cannot be identified in the continuous case. Approximate discrete soliton solutions, which are in excellent agreement with the numerically-found ones, are also obtained. These solutions are valid when the discrete solitons are quite broad (occupying many lattice sites), or narrow, i.e., when most of the power is essentially located in a few lattice sites.

2 Formulation

In general, the cubic-quintic DGL equation is given by

$$\mathrm{i}\dot{u}_n - \mathrm{i}\epsilon u_n + \alpha(u_{n+1} + u_{n-1}) + p|u_n|^2 u_n + q|u_n|^4 u_n = 0 \, , \qquad (1)$$

where $p = p_r + \mathrm{i}p_i$, $q = q_r + \mathrm{i}q_i$, $\alpha = \alpha_r + \mathrm{i}\alpha_i$, ϵ is a real parameter and $\dot{u}_n$ is the partial derivative of u_n with respect to space or time (depending on the specific problem). Here, without loss of generality, we will assume that $\dot{u}_n = \mathrm{d}u_n/\mathrm{d}z$. Physically, the discretization in (1) occurs by applying the tight-binding approximation (or coupled mode theory) [21, 30]. The original continuous system, which is periodic in space, is expanded into local modes whose amplitudes are described by the corresponding discrete model. In (1), α_r accounts for the energy tunneling between adjacent elements of the lattice, while its imaginary part stands for gain/loss due to coupling. The real parts of p and q represent the strengths of the cubic and quintic nonlinearities of the system, while ϵ, p_i, q_i are the linear and nonlinear gain/loss coefficients.

Within the context of nonlinear optics, the DGL equation can arise in the description of semiconductor laser arrays [21] and in semiconductor optical amplifiers [31] (due to applying a refractive index modulation along the crystal), where the quintic term can account for the gain/nonlinearity saturation of the lasing medium. The DGL equation can also describe the dynamics of an open Bose-Einstein condensate. In this case, the lattice potential is created by the interference of two optical standing waves [32] and, in this situation, solitons of the discrete nonlinear Schrödinger type are known to exist [33]. The dissipation of the Bose-Einstein condensate naturally occurs in an open condensate, while gain can result from the interaction between the condensed and the uncondensed atoms [34, 35]. Note that, in all of these cases, it is common practice to write the original saturable nonlinearity of the system in terms of a cubic-quintic expansion, and this, in turn, conveys the fundamental properties of the original model [21, 35].

3 Linear Properties

3.1 Discrete Diffraction

We begin our analysis by considering the linear dispersion properties of the DGL equation. To do so, we write

$$u_n \propto \exp(\mathrm{i}kz - \mathrm{i}\theta n) \, ,$$

where $k = k_r + ik_i$ is the complex propagation wave-number and θ is the wave momentum inside the lattice. The real and imaginary parts of (1) are then found to satisfy

$$k_r = 2\alpha_r \cos\theta \ , \tag{2}$$

$$k_i = -\epsilon + 2\alpha_i \cos\theta \ . \tag{3}$$

In Fig. 1, k_r and k_i are plotted as functions of θ. Equation (2) describes the dispersive properties of the lattice within the Brillouin zone, as defined in the region $|\theta| \leq \pi$. When $|\theta| < \pi/2$ and $\alpha_r > 0$, the curvature of the dispersion relation [(2)] implies that the effective diffraction of the array is normal. On the other hand, when $\pi/2 < |\theta| < \pi$ (and $\alpha_r > 0$), the effective diffraction of the array becomes anomalous, and, of course, these regimes are reversed for $\alpha_r < 0$.

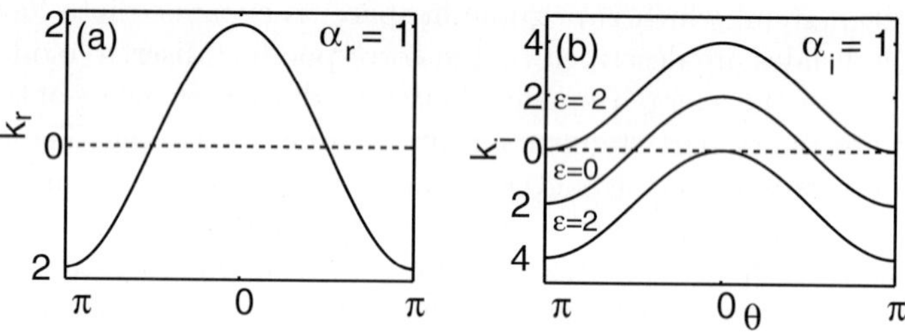

Fig. 1. (**a**) The dispersion curve for the array when $\alpha_r = 1$, and (**b**) k_i (associated with the instability growth rate) as a function of θ for $\alpha_i = 1$ and $\epsilon = -2, 0, 2$

The imaginary part of the propagation wave-number [(2)] is directly related to the growth rate of the perturbations of the "zero amplitude" solution. In particular, any perturbation frequency, θ, that satisfies the condition $k_i < 0$, will grow exponentially with a growth rate

$$g_d(\theta) = \epsilon - 2\alpha_i \cos\theta \ . \tag{4}$$

We emphasize that this growth rate is a periodic function of θ with period 2π, i.e., $g_d(\theta + 2\pi n) = g_d(\theta)$, where n is an integer. From all the possible frequencies, θ, within the Brillouin zone, only those that satisfy the inequality $\epsilon > 2\alpha_i \cos\theta$ will eventually develop instabilities. Therefore, the "zero amplitude" solution is absolutely stable for

$$\epsilon < -2|\alpha_i| \ . \tag{5}$$

On the other hand, when $\epsilon > 2|\alpha_i|$, every frequency is amplified, and the maximum growth rate (which occurs either at the base or at the edge of the Brillouin zone) is given by $\epsilon + 2|\alpha_i|$. In the regime between the two aforementioned cases, i.e., when $-2|\alpha_i| < \epsilon < 2|\alpha_i|$, only a subset of the frequencies (i.e. those satisfying $\epsilon > 2\alpha_i \cos\theta$) will be amplified, while the rest of them will decay. Note that the instability behavior of the "zero amplitude" solution of the DGL is fundamentally different from that arising in the continuous GL limit; it is described by (Sect. 4.1)

$$g_c(\theta) = \epsilon' + \alpha_i \theta^2 \ . \tag{6}$$

Apparently, in the continuous limit, the stability properties are strongly affected by the sign of the diffusion coefficient, α_i, and, in fact, for $\alpha_i > 0$, the "zero amplitude" solution will always be unstable, regardless of the value of ϵ. Thus, in order to stabilize the background of a self-localized state, it is essential to have $\alpha_i < 0$. On the other hand, the DGL model has the interesting property that the background can be stabilized for both signs of α_i by choosing the linear gain ϵ of the system in an appropriate manner.

The linear part of (1) also supports exponentially-decaying/growing solutions of the form

$$u_n \propto \exp(\pm sn + \mathrm{i}kz) \ , \tag{7}$$

where the parameters of (7) obey

$$k = 2\alpha_r \cosh s_r \cos s_i - 2\alpha_i \sinh s_r \sin s_i \ , \tag{8}$$

$$\epsilon = 2\alpha_r \sinh s_r \sin s_i + 2\alpha_i \cosh s_r \cos s_i \ . \tag{9}$$

In the above equations k is taken to be real and $s = s_r + is_i$ to be complex. We would like to mention that, in the linear regime, (1) can be solved analytically. When only one lattice element is initially excited (say $n = 0$ at $z = 0$), the field profile at z is given by

$$u_n(z) = u_0 J_n(2\alpha z) \mathrm{e}^{\mathrm{i}\pi n/2} \mathrm{e}^{\epsilon z} \ , \tag{10}$$

where $J_n(x)$ (with complex argument) is a Bessel function of the first kind with integer order n. The evolution of more involved initial field patterns can be readily obtained by simple superposition from this impulse response. In Fig. 2a–d, the impulse response of the lattice at $z = 3$ is depicted for $\alpha = 1, 1+i, i, -i$, respectively. In Fig. 2c, the out-of-phase mode is preferentially amplified, while in Fig. 2d, the mode is in-phase at the output.

3.2 Bloch Oscillations

Bloch was the first to show that an electron wavefunction has the tendency to periodically revive when an external potential is applied to the periodic lattice of a solid [36]. In optics, Bloch oscillations were predicted [27] and have been experimentally demonstrated in waveguide arrays [28, 29]. Let us introduce a linear refractive index modulation in a waveguide array or an additional linear potential in a Bose-Einstein condensate. In this case, the evolution of a low power field will obey

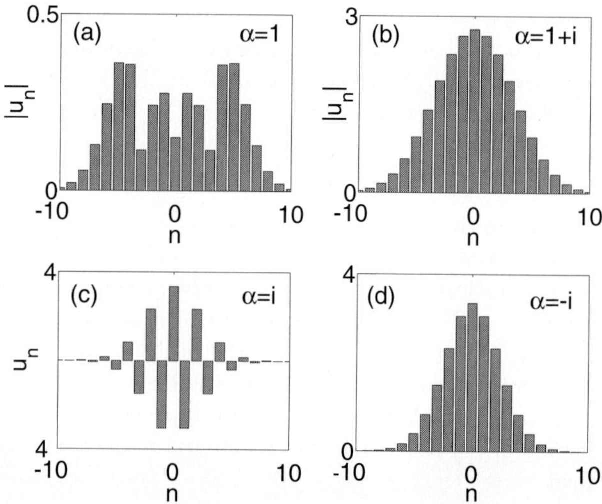

Fig. 2. Field profile of linear impulse response of the lattice at $z = 3$. In (**a**), $\epsilon = 0$, while in (**b**)–(**d**), $\epsilon = -1$

$$\mathrm{i}\dot{u}_n - \mathrm{i}\epsilon u_n + \alpha(u_{n+1} + u_{n-1}) + \beta n u_n = 0 \, , \tag{11}$$

where β is proportional to the modulation of the lattice. Since α_r represents the nearest-neighbor element coupling coefficient among the fundamental (zero-node) local modes, it is natural to assume that $\alpha_r > 0$.

We consider the possibility of Bloch oscillations in such a dissipative lattice. Extending the analysis of [27] to the complex domain, we find an exact solution for the impulse response of (11):

$$u_n(z) = J_n \left[\frac{4\alpha}{\beta} \sin\left(\frac{\beta z}{2}\right)\right] \exp\left[\frac{\mathrm{i}n}{2}(\beta z + \pi)\right] \mathrm{e}^{\epsilon z} \, . \tag{12}$$

Here, $J_n(x)$ is a Bessel function with a complex argument. Apart from the last exponential term in (12), the intensity of this wavefunction repeats itself after

$$Z_0 = 2\pi/\beta \, . \tag{13}$$

Apparently, the evolution of more complicated wavepackets will be given from superpositions of (12). The last term in (12) indicates that, on average, the power

$$P = \sum_n |u_n|^2 \tag{14}$$

will increase or decrease according to

$$\langle P \rangle \sim \exp(2\epsilon z) \, . \tag{15}$$

Notice that the average in (15) is taken over one period of the oscillations, Z_0:

$$\langle P \rangle = \frac{1}{Z_0} \int_z^{z+Z_0} P(z) dz \; . \tag{16}$$

In order to better understand the effect of the linear tilt displayed in (11), we will compare these results with the zero-tilt case. If $\beta = 0$, (11) has a complex dispersion curve allowing plane-wave solutions, as shown in the previous section. Obviously, under such conditions, where some modes are amplified more than others, periodic revivals of an input intensity pattern cannot occur.

If β is non-zero, the behavior of the system is very different. In Fig. 3, we can see the intensity evolution of a single waveguide excitation. The conservative case depicted on the left-hand-side is identical to the one discussed in [27]. Since there are no dissipation processes occurring in this case, the power is conserved with z. For the right-hand-side diagram, the coefficients are complex, i.e., $\alpha = 1 \pm \mathrm{i}0.4$ and $\epsilon = 0$, and so the total power is not locally conserved. The total power of the beam initially increases and attains a maximum at $z = Z_0/2$, but after that it decreases and the initial waveform recurs at $z = Z_0$. If ϵ is non-zero, the input beam will still demonstrate a similar behavior, apart from an average growth in its power, given by (15).

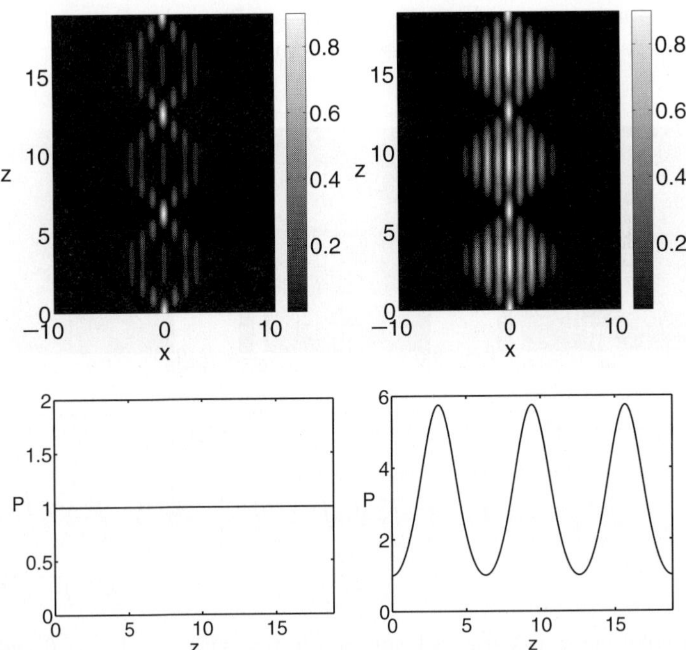

Fig. 3. Evolution of a single waveguide excitation when $\epsilon = 0$, $\beta = 1$. For the *left-hand-side* diagram, $\alpha = 1$, while for the *right-hand-side* one, $\alpha = 1 \pm \mathrm{i}0.4$

Qualitatively, this behavior can be understood on the basis that β causes the Bloch momentum θ to periodically scan the complex dispersion curves. As θ periodically oscillates in the Brillouin zone, the input beam experiences both normal and anomalous dispersion. In a similar way, due to the change in the value of θ, the growth rate is also periodically modulated. Interestingly enough, and since the impulse response excites the whole spectrum equally, the intensity profile is identical for both signs of α_i.

Another example is depicted in Fig. 4, where an initially-in-phase Gaussian beam of the form

$$u_n = \exp(-((n - n_0)/A)^2) \tag{17}$$

is launched into the lattice with $A = 2$. In this example, $\epsilon = 0$. On the left-hand-side, $\alpha = 1$ and so the power, P, remains invariant along z. In the middle plots, $\alpha = 1 + \mathrm{i}0.3$. Again, the $P - z$ diagram can be qualitatively explained. Since the initial configuration is in-phase, and $\alpha_i > 0$, the beam will initially experience loss. After some propagation length, the out-of-phase pattern grows and the power starts to increase with z. After approximately half a period, the phase of the pattern will shift and will have a tendency to become in-phase, thus experiencing losses. After one period, the input wavepacket is restored. The situation is different in the right-hand-side diagram, where $\alpha = 1 - \mathrm{i}0.3$. Since $\alpha_i < 0$, the in-phase patterns will have a tendency to gain power. Thus, as we can see from the $P - z$ diagram, the power increases with z when the beam is launched.

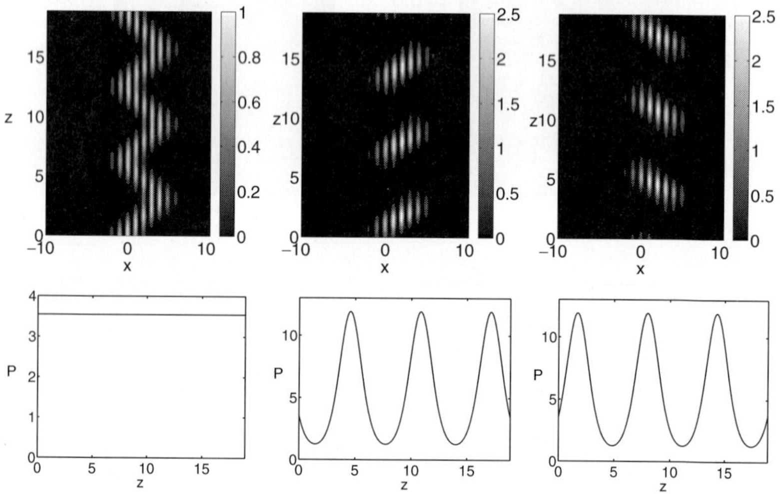

Fig. 4. Evolution of a Gaussian beam in a lattice with $\beta = 1$, $\epsilon = 0$ and $A = 2$, when (**a**) $\alpha = 1$ (*left-hand-side*), (**b**) $\alpha = 1 + \mathrm{i}3$ (*middle column*), and (**c**) $\alpha = 1 - \mathrm{i}3$ (*right-hand-side*)

This refractive index ladder also affects the stability of the zero background. Such an investigation is of interest when studying the stability of self-confined dissipative solitons. If $\beta = 0$, then the zero background should be stable for all values of θ, and so the necessary and sufficient stability condition is $\epsilon < -2|\alpha_i|$ [(5)]. However, if β differs from zero, then, due to the periodic variations of θ inside the Brillouin zone, stability can be established when the average growth is negative,

$$g_{\text{avg}} = \int_0^{\pi} g(\theta) d\theta = 2\pi\epsilon < 0 , \tag{18}$$

or

$$\epsilon < 0 . \tag{19}$$

By comparing the stability conditions, (5) and (19), it becomes apparent that a refractive index ladder improves the stability of the background.

4 Nonlinear Properties

4.1 Long Wavelength Regime

When a nonlinear wave is broad enough, i.e., when its envelope varies slowly with the index n, the long wavelength approximation can be applied. By studying the resulting continuous equation, many of the properties of the original discrete model, and especially those related to the Hopf bifurcation of the DS with the "zero amplitude" solution, can be revealed. Initially, we apply a phase transformation to the field,

$$u_n = v_n \exp(\mathrm{i}\theta n) , \tag{20}$$

resulting in

$$\mathrm{i}\dot{v}_n - \mathrm{i}\epsilon v_n + \alpha(u_{n+1}\mathrm{e}^{\mathrm{i}\theta} + u_{n-1}\mathrm{e}^{-\mathrm{i}\theta}) + p|u_n|^2 u_n + q|u_n|^4 u_n = 0 , \tag{21}$$

where θ is the linear phase difference between adjacent elements of the lattice. We can associate the discrete field, v_n, with a continuous function $v(x)$:

$$x_n = n\Delta x , \quad v(x_n) = v_n . \tag{22}$$

Without loss of generality, we will assume that $\Delta x = 1$. Of course, the solution of the original problem (21) can be reconstructed by sampling the continuous field $v(x)$ at the discrete points x_n. The field at lattice elements adjacent to n can be found be using the following Taylor expansion:

$$v_{n\pm1} = \sum_{n=0}^{\infty} \frac{(\pm 1)^n}{n!} \frac{\partial^n v(x)}{\partial x^n} . \tag{23}$$

Substituting (23) into (21), we find

$$\mathrm{i}\frac{\partial v}{\partial z} - \mathrm{i}\epsilon v + 2\alpha \sum_{n=0}^{\infty} \left[\frac{\cos\theta}{(2n)!} \frac{\partial^{2n} v}{\partial x^{2n}} + \frac{\mathrm{i}\sin\theta}{(2n+1)!} \frac{\partial^{2n+1} v}{\partial x^{2n+1}} \right] + p|v|^2 v + q|v|^4 v = 0 \,. \tag{24}$$

Equation (24) is equivalent to the following integro-differential equation:

$$\mathrm{i}\frac{\partial v}{\partial z} - \mathrm{i}\epsilon v + \frac{\alpha}{\pi} \int_{-\infty}^{\infty} dx' \int_{-\infty}^{\infty} dq \cos(\theta - q) e^{iq(x'-x)} v(x') + p|v|^2 v + q|v|^4 v = 0 \,. \tag{25}$$

We would like to mention that (25) is quite involved, and very little information about the dynamics of the DGL system can easily be derived from it. For this reason, we will further simplify (24) by employing the long-wavelength approximation.

In this latter regime, we assume that the discrete field varies quite slowly with the lattice index, i.e.,

$$|v_{n\pm1} - v_n| \ll 1 \,. \tag{26}$$

Utilizing (23) allows us to write the corresponding condition for the envelope:

$$\left| \sum_{n=1}^{\infty} \frac{(\pm 1)^n}{n!} \frac{\partial^n v(x)}{\partial x^n} \right| \ll 1 \,. \tag{27}$$

Assuming that v varies slowly with x as $v = v(\sigma x)$, where $\sigma \ll 1$, keeping terms up to the second order in σ, and applying the transformation $v(x,z) = U(x,z)\exp(i\alpha_r x)$, we arrive at:

$$\mathrm{i}\left(\frac{\partial U}{\partial z} + c\frac{\partial U}{\partial x} \right) - \mathrm{i}\epsilon' U + \alpha' \frac{\partial^2 U}{\partial x^2} + p|U|^2 U + q|U|^4 U = 0 \,, \tag{28}$$

where

$$\epsilon' = \epsilon - 2\alpha_i \cos\theta \,, \tag{29}$$

$$c = 2\alpha \sin\theta \,, \tag{30}$$

end

$$\alpha' = \alpha \cos\theta \,. \tag{31}$$

Changing the co-ordinate system to

$$X = x - cz, \quad Z = z \,, \tag{32}$$

we obtain the following cubic-quintic Ginzburg-Landau equation:

$$\mathrm{i}\frac{\partial U}{\partial Z} - \mathrm{i}\epsilon' U + \alpha' \frac{\partial^2 U}{\partial X^2} + p|U|^2 U + q|U|^4 U = 0 \,. \tag{33}$$

The solutions of Equation (33) will be only considered either at the base of the Brillouin zone ($\theta = 0$) or at the edges ($\theta = \pm\pi$) where the group velocity c of equation (30) is zero. Equation (33) can help us understand the Hopf bifurcations of the soliton solutions with zero background. Exact soliton solutions of (33) were first found in [10]. Since we are interested in low-intensity solutions, it is natural to assume that the quintic term is negligible. Under this assumption, the bright soliton solution of (33) reduces to [7]

$$U(X, Z) = A\,[\mathrm{sech}(\eta X)]^{1+\mathrm{i}\mu}\,\mathrm{e}^{\mathrm{i}kZ}\;, \tag{34}$$

where

$$\mu = \frac{3b \pm \sqrt{9b^2 + 8a^2}}{2a}\;, \tag{35}$$

$$\eta^2 = \frac{\epsilon}{2\mu\alpha_r + \alpha_i(1-\mu^2)}\;, \tag{36}$$

$$k = \eta^2[\alpha_r(1-\mu^2)2\mu\alpha_i]\;, \tag{37}$$

$$A^2 = -\frac{3\mu\left(\alpha_r^2 + \alpha_i^2\right)}{a}\eta^2\;, \tag{38}$$

$$a = p_r\alpha_i - p_i\alpha_r\;, \quad b = p_r\alpha_r - p_i\alpha_i\;. \tag{39}$$

Note that the plus/minus signs in (35) correspond to opposite signs of μ, of which only one satisfies the constraints $\eta^2 > 0$ and $A^2 > 0$. Using (38), we can see that the pulse width, $1/\eta$, of a soliton is inversely proportional to its amplitude A, and this, in turn, confirms our original assumption. Again, we would like to mention that the expressions derived here are valid for solutions that are sufficiently broad.

4.2 Highly-Confined Discrete Solitons

At the other extreme, i.e., when the discrete solitons are highly confined inside the lattice, these solutions are accurately described by

$$u = u_0 \exp(-s|n| + \mathrm{i}\lambda z)\;, \tag{40}$$

where the parameters of the solution satisfy

$$\lambda + \mathrm{i}\epsilon = 2\alpha\cosh s \tag{41}$$

and

$$\sinh s = u_0^2\left(p + qu_0^2\right)/(2\alpha)\;. \tag{42}$$

In addition, one can obtain valuable information from the DS tails, since locally, $u_n \propto \exp(i\lambda z - s|n|)$ is satified. In this case, the relations/constraints of (8)–(9) hold true.

Using a perturbation method for the cubic model, one can accurately estimate:

$$u_0^2 = \left(\sqrt{\frac{\epsilon}{p_i}} - \sqrt{\frac{p_i}{\epsilon}}\frac{2p_r\alpha_r\alpha_i + p_i(\alpha_i^2 - \alpha_r^2)}{\epsilon(p_r^2 + p_i^2)}\right)^2, \tag{43}$$

$$k = \frac{p_r\epsilon}{p_i} - 4\frac{\alpha_r\alpha_i(p_r^2 - p_i^2) + p_r p_i(\alpha_i^2 - \alpha_r^2)}{\epsilon(p_r^2 + p_i^2)} \tag{44}$$

$$u_{\pm 1} = \frac{\alpha}{p}\sqrt{\frac{p_i}{\epsilon}} \tag{45}$$

while similar expressions can be obtained for the quintic model. Again, for asymptotically-large values of the maximum intensity, $|u_0|^2$ is linearly-related to ϵ.

4.3 Discrete Solitons and Their Bifurcations

We will now investigate the structure, as well as some of the basic properties, of DS states existing in the GL lattice. To do so, we look for stationary localized modes, $u_n = \exp(i\lambda z)v_n$, of (1). Then the resulting algebraic system is solved numerically using the Newton iteration method. Note that (1) is subject to certain symmetries that can be used to reduce the parameter space of the system under study. We note that, by employing the phase transformation $u_n \to u_n \exp(i\pi n)$, along with $\alpha \to -\alpha$, (1) remains invariant. This, in turn, allows the one-to-one mapping $\alpha \leftrightarrow -\alpha$. Here, without any loss of generality, we assume that $\alpha_r > 0$. A second symmetry also exists, *viz.* $z \to -z$, and $u_n \to \exp(i\pi n)u_n$, $p \to -p$, $q \to -q$, which is used by applying an additional constraint, $p_r > 0$, to the system. In doing so, we consider immobile GL discrete solitons that reside either at the base ($\theta = 0$), or at the edge ($\theta = \pi$), of the Brillouin zone [37]. For $\theta = 0$, the DS bifurcates from the zero solution at $\epsilon = 2\alpha_i$, while in the case $\theta = \pi$, the DS bifurcates at $\epsilon = -2\alpha_i$. It is important to note that the existence of these two bifurcation points is a result of the periodicity introduced by the lattice model, and is in clear contrast to the continuous GL equation, where only one bifurcation occurs when the linear gain is zero.

In Fig. 5, typical co-dimension 1 bifurcation diagrams of the cubic DGL model are depicted. Figures 5a and c show bifurcations for DS located at the base of the Brillouin zone, whereas the curves of Fig. 5b and d correspond to the edge of the zone. As in the continuous GL case, these bifurcations are super-critical when $p_i > 0$ and sub-critical when $p_i < 0$. In all the bifurcation figures, solid and dash-dotted curves represent stable and unstable branches, respectively.

Close to the bifurcation point, the DS are quite broad, and, as a result, the numerically-found curves shown in Fig. 5 (with either $\theta = 0$ or $\theta = \pi$) are well-approximated by (33). Also, when the maximum intensity becomes relatively high, the solutions residing in the normal diffraction regime become highly localized inside the lattice.

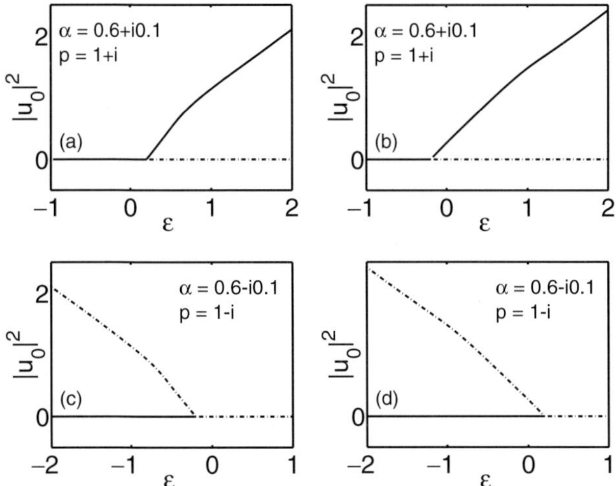

Fig. 5. Typical co-dimension 1 bifurcation diagrams of the cubic DGL model. In all the bifurcation figures, *solid* and *dash-dotted* curves represent stable and unstable branches, respectively

We next investigate DS solutions of the cubic-quintic DGL. Figure 6a depicts such a highly-confined DS (at the base of the Brillouin zone), which is in excellent agreement with the analytical results of (40). On the other hand, in the anomalous discrete diffraction regime (at the edge of the Brillouin zone), a rather peculiar feature arises, *viz.* that the amplitude profile becomes broader and flatter with stronger chirp. We attribute this property to the rather involved energy flow within the GL DS under anomalous diffraction conditions. As an example, Fig. 6b shows the field of a high intensity DS in the anomalous diffraction regime that extends over seven lattice points. Similar types of solutions (flat-top) can also be found in the continuous Ginzburg-Landau model [11, 12]. On the other hand, these discrete flat-top solutions exist when the maximum intensity of the solutions is above a certain threshold, and their stability properties may be relevant to the modulational instability of the corresponding continuous-wave solutions. However, the possible bifurcation of these solutions is an issue that merits further investigation.

In general, the stability properties of the cubic DGL solitons can be identified from the corresponding bifurcation diagrams. A sub-critical DS will be unstable, whereas an unstable background can destabilize a super-critical DS. The properties of the bifurcation diagrams can be modified when the quintic term in (1) is non-zero. In fact, a saddle-node bifurcation emerges if the condition $p_i q_i < 0$ is satisfied. It is then of interest to determine the parameter space where absolutely stable DS exist. To realize this, it is necessary that the zero background, on which these discrete modes reside, is stable for any perturbation frequency, i.e., $\epsilon < -2|\alpha_i|$. By taking this into account,

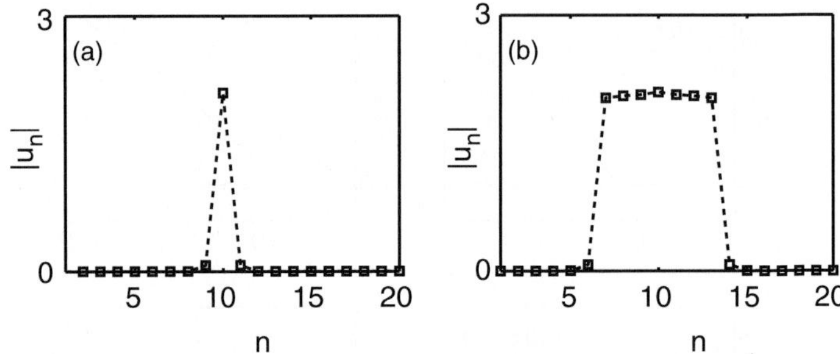

Fig. 6. (**a**) A highly-confined DS solution in the normal diffraction regime, and (**b**) the corresponding DS that resides at the edge of the Brillouin zone, when $\alpha = 0.1 - 0.2\mathrm{i}$, $p = 1 - \mathrm{i}0.8$, $q = 0.1 + 0.1\mathrm{i}$, $u_0 = 2.1$

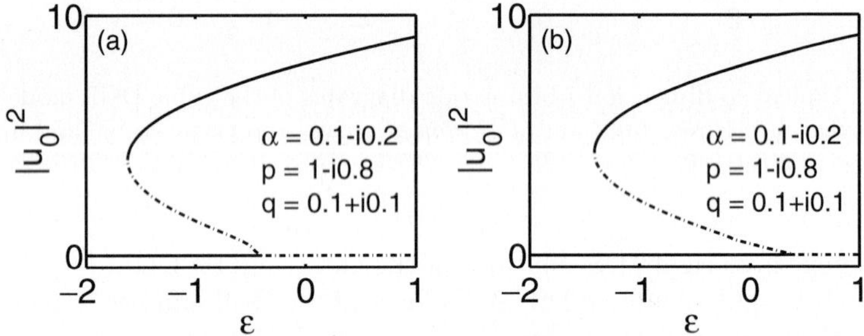

Fig. 7. Bifurcation diagrams of the quintic DGL equation (**a**) at the base and (**b**) at the edge of the Brillouin zone

and knowing, from standard bifurcation theorems, that stable and unstable manifolds alternate, one can then conclude that necessary conditions for DS stability are $p_i < 0$ and $q_i > 0$. In Fig. 7, such bifurcation diagrams of the quintic model, satisfying the necessary stability conditions, are presented. The curves shown in Fig. 7a and b correspond to the base and the edge of the Brillouin zone, respectively. The stability of these solutions has been investigated by performing numerical simulations. The DS shown in Fig. 6a lies on the upper branch of the bifurcation diagram and its stability has also been verified numerically. On the other hand, the solution shown in Fig. 6b happens to be unstable.

We would also like to mention that, for a certain range of parameters, the DS solutions of the DGL equation can exhibit interesting behavior – the tails of the DS become very broad (occupying many lattice sites), whereas the central part of it is confined and displays a cusp-like feature, as shown in Fig. 8. Note that no such cusp soliton structures are possible in either the continuous GL regime or in DNLS lattices. To understand this behavior, one may use the relation $\lambda + \mathrm{i}\epsilon = 2\alpha \cosh s$, whence the rate of decay, s_r, of the

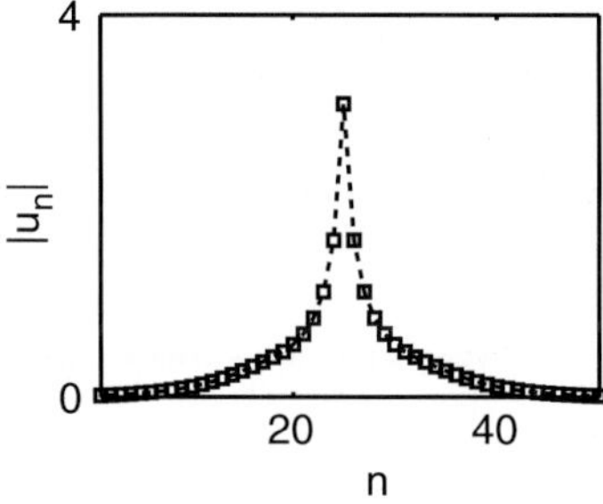

Fig. 8. A cusp-like DS solution for $\alpha = 0.4 - 2\mathrm{i}$, $p = 0.1 - 0.4\mathrm{i}$, $q = 0.43\mathrm{i}$ and $\epsilon = -0.4$

soliton tails in (40) can be determined. In the case of Fig. 8, $s_r = 0.2$, which indeed justifies the slow field decay in the tails.

More complicated bifurcation diagrams that have no analogs in the continuous GL case also appear in the discrete model. For example, in Fig. 9 (normal diffraction regime), we can observe a sub-critical bifurcation that is followed by three successive saddle-node bifurcations. As a result, four different branches of non-zero solutions exist, allowing up to two stable DS for the same value of ϵ.

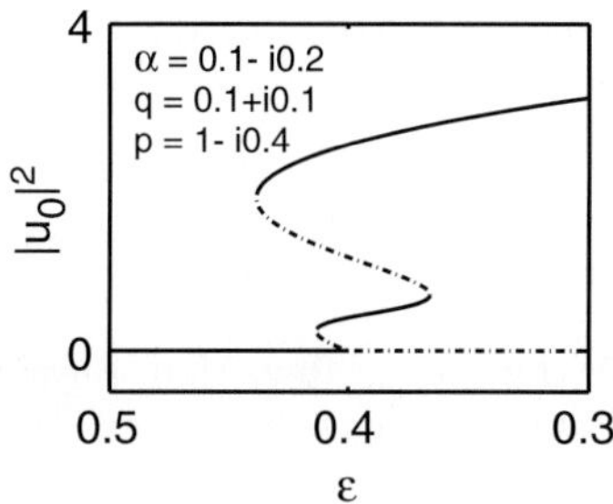

Fig. 9. A sub-critical bifurcation, followed by three successive saddle-node bifurcations at $\theta = 0$

Except for the DS that are centered on a single lattice point (on-site), two different types of DS that are centered between two lattice points (intra-site) exist. These are characterized by the phase difference between the two central lattice points, and this can be either 0 or π when the solutions are highly confined. In the DNLS regime, the π-out-of-phase intra-site DS are known to be stable [38, 39, 40] for relatively strong nonlinearities, whereas, the in-phase intra-site DS are always unstable because of oscillatory instabilities. However, in the DGL lattice system, we have identified regimes where both types of DS solutions are stable. In Fig. 10, such a stable intra-site

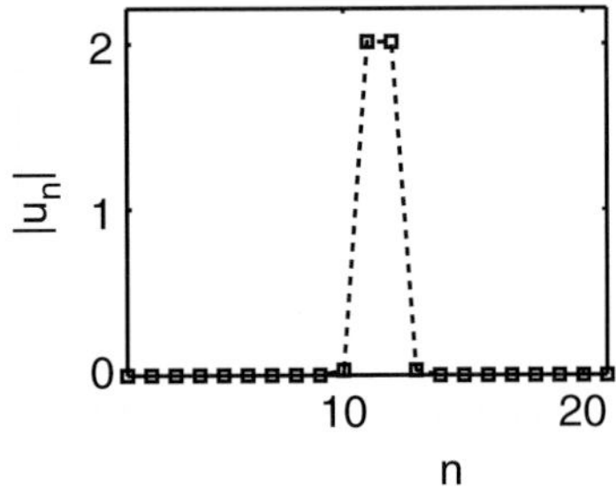

Fig. 10. In-phase stable intra-site DS for $\alpha = 0.1 - \mathrm{i}0.1$, $p = 0.2 - \mathrm{i}6$, $q = 0.2 + \mathrm{i}$, and $\epsilon = -8$

in-phase state is depicted. For these values of the parameters, both the in-phase and the out-of-phase DS are stable. Their intensity profiles are almost identical, and the main difference between them is the relative phase between the high-intensity lattice sites, which is 0 or π. The stability of these DS has been checked dynamically against symmetry-breaking perturbations. A more detailed stability analysis of these DS will be presented elsewhere.

Acknowledgements

This work was supported by an ARO MURI and by the Pittsburgh supercomputer center.

References

1. M. C. Cross and P. C. Hohenberg, Rev. Mod. Phys. **65**, 851 (1993).
2. Y. Kuramoto, *Chemical Oscillations, Waves and Turbulence*, (Springer, Berlin, 1984).
3. I. S. Aranson and L. Kramer, Rev. Mod. Phys. **74**, 99 (2002).
4. N. Akhmediev and A. Ankiewicz, *Solitons, Nonlinear pulses and beams*, (Chapman and Hall, London, 1997).
5. P. Manneville, *Dissipative Structures and Weak Turbulence*, (Academic, San Diego, 1990).
6. G. Nicolis *Introduction to Nonlinear Science*, (Cambridge University Press, Cambridge, 1995).
7. L. M. Hocking and K. Stewartson, Proc. R. Soc. London, Ser. A **326**, 289 (1972); N. R. Pereira and L. Stenflo, Phys. Fluids **20**, 1733 (1977).
8. K. Nozaki and N. Bekki, J. Phys. Soc. Jpn. **53**, 1581 (1984).
9. W. van Saarloos and P. C. Hohenberg, Physica D **56**, 303 (1992).
10. R. Conte and M. Musette, Physica D **69**, 1 (1993); P. Marcq, H. Chaté, and R. Conte, Physica D **73**, 305 (1994).
11. N. N. Akhmediev, V. V. Afanasjev and J. M. Soto Crespo, Phys. Rev. E **53**, 1190 (1996).

12. L. C. Crasovan, B. A. Malomed and D. Mihalache, Phys. Lett. A **289**, 59 (2001).
13. J. M. Soto-Crespo, N. Akhmediev and A. Ankiewicz, Phys. Rev. Lett. **85**, 2937 (2000).
14. P. S. Hagan, SIAM J. Appl. Math. **42**, 762 (1982).
15. L. C. Crasovan, B. A. Malomed and D. Mihalache, Phys. Rev. E **63**, 016605 (2001).
16. D. N. Christodoulides and R. I. Joseph, Opt. Lett. **13**, 794 (1988).
17. D. N. Christodoulides, F. Lederer and Y. Silberberg, Nature **424**, 817 (2003).
18. H. S. Eisenberg, Y. Silberberg, R. Morandotti, A. R. Boyd and J. S. Aitchison, Phys. Rev. Lett. **81**, 3383 (1998).
19. H. Willaime, O. Cardoso and P. Tabeling, Phys. Rev. Lett. **67**, 3247 (1991).
20. S. S. Wang and H. G. Winful, Appl. Phys. Lett. **52**, 1774 (1988); H. G. Winful and S. S. Wang, Appl. Phys. Lett. **53**, 1894 (1988).
21. K. Otsuka, *Nonlinear dynamics in optical complex systems*, (KTK Scientific Publishers, Tokyo, 1999).
22. K. Otsuka, Phys. Rev. Lett. **65**, 329 (1990); H. G. Winful and L. Rahman, Phys. Rev. Lett. **65**, 1575 (1990).
23. N. K. Efremidis and D. N. Christodoulides, Phys. Rev. E **67**, 026606 (2003).
24. K. Maruno, A. Ankiewicz and N. Akhmediev, Opt. Comm. **221**, 199 (2003).
25. K. Staliunas, Phys. Rev. Lett. **91**, 053901 (2003).
26. N. K. Efremidis and D. N. Christodoulides, Opt. Lett. **29**, 2485 (2004).
27. U. Peschel, T. Pertsch and F. Lederer, Opt. Lett., **23**, 1701 (1998).
28. T. Pertsch, P. Dannberg, W. Elflein, A. Brüer and F. Lederer, Phys. Rev. Lett. **83**, 4752 (1999).
29. R. Morandotti, U. Peschel, J. S. Aitchison, H. S. Eisenberg and Y. Silberberg, Phys. Rev. Lett. **83**, 4756 (1999).
30. C. Kittel, *Introduction to Solid State Physics* (Wiley, New York, 1986).
31. E. A. Ultanir, G. I. Stegeman, and D. N. Christodoulides, Opt. Lett. **29**, 845 (2004).
32. B. P. Anderson and M. A. Kasevich, Science **282**, 1686 (1998).
33. A. Trombettoni and A. Smerzi, Phys. Rev. Lett. **86**, 2353 (2001).
34. B. Kneer, T. Wong, K. Vogel, W. P. Schleich and D. F. Walls, Phys. Rev. A **58**, 4841 (1998).
35. F. T. Arecchi, J. Bragard and L. M. Castellano, In: *Bose-Einstein Condensates and Atom Lasers*, edited by S. Martellucci, A. N. Chester, A. Aspect, and M. Inguscio. (Kluver, New York, 2002).
36. F. Bloch, Z. Phys., **52**, 555 (1928).
37. Y. S. Kivshar, Opt. Lett. **18**, 1147 (1993).
38. S. Darmanyan, A. Kobyakov and F. Lederer, Zh. ksp. Teor. Fiz. **113**, 1253 (1998) [Sov. Phys. JETP **86**, 682 (1998)].
39. P. G. Kevrekidis, A. R. Bishop and K. O. Rasmussen, Phys. Rev. E **63**, 036603 (2001).
40. N. K. Efremidis and D. N. Christodoulides, Phys. Rev. E **65**, 056607 (2002).

Discrete Dissipative Solitons

F.Kh. Abdullaev

Physical-Technical Institute of the Uzbek Academy of Sciences, 700084, Tashkent-84, G.Mavlyanov str.2-b, Uzbekistan
fatkh@physic.uzsci.net

Abstract. The existence of *discrete* dissipative solitons in a nonlinear lattice is studied. The Ablowitz-Ladik (AL) model with linear damping, nonlinear cubic amplification, quintic damping and complex second difference, representing the discrete analogue of a filter, is investigated. The parameters of the discrete dissipative soliton are calculated using a perturbation theory for the AL model. Analytic predictions are confirmed by numerical simulations of the AL model with non-conservative perturbations. We also discuss modulational instability (MI) in the discrete complex Ginzburg-Landau (DCGL) equation and the existence the exact localized solutions.

1 Introduction

Properties of dissipative solitons in continuum nonlinear media with dissipation and nonlinear amplification are now well-explored [1]. They were predicted by Pereira-Stenflo in [2] (see also [3]), where the exact soliton solution of the one-dimensional cubic Ginzburg-Landau (GL) equation with filter was obtained for the first time. Different types of dissipative solitons have been found in one and two dimensions, namely flat-top solitons [4], erupting and creeping solitons, and spiraling solitons. Recently, these ideas have been applied to more complicated dissipative nonlinearities, and to a higher-dimensional GL equation [5]. There is an important difference between soliton solutions of Hamiltonian systems and dissipative soliton solutions in non-conservative systems. In Hamiltonian systems, soliton solutions appear as a result of a balance between nonlinearity and dispersion, and these solutions usually form a one-parameter family. The generation of dissipative solitons is possible when two conditions – an equilibrium between nonlinearity and dispersion and an equilibrium between dissipation and amplification – are met. The solution satisfying these requirements simultaneously exists for fixed parameters defined by the parameters of the equation.

For envelope waves, dissipative solitons (auto-solitons) are described by the complex Ginzburg-Landau equation, which in some limit can be transformed to the nonlinear Schrödinger equation with complex perturbation terms.

An exact solution exists for linear amplification and nonlinear damping and filtering [2]. However, this solution is unstable due to the instability of

F.Kh. Abdullaev: *Discrete Dissipative Solitons*, Lect. Notes Phys. **661**, 327–341 (2005)
www.springerlink.com

the zero mode, $u = 0$. A stable solution is possible for linear damping and cubic gain plus quintic damping [1, 6].

Recently, attention has been attracted to discrete solitons in dissipative lattices, an important example of which is the perturbed Ablowitz-Ladik (AL) model [7]. It was shown that gain via an a.c. driver leads to the compensation of dissipative and bistability effects. AL solitons for local dissipative impurities and a complex second difference (i.e. the discrete analogue of a filter), and also a shock-wave solution, were studied in [8, 9]. It is interesting to consider the discrete analogue of the complex Ginzburg-Landau equation and the existence of auto-solitons in such systems.

In this work, we study the discrete analogue of the complex GL equation, based on the AL model with non-conservative terms, including effects of linear and quintic damping, nonlinear (cubic) amplification and discrete filtering [10].

The merit of the model is that the unperturbed system is an integrable one and that it has many remarkable properties. One of them is that it admits moving discrete one and multi- solitonic solutions. Also, this system has an infinite number of conserved quantities. While the system has only little relevance to real systems at present, such a consideration makes sense. Firstly, in some cases, real models like the discrete nonlinear Schrödinger equation can be considered as perturbed AL models [11, 12]. Secondly, in modern optical systems like arrays of waveguides, the coupling between waveguides can be viewed as having a nonlocal (AL type) character, as they are embedded in the nonlinear medium.

Using perturbation theory for the AL model, based on inverse scattering theory (IST), we analyze the properties of *discrete* dissipative solitons.

We note that, recently, some results have been obtained for standing localized modes (discrete breathers) for the complex DNLS equation by [13]. The discrete breather solutions which were found do not have counterparts in the continuum complex GL equation. Exact solutions for the non-conservative AL system have been found in [14, 15]. In spite of the fact that the solutions are unstable, they can be useful for understanding the properties of discrete dissipative solitons, and we will give a short description of these solutions below.

2 Model and Basic Equations

The model considered in this work is the Ablowitz-Ladik model [7] with non-conservative perturbations describing the effects of linear damping and nonlinear amplification. Thus it can be considered as a *discrete* generalization of the complex Ginzburg-Landau equation.

$$\begin{aligned} \mathrm{i}u_{nt} &+ (u_{n+1} + u_{n-1})(1 - \lambda|u_n|^2) \\ &= -\mathrm{i}\delta u_n + \mathrm{i}\gamma|u_n|^2 u_n + \mathrm{i}\beta(u_{n+1} + u_{n-1} - 2u_n) \\ &\quad -\mathrm{i}\kappa|u_n|^4 u_n = \epsilon f(u_n) , \end{aligned} \tag{1}$$

where $\delta, \gamma, \kappa, \beta$ are the linear dissipation, cubic nonlinear amplification, quintic damping and filter coefficients, respectively, and $\lambda = \pm 1$ for defocusing and focusing cases, respectively. Below, we will drop the term $-2i\beta u_n$, since it can be included in the renormalization of δ. When $\delta = \gamma = \kappa = \beta = 0$, the model reduces to the well-known Ablowitz-Ladik model, which is integrable. Thus, when the perturbations are small, $\delta, \gamma, \kappa, \beta \ll 1$, we can use the IST perturbation theory for discrete solitons.

2.1 Modulational Instability in Non-Conservative AL Model

One possible mechanism for the generation of discrete dissipative solitons is modulational instability. Thus it is interesting to analyze the parameter values where the nonlinear plane waves solution is unstable against modulations. For continuum nonlinear media, this phenomenon results from the interplay between dispersive and nonlinear effects. In a discrete system, the discreteness introduces new peculiarities. In particular, in contrast to the continuum case, above a certain threshold in the amplitude of the wave, a carrier wave at small wave-numbers is unstable to all possible modulations [16]. Here we will study modulational instability for the non-conservative AL equation. Below, we consider the particular case where $\delta, \gamma \neq 0$. The plane-wave solution for (1) is

$$u = u_0 \mathrm{e}^{\mathrm{i}\theta_n} , \quad \theta_n = qn - \omega t , \tag{2}$$

where $u_0 = \sqrt{\delta/\gamma}$. The frequency, ω, satisfies the dispersion law

$$\omega + 2\cos(q)\left(1 + \frac{\delta}{\gamma}\right) = 0 . \tag{3}$$

To analyze the linear stability of the plane-wave solution, we consider a solution of the form (2).

$$u_n = (u_0 + b_n(t))\mathrm{e}^{\mathrm{i}\theta_n + \mathrm{i}\psi_n(t)} , \tag{4}$$

where b_n, ψ_n are assumed to be small compared to u_0, q, ω. Substituting (4) into (1) and keeping linear terms in b_n, ψ_n, we obtain the equations

$$\begin{aligned} b_{n,t} - 2\delta_n + (1 + u_0^2)(b_{n+1} - b_{n-1})\sin(q) + u_0(1 + u_0^2)(\psi_{n+1} \\ + \psi_{n-1} - 2\psi_n)\cos(q) = 0 , \end{aligned} \tag{5}$$

$$\begin{aligned} -u_0\psi_{n,t} + 2\cos(q)(u_0^2 - 1)b_n + (1 + u_0^2)(b_{n-1} + b_{n-1})\cos(q) \\ -u_0(1 + u_0)^2(\psi_{n+1} - \psi_{n-1})\sin(q) = 0 . \end{aligned} \tag{6}$$

Then we have the dispersion relation for the plane wave modulation with wave-number Q and frequency Ω:

$$\left[\Omega - 2(1+u_0^2)\sin(Q)\sin(q) - \delta\right]^2 = 16(1+u_0^2)\sin^2\left(\frac{Q}{2}\right)\cos^2(q)$$
$$\left[(1+u_0^2)\sin^2\left(\frac{Q}{2}\right) - u_0^2\right] . \qquad (7)$$

For small Q, q, corresponding to the continuum approximation, we obtain the dispersion relation

$$\left[\Omega - 2(1+u_0^2 Qq - \delta)\right]^2 = (1+u_0^2)Q^2\left[(1+u_0^2)Q^2 - 4u_0^2\right] , \qquad (8)$$

which agrees with the dispersion relation for the non-conservative continuous NLS equation [17]. Comparing (7) with the well-known result for MI of the unperturbed AL equation, we can conclude that the region of MI in Q remains unchanged. The critical wave number where the instability appears is defined by

$$Q = 2\text{arcsin}\left(\frac{\delta}{\sqrt{\delta+\gamma}}\right).$$

We note that only the increment of MI is changed. So, it is reasonable to think that localized solutions can exist for the AL equation with dissipation and damping. In the next section, we will study localized moving dissipative solutions using a perturbation theory for the AL equation.

2.2 The Ablowitz-Ladik Model and Solitonic Solutions

Let us describe briefly the Ablowitz-Ladik model and its solutions. The unperturbed equation follows from the Hamiltonian

$$H = -\sum_{n=-\infty}^{\infty}(u_n u_{n+1}^* + u_n^* u_{n+1}) + \frac{2}{\lambda}\sum_{n=-\infty}^{\infty}\ln(1-\lambda|u_n|^2) , \qquad (9)$$

with non-standard Poisson brackets

$$\{u_n, u_m^*\} = i(1-\lambda|u_n|^2)\delta_{mn}, \quad \{u_m, u_n\} = \{u_m^*, u_n^*\} = 0 . \qquad (10)$$

The system is integrable and has an infinite number of integrals of motion. Some of them are the norm N and the energy E:

$$N = -\frac{1}{\lambda}\sum_n \ln(1-\lambda|u_n|^2), \quad E = -\sum_{n=-\infty}^{\infty}(u_n u_{n+1}^* + u_n^* u_{n+1}) . \qquad (11)$$

Below, we will consider the focusing (attractive) interaction with $\lambda = -1$.

The AL equation has a moving 1-soliton solution of the form

$$u_s(n,t) = \frac{\sinh \mu}{\cosh[\mu(n - vt - x_0)]} \exp[ik(n - x(t)) + i\alpha(t)] , \qquad (12)$$

where

$$\alpha(t) = \omega t + \phi_0 , \quad \omega = 2(1 - \cosh \mu \cos k) + v , \quad v = \frac{2 \sinh \mu \sin k}{\mu} . \qquad (13)$$

The limiting velocity is equal to $v_{max} = 2 \sinh \mu / \mu$. The parameter μ determines the maximal amplitude of the soliton, k is its frequency, and α is the soliton phase. The discreteness of the system is evidenced by the existence of a maximum velocity $v_{max} = 2 \sinh(\mu)/\mu$. The energy and the norm of the one-soliton solution are

$$E = 4 \sinh(\mu) \cos(k) , \quad N = 2\mu . \qquad (14)$$

Let us briefly describe the information from the inverse scattering transform that it is necessary for the derivation of the perturbation equations for the AL model [7, 18].

The linear spectral problem associated with the AL equation is

$$\begin{pmatrix} 1 & 0 \\ 0 & 1 \end{pmatrix} f_{n+1} + \begin{pmatrix} 0 & q_n \\ -q_n^* & 0 \end{pmatrix} f_n(z) = \begin{pmatrix} z & 0 \\ 0 & z^{-1} \end{pmatrix} f_n(z) \qquad (15)$$

and

$$f_n(z) = \begin{pmatrix} f_{1,n}(z) \\ f_{2,n}(z) \end{pmatrix} .$$

Here z is the spectral parameter. If q_n decays rapidly as $|n| \to \infty$, then $\phi_n(z) z^{-n}$ and $\phi_n(z) z^n$ can be analytically continued for $|z| > 1$. Let us introduce $\bar{f}$ associated with the solution of (15) f, which is the solution of (15) with the same eigenvalues.

$$\bar{f}_n(z) = \begin{pmatrix} f_{2,n}^*(1/z^*) \\ -f_{1,n}^*(1/z^*) \end{pmatrix} .$$

Jost functions satisfy the boundary conditions

$$\begin{aligned} \phi_n(z) &\sim \hat{e}_1 z^n , \quad n \to -\infty , \\ \psi_n(z) &\sim \hat{e}_2 z^{-n} , \quad n \to \infty , \end{aligned} \qquad (16)$$

where $\hat{e_1} = (1,0)^T, \hat{e_2} = (0,1)^T$ and $\bar{\phi} z^n, \bar{\psi} z^{-n}$ are analytic for $|z| < 1$. Their Wronskian is non-zero:

$$W(\psi_n \bar{\psi_n}) = \psi_{1,n} \bar{\psi}_{2,n} - \psi_{2,n} \bar{\psi}_{1,n} = - \prod_{m=n}^{\infty} (1 + |q_n|^2)^{-1} , \qquad (17)$$

so the Jost functions $\psi, \bar{\psi}$ are linearly-independent. We also have the representation

$$\phi_n(z) = a(z)\bar{\psi}_n(z) + b(z)\psi_n(z) \,, \tag{18}$$
$$\phi_n(z) = a(z)\psi_n(z) + \bar{b}(z)\bar{\psi}_n(z) \,, \tag{19}$$

where $a(z), b(z)$ are the Jost coefficients. From these equations, we can find

$$a(z) = \frac{W(\phi_n(z), \psi_n(z))}{W(\bar{\psi}_n(z), \psi_n(z))} \,,$$
$$b(z) = \frac{W(\phi_n(z), \bar{\psi}_n(z))}{W(\bar{\psi}_n(z)\ , \psi_n(z))} \,, \tag{20}$$

and

$$|a(z)|^2 + |b(z)|^2 = \prod_{n=-\infty}^{\infty} (1 + |q_n|^2) \,.$$

The discrete spectrum is characterized by exponential decay of ϕ and ψ as $|n| \to \infty$. The zeros of $a(z)$ come in pairs. The spectral problem scattering data (15) are in the set $a(z), b(z), z_r, c_r, a'(z_r)$.

The Jost functions for the one-soliton solution are:

$$\Phi_{n,s}(z) = \frac{z^n}{2\cosh(\mu(n-1-x))} \begin{pmatrix} a_s(z)\mathrm{e}^{\mu(n-1-x)} + \mathrm{e}^{-\mu(n-1-x)} \\ (1 - a_s(z))z\mathrm{e}^{-\mathrm{i}k(n-x-\mathrm{i}\alpha)} \end{pmatrix} \,, \tag{21}$$

$$\Psi_{n,s}(z) = \frac{z^{-n}}{2\cosh(\mu(n-1-x))} \begin{pmatrix} (1 - \mathrm{e}^{-2\mu})a_s(z)z^{-1}\mathrm{e}^{\mathrm{i}k(n-x)+\mathrm{i}\alpha} \\ \mathrm{e}^{\mu(n-1-x)} + a_s(z)\mathrm{e}^{-\mu(n+1-x)} \end{pmatrix} \,. \tag{22}$$

The soliton solution is associated with the scattering data

$$a_s(z) = \frac{z^2 - \mathrm{e}^{\mathrm{i}k+\mu}}{z^2 - \mathrm{e}^{\mathrm{i}k-\mu}} \,, \quad b_s = 0 \,. \tag{23}$$

and

$$c_{rs} = z_{rs} \exp(\mathrm{i}kx + \mu(x+1) - \mathrm{i}\alpha) \,, \quad r = 1, 2 \,. \tag{24}$$

Let us uncover the evolution of the single soliton under perturbation, using the equations for the scattering data. The system of evolution equations for the scattering data are [12]

$$\frac{\partial a}{\partial t} = -\mathrm{i}\epsilon \sum_n \frac{W(\psi_{n+1}(z), \hat{F}(q_n)\phi_n(z))}{W(\bar{\psi}_{n+1}(z), \psi_{n+1}(z))} \,, \tag{25}$$

$$\frac{\partial b}{\partial t} = \mathrm{i}\omega(z)b + \mathrm{i}\epsilon \sum_n \frac{W(\bar{\psi}_{n+1}(z), \hat{F}(q_n(z))\phi_n(z))}{W(\bar{\psi}_{n+1}(z), \psi_{n+1}(z))} \,, \tag{26}$$

where $\omega(z) = 2 - z^2 - z^{-2}$ is the linear dispersion relation for the discrete linear Schrödinger equation.

$$\frac{dz_r}{dt} = \frac{i\epsilon}{a'(z_r)} \sum_{n=-\infty}^{\infty} \frac{W(\psi_{n+1}, \hat{F}(q_n, q_n^*)\phi_n)}{W(\bar{\psi}(\bar{z}), \psi_{n+1}(z))} , \tag{27}$$

$$\frac{dc_r}{dt} = i(2 - z_r^2 - z_r^{-2})c_r + \frac{i\epsilon}{a'(z_r)} \sum_{-\infty}^{\infty} \frac{W(\phi'_{n+1}(z_r) - c_r\psi'_{n+1}(z_r), \hat{F}(q_n, q_n^*)\phi_n(z_r))}{W(\bar{\psi}_{n+1}(z_r), \psi_{n+1}(z_r))} , \tag{28}$$

where

$$\hat{F} = \begin{pmatrix} 0 & f(q_n) \\ f^*(q_n) & 0 \end{pmatrix}, \quad a' = \left(\frac{d}{dz}a(z)\right)_{z=z_r} .$$

To find the perturbations equations for the soliton parameters, it is sufficient to substitute (12), (13), (21)–(24) into (27), (28). The calculations lead to the equations for the 1-soliton parameters:

$$x_t = \frac{2\sinh(\mu)}{\mu}\sin(k) + \epsilon\frac{\sinh(\mu)}{\mu} \sum_{n=-\infty}^{\infty} \frac{(n-x)\cosh[\mu(n-x)]}{\cosh[\mu(n+1-x)]\cosh[\mu(n-1-x)]} \cdot Im\{f(u_{ns})\exp[-ik(n-x) - i\alpha]\} , \tag{29}$$

$$k_t = -\epsilon\sinh(\mu) \sum_{n=-\infty}^{\infty} \frac{\cosh[\mu(n-x)]\tanh[\mu(n-x)]}{\cosh[\mu(n+1-x)]\cosh[\mu(n-1-x)]} \cdot Re\{f(u_{ns})\exp[-ik(n-x) - i\alpha]\} , \tag{30}$$

$$\mu_t = \epsilon\sinh(\mu) \sum_{n=-\infty}^{\infty} \frac{\cosh[\mu(n-x)]}{\cosh[\mu(n+1-x)]\cosh[\mu(n-1-x)]} \cdot Im\{f(u_{ns})\exp[-ik(n-x) - i\alpha]\} . \tag{31}$$

Here, the equation for α is not given, because it is decoupled from the others and will not be used. The function f describes a perturbation to the AL equation, (1) and ϵ is a small parameter.

2.3 Evolution of a Single Soliton Under Non-Conservative Perturbations

Substituting the expressions for the perturbations, via the 1-soliton solution, into (29)–(31), we find the system of equations for the soliton parameters. Using the Poisson summation formula

$$\sum_{n=-\infty}^{\infty} F(na) = \int_{-\infty}^{\infty} \frac{dx}{a} F(x) \left[1 + 2\sum_{s=1}^{\infty} \cos\left(\frac{2\pi s x}{a}\right)\right] \tag{32}$$

and retaining the first harmonic, we obtain, with an accuracy up to terms $\sim\exp(-\pi^2/\mu)$, the system [10]

$$\mu_t = -2\delta\tanh(\mu) + 2\gamma\frac{\sinh^2(\mu)}{\mu}\left(1 - \frac{2\mu}{\sinh(2\mu)}\right) \tag{33}$$

$$+2\beta\cos(k)\frac{\sinh(\mu)}{\mu}(\mu + \tanh(\mu) - \mu\tanh^2(\mu))$$

$$-2\kappa\frac{\sinh^4(\mu)}{\mu}\left(\frac{2}{3} - \frac{1}{\sinh^2(\mu)} + \frac{2\mu}{\sinh^2(\mu)\sinh(2\mu)}\right),$$

$$k_t = -2\beta\frac{\sinh(\mu)}{\mu}\sin(k)\left(1 - \frac{\mu}{2\sinh(2\mu)}\right), \tag{34}$$

$$x_t = \frac{2\sinh(\mu)}{\mu}\sin(k)\,. \tag{35}$$

Here the equation (35) is decoupled from the others.

Let us consider in more detail the cases where:

(a) $k = 2m\pi, \sin(k) = 0, \cos(k) = 1, m = 0, 1, 2\ldots.$

In this case, the system of equations (33)–(35) has the form

$$\mu_t = -2\delta\tanh(\mu) + 2\gamma\frac{\sinh^2(\mu)}{\mu}\left(1 - \frac{2\mu}{\sinh(2\mu)}\right) \tag{36}$$

$$+2\beta\frac{\sinh(\mu)}{\mu}(\mu + \tanh(\mu) - \mu\tanh^2(\mu))$$

$$-2\kappa\frac{\sinh^4(\mu)}{\mu}\left(\frac{2}{3} - \frac{1}{\sinh^2(\mu)} + \frac{2\mu}{\sinh^2(\mu)\sinh(2\mu)}\right),$$

$$k_t = 0\,, \tag{37}$$

$$x_t = 0\,. \tag{38}$$

Thus the soliton dynamics is described by just one ordinary differential equation for amplitude. The fixed points of equation (36), when $\mu_t = 0$, can easily be found graphically. These fixed points give us the parameters of the auto-soliton. The value of μ_{cr} for the fixed point, if it exists, can be expressed by the following approximate solution $(1 < \mu < \pi)$

$$\mu_{cr} = \sinh^{-1}\left[\left(\frac{3(\kappa+\gamma)}{2\kappa}\right)^{1/2}\right]. \tag{39}$$

In this region of soliton amplitude μ, cubic amplification and quintic damping dominates, and contributions from linear damping δ and the discrete filter β are small.

If we consider the range of parameters when $\mu \ll 1$, corresponding to a broad soliton, then the equation (36) can be simplified and written in the form

$$\mu_t = 2\mu(-\delta + 2\beta) + \frac{2}{3}\mu^3(\delta + 2\gamma - 3\beta) . \quad (40)$$

The equation (40) is valid only for $\mu \ll 1$, but, from analysis of this equation, one can obtain information about possible global scenarios of the evolution of μ, even when it becomes quite large (i.e. a narrow discrete soliton). Note that, when compared with the continuous case considered in [1], we have additional terms of the order of μ^3 in δ and β. These terms appear due to the discreteness of the model. Let us present the results of such analysis. From equation (40) in the linear approximation, one can deduce that if $2\beta > \delta$, then μ_t, as a function of μ, initially increases, but then at some point it crosses zero, going from positive to negative values, and this point corresponds to a stable auto-soliton. So, in this case, the auto-soliton exists for one fixed value of amplitude. If we start with a soliton with some value of amplitude, it will evolve to an auto-soliton with parameters determined by the fixed point.

In the opposite case, when $2\beta < \delta$, it is easy to see that there are two possibilities. The first one is that μ_t is negative for all values of μ, and an auto-soliton does not exist. All initial pulses disappear.

According to the second possibility, μ_t initially decreases to negative values, then starts to increase, and at some point crosses zero, going from negative to positive values, and this point corresponds to an unstable auto-soliton. If we increase μ, then μ_t again changes its sign and crosses zero, going from positive to negative values, and this point corresponds to a stable auto-soliton. So, in this case, there are two fixed points, with one being unstable and the other one stable. An initial pulse may evolve to a stable auto-soliton or disappear, depending on the initial value.

(b) Let $k = (2m + 1)\pi$, $\sin(k) = 0$, $\cos(k) = -1$. In this case, results can be obtained in the same way as the mentioned above.

In Fig. 1, we show the phase portrait of the system (33)–(34), with parameters corresponding to the second case, where two fixed points exist. The upper fixed point is a sink which defines the parameters of the stable approximate auto-soliton of (1). Any soliton-like initial condition in close proximity to a fixed point will converge to a stable stationary solution. The analytic results described above have been checked by direct numerical solution of the perturbed AL equations (1). In Fig. 2, an example of the evolution of an initial pulse to an auto-soliton is presented; it confirms the result obtained from perturbation theory.

We clearly see the oscillations of the soliton profile that can be attributed to the non-zero chirp. To calculate this chirp, we should also calculate the contribution of the continuum spectrum induced by the perturbation.

The results of numerical simulations, and a comparison with perturbation theory (33)–(34) results and with the formula (40) are presented in Figs. 3–5. In Fig. 3, we show the evolution of the amplitude of a soliton for values of parameters of equation (1) where inequality $2\beta > \delta$ is satisfied, so an auto-soliton should exist. This prediction is confirmed by the numerical calculations. From Fig. 3, it is clear that the amplitude of the initial pulse

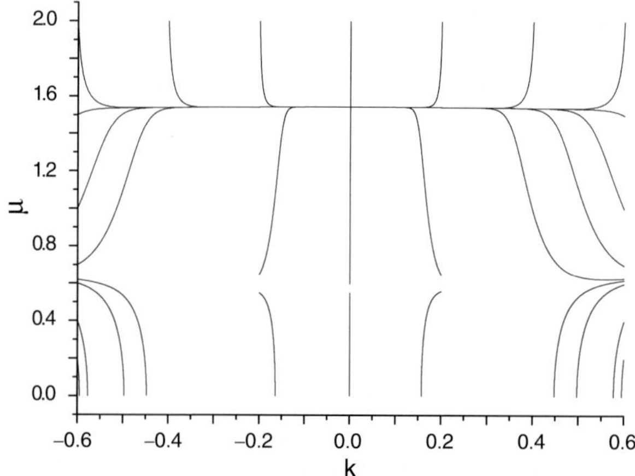

Fig. 1. The phase portrait of the dynamical system (33) and (34) for $\delta = 0.1, \beta = 0.07, \gamma = 0.2, \kappa = 0.05$

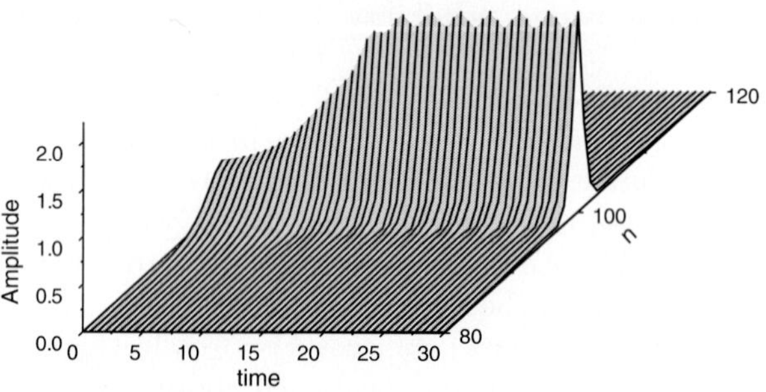

Fig. 2. Evolution of soliton amplitude for $\delta = 0.1, \beta = 0.06, \gamma = 0.1, \kappa = 0.03$, and initial conditions $\mu = 0.5, k = 0$

reaches a stationary value, but there are also oscillation of amplitude which are not described by perturbation theory. This result was obtained for various initial pulses with different amplitudes, but the resulting amplitude of the auto-soliton was the same.

In Figs. 4 and 5, we present examples of pulse evolution where the inequality $2\beta < \delta$ is satisfied. In Fig. 4, we show the evolution of the pulse amplitude, where its initial value is smaller than that of the unstable auto-soliton, so that it disappears.

In Fig. 5, the results of a simulation are presented, with the initial amplitude being above the amplitude of the unstable soliton. One can see that, in this case, a stable auto-soliton exists. The stationary amplitudes obtained from the solution of the system (33)–(34) are always slightly higher than those obtained from the solution of the full perturbed AL equations, and this can be explained as a result of the approximations used.

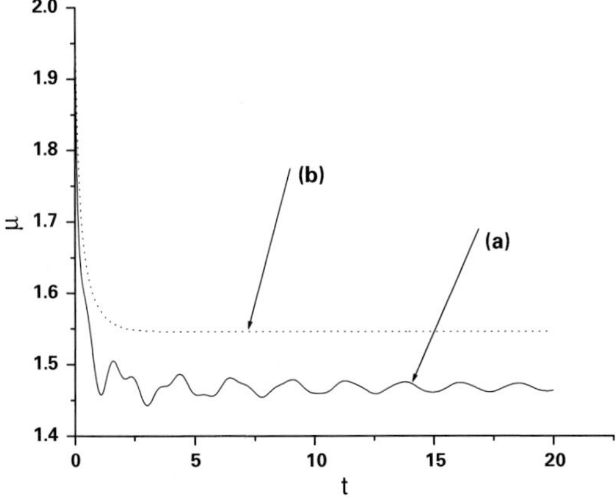

Fig. 3. Evolution of soliton maximum amplitude μ for $\delta = 0.1, \beta = 0.03, \gamma = 0.2, \kappa = 0.05$. (**a**) Numerical solution of perturbed AL equations (1). (**b**) Numerical solution of system (33)–(34). Initial conditions are: $\mu = 2, k = 0$

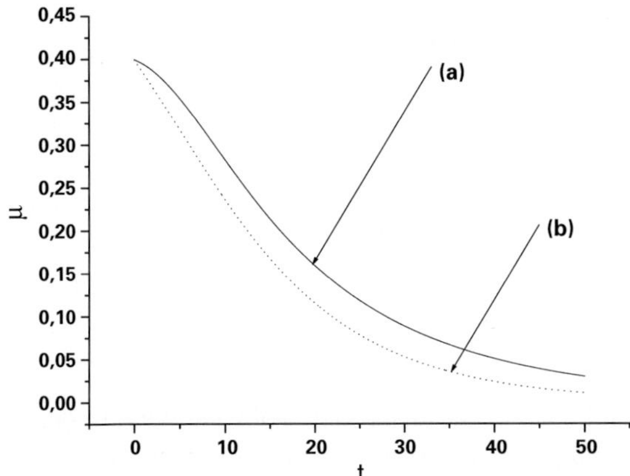

Fig. 4. Evolution of soliton maximum amplitude μ for $\delta = 0.1, \beta = 0.07, \gamma = 0.2, \kappa = 0.05$. (**a**) Numerical solution of perturbed AL equations (1). (**b**) Numerical solution of system (33)–(34). Initial conditions are: $\mu = 0.4, k = 0$

3 Exact Localized Solutions of DCGL Equation

There has been recent interest in finding the exact solutions of the DCGL equation, thus going beyond the assumption of weak gain and dissipation [14]. We take a perturbed AL equation of the form

$$\mathrm{i}u_{n,t} + \left(\frac{D}{2} - \mathrm{i}\beta\right)(u_{n+1} - u_{n-1} - 2u_n) + (1 + \mathrm{i}\epsilon)(u_{n+1} + u_{n-1})|u_n|^2 = 0 \,. \tag{41}$$

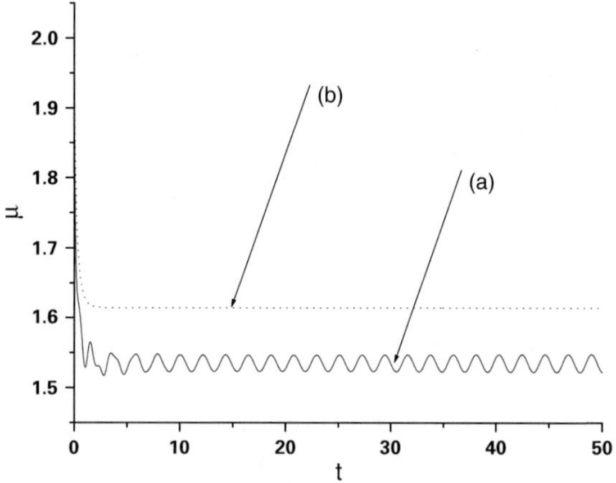

Fig. 5. Evolution of the soliton maximum amplitude μ for $\delta = 0.1, \beta = 0.07, \gamma = 0.2, \kappa = 0.05$. (**a**) Numerical solution of perturbed AL equations (1). (**b**) Numerical solution of the system (33)–(34), with initial conditions $\mu = 2, k = 0$

Let us apply the Hirota method. Following this method, we seek the solution of the form

$$u_n(t) = \phi_n e^{-i\omega t} , \quad \phi_n = \frac{g_n}{f_n} , \tag{42}$$

with f_n real. Substituting into (41), we obtain the multi-linear form

$$\left(\omega - \mathrm{i}\delta - 2\left(\frac{D}{2} - \mathrm{i}\beta\right)\right) f_{n+1} f_{n-1} g_n + \left(\frac{D}{2} - \mathrm{i}\beta\right) f_{n+1} f_n^2 g_{n-1} \\ +(1 - \mathrm{i}\epsilon) f_{n-1} g_{n+1} g_n g_n^* + (1 - \mathrm{i}\epsilon) f_{n+1} g_n g_n^* g_{n-1} = 0 . \tag{43}$$

Let us consider the case where the solution has a constant phase along the chain. The following constraint on the equation parameters is applicable:

$$D = \frac{2\beta}{\epsilon} . \tag{44}$$

The frequency ω is given by

$$\omega = \frac{\delta}{\epsilon} . \tag{45}$$

Applying the Hirota method, we find the solution for the *bright soliton* to be

$$\phi_n = \frac{1}{2}\sqrt{\frac{\delta}{\epsilon}\left(\frac{\delta}{\beta} - 4\right)} \operatorname{sech}\left[n \operatorname{arccosh}\left(1 - \frac{\delta}{2\beta}\right) + n_a\right] . \tag{46}$$

This solution requires $\delta/\beta < 0$ and opposite signs for δ and ϵ. The amplitude and the width of the discrete dissipative soliton are fixed, and are defined by the parameters of the equation. Another interesting property is that the constant n_a is the free parameter. This indicates translational invariance along the lattice. The result shows that translational invariance is restored in the dissipative system.

Another type of bright soliton is the *alternating-phase bright soliton* solution. It can be obtained by multiplying the solution (46) by the factor $(-1)^n$. In this solution, each site is out-of-phase with each of its neighbours. Such a solution has no analogue in the continuous NLS equation.

Oscillatory "sec" *solution.*

This solution represents a solution with a quasi-periodic oscillation. It requires the condition $0 < \delta/\beta < 4$ to be fulfilled, as well as $\delta/\epsilon < 0$. The solution has the form

$$\phi_n = \frac{1}{2}\sqrt{\frac{\delta}{\epsilon}\left(\frac{\delta}{\beta} - 4\right)} \sec\left[n \ \arccos\left(1 - \frac{\delta}{2\beta}\right) + n_a\right] . \tag{47}$$

The particular solution for $\delta = -3, \beta = -2, \epsilon = 5$ is

$$\phi_n = \frac{1}{2}\sqrt{\frac{3}{2}} \sec\left[n \ \arccos\left(\frac{1}{4}\right)\right] . \tag{48}$$

A solution corresponding to the *dark soliton* also exists:

$$\phi_n = \sqrt{\frac{-\delta}{2\epsilon}} \ \tanh\left[n \ \operatorname{arctanh}\left(\sqrt{\frac{\delta}{2\beta}}\right) + n_b\right] , \tag{49}$$

where n_b is an arbitrary constant, indicating translational invariance of the solution. For more details, we refer the reader to the article [15].

In conclusion, we note the important properties of the exact solutions. Firstly, dissipative discrete solitons admit translational invariance, like the solitons of the unperturbed AL equation. While the lattice lacks itself translational invariance, the solitons have this additional degree of freedom. This is an important distinguishing feature from discrete solitons in a Hamiltonian system, where the positions of the solitons relative to the lattice are fixed. The second property is the existence of "sec" type solutions, the continuum counterparts of which are singular. Here, the points of singularities can occur between the sites and the problem can be avoided. As seen in numerical simulations, the solutions of such type are not stable. If quintic terms of the form $(\nu - i\kappa)|u|_n^4 u_n$ are included in the system, stability can be obtained (see previous sections). Note that exact solutions for such a form of perturbed AL model have not been obtained.

4 Conclusion

In this work, we have studied the dynamics of an initial pulse in the perturbed AL system which is the discrete analogue of the quintic complex Ginzburg-Landau equation. Investigation of this equation is interesting, since it is a model which provides for a discrete auto-soliton, and also has applications in arrays of optical waveguides [19], nonlinear electrical chains [20], laser arrays [21, 22] and arrays of Bose-Einstein condensates [23, 24] and biological systems. The motivation for this study was to understand the general properties of discrete auto-soliton formation and its stability, where they may be applicable in various discrete one-dimensional systems. The modulational instability of the AL equation with non-conservative terms has been examined. The value of the critical wave-number for modulations for MI has been calculated. The process of MI can be used to generate discrete dissipative solitons. Perturbation theory has been applied to consider evolution of soliton parameters, and it was shown that discrete dissipative solitons may appear, depending on the range of parameters, as a result of the balance between gain and loss. The conditions where a soliton exists were given and an approximate expression has been found for the amplitude of a stationary solution. The results of perturbation analysis have been checked by direct numerical solution of the perturbed AL equation and sufficiently good agreement was found. We also presented recent results for exact discrete solitonic solutions of the non-conservative AL equation. We note that dissipative solitons can appear in the fully discrete regime, where only a few sites of the chain are exited. It can be expected that a similar mechanism should be valid in the corresponding discrete NLS equations with linear and nonlinear amplification and/or damping.

References

1. N. Akhmediev and A. Ankiewicz, Solitons of the complex Ginzburg-Landau equation, pp. 311–339 in *Spatial solitons*, editors: Trillo, S., Torruellas, W., (Springer, Berlin Heidelberg New York, 2001).
2. N. R. Pereira and L. Stenflo, Phys. Fluids **20**, 1733 (1976).
3. L. A. Ostrovsky, In *Nonlinear deformation waves* IUTAM Symposium, Tallin, U. Nigul, J. Engelbrecht (Eds.), (Springer, Berlin Heidelberg New York 1983), pp. 32–43.
4. N. Akhmediev, V. V. Afanas'ev and J. M. Soto-Crespo, Phys. Rev. E **53**, 1190 (1996).
5. V. Filho, F. Kh. Abdullaev, A. Gammal and L. Tomio, Phys. Rev. A **63**, 3603 (2001).
6. D. Cai, A. R. Bishop, N. G. Gronbech-Jensen and B. A. Malomed, Phys. Rev. E **50**, R694 (1994).
7. M. J. Ablowitz and J. F. Ladik, J. Math. Phys. **17**, 1011 (1976).
8. V. V. Konotop, D. Cai, M. Salerno, A. R. Bishop and N. Gronbech-Jensen, Phys. Rev. E **53**, 6476 (1996).
9. M. Salerno, B. A. Malomed and V. V. Konotop, Phys. Rev. E **62**, 8651 (2000).
10. F. K. Abdullaev, A. A. Abdumalikov and B. A. Umarov, Phys. Lett. A **305**, 371 (2002).
11. Y. Kivshar and D. K. Campbell, Phys. Rev. A **48**, 3077 (1993).
12. A. A. Vakhnenko and Yu. B. Gaididei, Teor. Math. Phys. **68**, 873 (1986).

13. N. K. Efremidis and D. N. Christodoulides, Phys. Rev. E **67**, 02606 (2003).
14. K. Maruno, A. Ankiewicz and N. Akhmediev, Opt. Commun. **221**, 199 (2003).
15. J. M. Soto-Crespo, N. Akhmediev and A. Ankiewicz, Phys. Lett. A **314**, 126 (2003).
16. Y. S. Kivshar and M. Peyrard, Phys. Rev. A **46**, 3198 (1992).
17. F. Kh. Abdullaev, S. A. Darmanyan and J. Garnier, Modulational instability of electromagnetic waves in inhomogeneous and in discrete media, In *Progress in Optics*, E. Wolf (Ed), **44** (Elsevier Publ., Amsterdam, 2002), pp. 303–365.
18. J. Garnier, Phys. Rev. E **63**, 026608 (2001).
19. A. B. Aceves, C. De Angelis, T. Peschel, R. Muschall, F. Lederer, S. Trillo and S. Wabnitz, Phys. Rev. E **53**, 1172 (1996).
20. M. Remoissenet, *Waves called solitons*, (Springer, Berlin Heidelberg New York 2000).
21. S. Wang and H. G. Winful, Appl. Phys. Lett. **52**, 1774 (1998).
22. K. Otsuka, *Nonlinear Dynamics in Optical Complex Systems* (KTK Scientific Publishers, Tokyo, 1999).
23. A. Trombettoni and A. Smerzi, Phys. Rev. Lett. **86**, 2353 (2001).
24. F. Kh. Abdullaev, B. B. Baizakov, S. A. Darmanyan, V. V. Konotop and M. Salerno, Phys. Rev. A **64**, 043606 (2001).

Nonlinear Schrödinger Equation with Dissipation: Two Models for Bose-Einstein Condensates

V.V. Konotop

Centro de Física Teórica e Computacional and Departamento de Física, Universidade de Lisboa, Complexo Interdisciplinar, Av. Prof. Gama Pinto 2, Lisbon 1649-003, Portugal
konotop@cii.fc.ul.pt

1 Introduction

Dissipation is a universal phenomenon which must be taken into account each time when one attempts to bring a theoretical description of a physical problem closer to the experimental situation. Dissipation can play either a principal role, i.e. determining the phenomenon itself (so that the latter disappears when dissipation is switched off), or a secondary role, i.e. affecting physical processes only by causing relatively small energy losses (so that the phenomenon persists in the absence of the dissipation). In the first case, it is customary to say that one deals with a dissipative system, while in the second case, one speaks of a dissipative perturbation of an originally conservative system.

Furthermore, there exists one more situation where dissipation appears in the description of a physical system. This is the case of effective dissipation, which does not exist in the original model, but appears after using a specific *ansatz* while looking for a solution. As an example, one can mention a nonlinear Schrödinger (NLS) equation with inhomogeneous coefficients of a specific form. In the present chapter, we study two particular realizations of this example. Although physical applications of the NLS equation are very wide (including optics, hydrodynamics, condensed matter physics and biophysics), we will be interested in an interpretation of the results obtained in terms of a specific problem – the dynamics of a Bose-Einstein condensate (BEC). More specifically, we will consider the effect of the Feshbach resonance on the dynamics of a BEC, in a parabolic trap and in an optical lattice, as well as modulational instability of a trapped BEC. From the practical point of view, these two problems represent examples of the management of a BEC by an external magnetic field.

V.V. Konotop: *Nonlinear Schrödinger Equation with Dissipation: Two Models for Bose-Einstein Condensates*, Lect. Notes Phys. **661**, 343–371 (2005)
www.springerlink.com

2 From a Three-Dimensional Gross-Pitaevskii Equation to the One-Dimensional Nonlinear Schrödinger Equation

The possibility of bosons condensing in the ground state at ultra-low temperatures was predicted in 1924 [1, 2] and its experimental realization in atomic vapors was achieved in 1995 [3, 4]. Since that time, the number of experimental and theoretical results reported in this area of physics has grown very quickly (see e.g. [5] for a review).

2.1 Mean-Field Approximation and Parameters of the Problem

From a theoretical point of view, there are two aspects of BECs to be taken into account. Firstly, this is a quantum phenomenon, and as such, its microscopic description must be based on the equation for the field operators (see e.g. [6]). Secondly, due to the macroscopic manifestation of the phenomenon, one can introduce an average characteristic of the condensate in order to describe it This is an order parameter, also called a macroscopic wave function, $\Psi(\mathbf{r},t)$. The equation governing the evolution of $\Psi(\mathbf{r},t)$ is obtained within the mean-field approximation and is called the Gross-Pitaevskii (GP) equation (due to the pioneering works [7, 8]). It reads

$$i\hbar\frac{\partial\Psi}{\partial t} = -\frac{\hbar^2}{2m}\Delta\Psi + V(\mathbf{r})\Psi + g_0|\Psi|^2\Psi \tag{1}$$

Here we use standard notation: $g_0 = 4\pi\hbar^2 a_s/m$ is the coupling constant characterizing two-body interactions, a_s is the s-wave scattering length, which can be either positive or negative, m is the atomic mass, Δ stands for the three-dimensional (3D) Laplacian, and $V(\mathbf{r})$ is an external potential.

We start with the simplified problem, where $V(\mathbf{r})$ is a harmonic trap potential, $V(\mathbf{r}) \equiv V_{trap}(\mathbf{r})$, created by an external magnetic field and well-approximated by a parabolic function

$$V_{trap}(\mathbf{r}) = \frac{m}{2}\left(\Omega_\perp^2\mathbf{r}_\perp^2 + \omega_0^2 x^2\right). \tag{2}$$

Here $\mathbf{r}_\perp = (y,z)$, $\Omega_\perp$ and ω_0 are the transverse and axial harmonic oscillator frequencies.

We limit ourselves to a cigar-shaped trap potential, i.e. a potential where the transverse linear oscillator length, $a_\perp = \sqrt{\hbar/m\Omega_\perp}$, is much smaller than the longitudinal one, $a_0 = \sqrt{\hbar/m\omega_0}$, $a_\perp \ll a_0$. Then it is natural to investigate the possibility of reducing the three-dimensional (3D) GP equation (1) to an effective 1D model. In addition to the characteristic scales, a_0 and $a_\perp$, the problem at hand possesses an additional one – the so-called "healing length" ξ defined as

$$\xi = \frac{1}{\sqrt{8\pi n|a_s|}} \tag{3}$$

where n is the mean particle density, which can be estimated from the relation $n \approx \frac{\mathcal{N}}{a_\perp^2 a_0}$ with $\mathcal{N}$ being the total number of atoms. This also defines the normalization of the order parameter:

$$\int |\Psi|^2 d\mathbf{r} = \mathcal{N} \tag{4}$$

Thus, the relation between the three characteristic scales $\{a_0, a_\perp, \xi\}$ defines the approximation to be used. We will concentrate on the case $a_\perp \ll a_0, \xi$, and define the small parameter of the problem

$$\epsilon = \frac{a_\perp}{\xi} \ll 1 \tag{5}$$

2.2 Effective One-Dimensional Equation

Now a reduction from the 3D GP equation to the 1D NLS equation with a parabolic potential can be made by means of the *multiple-scale expansion.* To this end, we introduce the dimensionless independent variables $T = \omega_\perp t/2$, $X = x/a_\perp$, $R_\perp = \mathbf{r}_\perp/a_\perp = (Y, Z)$, and a re-normalized order parameter $\tilde{\Psi}$ through the relation $\Psi = \frac{\epsilon}{a_\perp |a_s|^{1/2}}\tilde{\Psi}$, and re-write (1) in dimensionless form:

$$i\frac{\partial \tilde{\Psi}}{\partial T} = \mathcal{L}\tilde{\Psi} + 8\pi\epsilon^2\sigma|\tilde{\Psi}|^2\tilde{\Psi} \tag{6}$$

where $\sigma = \text{sign}(a_s)$,

$$\mathcal{L} = -\frac{\partial^2}{\partial X^2} - \frac{\partial^2}{\partial Y^2} - \frac{\partial^2}{\partial Z^2} + \nu^2 X^2 + Y^2 + Z^2$$

and $\nu = a_0^2/a_\perp^2$. In these units, the wave function is normalized to one, i.e.

$$\int | \tilde{\Psi} |^2 \, d\mathbf{R} = 1 \, . \tag{7}$$

Taking into account the fact that ϵ is a small parameter and that the nonlinearity disappears in the limit $\epsilon \to 0$, we construct a solution of (1) perturbatively as a modulation of a solution of the linear problem associated with (6):

$$\mathcal{L}\phi_{\mathbf{n}} = \mathcal{E}_{\mathbf{n}}\phi_{\mathbf{n}} \, . \tag{8}$$

Equation (8) is the equation for a linear quantum oscillator, whose solutions are very well-known (see e.g. [6]), while $\mathbf{n}$ refers to three quantum numbers.

The next calculation depends on the choice of the underlying linear mode $\phi_{\mathbf{n}}(X)$, i.e. on the particular statement of the problem. We concentrate on the most interesting case, from the practical point of view, where $\phi_{\mathbf{n}}$ is the ground state of the 3D linear oscillator (8), i.e. where $\mathbf{n} = (0,0,0)$ and

$$\phi_{\mathbf{n}} \equiv \phi_0(\mathbf{R}) = \frac{\nu^{1/4}}{\pi^{3/4}} \exp\left(-\frac{\nu X^2 + Y^2 + Z^2}{2}\right), \qquad \mathcal{E}_{\mathbf{n}} = \mathcal{E}_0 = 2 + \nu \quad (9)$$

Next, we look for solutions of (6) of the form

$$\tilde{\Psi} = \epsilon\psi_1 + \epsilon^2\psi_2 + \cdots , \quad (10)$$

where ψ_1 is the weakly-modulated linear ground state wavefunction mentioned above. It has the form

$$\psi_1 = A(x_1, t_1)\phi_0(\mathbf{r}_0)\mathrm{e}^{-\mathrm{i}(2+\nu)t_0} , \quad (11)$$

The modulating amplitude $A(x_1, t_1)$ is considered to be a function of a set of spatial and temporal variables $\{x_1, x_2, \ldots\}$ and $\{t_1, t_2, \ldots\}$, which are introduced as $x_n = \epsilon^n X$ and $t_n = \epsilon^n T$, and $\mathbf{r}_0 = (x_0, y_0, z_0)$. In the arguments of A, only the most "rapid" variables are explicitly shown (i.e., for example, $A(x_1, t_1)$ means that A depends on $\{x_1, x_2, \ldots\}$ and $\{t_1, t_2, \ldots\}$, but not on x_0 and t_0). The scaled variables x_j and t_j are considered to be independent, and hence time and co-ordinate derivatives are expanded as

$$\frac{\partial}{\partial T} = \sum_{\alpha=0} \epsilon^\alpha \frac{\partial}{\partial t_\alpha}, \quad \frac{\partial}{\partial X} = \sum_{\alpha=0} \epsilon^\alpha \frac{\partial}{\partial x_\alpha}, \quad \frac{\partial}{\partial Y} = \frac{\partial}{\partial y_0}, \quad \frac{\partial}{\partial Z} = \frac{\partial}{\partial z_0} \quad (12)$$

For the next part, it is important that we choose the basis $\phi_{\mathbf{n}}$ to be orthonormal

$$\int_{-\infty}^{\infty} \bar{\phi}_{\mathbf{n}}(\mathbf{r}_0)\phi_{\mathbf{n}'}(\mathbf{r}_0)d\mathbf{r}_0 = \delta_{\mathbf{n}\mathbf{n}'} \quad (13)$$

where $\delta_{\mathbf{n}\mathbf{n}'}$ is the Kronecker delta. An overbar, standing for complex conjugation, can be dropped, since $\phi_{\mathbf{n}}$ are real eigenfunctions of the linear oscillator.

Substituting (10)–(12) in (6) and collecting all terms of the same order in ϵ, we obtain, to first order:

$$\mathrm{i}\frac{\partial\psi_1}{\partial t_0} - \mathcal{L}\psi_1 \equiv 0 . \quad (14)$$

which is evidently satisfied by ψ_1 given by (11). Thus, ψ_1 is the kernel of the operator $i\partial/\partial t_0 - \mathcal{L}$.

To order ϵ^2, the following equation is obtained:

$$\mathrm{i}\frac{\partial\psi_2}{\partial t_0} - \mathcal{L}\psi_2 = -\mathrm{i}\frac{\partial\psi_1}{\partial t_1} - 2\frac{\partial^2\psi_1}{\partial x_0 \partial x_1} . \quad (15)$$

Its solution can be sought using the form

$$\psi_2 = \sum_{\mathbf{n}\neq\mathbf{0}} B_{\mathbf{n}}(x_1, t_1)\phi_{\mathbf{n}}(\mathbf{r}_0)e^{-i(\nu+2)t_0} \tag{16}$$

subject to the initial condition $\psi_2(x_0, 0) = 0$ at $t_\alpha = 0$. Note that the correction ψ_2 is orthogonal to ϕ_0 (and thus to ψ_1: $\int \psi_1\psi_2 d\mathbf{r}_0 = 0$), since this is a generic property of a perturbative expansion (see e.g. [6]). Including ψ_0 in the expansion (15) would result in the re-normalization of the initial condition for ψ_1. Multiplying (15) by $\bar{\phi}_0$ and integrating by x_0 over the real axis, one ensures that $\partial A/\partial t_1 = 0$, i.e. that A does not depend on t_1. This, in accord with accepted notation, can be written down as $A \equiv A(x_1, t_2)$. Next, applying $\int_{-\infty}^{\infty} d\mathbf{r}_0 \bar{\phi}_{\mathbf{n}}(\mathbf{r}_0)$, where $\mathbf{n} \neq \mathbf{0}$, one verifies that all $B_{\mathbf{n}}(x_1, t_1) \equiv 0$. In other words, we deduce that $\psi_2 \equiv 0$.

Finally, to third order in ϵ, we get

$$\begin{aligned} \mathrm{i}\frac{\partial\psi_3}{\partial t_0} - \mathcal{L}\psi_3 = &-\mathrm{i}\frac{\partial\psi_1}{\partial t_2} - \mathrm{i}\frac{\partial\psi_2}{\partial t_1} - 2\frac{\partial^2\psi_2}{\partial x_0\partial x_1} \\ &- \left(\frac{\partial^2}{\partial x_1^2} + 2\frac{\partial^2}{\partial x_0\partial x_2}\right)\psi_1 + 8\pi\sigma|\psi_1|^2\psi_1 \,. \end{aligned} \tag{17}$$

Requiring orthogonality between the right-hand-side of this equation and the kernel of the operator $i\partial/\partial t_0 - \mathcal{L}$, i.e. to ψ_1, and taking into account the expression for ψ_1 derived above and the fact that $\psi_2 \equiv 0$, we arrive at the following NLS equation:

$$\mathrm{i}\frac{\partial A}{\partial t_2} + \frac{\partial^2 A}{\partial x_1^2} - 2\sigma\sqrt{\frac{2\nu}{\pi}}|A|^2 A = 0 \,, \tag{18}$$

where we have used the fact that $\int |\phi_0|^4 d\mathbf{r}_0 = \nu^{1/2}/(2\pi)^{3/2}$. In order to reduce (18) to a standard form, we introduce $\psi = \left(\frac{2\nu}{\pi}\right)^{1/4} A$ and use x and t for the variables x_1 and t_2, respectively. Then we arrive at the NLS equation in the form:

$$\mathrm{i}\frac{\partial\psi(x,t)}{\partial t} + \frac{\partial^2\psi(x,t)}{\partial x^2} - 2\sigma|\psi(x,t)|^2\psi(x,t) = 0 \,, \tag{19}$$

and this will be the starting point for the next section.

Since $|\psi|^2$ describes the local density of atoms, condensate excitations governed by (19) can be called *matter waves.*

3 Periodic Solutions

3.1 Stationary Periodic Waves

Let us discuss periodic solutions of (19), i.e. solutions satisfying

$$\psi(x,t) = \psi(x+L,t) \,, \tag{20}$$

where L is the period. The simplest solutions of such type can be obtained in the form of a standing wave by means of the *ansatz*

$$\psi(x,t) = \mathrm{e}^{\mathrm{i}\omega t} u(x) , \tag{21}$$

where $u(x)$ is a real periodic function, $u(x) = u(x+L)$, and ω is the frequency of the wave. Indeed, in this case, (19) is reduced to the following ordinary differential equation:

$$- u_{xx} + \omega u + 2\sigma u^3 = 0 \tag{22}$$

In order to find a solution of (22), we multiply it by u_x and integrate once. Designating the integration constant as C, we then arrive at

$$u_x^2 = C + \omega u^2 + \sigma u^4 , \tag{23}$$

and hence

$$x - x_0 = \int_0^u \frac{du}{\sqrt{C + \omega u^2 + \sigma u^4}} , \tag{24}$$

where x_0 is a constant.

Before proceeding, let us mention that, due to the symmetry of the NLS equation, if a solution $u(x)$ of (23) is found, then it is not difficult to construct a solution of (19) moving with a constant velocity, V. Indeed, it will read

$$\psi(x,t) = \exp\left(\mathrm{i}\left(\omega - \frac{V^2}{4}\right)t + \mathrm{i}\frac{V}{2}x\right) u(x - Vt) , \tag{25}$$

as can be verified by direct substitution of (25) into (19). This point allows us to restrict our consideration in what follows to stationary waves only.

The remaining calculations depend on the particular choice of the parameters C and ω. Although only the simplest solutions of the form (24) will be our main interest, we outline below some aspects of a more general method – the inverse scattering technique (IST) (also referred to as the method of finite gap integration). For more details, see e.g. [9, 10]), which, in the general case provides an algorithm for solving the respective Cauchy problem for (19), and in particular, offers a very convenient parametrization of the problem, i.e. more convenient than the one given by C and ω defined by (23. This last point is especially important in the context of the perturbation theory developed below.

3.2 Some Elements of the Inverse Scattering Technique

The starting point of the IST is the fact that (19) appears as a compatibility condition for two systems of differential equations [9]:

$$|f\rangle_x = \mathbf{U}|f\rangle\,, \qquad |f\rangle_t = \mathbf{V}|f\rangle \tag{26}$$

where the ket-vector is defined as $|f\rangle = \mathrm{col}(f_1, f_2)$, $f_{1,2} = f_{1,2}(x,t)$, and matrices $\mathbf{U}$ and $\mathbf{V}$ are given by

$$\mathbf{U} = \begin{pmatrix} \frac{\lambda}{2\mathrm{i}} & \sqrt{\sigma}\bar{\psi} \\ \sqrt{\sigma}\psi & -\frac{\lambda}{2\mathrm{i}} \end{pmatrix} \text{ and } \mathbf{V} = \begin{pmatrix} -\frac{\lambda^2}{2\mathrm{i}} + \mathrm{i}\sigma|\psi|^2 & -\sqrt{\sigma}(\mathrm{i}\bar{\psi}_x + \lambda\bar{\psi}) \\ \sqrt{\sigma}(\mathrm{i}\psi_x - \lambda\psi) & \frac{\lambda^2}{2\mathrm{i}} - \mathrm{i}\sigma|\psi|^2 \end{pmatrix} \tag{27}$$

and the constant λ is called a spectral parameter. Note that t and x play the role of parameters in the first and second systems in (26). The first of the problems (26) is usually called a Zakharov-Shabat spectral problem. Differentiating these equations with respect to t and x respectively, and requiring (26) to be satisfied for any value of λ, one arrives at the compatibility condition:

$$\mathbf{U}_t - \mathbf{V}_x + [\mathbf{U}, \mathbf{V}] = 0\,, \tag{28}$$

and, as a consequence, to the NLS equation (19).

Let us now follow [27] and introduce a bra-vector $\langle f| = (-f_2, f_1)$, and define the internal product

$$\langle f|g\rangle = \det\begin{vmatrix} f_1 & g_1 \\ f_2 & g_2 \end{vmatrix} \tag{29}$$

and a projector matrix

$$\mathbf{P} \equiv |f\rangle\langle g| = \begin{pmatrix} -f_1 g_2 & f_1 g_1 \\ -f_2 g_2 & f_2 g_1 \end{pmatrix} \tag{30}$$

It is clear that, if $|f\rangle$ and $|g\rangle$ are two different solutions of each of the systems (26), then $\langle f|g\rangle$ is nothing but a Wronskian of each of them, and hence it depends neither on x nor on t, i.e. $\langle f|g\rangle = p(\lambda)$, where $p(\lambda)$ is a function of the spectral parameter only. Let us suppose that a solution $|f\rangle = \mathrm{col}(f_1, f_2)$ of (26) with some ψ is given. Then, the symmetry properties of the matrices $\mathbf{U}$ and $\mathbf{V}$ further imply that $|g\rangle = \mathrm{col}(\bar{f}_2, \sigma\bar{f}_1)$ is a solution of (26) with the same ψ. Using the pair of solutions $|f\rangle$ and $|g\rangle$ (we will refer to them as conjugate solutions), one immediately obtains:

$$p(\lambda) = \langle f|g\rangle = \sigma|f_1|^2 - |f_2|^2 \tag{31}$$

Next, by using straightforward algebra, one ensures that $\mathbf{P}$ solves equations

$$\mathbf{P}_x = [\mathbf{U}, \mathbf{P}]\,, \qquad \mathbf{P}_t = [\mathbf{V}, \mathbf{P}] \tag{32}$$

From this system, it follows that

$$\Delta P^2 - 4P_{12}P_{21} = P(\lambda) \tag{33}$$

where $\Delta P = P_{22} - P_{11}$, P_{ij} are entries of the matrix $\mathbf{P}$, and $P(\lambda)$ is a function of λ only (i.e. it does not depend on x and t). For a pair of conjugate solutions, one verifies that $P_{21} = \sigma \bar{P}_{12}$, and thus $P_{12}P_{21} = \sigma |f_1|^2|f_2|^2$ and $\Delta P = \sigma|f_1|^2 + |f_2|^2$. Consequently, $P(\lambda) = p^2(\lambda)$, i.e. it is a real positive function of λ.

Thus far, we have not specified the functional dependence of $P(\lambda)$ (or $p(\lambda)$). By imposing such a dependence, one restricts consideration to a specific class of solutions. We are interested in the simplest periodic ones. This justifies the choice of $P(\lambda)$ made in what follows. Specifically, we note that $\mathbf{U}$ and $\mathbf{V}$ are matrix polynomials in λ. This allows us to guess that the simplest non-trivial solution $|f\rangle$, and hence $\mathbf{P}$, should also be sought in the forms of polynomials with respect to λ. As long as $\mathbf{V}$ is quadratic with respect to λ, one has to consider $|f\rangle$ to be a second (or higher) order polynomial with respect to λ. Since $\mathbf{P}$ is quadratic with respect to the elements of $|f\rangle$, one concludes that the simplest non-trivial form of $P(\lambda)$ must also be a polynomial of fourth degree with respect to λ. Also taking into account the fact that $P(\lambda)$ is real, one arrives at the following representation

$$\begin{aligned} P(\lambda) &= (\lambda - \lambda_1)(\lambda - \lambda_2)(\lambda - \lambda_3)(\lambda - \lambda_4) \\ &= \lambda^4 + c_3\lambda^3 + c_2\lambda^2 + c_1\lambda + c_0 \end{aligned} \tag{34}$$

where $\lambda_j (j = 1, \ldots, 4)$ are complex numbers to be chosen in such a manner that all the c_j are real. (They are expressed through λ_j by the obvious relations). The real $\xi_j = \mathrm{Re}\lambda_j$ and imaginary $\eta_j = \mathrm{Im}\lambda_j$ parts of λ_j are the parameters characterizing respective solutions of the nonlinear problem (19). The requirement that all c_j $(j = 0, \ldots, 4)$ be real imposes four constraints on the parameters, and this means that we have only four independent parameters. Indeed, considering $\bar{P}(\lambda)$, one concludes that: (i) all λ_j are real, i.e. all $\eta_j = 0$ and ξ_j parametrize the problem, or (ii) two of them are real and two are complex conjugates, say $\mathrm{Im}\lambda_1 = \mathrm{Im}\lambda_2 = 0$ and $\lambda_3 = \bar{\lambda}_4$, in which case the parameters are $\{\xi_1, \xi_2, \xi_3, \eta_3\}$, or (iii) there are two pairs of complex conjugate parameters, say $\lambda_1 = \bar{\lambda}_2$ and $\lambda_3 = \bar{\lambda}_4$, i.e. the parameters are $\{\xi_1, \xi_3, \eta_1, \eta_3\}$.

Formulae (33) and (34) suggest that P_{ij} should also be sought in the form of polynomials (at least of second order) with respect to the parameter λ. This is why we make the substitution

$$\Delta P = \lambda^2 + \lambda\Delta_1 + \Delta_0 \,, \qquad P_{12} = \lambda P_{12}^{(1)} + P_{12}^{(0)} \tag{35}$$

and look for explicit forms of the coefficients Δ_j and $P_{12}^{(j)}$. Comparing powers of λ in (33), we obtain four relations:

$$\Delta_1 = \frac{c_3}{2} \,, \qquad \Delta_0 = 2\sigma|P_{12}^{(1)}|^2 + \frac{1}{2}\left(c_2 - \frac{1}{4}c_3^2\right) \tag{36}$$

$$c_3\Delta_0 - 4\sigma(\bar{P}_{12}^{(1)}P_{12}^{(0)} + \bar{P}_{12}^{(0)}P_{12}^{(1)}) = c_1 \tag{37}$$

$$\Delta_0^2 - 4\sigma|P_{12}^{(0)}|^2 = c_0 \tag{38}$$

From the first of (32), we compute the equation for $\partial P_{12}/\partial x$ whose consistency with representations (35) and (36) results in the following connecting formulae:

$$\bar{\psi} = \frac{\mathrm{i}}{\sqrt{\sigma}} P_{12}^{(1)} , \tag{39}$$

$$\bar{\psi}_x = \frac{1}{\sqrt{\sigma}} P_{12}^{(0)} + \frac{\mathrm{i}}{2} c_3 \tag{40}$$

$$\bar{\psi}_{xx} = \bar{\psi} \Delta_0 \tag{41}$$

Comparing (41) with (19), one easily verifies that the two equations coincide, subject to the conditions

$$c_2 = \sum_{\substack{i,j=1 \\ i \neq j}}^{4} \lambda_i \lambda_j = 2\omega \tag{42}$$

$$c_3 = \lambda_1 + \lambda_2 + \lambda_3 + \lambda_4 = 0 \tag{43}$$

(It worth pointing out here that in the case of the more general solution (25), the sum of the parameters λ_j, i.e. coefficient c_3, determines the velocity V, and thus (43) expresses the fact that the velocity of the solution we are seeking is equal to zero). Then, it follows from (39) that $\bar{P}_{12}^{(1)}$ is a solution of (19). If one also takes into account the particular form of the solution (21), one immediately concludes from (37), (39), (40), and (43) that $c_1 = 0$.

Finally, we compare (38) with (23) taking into account (21). This gives

$$C = \frac{\sigma}{16}(c_2^2 - 4c_0) = \frac{\sigma}{4}(\omega^2 - \lambda_1\lambda_2\lambda_3\lambda_4) \tag{44}$$

The last condition (43) leaves us with only two parameters, i.e. in other words we are dealing with the simplest two-parameter solutions. In Fig. 1, we show the possible locations of λ_j, in the complex plane, which satisfy (43) and result in periodic solutions – they correspond to cases (i) Fig. 1a and (ii) Fig. 1c, e.

Let us now try to understand why the parametrization given by λ_j is convenient for the construction of the perturbation theory for periodic waves. To this end, we explicitly write down three types of periodic solutions, corresponding to the Figs. 1a, c, and e.

3.3 *sn*-Wave in a BEC with a Positive Scattering Length

We start with a periodic solution of (19), with $\sigma = 1$ and parameters as shown in Fig. 1a. Then, as follows from (42) and (44), $\omega = (\xi_1^2 + \xi_2^2)/2$ and $C = (\xi_1^2 - \xi_2^2)^2/4$. Substituting these expressions into (24), and computing the integral (see e.g. [11]), one obtains

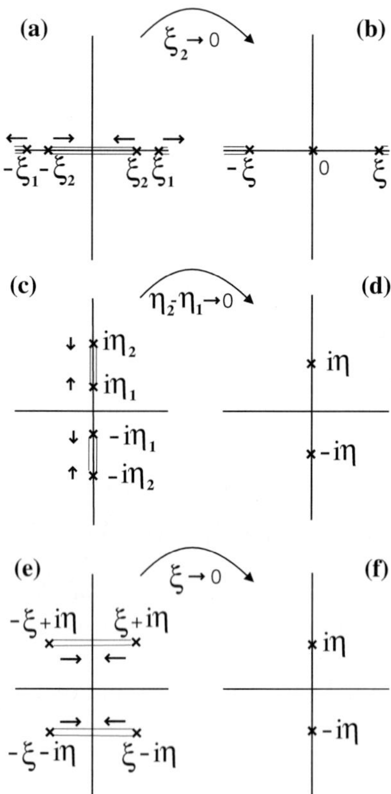

Fig. 1. The Riemann surface of the function $\sqrt{P(\lambda)}$ consists of two complex planes, one of which is shown in diagram (**a**) for the sn-wave: $\lambda_1 = -\lambda_3 = \xi_1$, $\lambda_2 = -\lambda_4 = \xi_2$, (**c**) for the dn-wave: $\lambda_1 = \bar{\lambda}_3 = i\eta_1$, $\lambda_2 = \bar{\lambda}_4 = i\eta_2$, and (**e**) for the cn-wave: $\lambda_1 = -\bar{\lambda}_3 = -\lambda_2 = -\bar{\lambda}_4 = \xi + i\eta$. The passage between (**a**) and (**b**) illustrates the motion of the branching points $\xi_{1,2}$ (indicated by arrows) corresponding to the deformation of the sn-wave into a dark soliton. The branching points $\pm\xi_2$ collapse onto one point, corresponding to the discrete spectrum, while $\pm\xi_1$ go to the edges of the continuum spectrum of the ZS spectral problem for a finite-density. (This is the case $\sigma = 1$ and $\lim_{x\to\pm\infty}\psi = \rho e^{\pm i\varphi}$, where ρ and φ are real constants, see e.g. [9] for details). The passage between (**c**) and (**d**) illustrates the motion of the branching points, resulting in a transformation of the dn-wave into a static bright soliton. The passage between (**e**) and (**f**) illustrates the motion of the branching points, resulting in the transformation of the cn-wave into a static bright soliton. In these last two cases, the branching points collapse onto the discrete spectrum of the ZS spectral problem (the case $\sigma = -1$, $\lim_{x\to\pm\infty}|\psi| = 0$)

$$\psi_{sn} = \frac{\xi_1 - \xi_2}{2} \exp\left(-\frac{i}{2}(\xi_1^2 + \xi_2^2)t\right) \mathrm{sn}\left(\frac{\xi_1 + \xi_2}{2} x \mid m\right) \tag{45}$$

where "sn" (also "cn" and "dn" below) stands for a Jacobi elliptic function and m, the elliptic parameter, is given by

$$m = \frac{(\xi_1 - \xi_2)^2}{(\xi_1 + \xi_2)^2} \ . \tag{46}$$

The above solution is a periodic matter wave (see Fig. 2 below) in a BEC with a positive scattering length. (Hereafter, for the elliptic modulus, periods and number of particles for different solutions, we use the same notations m, L and N. This does not lead to confusion, because each solution is considered separately.)

With any periodic wave, $\psi(x,t)$, one can also associate the number of particles belonging to one period (below, this is simply the number of particles, which, however, should not be confused with the real number of particles $\mathcal{N}$ introduced in (4)):

$$N(\lambda_1(t), \lambda_2(t)) = \int_0^L |\psi|^2 dx \ . \tag{47}$$

Then, for the sn-wave (45), one has

$$L = \frac{8K(m)}{\xi_1 + \xi_2} \ , \qquad N = 2(\xi_1 + \xi_2)(K(m) - E(m)) \ , \tag{48}$$

where $K(m)$ and $E(m)$ are standard notations for the complete elliptic integrals of the the first and second kinds. (For the properties of Jacobi elliptic functions and their integrals, see e.g. [11, 12]).

Let us suppose now that one is interested in the parametric transformation of the periodic sn-wave into a dark soliton. Then, as is shown in Fig. 1a, b, one has to shift (by some means) the branching points $\pm\xi_2$ towards zero, providing the limiting transition $\xi_2 \to 0$, while $\xi_1 \to \xi$, where ξ is a finite value. Assuming that this limit is achieved, one obtains, from (46) the fact that $m \to 1$. Then, using the well-known limits of the Jacobi elliptic functions (see e.g. [11]), one obtains, from (45), the limit

$$\psi_{sn} \stackrel{m\to 1}{\longrightarrow} \psi_{ds} \ , \tag{49}$$

where

$$\psi_{ds} = \frac{\xi}{2} \exp\left(-\frac{\mathrm{i}}{2}\xi^2 t\right) \tanh\left(\frac{\xi}{2} x\right) \tag{50}$$

is nothing but the dark soliton of the NLS equation.

We also mention here that another limit, $\xi_{1,2} \to \xi$, i.e. $2\Delta\xi = \xi_1 - \xi_2 \to 0$ and $m \to 0$, yields

$$\psi_{sn} \stackrel{m\to 0}{\longrightarrow} \Delta\xi \mathrm{e}^{-\mathrm{i}\xi^2 t} \sin(\xi x) \ , \tag{51}$$

which corresponds to a linear wave.

3.4 *dn*-Wave in a BEC with a Negative Scattering Length

Let us now turn to the case of a BEC with a negative scattering length, $\sigma = -1$, and consider deformations of the periodic dn-wave, as illustrated in Figs. 1 c,d. In order to find an explicit form for the periodic dn-wave, one takes into account that in the case at hand $\omega = (\eta_1^2 + \eta_2^2)/2$ and $C = (\eta_1^2 - \eta_2^2)^2/16$. Then, computing the integral in (24) gives

$$\psi_{dn} = \frac{\eta_1 + \eta_2}{2} \exp\left(\frac{\mathrm{i}}{2}(\eta_1^2 + \eta_2^2)t\right) \mathrm{dn}\left(\frac{\eta_1 + \eta_2}{2} x \mid m\right) \tag{52}$$

where the parameter m is given by

$$m = \frac{4\eta_1\eta_2}{(\eta_1 + \eta_2)^2} \ . \tag{53}$$

The two important parameters of this solution – the period L and the number of particles N – are now given by

$$L = \frac{4K(m)}{\eta_1 + \eta_2}\ , \qquad N = (\eta_1 + \eta_2)E(m) \tag{54}$$

If the respective NLS equation (19) is considered subject to the zero boundary conditions, it possesses a solution in the form of a bright soliton (see e.g. [9])

$$\psi_{bs} = \eta e^{\mathrm{i}\eta^2 t} \mathrm{sech}(\eta x) \ . \tag{55}$$

The complex plane of the corresponding ZS spectral problem is shown in Fig. 1d. As is shown in the figure, the deformation should result in $\eta_{1,2} \to \eta$ (i.e. $|\eta_1 - \eta_2| \to 0$) when the branching points $\pm i\eta_{1,2}$ collapse into two poles, $\pm i\eta$, which determine the discrete spectrum of the corresponding ZS spectral problem. Then, the following transition occurs

$$\psi_{dn} \stackrel{m\to 1}{\longrightarrow} \psi_{bs} \ . \tag{56}$$

The linear limit of the solution (52) corresponds to $\eta_{1,2} \to 0$, where ψ_{dn} reduces to a constant: $\psi_{dn} \stackrel{m\to 0}{\longrightarrow} \eta$ $(\eta \to 0)$.

3.5 *cn*-Wave in a BEC with a Negative Scattering Length

Finally, we consider the situation shown in Figs. 1 e, f for a BEC with a negative scattering length, $\sigma = -1$. Then $\lambda_1 = \bar{\lambda}_2 = -\lambda_3 = -\bar{\lambda}_4 = \xi + \mathrm{i}\eta$ and we get a deformation of the periodic cn-wave into a bright soliton (55)

$$\psi_{cn} \stackrel{m\to 1}{\longrightarrow} \psi_{bs} \tag{57}$$

corresponds to $\xi \to 0$.

In order to find an explicit expression for the cn-wave, one first computes $\omega = \eta^2 - \xi^2$ and $C = \eta^2\xi^2$, and then the integral in (24); this gives

$$\psi_{cn} = \eta \exp\left(-i(\eta^2 - \xi^2)t\right) \operatorname{cn}\left(\sqrt{\xi^2 + \eta^2}x \mid m\right) \tag{58}$$

Now the parameter m is given by

$$m = \frac{\eta^2}{\xi^2 + \eta^2} \tag{59}$$

The period and the number of particles for the cn-wave are given by

$$L = \frac{4K(m)}{\sqrt{\xi^2 + \eta^2}}, \qquad N = 4\sqrt{\xi^2 + \eta^2}E(m) - \xi^2 L \tag{60}$$

The linear limit is described by $\eta \to 0$, where $m \to 0$ and

$$\psi_{cn} \stackrel{m\to 0}{\longrightarrow} \eta e^{2i\xi t} \cos(\xi x) \tag{61}$$

4 Management of Matter Waves Using the Feshbach Resonance

As has been shown above, in order to change the aspect ratio of the periodic wave, and in particular to transform it into a soliton (or into a sequence of solitons as will be done below), in practice one has to find a way of moving the branching points λ_j in accordance with the change in wave parameters. To this end, one has to perturb the evolution equation (19), but this must be done weakly enough, to avoid destroying the structure of the periodic wave. One way could be a change in the nonlinearity, which, in the case of a BEC, can be achieved by means of the Feshbach resonance.

The Feshbach resonance occurs in a scattering process of two atoms when a quasi-bound state and free states (i.e. asymptotic states before and after the scattering) of the atoms are resonantly coupled, so that the scattering process passes through the bound phase. Then, the scattering length (we recall that a_s was introduced in (1)) can be drastically changed (even its sign can be changed). The process can be controlled by an external magnetic field, $B(t)$, varying with time. A model simulating the process has the following mathematical expression [16]:

$$a_s(t) = a_s(0)g(t), \qquad g(t) = 1 + \frac{\Delta}{B_0 - B(t)}. \tag{62}$$

Here $a_s(0)$ is the asymptotic value of the scattering length far from resonance, B_0 is the resonant value of the magnetic field, and Δ is the width of the resonance. Feshbach resonances have been observed in Na at 853 and 907 G

[17], in ^{7}Li at 725 G [19], and in ^{85}Rb at 164 G with $\Delta = 11$G [18]. Also, a rapid variation in time of a_s has been used for generation [20, 19] and was predicted to be useful for management [21] of bright solitons in BECs.

Temporal dependence of a_s means a dependence on time of the nonlinear coefficient: $g_0 \to g_0 \cdot g(t)$. Assuming that the dependence of $g(t)$ on time is slow enough (i.e. that it is governed by $\epsilon^2 t$, where ϵ is given by (5) and t is the real physical time), one can repeat the arguments of Sect. 2 and arrive at the evolution equation (cf. (19))

$$\mathrm{i}\frac{\partial\psi(x,t)}{\partial t} + \frac{\partial^2\psi(x,t)}{\partial x^2} - 2\sigma g(t)|\psi(x,t)|^2\psi(x,t) = 0\,. \tag{63}$$

In what follows, we consider only the case where $g(t)$ is a positive function, so that the Feshbach resonace does not change the type of two-body interaction. Then, in order to study the condensate evolution within the framework of the model (63), we note that substitution

$$\psi(x,t) = v(x,t)/\sqrt{g(t)} \tag{64}$$

transforms (63) into a dissipative NLS equation (cf. (19))

$$\mathrm{i}v_t + v_{xx} - 2\sigma|v|^2 v = \mathrm{i}\gamma v, \tag{65}$$

where

$$\gamma(t) = \frac{1}{2g(t)}\frac{dg(t)}{dt} \tag{66}$$

Thus, for slowly-varying $g(t)$, the right-hand-side of (65) can be considered as a small perturbation: $|\gamma(t)| \ll 1$. If the nonlinearity increases ($\gamma(t) > 0$, the case we will study), then the perturbation describes growth; otherwise, when the nonlinearity is decaying ($\gamma(t) < 0$), the perturbation in (65) describes dissipation.

Since $g(t)$ has been defined so that $g(0) = 1$ is satisfied, it follows from (64), that one has, for the initial distribution, $v(x,0) = \psi(x,0)$.

We will consider the perturbed evolution of periodic waves within the framework of (65). Under the influence of the dissipative perturbation, the wave shape is changed in a way that can be described by variations of the parameters λ_j, assuming them to be slow functions of time t (the so-called *adiabatic approximation*). Equations which govern their evolution can be derived by the following simple method.

First of all, we recall that in all the examples considered in Sect. 3, the waves had two parameters, and hence we expect to obtain two equations for the adiabatic approximation. Let us temporarily change notation and use, in the rest of this subsection, $\lambda_{1,2}$ for the respective parameters (thus $\lambda_{1,2} = \xi_{1,2}$ for the sn-wave, $\lambda_{1,2} = \eta_{1,2}$ for the dn-wave, and $\lambda_1 = \eta$ and $\lambda_2 = \xi$ for the cn-wave). Since the coefficients in the dissipative model (65) are supposed to be independent of x, the perturbation does not result in any change of the period L of the nonlinear wave; this can be expressed as follows:

$$\frac{d}{dt}L(\lambda_1,\lambda_2) = 0 \,. \tag{67}$$

Next, we use straightforward algebra to show that the number of particles defined by (47), of a periodic solution of (65), evolves according to

$$\frac{dN(\lambda_1,\lambda_2)}{d\zeta} = N(\lambda_1,\lambda_2) \tag{68}$$

where

$$\zeta = \zeta(t) = 2\int_0^t \gamma(t')dt' = \ln g(t)\,, \qquad \zeta(0) = 0 \tag{69}$$

Then, if the expressions for L and N in terms of $\lambda_{1,2}$ are known, (67) and (68) reduce to a system of two quasi-linear differential equations of first order for $\lambda_{1,2}$. The form of this system depends, of course, on the choice of the parameters $\lambda_{1,2}$, and below it will be analyzed for the three types of the nonlinear periodic waves introduced in Sect. 3.

4.1 Adiabatic Dynamics of the Sn-Wave

Let us start with the adiabatic dynamics of a periodic matter wave in a BEC with a positive scattering length. Now, we seek a solution of (65) in the form (45), with the ξ_j being functions of time, or, more precisely, of $\zeta(t)$. Substituting (48) into (67) and (68), and rearranging the terms we obtain

$$\begin{aligned}\frac{d\xi_1}{d\zeta} &= \frac{\xi_1(\xi_1+\xi_2)(\mathrm{K}(m)-\mathrm{E}(m))}{2\xi_1\mathrm{K}(m)-(\xi_1+\xi_2)\mathrm{E}(m)}\\ \frac{d\xi_2}{d\zeta} &= \frac{\xi_2(\xi_1+\xi_2)(\mathrm{K}(m)-\mathrm{E}(m))}{2\xi_2\mathrm{K}(m)-(\xi_1+\xi_2)\mathrm{E}(m)}\end{aligned} \tag{70}$$

where m is defined through ξ_k in (46). Having solved this system, and recalling links (64) and (69), one can find the the local density of particles in the condensate through the formula

$$|\psi(x,t)|^2 = e^{-\zeta}\frac{[\xi_1(\zeta)-\xi_2(\zeta)]^2}{4}\mathrm{sn}^2\left(\frac{\xi_1(\zeta)+\xi_2(\zeta)}{2}x \mid m\right) \tag{71}$$

System (70) cannot be solved explicitly. In Fig. 2, we represent its numerical solution, illustrating how the sn-wave is transformed into a lattice of dark solitons by means of an increase in the magnetic field. (Henceforth, in numerical simulations, MAPLE codes are used for elliptic functions and to solve the ODEs).

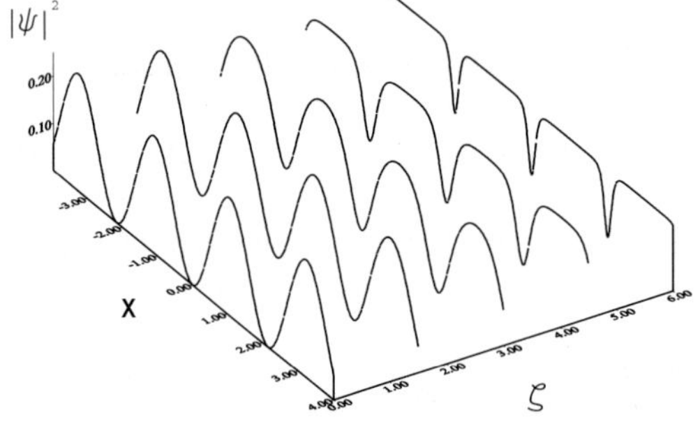

ζ	ξ_2	ξ_1
0	0.5	1
1.5	0.306	1.34
3	0.087	2.13
4.5	0.0036	3.87
6	$0.2{\cdot}10^{-5}$	7.63

Fig. 2. Adiabatic dynamics of the sn-wave, characterized by $L \approx 4.313$ and $N \approx 0.547$. The table shows the corresponding dynamics of the wave parameters. At $\zeta = 6$, the wave profile is a clearly-visible train of dark solitons

Analytic treatment of (70), however, is possible in the limit where $\xi_2 \to 0$, i.e. in the limit where the sn-wave is transformed into a train of dark solitons. To this end, we employ the well-known asymptotics of the complete elliptic integrals where $m \to 1$ (see e.g. [11]):

$$\begin{aligned} \mathrm{K}(m) &= \ln \frac{4}{\sqrt{1-m}} \left(1 + O(1-m)\right) \\ \mathrm{E}(m) &= 1 + O((1-m)\ln(1-m)) \end{aligned} \tag{72}$$

Then, to leading order (when $\xi_1 \to \infty$ and $\xi_2 \to 0$), the system (70) reduces to

$$\frac{d\xi_1}{d\zeta} = \frac{\xi_1}{2}, \qquad \frac{d\xi}{d\zeta} = -\frac{\xi_2}{2} \ln \frac{\xi_1}{\xi_2} \tag{73}$$

The first of these equations can be solved trivially. To solve the second one, we observe that, using (73), one can deduce the relation

$$\frac{d\xi_2}{d\xi_1} = -\frac{\xi_2}{\xi_1} \ln \frac{\xi_1}{\xi_2}, \tag{74}$$

which is also readily solved, giving

$$\xi_2 = \exp(1 + \ln \xi_1 + C_0 \xi_1) \sim \mathrm{e}^{C_0 \xi_1}. \tag{75}$$

Here C_0 is an integration constant which can be found from the (asymptotic) condition of the period conservation computed from (48):

$$L \sim \frac{4}{\xi_1} \ln \frac{\xi_1}{\xi_2} \sim -4C_0. \tag{76}$$

Thus we arrive at the asymptotics:

$$\xi_1 \sim \mathrm{e}^{\zeta/2}\,, \qquad \xi_2 \sim \xi_1 \mathrm{e}^{-L\xi_1/4} \tag{77}$$

Thus one observes a scenario different from the one described by (49) for the deformation of a periodic sn-wave into a dark soliton. In fact, now ξ_1 goes to infinity, and we obtain a "train of dark solitons". This is related to the fact that the perturbation does not affect the period of the nonlinear wave, however, in order to obtain a single dark soliton, the period should go to infinity in the process of the wave deformation. Nonetheless, the shape of the train of dark solitons indeed approximates the shape of a set of dark solitons. This occurs due to the extremely rapid convergence of ξ_2 to zero. According to experimental estimations, a good model for the temporal dependence is given by the exponential function: $g(t) = \exp(t/\tau)$, where τ is a constant. Then $\zeta = t/\tau$ and $\xi_1 \sim t$ while $\xi_2 \sim t\exp(-\mathrm{const}\cdot t)$.

4.2 Adiabatic Dynamics of the Dn-Wave

Let us now consider the effect of the Feshbach resonance on the dn-wave in a BEC with a negative scattering length. Assuming that the form of the solution of (65) is given by (52) with time-dependent parameters, η_j, and substituting (54) into (67) and (68), we obtain a system of equations for the adiabatic approximation for η_1 and η_2:

$$\begin{aligned}\frac{d\eta_1}{d\zeta} &= -\frac{\eta_1(\eta_1+\eta_2)\mathrm{E}(m)}{(\eta_2-\eta_1)\mathrm{K}(m)-(\eta_1+\eta_2)\mathrm{E}(m)}\\ \frac{d\eta_2}{d\zeta} &= \frac{\eta_2(\eta_1+\eta_2)\mathrm{E}(m)}{(\eta_2-\eta_1)\mathrm{K}(m)+(\eta_1+\eta_2)\mathrm{E}(m)}\end{aligned} \tag{78}$$

where, for the sake of definiteness, it is supposed that $\eta_2 > \eta_1$, m is given by (53), and ζ is defined by (69). Now the density of particles is given by

$$|\psi(x,t)|^2 = \mathrm{e}^{-\zeta}\frac{[\eta_1(\zeta)+\eta_2(\zeta)]^2}{4}\mathrm{dn}^2\left(\frac{\eta_1(\zeta)+\eta_2(\zeta)}{2}x \mid m\right) \tag{79}$$

An example of a numerical solution of (78) is given in Fig. 3. As in the previous sub-section, one observes generation of a train of bright matter solitons from a weakly-modulated periodic wave, due to the conservation of the period of the wave.

In order to describe the asymptotic behavior at $|\eta_1 - \eta_2| \to 0$, i.e. at $m \to 1$, we introduce $\eta = (\eta_1+\eta_2)/2$ and $\Delta\eta = (\eta_1-\eta_2)/2$. Then, using (72), we obtain from (78):

$$\frac{d\eta}{d\zeta} = \eta\,, \qquad \frac{d\Delta\eta}{d\zeta} = -\Delta\eta \ln\frac{\eta}{\Delta\eta} \tag{80}$$

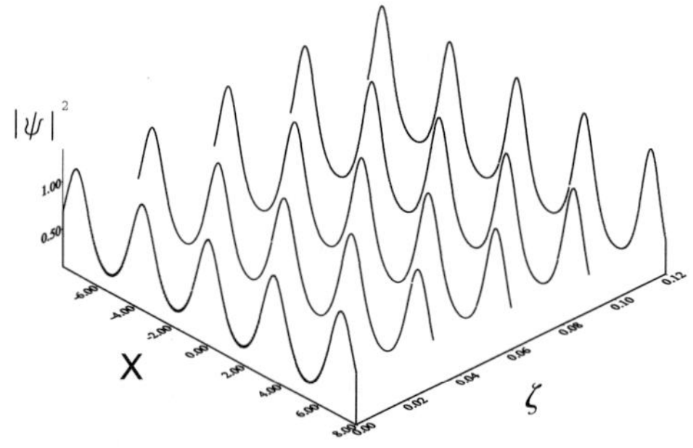

ζ	η_2	η_1
0.0	0.3	0.9
0.03	0.37	0.91
0.06	0.44	0.93
0.09	0.5	0.95
0.12	0.55	0.96

Fig. 3. Evolution of the dn-wave in a BEC with a negative scattering length. The table shows the respective change in the wave parameters. The situation considered corresponds to $L \approx 1.74$ and $N \approx 1.2$

Comparing these equations with (73), one observes the universality of the limiting behavior. Thus, using (77), one can immediately write down the asymptotic formulae (where, as before, the constant is found from the conservation of the wave period):

$$\eta \sim e^{\zeta} , \qquad \Delta\eta \sim \eta e^{-L\eta/4} \tag{81}$$

As in the case of the sn-wave, there is a peculiarity in the behavior of the dn-wave parameters in the limit $m \to 1$: both η_1 and η_2 tend to infinity. This is consistent with the fact that we cannot obtain a single bright soliton if we keep the wave period constant. At the same time, as we would expect, the difference $|\eta_1 - \eta_2|$ rapidly goes to zero, and as a result, the wave shape approaches the shape of a "train of bright solitons".

4.3 Adiabatic Evolution of the Cn-Wave

As the final example in this section, we consider the evolution of the periodic cn-wave, given by (58), within the framework of the dissipative NLS equation (65) with $\sigma = -1$. Substitution of the calculated L and N from (60) into (67) and (68) yields the system

$$\begin{aligned} \frac{d\eta}{d\zeta} &= \frac{[(\eta^2+\xi^2)\mathrm{E}(m) - \xi^2\mathrm{K}(m)]\mathrm{E}(m)\eta}{\eta^2\mathrm{E}^2(m) + \xi^2[\mathrm{K}(m) - \mathrm{E}(m)]^2} , \\ \frac{d\xi}{d\zeta} &= \frac{[\xi^2\mathrm{K}(m) - (\eta^2+\xi^2)\mathrm{E}(m)][\mathrm{K}(m) - \mathrm{E}(m)]\xi}{\eta^2\mathrm{E}^2(m) + \xi^2[\mathrm{K}(m) - \mathrm{E}(m)]^2} \end{aligned} \tag{82}$$

If the dependence of ξ and η on ζ is found from (82), then (64) gives the evolution of the periodic wave $\psi(x,t)$ with a slow change in the parameter ζ

related to time by (69). In particular, the density of particles in the condensate is now given by

$$|\psi(x,t)|^2 = 4\mathrm{e}^{-\zeta}\eta^2(\zeta)\,\mathrm{cn}^2(\sqrt{\eta^2(\zeta)+\xi_2^2(\zeta)}\,x|m) \tag{83}$$

In Fig. 4, we present an example of the evolution of the respective density distribution. The figure shows that, in the case of a negative scattering length given by (62), the increase in the scattering length results in compression of the atomic density and the formation of a lattice of matter solitons.

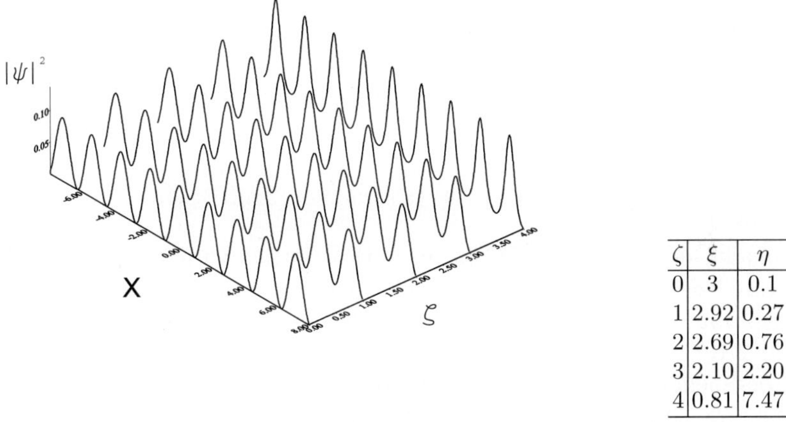

ζ	ξ	η
0	3	0.1
1	2.92	0.27
2	2.69	0.76
3	2.10	2.20
4	0.81	7.47

Fig. 4. Evolution of the cn-wave in a BEC with a negative scattering length ($\sigma = -1$) and a table of the respective change in the wave parameters. The situation considered corresponds to $L \approx 3.57$ and $N \approx 0.176$. At $\zeta \approx 0.81$, one already observes a "train" of clearly-defined bright solitons

As in the two previous examples, analytical treatment can be given for the limiting case $\xi \to 0$ and $\eta \to \infty$. Using (72), only straight-forward algebra is required to ensure that, for the case at hand, we arrive at previously-known equations of motion:

$$\frac{d\eta}{d\zeta} = \eta\,, \qquad \frac{d\xi}{d\zeta} = -\xi \ln\frac{\eta}{\xi} \tag{84}$$

and thus to the universal law of the limiting transition (cf. (77) and (81))

$$\eta \sim \mathrm{e}^{\zeta}\,, \qquad \xi \sim \eta\mathrm{e}^{-L\eta/4} \tag{85}$$

5 BEC in an Optical Lattice Controlled by the Feshbach Resonance

The approximation developed in Sect. 4 is not restricted to the cases considered there, and it does allow generalizations. One of them is considered in this section. More specifically, we consider an NLS equation with a periodic potential $V_{latt}(x)$, called a lattice, which allows exact solutions. This is the case for a potential in the form of an elliptic Jacobi function; we write it down in the form

$$V_{latt}(x) = 2V_0 \mathrm{sn}^2 (\kappa x | m_0) \ , \tag{86}$$

where $2V_0$ is a potential amplitude and κ and m_0 are parameters determining the aspect of the potential. As in Sect. 4, it will be assumed that the BEC is subject to the Feshbach resonance. Thus, we concentrate on periodic solutions of the equation

$$\mathrm{i}\frac{\partial \psi}{\partial t} = -\frac{\partial^2 \psi}{\partial x^2} + 2V_0 \mathrm{sn}^2 (\kappa x | m_0) \psi + 2\sigma g(t) |\psi|^2 \psi \tag{87}$$

This model, describing a BEC in an optical lattice with a slowly-varying scattering length a_s due to the Feshbach resonance, can be deduced from the GP equation (1) with the lattice potential (86) in the same wave, as was done in Sect. 2.1.

Experimentally, the periodic potential is created by a standing laser wave (see e.g. [28, 29]). Although an exact sn-potential (86) is then not easily achievable in the general case, for a large range of m_0, it is approximated very well by just the first few Fourier harmonics (see e.g. [11]). The use of the optical lattice is of great practical importance, since in any experimental setting, the preparation of the initial atomic distribution is the first step. It turns out that condensation of atoms in a lattice is one of the most natural controlled ways to generate a periodic matter wave.

The fact that potential (86) allows exact solutions is intimately related to the fact that the periodic wave solutions are expressed through elliptic functions (see Sect. 3) and that the nonlinearity is a squared function of the modulus of the field. The solutions mentioned, their diverse nature and linear stability have recently been studied in [22, 23], while here we address the problem of their deformation subject to the Feshbach resonance.

5.1 BEC with a Positive Scattering Length: Sn-Wave

Using the experience acquired in Sect. 3 while obtaining periodic soltions of the NLS equation, one immediately can verify that, for $g(t) \equiv 1$ (i.e. in absence of the Feshbach resonance), the function (see (45))

$$\psi_{sn-l} = \frac{\xi_{01} - \xi_{02}}{2} \exp\left(-\frac{\mathrm{i}}{2}(\xi_{01}^2 + \xi_{02}^2)t - \mathrm{i}V_0 t\right) \mathrm{sn}\left(\frac{\xi_{01} + \xi_{02}}{2} x \mid m_0\right) \tag{88}$$

with

$$\xi_{10} = \kappa + \sqrt{\kappa^2 m_0 - V_0}\,, \qquad \xi_{20} = \kappa - \sqrt{\kappa^2 m_0 - V_0} \tag{89}$$

is a solution of (87) with $\sigma = 1$.

When the Feshbach resonance is switched on, i.e. $dg(t)/dt \neq 0$, one can describe the evolution of the matter wave by assuming the functional form (88) and considering slow variations of the parameters $\xi_{0j} \mapsto \xi_j = \xi_j(t)$ and

$$m_0 \mapsto m = \frac{4V_0 + (\xi_1 - \xi_2)^2}{(\xi_1 + \xi_2)^2} \tag{90}$$

such that $\xi_j(0) = \xi_{j0}$, $m(t = 0) = m_0$, and the relation between ξ, m, κ and V_0 is given by (89). As in Sect. 4, we introduce a new dependent variable $v(x, t)$ through the relation (64). It reduces (87) to the form

$$\mathrm{i}v_t + v_{xx} - 2V_0 \mathrm{sn}^2(\kappa x|m_0)v - 2|v|^2 v = \mathrm{i}\gamma v \tag{91}$$

($\gamma(t)$ given by (69)). This equation preserves the period of the wave L and the number of particles N; they can be expressed through the parameters ξ_j:

$$L = \frac{8\mathrm{K}(m)}{\xi_1 + \xi_2}\,, \qquad N = 2\frac{(\xi_1 - \xi_2)^2(\xi_1 + \xi_2)}{(\xi_1 - \xi_2)^2 + 4V_0}\left[\mathrm{K}(m) - \mathrm{E}(m)\right] \tag{92}$$

(It worth mentioning that, at $V_0 = 0$, (92) is reduced to (48)).

Now, one can consider equations of the adiabatic approximation (67), (68). It turns out, however, that they assume their most compact forms in terms of $\xi_\pm = \xi_1 \pm \xi_2$:

$$\begin{aligned}
\frac{d\xi_+}{d\zeta} &= \frac{\xi_+\xi_-^2}{\Delta}(\mathrm{K}(m) - \mathrm{E}(m))[(4V_0 + \xi_-^2)\mathrm{K}(m) + \xi_+^2(\mathrm{E}(m) - \mathrm{K}(m))] \\
\frac{d\xi_-}{d\zeta} &= \frac{\xi_+^2\xi_-}{\Delta}(\mathrm{K}(m) - \mathrm{E}(m))(4V_0 + \xi_-^2)\mathrm{E}(m)
\end{aligned} \tag{93}$$

where

$$\begin{aligned}
\Delta = -\xi_+^2\mathrm{E}^2(m)(8V_0 + \xi_-^2) + 2\xi_+^2\mathrm{E}(m)\mathrm{K}(m)(4V_0 + \xi_-^2) \\
-\xi_-^2\mathrm{K}^2(m)(\xi_+^2 - \xi_-^2 - 4V_0)
\end{aligned}$$

One can distinguish two distinct cases, *viz.* positive and negative potential amplitude, V_0. Let us consider them separately (see also Fig. 5).

Case $V_0 < 0$*:*

Recalling that $0 \leq m_0 \leq 1$ and that both ξ_j are positive (as before we assume that $\xi_1 > \xi_2 \geq 0$), then from (89), one deduces that the solution (45) does not exist for all V_0, but for those satisfying the condition

$$|V_0| < \kappa^2(1 - m_0) . \tag{94}$$

An example of this case is illustrated in Fig. 5a. Recalling that $|\psi|^2$ describes the local density of the atoms, one concludes that, in the case $V_0 < 0$, particles are collected around the potential minima, which is the typical situation for the linear case, i.e. for non-interacting particles. Growth of the nonlinearity results in an increase of the defocusing effect, and as time goes on, an increasing number of atoms condenses in the classically forbidden region. The respective dark solitons can be viewed as density minima in the forbidden region. In this case, however there are constraints to be imposed on ξ_j. As one can verify from (90), the "linear" limit $m \to 0$ corresponds to $\xi_1 - \xi_2 \to 2\sqrt{V_0}$, while the solitonic limit $m \to 1$ corresponds to $\xi_2 \to 0$ and $\xi_1 \to \infty$.

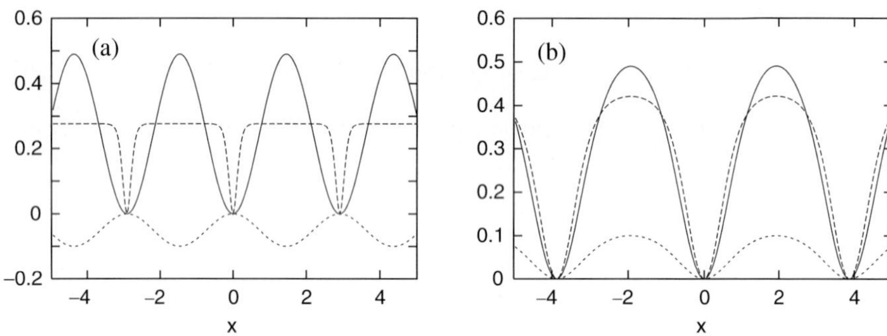

Fig. 5. Examples of sn-wave solutions at $\zeta = 0$ (*solid lines*) and $\zeta = \zeta_{fin}$ (*dashed lines*) in the periodic potential (*dotted lines*) for (**a**) $V_0 = -0.1$, $m \approx 0.074$, $L \approx 5.82$, $N \approx 1.44$, and $\zeta_{fin} = 5$ and (**b**) $V_0 = 0.1$, $m \approx 0.74$, $L \approx 7.75$, $N \approx 2.2$, and $\zeta_{fin} = 4$. In both cases $\xi_{01} = 0.9$, $\xi_{02} = 0.2$, and $\kappa = 0.55$

In order to better understand the role of the lattice potential, we simplify (93) using (72). This leads to the asymptotic behavior $\xi_1 \sim e^{\zeta}$, $\xi_2 \sim e^{-\zeta^2/2}$. Thus, while ξ_2 goes to zero faster than ξ_1 grows, the former tendency is slower than that in the absence of the lattice potential (see (77)). An explanation of this fact is that the lattice potential slows down creation of dark solitons because, in its presence, a number of atoms have to be collected in classically-forbidden regions.

Case $V_0 > 0$

Now the solution exists only for those values satisfying the condition

$$V_0 < \kappa^2 m_0 \tag{95}$$

(recall that $0 \le m_0 \le 1$). In order to understand this constraint, we note that atoms are now gathered at the maxima of the external potential. This is possible because self-action, due to the nonlinearity, creates an effective potential which can be written down in the form $V_0\mathrm{sn}^2(\kappa x|m) + 2g|\psi|^2$. This has minima where the pure lattice potential, V_{latt}, has maxima. If, however, the amplitude of the lattice potential is large enough, it becomes a dominating term in the expression for the effective potential, and the nonlinearity no longer changes the locations of the potential minima.

In the case at hand, the linear limit, $m \to 0$, can be achieved only when $\xi_{1,2} \to \infty$ and $\xi_1 - \xi_2$ is a constant. The nonlinear limit, $m \to 1$ is already achieved at finite ξ_j, such that $\xi_1\xi_2 = V_0$ (or $\xi_+^2 - \xi_-^2 = 4V_0$).

5.2 BEC with a Negative Scattering Length. Cn-Wave

Now we consider a deformation of the periodic cn-wave resulting in the creation of a train of bright matter solitons in a BEC with the negative scattering length placed in an optical lattice. This situation generalizes the results of Sect. 4.3 and is described by (87) with $\sigma = -1$. As before, by using the *ansatz* (64), it is reduced to

$$\mathrm{i}v_t + v_{xx} - 2V_0\mathrm{sn}^2(\kappa x|m_0)v + 2|v|^2 v = \mathrm{i}\gamma v \,. \tag{96}$$

The unperturbed solution of this equation is sought in the form (58), where, however, the link between the wave parameters is given by

$$\eta = \sqrt{m\kappa^2 - V_0}\,, \qquad \xi = \sqrt{(1-m)\kappa^2 + V_0}\,, \tag{97}$$

where η, ξ and m are functions of ζ, satisfying the initial conditions $\eta(0) = \eta_0$, $\xi(0) = \xi_0$, and $m(0) = m_0$. For analysis of the results, it is useful to write down an explicit form of the parameter m for the case at hand:

$$m = \frac{\eta^2 + V_0}{\xi^2 + \eta^2}\,. \tag{98}$$

Next we compute

$$L = \frac{4\mathrm{K}(m)}{\sqrt{\xi^2+\eta^2}}\,, \qquad N = \frac{\eta^2}{\eta^2+V_0}\left[4\sqrt{\xi^2+\eta^2}\mathrm{E}(m) - (\xi^2 - V_0)L\right] \tag{99}$$

Using these representations, the equations of the adiabatic approximation, (67) and (68) can be written down for $X_1 = \eta^2$ and $X_2 = \xi^2$:

$$\begin{aligned}
\frac{dX_1}{d\zeta} &= \frac{2X_1}{\Delta}\mathrm{E}(m)(X_1+V_0)[\mathrm{E}(m)(X_1+X_2) - \mathrm{K}(m)(X_2-V_0)] \\
\frac{dX_2}{d\zeta} &= \frac{2X_1}{\Delta}(X_2-V_0)[\mathrm{E}(m)-\mathrm{K}(m)][\mathrm{E}(m)(X_1+X_2) - \mathrm{K}(m)(X_2-V_0)]
\end{aligned} \tag{100}$$

where

$$\Delta = X_1^2\mathrm{E}^2(m) + X_1X_2(\mathrm{E}(m) - \mathrm{K}(m))^2 + 2V_0(X_1 + X_2)\mathrm{E}^2(m)$$
$$+2V_0\mathrm{K}(m)\mathrm{E}(m)(X_1 - X_2 + V_0) - X_1V_0\mathrm{K}^2(m)$$

An example of a numerical solution of the system (100) is presented in Fig. 6. Here we again have two different cases. If $V_0 > 0$, then there exists the constraint $\xi > V_0$. For fixed ξ, the linear limit, $m \to 0$, is impossible, while the nonlinear limit, $m \to 1$, is achieved when $\eta \to \infty$. The respective asymptotics are given by (85), where ξ^2 must be replaced by $\xi^2 - V_0$. In the opposite case, $V_0 < 0$, atoms are collected at the maxima of the potential. An important constraint now is that $\eta > |V_0|$. In the nonlinear limit, which corresponds to $\eta \to \infty$, the potential becomes an obstacle to the creation of bright solitons, thus slowing down the process. This happens because bright solitons can be created only in classically-forbidden regions.

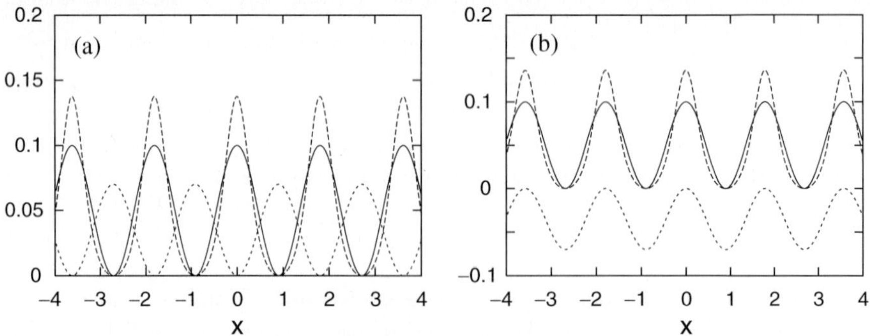

Fig. 6. Examples of the cn-wave solution at $\zeta = 0$ (*solid lines*) and $\zeta = 4$ (*dashed line*) in the periodic potential (*dotted lines*) for (**a**) $V_0 = 0.07$, $m \approx 0.097$, $L \approx 3.571$, and $N \approx 1.749$ and (**b**) $V_0 = -0.07$, $m \approx 0.055$, $L \approx 3.57$, and $N \approx 0.1778$. In both cases, $\eta_0 = 0.1$, $\xi_0 = 3.0$, and $\kappa = 1.76$

6 Modulation Instability of BEC in a Parabolic Trap

Now, following [24], we turn to another problem where an NLS equation with an inhomogeneous term is reduced to an effectively dissipative NLS equation. More specifically, we concentrate on the model

$$\mathrm{i}\frac{\partial\psi}{\partial t} + \frac{\partial^2\psi}{\partial t^2} - \sigma|\psi|^2\psi - \nu^2(t)x^2\psi = 0\ , \qquad (101)$$

which can be viewed as a quasi-1D BEC in a varying parabolic trap, i.e. in a parabolic trap formed by a time-dependent magnetic field. In particular, we consider modulational stability of the ground state.

6.1 Concept of Modulational Instability of the NLS Equation

Let us start by recalling well-known results (see e.g. [25]) on the modulational instability of the NLS equation with no external potential, i.e. when $\nu \equiv 0$. One considers the evolution of a small amplitude modulation of a plane wave, i.e. uses the *ansatz*:

$$\psi(x,t) = (\psi_0 + \varepsilon\psi_1)\exp[\mathrm{i}(qx - \omega t + \varepsilon\varphi(x,t))] \tag{102}$$

where

$$\psi_1(x,t) = \psi_{10}\mathrm{e}^{\mathrm{i}(Qx-\Omega t)}\,, \qquad \varphi(x,t) = \varphi_0\mathrm{e}^{\mathrm{i}(Qx-\Omega t)} \tag{103}$$

Here ψ_{10} and φ_0 are real constants, while Q and Ω play the roles of the wave-vector and frequency, respectively, of the perturbation of the plane wave whose amplitude, wave vector and frequency are ψ_0, q and ω, respectively.

Substituting (102) and (103) into (101) with $\nu = 0$, one finds

$$(\Omega - 2qQ)^2 = Q^2(Q^2 + 2\sigma\psi_0^2)\,. \tag{104}$$

Thus, if $\sigma < 0$ (recall that this corresponds to a BEC with a positive scattering length), then, for $Q^2 < 2\psi_0^2$, the frequency Ω acquires an imaginary part and this implies plane wave instability – it is called *modulational instability*. This phenomenon has been very well-known for a long time (see e.g. [30] for a recent review), and one of its practical applications is the generation of solitary pulses. (In the BEC context, this problem was considered in [31, 32]).

It should be mentioned here that, in comparison with the Feshbach resonance, an advantage of the use of modulational instability for soliton generation is the possibility of using an almost arbitrary initial profile of the condensate, while a disadvantage is that the soliton generation cannot be fully controlled. (Using the Feshbach resonance, one can create solitons with parameters specified a priori).

6.2 BEC with Negative Scattering Length in a Parabolic Trap

As has been mentioned in Sect. 2, when one deals with a BEC, the presence of a trap potential confining the atoms is crucial for many experimental settings. Let us now address the question of how the presence of the external parabolic potential affects the MI phenomenon. To this end, we transform (101) using the so-called *lens-transformation* (see e.g. [26]):

$$\psi(x,t) = \ell^{-1}\exp[\mathrm{i}f(t)x^2]v(\zeta,\tau) \tag{105}$$

where

$$\frac{d\tau}{dt} = \frac{1}{\ell^2}\,, \qquad \frac{d\ell}{d\tau} = 4f\ell^3\,, \qquad \frac{df}{dt} = -4f^2 - \nu^2(t)\,. \tag{106}$$

From the first of the these equations, one can express $\ell(t)$:

$$\ell(t) = \ell(0) \exp\left(4 \int_0^t f(t')dt'\right) \tag{107}$$

Then, substituting (105) into (101) for $\sigma = -1$, we obtain the equation for u:

$$\mathrm{i}v_\tau + v_{\zeta\zeta} + 2|v|^2 v = \mathrm{i}\gamma v\ , \tag{108}$$

where

$$\gamma = 2f\ell^2 \tag{109}$$

is the time-dependent dissipation in the case $f\ell^2 < 0$, and time-dependent growth if $f\ell^2 > 0$.

Here we only consider the case where the trap does not depend on time, i.e. where $\nu \equiv$ const. (See [24] for more details). In this case, one can solve (106) explicitly to give

$$f(t) = -\frac{\sqrt{\nu}}{2}\tan(2\sqrt{\nu}t)\ , \quad \ell(t) = \cos(2\sqrt{\nu}t)\ , \quad \tau(t) = \frac{1}{2\sqrt{\nu}}\tan(2\sqrt{\nu}t) \tag{110}$$

Here the constants of integration are chosen to make $\tau(0) = 0$ and $\ell(0) = 1$, so that the initial conditions for (101) and (108) coincide.

There are two basic features displayed by the solution (110). First, it shows periodic behavior. This is nothing but a manifestation of the well-known Ehrenfest theorem. Indeed, if we define a co-ordinate for the center of mass of the condensate:

$$X(t) = \frac{\int_{-\infty}^{\infty} x|\psi(x,t)|^2 dx}{\int_{-\infty}^{\infty} |\psi(x,t)|^2 dx} \tag{111}$$

then, differentiating $X(t)$ with respect to time twice, and using (101), we find that the center of mass is governed by the equation for a linear oscillator

$$\frac{d^2X}{dt^2} + 4\nu X = 0 \tag{112}$$

having frequency $2\sqrt{\nu}$.

Another feature of (110) is that they display divergence of $f(t)$ and thus, as follows from (105), the divergence of the phase of the condensate. This is also related to the oscillatory behavior of the condensate in a parabolic potential, and can be understood from the fact that the derivative of the phase determines the hydrodynamic velocity of the condensate, and this changes periodically due to oscillations in the parabolic potential.

Thus, in order to study the effect of the parabolic trap on modulational instability, one can use (108). In (Fig. 7), we present the results of a numerical study of the modulational instability of the plane wave, i.e. of the initial condition

$$u(x,0) = 1 + \delta \cos(qx) \; . \tag{113}$$

In both cases, for the GP equation, due to the presence of the potential, the condensate will gradually become peaked toward the center as time evolves. However, the development of the instability is clear from a comparison of the corresponding amplitudes of the oscillation of the field magnitude as a function of time.

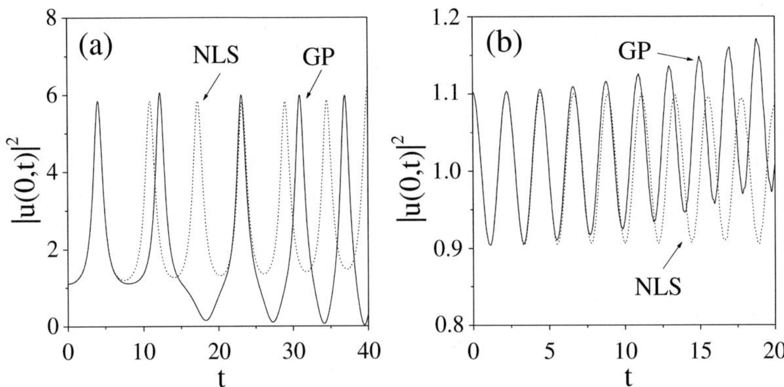

Fig. 7. Time evolution of the amplitude of the solution at the point $x = 0$ shown in (**a**) for $Q = 1$ and in (**b**) for $Q = 2$ for the GP equation with the parabolic potential $\nu = 0.01$ (*solid line*), and for the NLS equation with the same initial condition (*dotted line*). In each case $\delta = 0.05$

7 Concluding Remarks

The examples considered in the present chapter do not exhaust all applications of the approach developed. Let us list some related problems.

Starting from the very beginning, the consideration was restricted to cigar-shaped BECs and to 1D lattices. However, other configurations, as well as 2D and 3D lattices, are available experimentally (see e.g. [33, 33]). As is clear, an *ansatz* like (64) can be used to reduce a GP equation, in a 2D or 3D lattice with nonlinearity varying in time, to the corresponding dissipative NLS equations with periodic potentials.

Another limitation of the results obtained is related to the fact that, in all the examples, only one individual effect on the wave dynamics was considered. On the other hand, the interplay among different phenomena is of significant physical interest. In particular, in real experiments, the lattice optical potential is imposed simultaneously with a parabolic magnetic trap, or the experiment is carried out in a regime of free expanding condensate, where there is a periodic wave of finite width which increases due to the expansion.

Finally, we mention that the real dissipation gain could be additionally included in the consideration. Taking advantage of the fact that the integrals of even powers of Jacobi elliptic functions can be computed explicitly (see e.g. [12]), the interplay between linear gain and nonlinear dissipation, and between Feshbach resonance and dissipation, etc., can be investigated.

Acknowledgements

Different results reported in the present chapter were obtained in collaboration with F. Kh. Abdullaev, V. Brazhnyi, A. M. Kamchatnov, P. G. Kevrekidis and D. Frantzeskakis, who are greatly acknowledged for fruitful and stimulating discussions. The work was supported by the Programme "Human Potential-Research Training Networks", contract No. HPRN-CT-2000-00158.

References

1. S. N. Bose, Z. Phys, **26**, 178 (1924).
2. A. Einstein, Sitzber. Kgl. Preuss. Akad. Wiss. 261 (1924).
3. M. H. Anderson, J. R. Ensher, M. R. Matthews, C. E. Wieman, and E. A. Cornell, Science, **269**, 198 (1995).
4. K. B. Davis, M.-O. Mewes, M. R. Andrews, N. J. van Druten, D. S. Durfee, D. M. Kurn, W. Ketterle, Phys. Rev. Lett. **75**, 3969 (1995).
5. F. Dalfovo, S. Giorgini, L. P. Pitaevskii, S. Stringari: Rev. Mod. Phys. **71**, 463 (1999).
6. L. D. Landau, E. M. Lifshitz, *Quantum Mechanics: Non-relativistic Theory*, (Pergamon Press, New York 1977).
7. L. P. Pitaevskii: Sov. Phys. JETP, **13**, 451 (1961).
8. E. P. Gross, Nuovo Cimento, **20**, 454 (1961).
9. L. D. Faddeev, L. A. Takhtajan, *Hamiltonian Methods in the Theory of Solitons*, (Springer Verlag, Berlin, 1987).
10. E. D. Belokolos, A. I. Bobenko, V. Z. Enol'skjii, A. R. Its, V. B. Matveev, *Algebro-Geometric Approach to Nonlinear Integrable Equations*, (Springer-Verlag, Berlin, 1994).
11. M. Abramovitz, I. A. Stegun, *Handbook of Mathematical Functions*, (Dower, New York, 1965).
12. D. K. Lawden, *Elliptic Functions and applications*, (Springer-Verlag, New York Inc., 1989).

13. F. Kh. Abdullaev, A. M. Kamchatnov, V. V. Konotop, A. V. Brazhnyi, Phys. Rev. Lett. **90**, 230402 (2003).
14. S. Burger, K. Bongs, S. Dettmer, et al, Phys. Rev. Lett. **83**, 5198 (1999).
15. J. Denschlag et al, Science **287**, 97 (2000).
16. A. J. Moerdijk, B. J. Verhaar, A. Axelsson, Phys. Rev. A **51**, 4852 (1995).
17. S. Inouye, et al, Nature (London), **392**, 151 (1998).
18. P. Courteille et al, Phys. Rev. Lett., **81**, 69 (1998).
19. K. E. Strecker et al, Nature **417**, 150 (2002).
20. L. Khaykovich et al, Science **296**, 1290 (2002).
21. P. G. Kevrekidis, G. Theocharis, D. J. Frantzeskakis, and B. A. Malomed, Phys. Rev. Lett. **90**, 230401 (2003).
22. J. C. Bronski, L. D. Carr, B. Deconinck, J. N. Kutz, K. Promislow, Phys. Rev. E **63**, 036612 (2001).
23. J. C. Bronski, L. Carr, R. Carretero-Gonzalez, B. Deconinck, J. N. Kutz, K. Promislow, Phys. Rev. E **64**, 056615 (2001).
24. G. Theocharis, Z. Rapti, P. G. Kevrekidis, D. J. Frantzeskakis, and V. V. Konotop, Phys. Rev. A **67**, (2003).
25. L. D. Landau, E. M. Lifshitz, *Electromagnetics of Continuous Media*, (Pergamon Press, New York 1984).
26. C. Sulem, P. L. Sulem: *The Nonlinear Schrödinger Equation* (Springer-Verlag, New York, 1999).
27. T. Kawata, J. Phys. Soc. Japan, **51**, 3381 (1982).
28. B. P. Anderson and M. Kasevich, Science, **282**, 1686 (1998).
29. O. Morsch, E. Arimondo, *Ultracold atoms and Bose-Einstein condensates in optical lattices* arXiv:cond-mat/0209034 (2002).
30. F. Kh. Abdullaev, S. A. Darmanyan, J. Garnier, In *Progress in Optics*, vol 44, ed by E. Wolf (Elsvier, Amsterdam, 2002), p. 303
31. V. V. Konotop, M. Salerno, Phys. Rev. A **65**, 021602 (2002).
32. B. B. Baizakov, V. V. Konotop, M. Salerno, J. Phys. B: Atomic, Molecular & Optical Physics, **35**, 5105 (2002).
33. M. Greiner, O. Mandel, T. Esslinger, T. W. Hänsch, I. Bloch, Nature **415**, 6867 (2002).
34. M. Greiner, I. Bloch, O. Mandel, T. W. Hänsch, T. Esslinger, Phys. Rev. Lett. **87**, 160405 (2001).

Solitary Waves of Nonlinear Nonintegrable Equations

R. Conte[1] and M. Musette[2]

[1] Service de physique de l'état condensé (URA no. 2464), CEA–Saclay, 91191 Gif-sur-Yvette, Cedex, France
Conte@drecam.saclay.cea.fr

[2] Dienst Theoretische Natuurkunde, Vrije Universiteit Brussel, Pleinlaan 2, 1050 Brussels, Belgium
MMusette@vub.ac.be

Abstract. Our goal is to find closed form analytic expressions for the solitary waves of nonlinear nonintegrable partial differential equations. The suitable methods, which can only be non-perturbative, are divided in two classes. In the first class, which includes the well-known so-called truncation methods, one a priori assumes a given class of expressions (polynomials, etc.) for the unknown solution; the work involved can easily be done by hand but all solutions outside the given class are certainly missed. In the second class, instead of seeking an expression for the solution, one builds an intermediate equation with equivalent information, namely the *first-order* autonomous ODE satisfied by the solitary wave; in principle, no solution can be missed, but the work involved requires computer algebra. We present the application to the cubic and quintic complex one-dimensional Ginzburg-Landau equations, and to the Kuramoto-Sivashinsky equation.

1 Introduction

Many nonlinear partial differential equations (PDEs) encountered in physics are autonomous, i.e. they do not depend explicitly on the independent variables x (space) and t (time). In such cases, they admit a reduction, called a traveling wave reduction, to an autonomous nonlinear ordinary differential equation (ODE), defined in the simplest case by $u(x,t) = U(\xi), \xi = x - ct$, where c is a constant speed. A *solitary wave* is then defined as any solution of this nonlinear autonomous ODE. Physically relevant solitary waves must satisfy some decay condition when ξ goes to $\pm\infty$.

The distinctive feature of this chapter is that we explain the methods used to find closed form expressions for these solitary waves when the PDE and the reduced ODE are algebraic and nonintegrable. These solitary waves may have the topology of a front (for instance tanh), a pulse (for instance sech), a source, a sink, etc., but we will not discard an apparently physically uninteresting solution, because it might appear interesting in another field.

Why "algebraic"? This only excludes equations which cannot be converted into an algebraic form. For instance, the sine-Gordon equation $u_{xt} - \sin u = 0$ is not excluded because it is algebraic in e^{iu}.

R. Conte and M. Musette: *Solitary Waves of Nonlinear Nonintegrable Equations*, Lect. Notes Phys. **661**, 373–406 (2005)
www.springerlink.com

Why "nonintegrable"? Because the integrable ones (nonlinear Schrödinger (NLS), coupled NLS in the Manakov case, etc.) are "easy" to solve using powerful tools like the inverse spectral transform (IST) [1]. The difficulty with the nonintegrable equations is the absence of a general method to achieve the goal.

Why "autonomous"? Though irrelevant for the truncation methods (Sect. 6), this restriction is essential for the mathematical method of Sect. 7. Physically, this is not an important restriction, since many interesting PDEs are autonomous – see the examples below.

Why "closed form expressions"? Because a solution represented by a series can be misleading (*illusoire*, as Painlevé used to say). Consider, for instance, a chaotic deterministic dynamical system for which no analytic solution exists. Around a regular point, it admits a solution represented by a Taylor series, but one can conclude nothing before some analytic continuation has been performed. On the contrary, the Laurent series around a movable singularity (i.e. one whose location depends on the initial conditions) provides some constructive information (see Sect. 3.1) about the (global) integrability of the equation.

The methods described here are all based on the a priori singularities [26] of the solutions of the given ODE. In particular, we do not consider group theoretical methods [40]. These methods can be applied mainly to dissipative equations of importance in physics, nonlinear optics, mechanics, etc. Our specific examples are the following.

1. The one-dimensional cubic complex Ginzburg-Landau equation (CGL3)

$$\begin{gathered} \mathrm{i}A_t + pA_{xx} + q|A|^2 A - \mathrm{i}\gamma A = 0\,, \\ pq\gamma \neq 0\,, \quad \mathrm{Im}(p/q) \neq 0\,, \quad (A,p,q) \in \mathcal{C}\,, \quad \gamma \in \mathcal{R}\,, \end{gathered} \tag{1}$$

   (and its complex conjugate, i.e. a total differential order four), in which p, q, γ are constants, is a generic equation which describes many physical phenomena, such as the propagation of a signal in an optical fiber [2] and spatio-temporal intermittency in spatially-extended dissipative systems [22, 14, 34]. We will restrict ourselves to the proper CGL3 case with $\mathrm{Im}(p/q) \neq 0$.
   For analytic results on two coupled CGL3 equations, see [10].
2. The Kuramoto and Sivashinsky (KS) equation,

$$\begin{gathered} \varphi_t + \nu\varphi_{xxxx} + b\varphi_{xxx} + \mu\varphi_{xx} + \varphi\varphi_x = 0\,, \\ \varphi \in \mathcal{C}\,,\ (\nu, b, \mu) \in \mathcal{R}\,,\ \nu \neq 0\,. \end{gathered} \tag{2}$$

   in which ν, b, μ are constants. This PDE is obeyed by the variable $\varphi = \arg A$ of the above field A of the CGL3 under some limit [32, 20], hence its name – the "phase turbulence equation".
3. The one-dimensional quintic complex Ginzburg-Landau equation (CGL5),

$$\begin{aligned} &\mathrm{i}A_t + pA_{xx} + q|A|^2 A + r|A|^4 A - \mathrm{i}\gamma A = 0\,, \\ pr \neq 0\,,\quad &\mathrm{Im}(p/r) \neq 0\,,\quad (A,p,q,r) \in \mathcal{C}\,,\quad \gamma \in \mathcal{R}\,. \end{aligned} \tag{3}$$

4. The Swift-Hohenberg equation [37, 21]

$$\begin{aligned} \mathrm{i}A_t + bA_{xxxx} + pA_{xx} + q|A|^2 A + r|A|^4 A - \mathrm{i}\gamma A = 0\,, \\ br \neq 0\,,\quad (A,b,p,q,r) \in \mathcal{C}\,,\quad \gamma \in \mathcal{R}\,, \end{aligned} \tag{4}$$

in which b,p,q,r,γ are constants.

For the CGL3, KS, and Swift-Hohenberg equations (with one exception, namely the KS with $b^2 = 16\,\mu\nu$), all the solitary wave solutions $|A|^2 = f(\xi), \varphi = \Phi(\xi), \xi = x - ct$, which are known hitherto are just polynomials in $\tanh k\xi$.

So, two natural questions arise:

1. Is it possible or impossible that other solitary waves exist?
2. If other such solutions may exist, can one find not just a few more, but all of them?

From now on, let us denote the reduced ODE as

$$E(u^{(N)}, \ldots, u', u) = 0\,,\ ' = \frac{\mathrm{d}}{\mathrm{d}\xi}\,,\ \xi = x - ct, \tag{5}$$

and let us assume that it is also nonintegrable.

The chapter is organized as follows. In Sect. 2, one recalls the analytic expressions of the known solitary waves of the examples. This list, to be retrieved or augmented by the singularity-based methods, will allow us to rate the efficiency of the various methods. In Sect. 3, one investigates the amount of integrability of the equation, by applying the so-called "Painlevé test". More specifically, one checks the existence of particular solutions which admit a *local* representation as a Laurent series. This allows us to count the gap, strictly positive because of the assumed nonintegrability, between the differential order of the ODE and the maximal number of available integration constants. In Sect. 4, we discuss the choice of the suitable dependent variable to be used in the subsequent sections. In Sect. 5, one tries to obtain a *global* representation for the local information (Laurent series) previously found. One introduces a distinction between two main classes of methods, according to the following criteria: (i) computations easy enough to be carried out by hand, (ii) generality or particularity of the expected solution. The Sect. 6 is devoted to the first class of methods, which are known as "truncation methods". The input is a class of expressions for u (usually polynomials), which are chosen a priori, in some intermediate variable χ, which satisfies a given first-order ODE (e.g. Riccati, Weierstrass, Jacobi). Then, by a direct computation, easy to carry out by hand, one checks whether solutions in the given class indeed exist. The solutions which have a simple profile (such as tanh for a front, sech for a pulse), are easily found by this class of methods.

In the second class of methods [27], presented in Sect. 7, rather than directly looking for the unknown solution

$$u = f(\xi - \xi_0) , \tag{6}$$

in which ξ_0 is an arbitrary complex constant, then as an intermediate stage, one looks for the first-order nonlinear ODE

$$F(u, u') = 0 , \tag{7}$$

obtained by eliminating ξ_0 between (6) and its derivative, in which F is just as unknown as f. Indeed, provided that f is single-valued, then by a classical theorem recalled in the Appendix, there is an equivalence between knowledge of the solution f and that of the sub-equation F which it satisfies. The way to obtain the sub-equation F is to require that it be satisfied by the Laurent series obtained in a previous step.

The difference between the two classes of methods is the following. The solutions found by the first class of methods can only be a subset of those found by the second class. However, the computations involved can easily be performed by hand for the first class, while, for the second class, a computer algebra package is highly recommended.

2 The Known Solutions of the Examples

None of the expressions listed below represents the largest analytic solution which one could find, and their "distance" from this largest, yet unknown, analytic solution will be computed precisely in Sect. 3.

2.1 CGL3

The traveling wave reduction of (1)

$$A(x,t) = \sqrt{M(\xi)} \mathrm{e}^{\mathrm{i}(-\omega t + \varphi(\xi))} , \quad \xi = x - ct , \quad (c, \omega, M, \varphi) \in \mathcal{R} , \tag{8}$$

$$\frac{M''}{2M} - \frac{M'^2}{4M^2} + \mathrm{i}\varphi'' - \varphi'^2 + \mathrm{i}\varphi' \frac{M'}{M} - \mathrm{i}\frac{c}{2p}\frac{M'}{M} + \frac{1}{p}(c\varphi' + \omega) + \frac{q}{p}M - \frac{\mathrm{i}\gamma}{p} = 0 , \tag{9}$$

introduces two additional real constants (c, ω), and it is convenient to define the six real parameters $d_r, d_i, s_r, s_i, g_r, g_i$,

$$d_r + \mathrm{i}d_i = \frac{q}{p} , \quad s_r - \mathrm{i}s_i = \frac{1}{p} ,$$
$$g_r + \mathrm{i}g_i = \frac{\gamma + \mathrm{i}\omega}{p} + \frac{1}{2}c^2 s_r s_i + \frac{i}{4}c^2 s_r^2 . \tag{10}$$

In the proper CGL3 case, $d_i \neq 0$, to which we restrict ourselves here, only three solutions are currently known. The two real constants included in the complex equation

$$(-1 + \mathrm{i}\alpha)(-2 + \mathrm{i}\alpha)p + A_0^2 q = 0 , \tag{11}$$

are denoted by A_0^2 and α. Then these three solutions are as follows:

1. A heteroclinic source or propagating hole [4]

$$\begin{cases} A = A_0 \left[\frac{k}{2} \tanh \frac{k}{2}\xi - \frac{\mathrm{i}qp_i}{2(1-\mathrm{i}\alpha)p|p|^2 d_i} c \right] \\ \qquad \times \exp \left[\mathrm{i} \left(\alpha \operatorname{Log} \cosh \frac{k}{2}\xi + \frac{q_i}{2|p|^2 d_i} c\xi - \omega t \right) \right] , \\ \frac{i\gamma - \omega}{p} = \left(\frac{c}{2p} \right)^2 - (2 - 3\mathrm{i}\alpha)\frac{k^2}{4} , \end{cases} \tag{12}$$

in which the velocity c is arbitrary. Indeed, the real and imaginary parts of the last equation define the value of ω and a linear relation between c^2 and k^2 – see [9, (79)].

2. A homoclinic pulse or solitary wave [30]

$$\begin{cases} A = A_0(-\mathrm{i}k \operatorname{sech} kx)\mathrm{e}^{\mathrm{i}[\alpha \operatorname{Log} \cosh kx - \omega t]} , \\ \frac{\mathrm{i}\gamma - \omega}{p} = (1 - \mathrm{i}\alpha)^2 k^2 , \ c = 0 . \end{cases} \tag{13}$$

3. A heteroclinic front or shock [28]

$$\begin{cases} A = A_0 \frac{k}{2} \left[\tanh \frac{k}{2}\xi + \varepsilon \right] \mathrm{e}^{\mathrm{i}[\alpha \operatorname{Log} \cosh \frac{k}{2}\xi + \frac{3p_r + \alpha p_i}{6|p|^2} c\xi - \omega t]} , \\ \varepsilon^2 = 1 , \\ \frac{\mathrm{i}\gamma - \omega}{p} = \left(\frac{c}{2p} \right)^2 + \frac{k^2}{4} , \ \frac{k}{2} = \varepsilon \frac{p_i c}{6|p|^2} . \end{cases} \tag{14}$$

None of these three solutions requires any constraint on p, q, γ, and they depend on an additional sign resulting from the resolution of (11),

$$A_0^2 = \frac{3(3d_r + \varepsilon_1 \Delta)}{2d_i^2} , \ \alpha = \frac{3d_r + \varepsilon_1 \Delta}{2d_i} , \ \Delta = \sqrt{9d_r^2 + 8d_i^2} , \ \varepsilon_1^2 = 1 . \tag{15}$$

In all of them, M is a much simpler expression, namely a second degree polynomial in

$$\tau = (k/2) \tanh k\xi/2 , \ k^2 \in \mathcal{R} . \tag{16}$$

Therefore, if one wants to extend the three solutions given above, it is advisable to eliminate φ between the system of two real equations equivalent to (9),

$$\begin{cases} \dfrac{M''}{2M} - \dfrac{M'^2}{4M^2} - \varphi'^2 - s_i\left(\dfrac{cM'}{2M} + \gamma\right) + s_r\left(c\varphi' + \omega\right) + d_r M = 0\,, \\ \varphi'' + \varphi'\dfrac{M'}{M} - s_r\left(\dfrac{cM'}{2M} + \gamma\right) - s_i\left(c\varphi' + \omega\right) + d_i M = 0\,, \end{cases} \tag{17}$$

which results in

$$\varphi' = \frac{cs_r}{2} + \frac{G' - 2cs_iG}{2M^2(g_r - d_iM)},\ \left(\varphi' - \frac{cs_r}{2}\right)^2 = \frac{G}{M^2}\,, \tag{18}$$

$$(G' - 2cs_iG)^2 - 4GM^2(d_iM - g_r)^2 = 0\,, \tag{19}$$

$$G = \frac{1}{2}MM'' - \frac{1}{4}M'^2 - \frac{cs_i}{2}MM' + d_rM^3 + g_iM^2\,, \tag{20}$$

and to concentrate on the single third-order equation (19) for $M = |A|^2$.

2.2 KS

The traveling wave reduction is defined as

$$\varphi(x,t) = c + u(\xi),\ \xi = x - ct,\ \left[\nu u''' + bu'' + \mu u' + \frac{u^2}{2}\right]' = 0\,,$$
$$(\nu, b, \mu) \in \mathcal{R}\,,\ \nu \neq 0\,, \tag{21}$$

which integrates once as

$$\nu u''' + bu'' + \mu u' + \frac{u^2}{2} + A = 0\,, \tag{22}$$

where A is an integration constant. It exhibits chaotic behavior [22], and depends on two dimensionless parameters, $b^2/(\mu\nu)$ and $\nu A/\mu^3$. The known solutions are one elliptic solution, six trigonometric solutions and one rational solution.

The unique known elliptic solution exists for one constraint between the parameters ν, b, μ of the PDE [12, 18],

$$b^2 - 16\mu\nu = 0 : u = -60\nu\wp' - 15b\wp - \frac{b\mu}{4\nu}, g_2 = \frac{\mu^2}{12\nu^2}, g_3 = \frac{13\mu^3 + \nu A}{1080\nu^3}\,. \tag{23}$$

in which $\wp$ is the elliptic function of Weierstrass, defined by the ODE

$$\wp'^2 = 4\wp^3 - g_2\wp - g_3\,. \tag{24}$$

The six trigonometric solutions [19, 16], all of them rational in $e^{k\xi}$, exist at the price of one constraint between ν, b, μ and another one on A,

$$u = 120\nu\tau^3 - 15b\tau^2 + \left(\frac{60}{19}\mu - 30\nu k^2 - \frac{15b^2}{4\times 19\nu}\right)\tau + \frac{5}{2}bk^2$$
$$- \frac{13b^3}{32\times 19\nu^2} + \frac{7\mu b}{4\times 19\nu}\,, \tag{25}$$
$$\tau = \frac{k}{2}\tanh\frac{k}{2}(\xi - \xi_0)\,,$$

with the allowed values being listed in Table 1.

Table 1. The six known trigonometric solutions of KS, (22). They all have the form (25). The last line shows a degeneracy of the elliptic solution (23)

$b^2/(\mu\nu)$	$\nu A/\mu^3$	$\nu k^2/\mu$
0	$-4950/19^3,\ 450/19^3$	$11/19,\ -1/19$
144/47	$-1800/47^3$	1/47
256/73	$-4050/73^3$	1/73
16	$-18,\ -8$	$1,\ -1$

Finally, the only known rational solution

$$b = 0\,,\quad \mu = 0\,,\quad A = 0\ :\quad u = 120\nu(\xi - \xi_0)^{-3}\,, \tag{26}$$

is a limit of all the above solutions.

A nice property common to all these solutions is that they admit the representation

$$u = \mathcal{D}\,\mathrm{Log}\,\psi + \text{constant}\,,\quad \mathcal{D} = 60\nu\frac{\mathrm{d}^3}{\mathrm{d}\xi^3} + 15b\frac{\mathrm{d}^2}{\mathrm{d}\xi^2} + \frac{15(16\mu\nu - b^2)}{76\nu}\frac{\mathrm{d}}{\mathrm{d}\xi}\,, \tag{27}$$

in which ψ is an *entire* function (i.e. one without any singularity at a finite distance) whose ODE is easy to construct. Thus:

$$(-\mathrm{Log}\,\psi)'''^2 - 4(-\mathrm{Log}\,\psi)''^3 + g_2(-\mathrm{Log}\,\psi)'' + g_3 = 0\,, \tag{28}$$
$$\psi'' - \frac{k^2}{4}\psi = 0\,, \tag{29}$$
$$\psi'' = 0\,. \tag{30}$$

This linear operator $\mathcal{D}$, which captures the singularity structure, is called the *singular part operator.*

2.3 CGL5

The traveling wave reduction is quite similar to that of CGL3, so we do not repeat it. Again, A_0^2 and α denote two real constants appearing in the complex equation

$$(-1/2 + \mathrm{i}\alpha)(-3/2 + \mathrm{i}\alpha)p + A_0^4 r = 0 \,, \tag{31}$$

and the following constants are convenient

$$\begin{aligned} e_r + \mathrm{i}e_i &= \frac{r}{p} \,, \quad s_r - \mathrm{i}s_i = \frac{1}{p} \,, \\ g_r + \mathrm{i}g_i &= \frac{\gamma + \mathrm{i}\omega}{p} + \frac{1}{2}c^2 s_r s_i + \frac{i}{4}c^2 s_r^2 \,. \end{aligned} \tag{32}$$

In the proper CGL5 case $e_i \neq 0$ to which we restrict ourselves here, only two solutions are currently known:

1. A heteroclinic front or shock [35],

$$\begin{cases} A = A_0 \left((k/2)(\tanh k\xi/2 + \varepsilon)\right)^{1/2} \mathrm{e}^{\mathrm{i}[\alpha \operatorname{Log} \cosh k\xi/2 + K\xi - \omega t]} \,, \\ \varepsilon^2 = 1 \,, \\ (-1/2 + \mathrm{i}\alpha)\left[\mathrm{i}(c - 2pK) + 2\varepsilon(-2 + \mathrm{i}\alpha)pk\right] p - A_0^2 q = 0 \,, \\ \dfrac{\mathrm{i}\gamma - \omega}{p} = \left(\dfrac{c}{2p}\right)^2 - \left(K - \dfrac{c}{2p} + \varepsilon(1 - \mathrm{i}\alpha)k/2\right)^2 . \end{cases} \tag{33}$$

   The number of constraints restricting (p, q, r, γ) is either two (case $s_i = 0$, c arbitrary), or zero (case $s_i \neq 0$, with a fixed velocity).
2. A homoclinic source or sink [23],

$$\begin{cases} A = A_0 \left(\dfrac{k \sinh ka}{\cosh k\xi + \cosh ka} + r_0\right)^{1/2} \\ \qquad \times \exp\left[\mathrm{i}[\alpha \operatorname{Log}(\cosh k\xi + \cosh ka) + K\xi - \omega t]\right] \,, \\ c - 2pK = 0, \quad \text{which implies } cp_i = 0 \,, \\ (-1/2 + \mathrm{i}\alpha)\left[-2k\mu_0(1 - \mathrm{i}\alpha) + 2(-2 + \mathrm{i}\alpha)r_0\right] p - A_0^2 q = 0 \,, \\ \dfrac{\mathrm{i}\gamma - \omega}{p} = \left(\dfrac{c}{2p}\right)^2 + (1/2 - \mathrm{i}\alpha)^2 k^2 + (3 - 10\mathrm{i}\alpha - 4\alpha^2)k\mu_0 r_0 \\ \qquad + (3 - 8\mathrm{i}\alpha - 2\alpha^2)r_0^2/2 \,, \\ r_0\left(r_0^2 + 2k\mu_0 r_0 + k^2\right) = 0 \,. \end{cases} \tag{34}$$

   in which $K, k^2, r_0, \mu_0 = \coth ka$ are real constants. The number of constraints restricting (p, q, r, γ) is either two (case $s_i = 0$, with c arbitrary), or one (case $s_i \neq 0, c = 0$).

Each of these solutions depends on two additional signs arising from the resolution of (31)

$$\begin{aligned} A_0^2 &= \varepsilon_2 \sqrt{\frac{2e_r + \varepsilon_1 \Delta}{e_i^2}} \,, \quad \alpha = \frac{2e_r + \varepsilon_1 \Delta}{2e_i} \,, \\ \Delta &= \sqrt{4e_r^2 + 3e_i^2} \,, \quad \varepsilon_1^2 = \varepsilon_2^2 = 1 \,. \end{aligned} \tag{35}$$

2.4 Swift-Hohenberg

Now, A_0^4 and α denote two real constants appearing in

$$(-1+\mathrm{i}\alpha)(-2+\mathrm{i}\alpha)(-3+\mathrm{i}\alpha)(-4+\mathrm{i}\alpha)b + A_0^4 r = 0 \,. \tag{36}$$

In the case $\mathrm{Im}(r/b) \neq 0$ to which we restrict ourselves, only two solutions seem to be currently known.

1. A stationary front [24, (127)]

$$\begin{cases} A = A_0(k/2)\tanh kx/2 \; \mathrm{e}^{\mathrm{i}[\alpha \,\mathrm{Log}\cosh kx/2 - \omega t]} \,, \\ k^2 = \dfrac{2}{5(2-i\alpha)b}\left[\dfrac{A_0^2 q}{(1-\mathrm{i}\alpha)(2-\mathrm{i}\alpha)} + p\right] , \\ \mathrm{i}\gamma - \omega = \dfrac{16 - 30\mathrm{i}\alpha - 15\alpha^2}{16} bk^4 + \dfrac{-2+3\mathrm{i}\alpha}{4} pk^2 \,. \end{cases} \tag{37}$$

2. A stationary pulse [24, (119)]

$$\begin{cases} A = A_0(-\mathrm{i}k \,\mathrm{sech}\, kx)\mathrm{e}^{\mathrm{i}[\alpha \,\mathrm{Log}\cosh kx - \omega t]} \,, \\ k^2 = \dfrac{-1}{2(5-4\mathrm{i}\alpha-\alpha^2)b}\left[\dfrac{A_0^2 q}{(1-\mathrm{i}\alpha)(2-\mathrm{i}\alpha)} + p\right] , \\ i\gamma - \omega = (1-\mathrm{i}\alpha)^2\left[(1-\mathrm{i}\alpha)^2 bk^4 + pk^2\right] \,. \end{cases} \tag{38}$$

In both solutions, the number of constraints on (b,p,q,r,γ) is two (these are defined by the vanishing of the imaginary part of the relations for k^2 and $\mathrm{i}\gamma - \omega$).

3 Investigation of the Amount of Integrability

3.1 Counting Arguments Based on Singularity Analysis

Because the ODE (5) is assumed nonintegrable, the number of integration constants which can be present in any closed form solution is strictly smaller than the differential order of the ODE. Let us first precisely compute this difference, which is an indicator of the amount of integrability of the ODE. The technique to do this is just the Painlevé test (see [7] for the basic vocabulary of this technique). Let us present it for the KS example (22).

Looking for local algebraic behaviour near a movable singularity x_0 (*movable* means that it depends on the initial conditions),

$$u \underset{x \to x_0}{\sim} u_0 \chi^p, \; u_0 \neq 0, \; \chi = x - x_0 \,, \tag{39}$$

one first obtains the usual balancing conditions (here, between the highest derivative and the nonlinearity)

$$p-3=2p,\ p(p-1)(p-2)\nu u_0+\frac{u_0^2}{2}=0\ , \tag{40}$$

easily solved to give

$$p=-3,\ u_0=120\nu\ , \tag{41}$$

which yields the Laurent series,

$$\begin{aligned} u^{(0)} &= 120\nu\chi^{-3}-15b\chi^{-2}+\frac{15(16\mu\nu-b^2)}{4\times 19\nu}\chi^{-1} \\ &+\frac{13(4\mu\nu-b^2)b}{32\times 19\nu^2}+O(\chi^1)\ , \end{aligned} \tag{42}$$

from which two out of the three arbitrary constants are missing. These two constants appear in a perturbation [11],

$$u=u^{(0)}+\varepsilon u^{(1)}+\varepsilon^2 u^{(2)}+\dots\ , \tag{43}$$

in which the small parameter ε is not in the ODE (22). The linearized equation around $u^{(0)}$,

$$\left(\nu\frac{\mathrm{d}^3}{\mathrm{d}x^3}+b\frac{\mathrm{d}^2}{\mathrm{d}x^2}+\mu\frac{\mathrm{d}}{\mathrm{d}x}+u^{(0)}\right)u^{(1)}=0\ , \tag{44}$$

is then of the Fuchsian type near $x=x_0$, with an indicial equation ($q=-6$ denotes the singularity degree of the l.h.s. E of (22))

$$\lim_{\chi\to 0}\chi^{-j-q}(\nu\partial_x^3+u_0\chi^p)\chi^{j+p} \tag{45}$$

$$=\nu(j-3)(j-4)(j-5)+120\nu=\nu(j+1)(j^2-13j+60) \tag{46}$$

$$=\nu(j+1)\left(j-\frac{13+\mathrm{i}\sqrt{71}}{2}\right)\left(j-\frac{13-\mathrm{i}\sqrt{71}}{2}\right)=0\ . \tag{47}$$

The local representation of the general solution,

$$\begin{aligned} u(x_0,\varepsilon c_+,\varepsilon c_-) = 120\nu\chi^{-3}\{&\mathrm{Taylor}(\chi) \\ &+\varepsilon[c_+\chi^{(13+i\sqrt{71})/2}\mathrm{Taylor}(\chi) \\ &+\ c_-\chi^{(13-i\sqrt{71})/2}\mathrm{Taylor}(\chi)]+\mathcal{O}(\varepsilon^2)\}\ , \end{aligned}$$

in which "Taylor" denotes a converging series of χ, does depend on three arbitrary constants $(x_0,\varepsilon c_+,\varepsilon c_-)$ (the Fuchs index -1 only represents a shift of x_0). The dense movable branching arising from the two irrational indices characterizes [38] the chaotic behaviour, and the only way to remove it is to require $\varepsilon c_+=\varepsilon c_-=0$, i.e. $\varepsilon=0$, thus restricting the analytic part of the solution to a single arbitrary constant.

To summarize, let us introduce two notions.

The first one is trivial. One calls *irrelevant* any integration constant which, because of some symmetry, is always present in any solution. The KS ODE has one such irrelevant integration constant, the origin ξ_0 of ξ, and we will systematically omit including it. The traveling wave reduction of CGL3 has two irrelevant integration constants, *viz* the origins of ξ and φ. The second notion is quite an important property of the equation. We will call *unreachable* any constant of integration which cannot participate in any closed form solution. The KS ODE has two unreachable integration constants. Furthermore, the closed form solution which depends on the maximal possible number of reachable integration constants will be described as the *general analytic solution*, and our goal is precisely to exhibit a closed form expression for this general analytic solution, whose local representation is a Laurent series like (42).

The above notions (irrelevant, unreachable) are attached to an equation, not to a solution. Let us similarly introduce two integer numbers, attached to a solution, allowing one to quantify how far this solution is from the general analytic solution. The number of reachable integration constants, excluding the irrelevant ones, which are missing in this solution will be called the *deficiency* of a closed form solution. In the KS, for instance, the elliptic solution has zero deficiency (as A is arbitrary), and all the trigonometric solutions have deficiency one (as A is fixed).

Let us finally define the *co-dimension* of a closed form solution of an equation as the number of constraints on the fixed parameters. (Here *fixed* means those which occur in the definition of the equation.) Thus, the elliptic and trigonometric solutions of KS have co-dimension one, and the rational solution has co-dimension two.

3.2 Evidence for Unknown Solutions

Computer simulations, as well as real experiments, (for a recent review, see [34]) sometimes display regular patterns in the (x, t) plane, and some of them are indeed described by some analytic solution. For the remaining patterns, we can guess that analytic expressions corresponding to these patterns should exist, so they are “to be found”. For the KS equation (2), one thus observes a homoclinic solitary wave [39, Fig. 7] $\varphi = f(\xi), \xi = x - ct$, while all solutions known to date are heteroclinic. For the CGL3 equation (1), [13] has predicted the existence of a fourth physically interesting solution, which is a co-dimension-one homoclinic hole solution with an arbitrary velocity c. Table 2 collects the current state of the known solutions for the various non-integrable equations considered in this chapter. In [14], another count, based on the various possible topological structures, is made for the CGL3, and provides the same results.

Table 2. Integers rating the particular solutions of a nonintegrable equation. The vocabulary (irrelevant, unreachable, deficiency, co-dimension) is defined in Sect. 3.1. The column "Available" indicates the number of relevant, reachable integration constants, which is the algebraic sum "Order" – "Irrelevant" – "Unreachable". The last two columns indicate the properties of the solutions in their order of appearance in Sect. 2. The best solution would be one with deficiency and co-dimension both equal to zero, with the reduction parameters c and ω being arbitrary

Equation	Order	Irrelevant	Unreachable	Available	Deficiency	Co-dimension
CGL3	4	$2 = \xi_0, \varphi_0$	2	0	0, 0, 0	1, 2, 2
CGL5	4	$2 = \xi_0, \varphi_0$	2	0	0, 0	1, 1
KS	4	$1 = \xi_0$	2	1	$0, 1, 1, 1,$ $1, 1, 1, 1$	$1, 1, 1, 1,$ $1, 1, 1, 2$
SH	8	$2 = \xi_0, \varphi_0$	6	0	0, 0	2, 2

4 Selection of Possibly Single-Valued Dependent Variables

Whatever the class of methods to be applied, a pre-requisite is to determine a variable whose dominant behaviour is single-valued and which satisfies some algebraic ODE (or more generally, PDE). This is the case for the KS, since the solution p of (40) is an integer, but in the CGL3, the CGL5, or the SH, this is the case for neither $(A, \bar{A})$ nor $\arg A$, and for the CGL5, it is not even the case for $|A|$. Indeed, considering the CGL3 for instance, the dominant terms are

$$pA_{xx} + q|A|^2 A \,, \tag{48}$$

and one can easily check that $|A|$ generically behaves like simple poles. Let us therefore define the dominant behaviour of the two fields $(A, \overline{A})$ as

$$A \sim A_0 \chi^{-1+\mathrm{i}\alpha} \,, \quad \overline{A} \sim \overline{A_0} \chi^{-1-\mathrm{i}\alpha} \,, \quad A_0 \in \mathcal{C} \,, \quad \alpha \in \mathcal{R} \,, \tag{49}$$

in which (A_0, α) are constants to be determined. The resulting complex equation (equivalent to two real equations)

$$(-1 + \mathrm{i}\alpha)(-2 + \mathrm{i}\alpha)p + |A_0|^2 q = 0 \,, \tag{50}$$

is precisely the one artificially introduced earlier as (11) and solved in (15), with the convention that A_0 is real being allowed by the phase invariance of the CGL3. The same applies for the CGL5

$$A \sim A_0 \chi^{-1/2+\mathrm{i}\alpha} \,, \quad \overline{A} \sim A_0 \chi^{-1/2-\mathrm{i}\alpha} \,, \tag{51}$$

see (31) and (35), and for the SH,

$$A \sim A_0 \chi^{-1+\mathrm{i}\alpha} , \quad \overline{A} \sim A_0 \chi^{-1-\mathrm{i}\alpha} , \tag{52}$$

see (36).

In all three examples (CGL3, CGL5 and SH), the variable $M = |A|^2$ satisfies an algebraic ODE, which can be constructed by elimination of $\arg A$, and it has single-valued dominant behaviour (respectively, movable double poles, simple poles and double poles). Moreover, again for the CGL3, CGL5 and SH, for all the solitary wave solutions which are known to date (see Sect. 2), this variable, M, is represented by quite simple mathematical expressions, namely either polynomials in one elementary variable τ, (16), which satisfies a Riccati equation

$$\frac{\mathrm{d}}{\mathrm{d}z}\tau(z) = 1 - \tau^2 , \quad \tau = \tanh z , \tag{53}$$

or (source solution of the CGL5, (34)) polynomials in two elementary variables (σ, τ) which satisfy a projective Riccati system [9]

$$\frac{\mathrm{d}}{\mathrm{d}z}\tau = 1 - \tau^2 - \mu_0 \sigma , \quad \frac{\mathrm{d}}{\mathrm{d}z}\sigma = -\sigma\tau , \quad \sigma^2 - \tau^2 - 2\mu_0\sigma + 1 = 0 , \tag{54}$$

in which μ_0 is a constant, and whose solution can be expressed as

$$\tau = \frac{\sinh z}{\cosh z + \cosh ka} , \quad \sigma = \frac{\sinh ka}{\cosh z + \cosh ka} , \quad \mu_0 = \coth ka . \tag{55}$$

When $\mu_0(\mu_0^2 - 1) = 0$, the class of polynomials in (σ, τ) degenerates to the class of polynomials in (sech, tanh).

Therefore $M = |A|^2$ will be our best choice to search for closed form solutions of the CGL3, CGL5 and SH.

Remark. Despite the multi-valued dominant behaviour of the complex amplitude A of the CGL3 and SH, one can define two variables with single-valued dominant behaviour. In this complex modulus representation [9]

$$A = A_0 Z(\xi) e^{i[\Phi(\xi) - \omega t]} , \quad \overline{A} = A_0 \overline{Z}(\xi) e^{-i[\Phi(\xi) - \omega t]} , \tag{56}$$

with Z complex and Φ real, the dominant behaviour is

$$Z \sim \chi^{-1} , \quad \overline{Z} \sim \chi^{-1} , \quad \Phi' \sim \alpha\chi^{-1} , \tag{57}$$

and the truncation of $(Z, \overline{Z}, \Phi')$ might prove to be much more economical than that of M. All the solutions listed in Sect. 2 for the CGL3, CGL5 and SH have been written in this representation.

5 On the Cost of Obtaining Closed Form Expressions

Let us now give some details on the distinction between the two main classes of methods outlined in the introduction.

In the **first class of methods**, one gives, as an input, some class of expressions $f(\xi)$ (for instance polynomials in $\operatorname{sech} k\xi$ and $\tanh k\xi$), and, by a direct computation, one checks whether there are indeed some solutions in the given class. For brevity, we will call these methods *sufficient*, because they certainly miss any solution outside the given class, e.g. for the ODE

$$M'^2 + \left(12M^2 - \frac{3}{2}\right) M' + 36M^4 - \frac{17}{2}M^2 + \frac{1}{2} = 0\,, \tag{58}$$

a polynomial would miss its solution, since it is rational in $\tanh k\xi$:

$$M = \frac{\tanh(\xi - \xi_0)}{2 + \tanh^2(\xi - \xi_0)}\,. \tag{59}$$

In the **second class of methods**, the search for first-order autonomous sub-equations (7) requires no a priori assumption at all, and, from the classical results recalled in the Appendix, knowledge of the first-order sub-equation is indeed equivalent to knowledge of the explicit expression (6). In contrast to the previous methods, which are "sufficient", as stated above, the proposed method can be qualified as "necessary".

The difference between the two classes of methods is obvious: the class of expressions $f(\xi)$ is an output of the second method, while it is an input of the first one. This is why the second method can find, not just some, but *all* the solutions which are elliptic or trigonometric, if they exist.

Remark. The "cost" of the method of first-order autonomous sub-equations is an increasing function of the positive integer m appearing in (168), but m, which is an input of the method, is not bounded. Indeed, any rational function $u = P_N(\tanh(\xi - \xi_0))/P_D(\tanh(\xi - \xi_0))$ satisfies an ODE (7) of order one and degree $\max(N, D)$. By considering only some differential consequence of this ODE, one cannot guess the correct value of m in advance.

6 First Class of Methods: Truncations

After having selected, as indicated in Sect. 4, dependent variables with single-valued leading behaviour, we note that the methods called truncations involve defining, for each such dependent variable, some single-valued closed form class of expressions, and then checking whether there exist solutions in that class.

The class of expressions to be chosen as the input depends on the number of *families of movable singularities* of the considered dependent variable. Thus, the field u of the KS has only one family, i.e. one value of u_0, while the field $M = |A|^2$ of the CGL3, CGL5 or SH has respectively two, four and four families. Let us start with the simplest class of expressions.

6.1 Polynomials in tanh (One-Family Truncation)

The class of polynomials in $\tanh(k/2)\xi$ is the most frequently encountered class of closed form solutions of autonomous PDEs. This fact is a direct consequence of a quite remarkable property. Indeed, from a result of Painlevé, the variable τ in (53) is the unique variable that is, at the same time, single-valued and closed by differentiation: if u is such a polynomial,

$$u = \sum_{j=0}^{-p} u_j \chi^{j+p} , \quad \chi^{-1} = \frac{k}{2} \tanh \frac{k}{2}\xi , \quad \xi = x - ct , \tag{60}$$

the l.h.s. $E(u, x, t)$ of the equation of the PDE is also such a polynomial,

$$E = \sum_{j=0}^{-q} E_j \chi^{j+q} , \tag{61}$$

and its identification with the null polynomial

$$\forall j : E_j = 0 , \tag{62}$$

generates the smallest possible number of *determining equations* $E_j = 0$. As compared with the Laurent series (42), the series (60) terminates, hence it is named *truncation.*

The truncation (60) involves only one value of u_0, so, for this reason it is called a *one-family truncation.* Let us give a few examples.

One-Family Truncation of the KS Equation

The symbols u_0 and p denote the leading behaviour of the ODE (22), and the truncation (60) defines $-q + 1 = 7$ determining equations, the first four being

$$E_0 \equiv -60\nu u_0 + \frac{u_0^2}{2} = 0 , \tag{63}$$

$$E_1 \equiv 12bu_0 + (u_0 - 24\nu)u_1 = 0 , \tag{64}$$

$$E_2 \equiv -3\mu u_0 + \frac{57}{2}k^2\nu u_0 + 6bu_1 + \frac{1}{2}u_1^2 + (u_0 - 6\nu)u_2 = 0 , \tag{65}$$

$$E_3 \equiv -\frac{9}{2}bk^2 u_0 - 2\mu u_1 + 10k^2\nu u_1 + 2bu_2 + u_1u_2 + u_0u_3 = 0 . \tag{66}$$

The structure of these types of algebraic determining equations is always the same: one algebraic equation for u_0 ($j = 0$), followed by $-p$ equations, which are linear in $u_j, j = 1, \ldots, -p$. Equation $j = 0$ has already been solved – see (41) – and the next equations, $j = 1, \ldots, -p$, have the same solution u_j as in the infinite Laurent series (42). The truncated expansion (60) then evaluates to

$$u = \mathcal{D} \operatorname{Log} \psi + \text{constant} , \tag{67}$$

in which $\mathcal{D}$ is the singular part operator defined in (27) from the Laurent series, and ψ is the logarithmic primitive of χ^{-1}, an entire function defined by

$$\psi'' - \frac{k^2}{4}\psi = 0 , \tag{68}$$

whose value can be chosen, without loss of generality, as

$$\psi = \cosh \frac{k}{2}\xi . \tag{69}$$

After the operator $\mathcal{D}$ has been computed, the two equations (67) and (68) are an equivalent way of defining a one-family truncation, which is much more elegant than with (60) and (53).

The remaining $-q+p$ equations, $j = -p+1, \dots, -q$, are algebraic in k^2 and the parameters appearing in the definition of the equation (22) (one says the *fixed* parameters),

$$\begin{aligned} E_4 \equiv & -\frac{5}{2}b^2k^2 + \frac{44}{19}\mu^2 + \frac{131}{304}b^4\nu^{-2} - \frac{87}{38}b^2\mu\nu^{-1} \\ & +40k^2\mu\nu - 76k^4\nu^2 = 0 , \end{aligned} \tag{70}$$

$$E_5 \equiv b(b^2 - 16\mu\nu)\left(5k^2 + \frac{13}{152}b^2\nu^{-2} - \frac{7}{19}\mu\nu^{-1}\right) = 0 , \tag{71}$$

$$E_6 \equiv 32A + 3\nu u_0 k^6 + 4(bu_1 - \nu u_2)k^4 + 8k^2\mu u_2 + 16u_3^2 = 0 , \tag{72}$$

and they admit only the six solutions listed in Table 1.

One-Family Truncation of the Real Modulus of CGL3

Whatever representation is chosen (couple (M, φ), $(Z, \overline{Z}, \Phi)$, etc), the CGL3 equation has more than one family – see (15) – therefore any one-family truncation only captures part of the whole singularity structure, and cannot yield the general analytic solution. Nevertheless, as already noted in Sect. 2.1, (16), the one-family truncation of $M = |A|^2$ must provide at least the three known solutions. Let us carry it out.

The field M has two families of movable double pole-like singularities

$$M = \frac{3(3d_r \pm \Delta)}{2d_i^2}\chi^{-2}\left(1 + \frac{cs_i}{3}\chi + O(\chi^2)\right) , \tag{73}$$

with singular part operators $\mathcal{D}_\pm$ equal to

$$\mathcal{D}_\pm = \frac{3(3d_r \pm \Delta)}{2d_i^2}\left(-\partial_x^2 + \frac{cs_i}{3}\partial_x\right) . \tag{74}$$

In its elegant definition, the one-family truncation,

$$\begin{cases} M = \mathcal{D}_{\pm} \operatorname{Log} \psi + m \,, \\ \psi'' + \dfrac{S}{2}\psi = 0, \; S = -\dfrac{k^2}{2} = \text{ constant} \,, \end{cases} \tag{75}$$

transforms (19) into the truncated Laurent series

$$\sum_{j=0}^{14} E_j \chi^{j-14} = 0 \,, \tag{76}$$

and one must solve the 15 real determining equations, $E_j = 0$, for the two constant unknowns (S, m) and the five parameters d_r, d_i, g_r, g_i, cs_i appearing in (19). By the construction of $\mathcal{D}_{\pm}$, equations $E_j = 0, j = 0, 1$, are identically zero.

To avoid carrying complicated expressions, let us make the following non-restrictive simplification. Out of the five parameters d_r, d_i, g_r, g_i, cs_i of the ODE (19), only three are essential (g_r, g_i, c, equivalent to γ, ω, c). Indeed, p and q (i.e. $d_r + id_i$ and $s_r - is_i$) can be re-scaled to convenient numerical values, such as

$$\begin{aligned} & p = -1 - 3i \,, \quad q = 4 - 3i \,, \\ & d_r = \frac{1}{2} \,, \quad d_i = \frac{3}{2} \,, \quad s_r = -\frac{1}{10} \,, \quad s_i = -\frac{3}{10} \,, \quad \Delta = \frac{9}{2} \,. \end{aligned} \tag{77}$$

Choosing the + sign in (75), one has $\mathcal{D}_{+} = 4(-\partial_x^2 - (c/10)\partial_x)$. As seen from the first few determining equations,

$$E_2 \equiv \frac{57}{100}c^2 + 156c_2 + 13k^2 + 4g_i + 16g_r = 0 \,, \tag{78}$$

$$E_3 \equiv \left(-\frac{39}{25}c^2 - 432c_2 - 28k^2 - 16g_i - 48g_r \right) c = 0 \,, \tag{79}$$

the resolution presents no difficulty. In particular, after solving the equations numbered $j = 0, \ldots, 6$, all the remaining equations are identically zero. This fact indicates a high redundancy in these determining equations, which are, therefore, not at all optimal. In the proper CGL3 case, $\operatorname{Im}(p/q) \neq 0$, for each sign in (74), one obtains three solutions,

$$\begin{cases} M = -2\left[\left(\tau - \dfrac{c}{20}\right)^2 + \left(\dfrac{c}{10}\right)^2 \right], \\ \varphi' - \dfrac{cs_r}{2} = -\tau - \dfrac{c}{20} - \dfrac{c}{5M}\left(\tau^2 - \dfrac{k^2}{4}\right), \\ k^2 = -7\left(\dfrac{c}{10}\right)^2 - \dfrac{4}{3}g_r \,, \quad 3g_i + 2g_r + \dfrac{3c^2}{50} = 0 \,, \end{cases} \tag{80}$$

$$\begin{cases} M = -2\left(\tau^2 - \dfrac{k^2}{4}\right), & \varphi' - \dfrac{cs_r}{2} = -\tau, \\ k^2 = 2g_r\,, \quad c = 0\,, \quad g_i = 0\,, \end{cases} \tag{81}$$

$$\begin{cases} M = -2\left(\tau \pm \dfrac{k}{2}\right)^2, & \varphi' - \dfrac{cs_r}{2} = -\tau + \dfrac{c}{20}, \\ k^2 = \left(\dfrac{c}{10}\right)^2, & g_r = 0,\ g_i - \dfrac{c^2}{50} = 0\,, \end{cases} \tag{82}$$

and

$$\begin{cases} M = 4\left[\left(\tau - \dfrac{c}{20}\right)^2 + \left(\dfrac{c}{20}\right)^2\right], \\ \varphi' - \dfrac{cs_r}{2} = 2\tau - \dfrac{c}{20} - \dfrac{c}{5M}\left(\tau^2 - \dfrac{k^2}{4}\right), \\ k^2 = -\left(\dfrac{c}{10}\right)^2 + \dfrac{2}{3}g_r\,, \quad 3g_i - g_r + \dfrac{3c^2}{80} = 0\,, \end{cases} \tag{83}$$

$$\begin{cases} M = 4\tau^2,\ \varphi' - \dfrac{cs_r}{2} = 2\tau, \\ k^2 = \dfrac{2}{3}g_r\,, \quad c = 0,\ 3g_i - g_r = 0\,, \end{cases} \tag{84}$$

$$\begin{cases} M = 4\left(\tau \pm \dfrac{k}{2}\right)^2, & \varphi' - \dfrac{cs_r}{2} = 2\tau - \dfrac{c}{10}, \\ k^2 = \left(\dfrac{c}{10}\right)^2, & g_r = 0\,, \quad g_i - \dfrac{c^2}{50} = 0. \end{cases} \tag{85}$$

These solutions are identical to those listed, in the same order, in Sect. 2.1.

One-Family Truncation of CGL3 in the Complex Modulus Representation

As already outlined at the end of Sect. 4, the one-family truncation of $(Z, \overline{Z}, \Phi')$

$$\begin{cases} Z = \chi^{-1} + X + iY\,, \\ \overline{Z} = \chi^{-1} + X - iY\,, \\ \Phi = \alpha \operatorname{Log}\psi + K\xi\,, \end{cases} \tag{86}$$

with the gradient definitions

$$\begin{cases} (\operatorname{Log}\psi)' = \chi^{-1}\,, \\ \chi' = 1 - \dfrac{k^2}{4}\chi^2\,, \end{cases} \tag{87}$$

puts the l.hs. of (1) in the form

$$\sum_{j=0}^{3} E_j \chi^{j-3} = 0\,, \tag{88}$$

thus generating four complex determining equations, $E_j = 0$, (i.e. eight real, to be compared with the fifteen of Sect. 6.1). These equations must first be solved as a *linear* system on $\mathcal{C}$, as follows [9, Appendix A]. The first equation, $E_0 = 0$, identical to (11), is linear in p and q. Let us solve it for q:

$$q = -(1 - \mathrm{i}\alpha)(2 - \mathrm{i}\alpha)A_0^{-2} p \,. \tag{89}$$

The next equation, $j = 1$, is then linear in K, X, Y, c; let us solve it, for instance, for K:

$$K = (3\mathrm{i} + \alpha)X - Y + \frac{c}{2p} \,. \tag{90}$$

The equation $j = 2$, linear in γ, ω, k^2, is solved for $(i\gamma - \omega)/p$:

$$\begin{aligned} \frac{\mathrm{i}\gamma - \omega}{p} &= \left(\frac{c}{2p}\right)^2 + [X - (1 - \mathrm{i}\alpha)iY]^2 \\ &\quad -(2 - 3\mathrm{i}\alpha)\left[\frac{k^2}{4} - (X + \mathrm{i}Y)^2\right] . \end{aligned} \tag{91}$$

The advantage of this "pivoting" elimination is that the last equation, $j = 3$, which does not depend on (q, K, γ) by construction, is also independent of (p, c, ω, A_0). It only depends on (X, Y, α, k^2), and it factorizes as

$$E_3 \equiv (2X - \alpha Y)(4(X + iY)^2 - k^2) = 0 \,, \tag{92}$$

thus defining two solutions on $\mathcal{C}$.

Finally, now considering the system (89)–(92) for the real unknowns or parameters $(A_0^2, \alpha, K, c, X, Y, \gamma, \omega, k^2)$, it is quite easy to obtain the three solutions listed in Sect. 2.1.

6.2 Polynomials in tanh and sech (Two-Family Truncation)

The class of polynomials in tanh and sech

$$u = \left(\sum_{j=0}^{-p} a_j \tanh k\xi\right) + \left(\sum_{j=0}^{-p-1} b_j \tanh k\xi\right) \operatorname{sech} k\xi \,, \tag{93}$$

can equivalently be represented by the class of powers of tanh ranging from p to $-p$ [31],

$$u = \sum_{j=0}^{-2p} u_j \chi^{j+p} \,,\ \chi^{-1} = \frac{k}{2} \tanh \frac{k}{2}\xi \,,\ \xi = x - ct \,,\ u_0 u_{-2p} \neq 0 \,, \tag{94}$$

because of the elementary identities [9]

$$\begin{aligned} \tanh z - \frac{1}{\tanh z} &= -2\mathrm{i}\,\mathrm{sech}\left[2z + \mathrm{i}\frac{\pi}{2}\right], \\ \tanh z + \frac{1}{\tanh z} &= 2\tanh\left[2z + \mathrm{i}\frac{\pi}{2}\right]. \end{aligned} \tag{95}$$

A solution in this class can only exist for ODEs admitting at least two families with the same p. Indeed, if, for this p, there exists only one value of u_0, then only the second combination $\tanh + 1/\tanh = 2\tanh$ can contribute. For instance, the KS equation cannot admit such a solution.

More generally, the class of polynomials in τ and σ defined in (55),

$$u = \left(\sum_{j=0}^{-p} a_j \tau^j\right) + \left(\sum_{j=0}^{-p-1} b_j \tau^j\right)\sigma\,, \quad (a_{-p}, b_{-p-1}) \neq (0,0)\,, \tag{96}$$

is equivalently defined as [9, Appendix A]

$$\begin{cases} u = \mathcal{D}_1 \operatorname{Log} \psi_1 + \mathcal{D}_2 \operatorname{Log} \psi_2 + m\,, \quad m = \text{const}\,, \\ \psi_1'' + \dfrac{S}{2}\psi_1 = 0\,, \quad \psi_2'' + \dfrac{S}{2}\psi_2 = 0\,, \quad S = -\dfrac{k^2}{2} = \text{constant}\,, \\ \dfrac{\psi_1'}{\psi_1}\dfrac{\psi_2'}{\psi_2} = -\dfrac{S}{2} - \dfrac{k}{2}\mu_0\left(\dfrac{\psi_1'}{\psi_1} - \dfrac{\psi_2'}{\psi_2}\right). \end{cases} \tag{97}$$

In this form, which is the natural extension of (75) to two families, the linear operators $\mathcal{D}_1$ and $\mathcal{D}_2$ are the singular part operators of two *different* families, and the entire functions ψ_1 and ψ_2 obey the same second-order linear equation, but with a different choice of the integration constants,

$$\psi_1 = \cosh\frac{k}{2}(\xi + a)\,, \quad \psi_2 = \cosh\frac{k}{2}(\xi - a)\,, \quad \mu_0 = \coth ka\,. \tag{98}$$

The case $\mu_0(\mu_0^2 - 1) = 0$ reduces to the class of polynomials in tanh and sech.

The practical implementation is the following.

1. For the class of polynomials in tanh and sech, one puts the l.h.s. $E(u)$ of the nonlinear ODE into the same form as u,

$$\begin{cases} u = \displaystyle\sum_{j=0}^{-2p} u_j \chi^{j+p}\,, \quad u_0 u_{-2p} \neq 0\,, \\ \chi' = 1 + \dfrac{S}{2}\chi^2\,, \quad S = -\dfrac{k^2}{2}\,, \\ E = \displaystyle\sum_{j=0}^{-2q} E_j \chi^{j+q}\,, \\ \forall j : E_j = 0\,. \end{cases} \tag{99}$$

and one solves the set of $-2q+1$ determining equations, $E_j = 0$.

(100)

2. For the class of polynomials in τ and σ defined in (55), under the assumption (97), the l.h.s. $E(u)$ is first expressed as a polynomial of the two variables $\psi_j'/\psi_j, j=1,2$

$$\sum_{k=0}^{-q}\sum_{l=0}^{-q-k} E_{k,l}\left(\frac{\psi_1'}{\psi_1}\right)^k\left(\frac{\psi_2'}{\psi_2}\right)^l=0, \tag{101}$$

which further reduces, thanks to the third line of (97), to the sum of two polynomials of one variable,

$$E_0+\left(\sum_{j=1}^{-q} E_j^{(1)}\left(\frac{\psi_1'}{\psi_1}\right)^j\right)+\left(\sum_{j=1}^{-q} E_j^{(2)}\left(\frac{\psi_2'}{\psi_2}\right)^j\right)=0\,. \tag{102}$$

One then requires the vanishing of the $-2q+1$ determining equations

$$E_0=0,\ E_j^{(1)}=0,\ E_j^{(2)}=0,\ j=1,\dots,-q\,. \tag{103}$$

As an example, let us apply this to the ODE

$$E(u)\equiv\left(\frac{\mathrm{d}u}{\mathrm{d}\xi}\right)^2-\alpha^2(u^2-b^2)^2+c=0\,. \tag{104}$$

It admits two families, with singular part operators $\mathcal{D}_1=\alpha^{-1}\partial_\xi, \mathcal{D}_2=-\alpha^{-1}\partial_\xi$. The relation $\mathcal{D}_2=-\mathcal{D}_1$ implies $E_j^{(1)}+(-1)^jE_j^{(2)}\equiv 0,\ j=1,2,3,4,$ and only 5 out of the 9 determining equations are linearly independent. Moreover, from the construction of the singular part operators, the two equations, $j=4$, are identically satisfied. The next equation, $j=3$,

$$E_3^{(1)}\equiv -2\alpha^{-2}k\mu_0-4\alpha^{-1}m=0\,, \tag{105}$$

is solved for m. Then, the equation $j=2$

$$E_2^{(1)}\equiv 2b^2+\alpha^{-2}k^2-\frac{3}{2}\alpha^{-2}(k\mu_0)^2=0\,, \tag{106}$$

is solved for k^2, viewing $k\mu_0$ as a single variable. The remaining system

$$E_1^{(1)}\equiv\left(-2b^2+\frac{1}{2}\alpha^{-2}(k\mu_0)^2\right)(k\mu_0)=0\,, \tag{107}$$

$$E_0\equiv-\alpha^2b^4+c+\frac{1}{2}b^2(k\mu_0)^2-\frac{1}{16}\alpha^{-2}(k\mu_0)^4=0\,, \tag{108}$$

admits two solutions.

The first one, $c=0, (k\mu_0)^2=(2\alpha b)^2$, corresponds to a factorization of the equation $E(u)=0$ into two Riccati equations, and therefore must be rejected. The second one

$$k\mu_0 = 0\,, \quad \alpha^2 b^4 - c = 0\,, \tag{109}$$

defines a solution, provided the indicated constraint on the fixed parameters (α, b, c) is satisfied. This solution

$$\begin{aligned} \mu_0 = 0\,, \quad m = 0\,, \quad k^2 = -2(\alpha b)^2\,, \\ u = \alpha^{-1} \frac{\mathrm{d}}{\mathrm{d}\xi} \operatorname{Log} \frac{\cosh(k/2)(\xi + a)}{\cosh(k/2)(\xi - a)} \end{aligned} \tag{110}$$

is nothing other than $u = i(k/\alpha) \operatorname{sech} k\xi$, using the relation $\mu_0 = \coth ka$.

Indeed, in contrast to the function tanh, which satisfies an ODE admitting only one family of movable singularities (the Riccati equation), the function sech (or more generally its homographic transform σ) satisfies a first-order second degree ODE

$$\operatorname{sech}'^2 + \operatorname{sech}^4 - \operatorname{sech}^2 = 0\,, \tag{111}$$

which admits two families of movable simple poles with opposite residues

$$\operatorname{sech}(\xi - \xi_0) \sim \pm \mathrm{i}(\xi - \xi_0)^{-1}\,. \tag{112}$$

Two-Family Truncation of the Real Modulus of CGL3

M admits exactly two families, so its two-family truncation is quite appropriate. The two singular part operators are, for each family, defined in (74), with $\mathcal{D}_1 = \mathcal{D}_+, \mathcal{D}_2 = \mathcal{D}_-$. The assumption (97), with $p = -2, q = -14$, transforms (19) into the sum (102) of two polynomials of one variable, and the four equations $j = 14, 13$ are identically zero, using the definition of $\mathcal{D}_\pm$.

For p, q, γ arbitrary, the resolution of the 25 remaining determining equations is impossible to carry out by hand. However, a hand computation becomes possible with the generic numerical values (77). First, the system $j = 12$

$$\begin{aligned} E_{12}^{(1)} \equiv \frac{57}{5} c^2 - 780m - 520k^2 + 78ck\mu_0 \\ +390(k\mu_0)^2 + 80g_i + 320g_r = 0\,, \end{aligned} \tag{113}$$

$$\begin{aligned} E_{12}^{(2)} \equiv \frac{3}{5} c^2 + 120m - 40k^2 - 24ck\mu_0 \\ +120(k\mu_0)^2 + 20g_i - 40g_r = 0\,, \end{aligned} \tag{114}$$

is solved as a linear system for m and k^2. The next system, $j = 11$

$$\begin{aligned} E_{11}^{(1)} \equiv -\frac{39}{125} c^3 - \frac{254}{25} c^2 k\mu_0 + \frac{468}{5} c(k\mu_0)^2 - 312(k\mu_0)^3 \\ -\frac{156}{5} cg_i - 168k\mu_0 g_i - \frac{104}{5} cg_r - 48k\mu_0 g_r = 0\,, \end{aligned} \tag{115}$$

$$\begin{aligned} E_{11}^{(2)} \equiv -\frac{177}{500} c^3 + \frac{218}{25} c^2 k\mu_0 - 39c(k\mu_0)^2 + 156(k\mu_0)^3 \\ -\frac{87}{5} cg_i + 84k\mu_0 g_i - 2cg_r + 24k\mu_0 g_r = 0\,, \end{aligned} \tag{116}$$

is linear in (g_r, g_i), with a Jacobian $J = c(3c - 5k\mu_0)$. For the first sub-case, $J \neq 0$, after solving for (g_r, g_i) as functions of $(c, k\mu_0)$, the next system $j = 10$ only depends on $k\mu_0/c$, and it admits no solution. A discussion of the second sub-case, $J = 0$, using only the next system, $j = 10$, leads to the conclusion that no solution exists. An identical result is achieved for arbitrary values of (p, q, γ) using computer algebra.

This unfortunate situation is exceptional, and only reflects the difficulty of the CGL3. Should such a solution exist, it would have the form

$$M = \left(\frac{\Delta}{2d_i^2}\tanh + c_1\right)\operatorname{sech} + \frac{9d_r}{2d_i^2}\tanh^2 + c_3\tanh + c_4\;, \tag{117}$$

and the constraint $c_3 = 0$ could define a homoclinic hole solution, just like the (yet analytically unknown) one of van Hecke [13].

Two-Family Truncation of the CGL3 in the Complex Modulus Representation

Let us denote two different solutions of (15) as (A_0, α) and (A_2, α_2) .

In the complex modulus representation (56), the two-family truncation of $(Z, \overline{Z}, \Phi)$ is defined as [9, Appendix A],

$$\begin{cases} A = (A_0(\partial_\xi \operatorname{Log}\psi_1(\xi) + X + \mathrm{i}Y) + A_2\partial_\xi \operatorname{Log}\psi_2(\xi))\,\mathrm{e}^{\mathrm{i}[-\omega t + \Phi(\xi)]}, \\ \overline{A} = (A_0(\partial_\xi \operatorname{Log}\psi_1(\xi) + X - \mathrm{i}Y) + A_2\partial_\xi \operatorname{Log}\psi_2(\xi)) \\ \qquad\qquad \times \mathrm{e}^{-\mathrm{i}[-\omega t + \Phi(\xi)]}, \\ \Phi = \alpha \operatorname{Log}\psi_1 + \alpha_2 \operatorname{Log}\psi_2 + K\xi\;, \end{cases} \tag{118}$$

with the definitions for the derivatives of (ψ_1, ψ_2) given by the last two lines of (97). The l.h.s. of (1) then takes the form (102) with $q = -3$, and one solves the seven complex determining equations as a linear system on $\mathcal{C}$, similarly to what has been done in Sect. 6.1.

From the two equations $j = 3$,

$$E_3^{(1)} \equiv A_0\left((1 - \mathrm{i}\alpha)(2 - \mathrm{i}\alpha)p + A_0^2 q\right) = 0\;, \tag{119}$$

$$E_3^{(2)} \equiv A_2\left((1 - \mathrm{i}\alpha_2)(2 - \mathrm{i}\alpha_2)p + A_2^2 q\right) = 0\;, \tag{120}$$

and the two relations implied by (15),

$$\alpha = \frac{d_i}{3}A_0^2\;, \quad \alpha_2 = \frac{d_i}{3}A_2^2\;. \tag{121}$$

one first proves that the only possibility is $A_2 = -A_0$. Therefore, the two complex equations, $j = 3$, are solved as

$$A_2 = -A_0\;, \quad \alpha_2 = \alpha\;, \quad q = -(1 - \mathrm{i}\alpha)(2 - \mathrm{i}\alpha)A_0^{-2}p\;. \tag{122}$$

At the level $j = 2$, the symmetric combination

$$E_2^{(1)} + E_2^{(2)} \equiv p(1 - \mathrm{i}\alpha)\left[(\mathrm{i}\alpha - 3)X - \mathrm{i}Y + \left(\mathrm{i}\alpha - \frac{3}{2}\right) k\mu_0\right] = 0\,, \tag{123}$$

is solved for the two pieces of information

$$X = -\frac{1}{2}k\mu_0\,, \quad Y = \frac{1}{2}\alpha k\mu_0\,, \tag{124}$$

and then the anti-symmetric combination

$$E_2^{(1)} - E_2^{(2)} \equiv p(1 - \mathrm{i}\alpha)\left[\frac{c}{2p} - K\right] = 0\,, \tag{125}$$

is solved as

$$K = \frac{c}{2p}\,. \tag{126}$$

At the level $j = 1$, the symmetric combination is identically zero, and the anti-symmetric combination is solved for $(i\gamma - \omega)/p$ (we omit the expression). The remaining equation

$$E_0 \equiv \mu_0\left[(2 + (\alpha^2 - 2)\mu_0^2) + \mathrm{i}\alpha(-4 + (\alpha^2 + 4)\mu_0^2)\right] = 0 \tag{127}$$

admits $\mu_0 = 0$ as the only solution. Therefore, one obtains the unique solution

$$A_2 = -A_0\,, \quad \alpha_2 = \alpha\,, \quad q = -(1 - \mathrm{i}\alpha)(2 - \mathrm{i}\alpha)A_0^{-2}p\,, \tag{128}$$

$$X = 0\,, \quad Y = 0\,, \quad \mu_0 = 0\,, \quad K = \frac{c}{2p}\,, \tag{129}$$

$$\frac{\mathrm{i}\gamma - \omega}{p} = \left(\frac{c}{2p}\right)^2 + (1 - \mathrm{i}\alpha)^2 k^2\,. \tag{130}$$

The value of K implies $cp_i = 0$, and the case $c = 0$ represents the homoclinic pulse (13).

6.3 Polynomials in $\wp$ and $\wp'$

A class of polynomial elliptic functions can be defined, for instance, with the Weierstrass function $\wp$ and its derivative [17, 33],

$$u = \left(\sum_{j=0}^{[-p/2]} a_j \wp(\xi)^j\right) + \left(\sum_{j=0}^{[(-p-3)/2]} b_j \wp(\xi)^j\right) \wp'(\xi)\,. \tag{131}$$

Since $\wp$ admits only one family, such a solution may exist for any ODE. It will be quite useful to take advantage of the value of the singular part operator of $\wp(\xi)$,

$$\mathcal{D} = -\frac{\mathrm{d}^2}{\mathrm{d}\xi^2} \, . \tag{132}$$

For the KS, the assumption of seeking solutions in the above class

$$u = c_0 \wp' + c_1 \wp + c_2, \; c_0 \neq 0 \, , \tag{133}$$

together with knowledge of the singular part operators (27) and (132), first yields the correct values of c_0 and c_1,

$$u = -60\nu\wp' - 15b\wp + c_2 \, , \tag{134}$$

and then a truncation which defines the four determining equations

$$\begin{cases} b^2 - 16\mu\nu = 0 \, , \\ b\mu + 4\nu c_2 = 0 \, , \\ bc_2 - 720\nu^2 g_2 = 0 \, , \\ A + \frac{1}{2}c_2^2 + \frac{15}{2}b^2 g_2 + 30\mu\nu g_2 - 1080\nu^2 g_3 = 0 \, . \end{cases} \tag{135}$$

Their unique solution is (23).

For the CGL3, the assumption that M be in this class

$$M = a_2\wp + c_2, \; a_2 \neq 0 \tag{136}$$

generates 10 determining equations, which are in the parameters and unknowns $(a_2, c_2, g_2, g_3; d_r, d_i, s_r, cs_i, g_r, g_i)$. The equation with the highest singularity degree

$$((a_2 d_i)^2 - 9a_2 d_r - 18)(2 + a_2 d_r) = 0 \, , \tag{137}$$

in which the vanishing of the second factor is forbidden, is first solved as a linear equation for d_r:

$$d_r = \frac{(a_2 d_i)^2 - 18}{9a_2} \, . \tag{138}$$

The next equation yields $cs_i = 0$. Then the next equations are successively solved for g_r, g_2, g_3, and the elliptic discriminant $g_2^3 - 27g_3^2$ is then divisible by the unique remaining determining equation. Therefore, one finds the pulse (13) as the unique solution.

Finally, let us mention the *Ansatz* made for the CGL5 [25, 3]

$$A = a(x)\mathrm{e}^{\mathrm{i}[-2\alpha \,\mathrm{Log}\, a(x) - \omega t]}, \; (\omega, \alpha, a) \in \mathcal{R} \, , \tag{139}$$

which sets an a priori constraint on the relation between the amplitude and the phase (similar to that made for the CGL3 in [5]), together with the assumption that a^2 obeys a first-order second degree elliptic equation. This allows one to retrieve (34) in the particular case $r_0 = 0, c = 0$.

7 Second Class of Methods: First-Order Sub-Equation

As noted in the Appendix, the requirement that the solution (6) be shared by the Nth-order ODE (5) and the first-order ODE (7) characterizes the single-valued expressions f as being elliptic or degenerate elliptic (i.e. trigonometric or rational), i.e. the class

$$u = R(\wp', \wp) \longrightarrow R(\mathrm{e}^{k\xi}) \longrightarrow R(\xi) \,, \tag{140}$$

in which R denotes rational functions and $\longrightarrow$ denotes the degeneracy. This class contains all the classes considered in previous sections (polynomials in tanh, in (tanh, sech), in (σ, τ) and in $(\wp, \wp')$), but it contains, in addition, expressions like (59).

The algorithm to obtain all the elliptic solutions combines two pieces of information:

1. a local one, a Laurent series representing the largest analytic solution of the N-th order ODE near a movable pole-like singularity,
2. a global one, the necessary form (168) of (7).

by requiring that the Laurent series satisfies the first-order sub-equation (168).

This provides the explicit form of the first-order sub-equation $F(u, u') = 0$. Then, one computes the solution $u = f(\xi - \xi_0)$ from this equation, $F = 0$.

The successive steps are [27, Sect. 5].

1. Choose a positive integer m and define the Briot and Bouquet first-order ODE

$$F(u, u') \equiv \sum_{k=0}^{m} \sum_{j=0}^{[(m-k)(p-1)/p]} a_{j,k} u^j u'^k = 0 \,, \quad a_{0,m} \neq 0 \,, \tag{141}$$

   in which $[z]$ denotes the "integer part" function. The upper bound on j implements the condition $m(p-1) \le jp + k(p-1)$, so that no term can be more singular than u'^m, and this is identically satisfied if $p = -1$. The polynomial F contains at most $(m+1)^2$ unknown constants $a_{j,k}$.
2. Compute J terms of the Laurent series, with J slightly greater than the number of unknown constants $a_{j,k}$.

$$u = \chi^p \left(\sum_{j=0}^{J} M_j \chi^j + \mathcal{O}(\chi^{J+1}) \right) \,, \quad \chi = \xi - \xi_0 \,, \tag{142}$$

   where p is -3 for the KS equation (22), -2 for the variable $|A|^2$ of the CGL3, etc.

3. Require the Laurent series to satisfy the Briot and Bouquet ODE, i.e. require the identical vanishing of the Laurent series for the l.h.s. $F(U,U')$ up to the order J

$$F \equiv \chi^D \left(\sum_{j=0}^{J} F_j \chi^j + \mathcal{O}(\chi^{J+1}) \right), \quad D = m(p-1) , \tag{143}$$

$$\forall j \ : \ F_j = 0 . \tag{144}$$

   If it has no solution for $a_{j,k}$, increase m and return to first step.
4. For every solution, integrate the first-order autonomous ODE (141).

Let us give two examples.

7.1 First-Order Autonomous Sub-Equations of the KS

The Laurent series of (22) is (42).

In the second step, the smallest integer, m, which allows a movable triple pole ($p=-3$) in (141) is $m=3$. With the normalization $a_{0,3}=1$, the sub-equation contains ten coefficients, which are first determined by the Cramer system of ten equations $F_j=0, j=0:6,8,9,12$. The first few are

$$F_0 \equiv -9a_{0,3} + 40\nu a_{4,0} = 0 , \tag{145}$$

$$F_1 \equiv 9ba_{0,3} + 12\nu a_{1,2} - 80b\nu a_{4,0} = 0 , \tag{146}$$

$$\begin{aligned} F_2 \equiv (2120b^2 + 2560\mu\nu)\nu a_{4,0} - (105b^2 + 144\mu\nu) a_{0,3} \\ -532b\nu a_{1,2} - 608\nu^2 a_{2,1} = 0 , \end{aligned} \tag{147}$$

$$\begin{aligned} F_3 \equiv (5b^2 + 72\mu\nu) ba_{0,3} + (137b^2 + 240\mu\nu)\nu a_{1,2} \\ -(442b^2 + 2656\mu\nu) b\nu a_{4,0} + 608b\nu^2 a_{2,1} + 608\nu^3 a_{3,0} = 0 . \end{aligned} \tag{148}$$

The remaining infinitely overdetermined nonlinear system for (ν, b, μ, A) contains, as its greatest common divisor (g.c.d.), $b^2 - 16\mu\nu$ (see (23)), which defines a first solution

$$\begin{aligned} &\frac{b^2}{\mu\nu} = 16 , \\ &\left(u' + \frac{b}{2\nu}u_s\right)^2 \left(u' - \frac{b}{4\nu}u_s\right) + \frac{9}{40\nu}\left(u_s^2 + \frac{15b^6}{1024\nu^4} + \frac{10A}{3}\right)^2 = 0 , \\ &u_s = u + \frac{3b^3}{32\nu^2} . \end{aligned} \tag{149}$$

After division by this g.c.d., the remaining system for (ν, b, μ, A) admits exactly four solutions (terminating the series at $j=16$ is sufficient to obtain the result), namely the first three lines of Table 1, with each solution defining the same kind of sub-equation:

$b=0$,

$$\left(u'+\frac{180\mu^2}{19^2\nu}\right)^2\left(u'-\frac{360\mu^2}{19^2\nu}\right)+\frac{9}{40\nu}\left(u^2+\frac{30\mu}{19}u'-\frac{30^2\mu^3}{19^2\nu}\right)^2=0\,, \quad (150)$$

$$b=0,\ u'^3+\frac{9}{40\nu}\left(u^2+\frac{30\mu}{19}u'+\frac{30^2\mu^3}{19^3\nu}\right)^2=0\,, \quad (151)$$

$$\frac{b^2}{\mu\nu}=\frac{144}{47},\ u_s=u-\frac{5b^3}{144\nu^2},\ \left(u'+\frac{b}{4\nu}u_s\right)^3+\frac{9}{40\nu}u_s^4=0\,, \quad (152)$$

$$\frac{b^2}{\mu\nu}=\frac{256}{73},\ u_s=u-\frac{45b^3}{2048\nu^2}\,,$$

$$\left(u'+\frac{b}{8\nu}u_s\right)^2\left(u'+\frac{b}{2\nu}u_s\right)+\frac{9}{40\nu}\left(u_s^2+\frac{5b^3}{1024\nu^2}u_s+\frac{5b^2}{128\nu}u'\right)^2=0\,. \quad (153)$$

In order to integrate the two sets of sub-equations (149), (150)–(153), one must first compute their genus[1], which is one for (149), and zero for (150)–(153). Therefore (149) has an elliptic general solution, listed above as (23), and initially found [12, 17] by other methods. As to the general solution of the four others (150)–(153), it is a third degree polynomial (25) in $\tanh\frac{k}{2}(\xi-\xi_0)$. These four solutions, obtained for the minimal choice of the sub-equation degree m, constitute all the analytic results known on (22). For $m=4$, no additional solution is obtained [41]. The computation for $m=5$ is in progress.

7.2 First-Order Autonomous Sub-Equations of the CGL3

We consider the variable $M=|A|^2$, i.e. $p=-2$. The smallest value of m is then 2. With the numerical values (77), the two Laurent series are

$$M_-=\chi^{-2}\left(-2+\frac{c}{5}\chi+\left(\frac{g_r}{3}-\frac{g_i}{6}-\frac{c^2}{200}\right)\chi^2+\mathcal{O}(\chi^3)\right), \quad (154)$$

$$M_+=\chi^{-2}\left(4-\frac{2c}{5}\chi+\left(\frac{16g_r}{39}+\frac{4g_i}{39}+\frac{19c^2}{1300}\right)\chi^2+\mathcal{O}(\chi^3)\right). \quad (155)$$

The existence of two Laurent series, rather than just one, is a feature which the sub-equation must also possess, and this has the effect of setting the lower bound to $m=4$, instead of 2. Indeed, the lowest degree sub-equations

$$F_2\equiv M'^2+M'(a_{1,1}M+a_{0,1})+a_{3,0}M^3+a_{2,0}M^2+a_{1,0}M+a_{0,0}=0\,, \quad (156)$$

$$F_3\equiv M'^3+M'^2(a_{1,2}M+a_{0,2})+M'(a_{3,1}M^3+a_{2,1}M^2+a_{1,1}M+a_{0,1})+a_{4,0}M^4+a_{3,0}M^3+a_{2,0}M^2+a_{1,0}M+a_{0,0}=0\,, \quad (157)$$

[1] For instance with the Maple command *genus* of the package *algcurves* [15], which implements an algorithm of Poincaré.

have the respective dominant terms ${M'}^2 + a_{3,0}M^3$ and ${M'}^3 + a_{3,1}M'M^3$, which define only one family of movable double poles.

Let us nevertheless start with $m = 2$, for which (156) can only be satisfied by one series, e.g. (154), thus preventing the full desired result from being obtained. The six coefficients $a_{j,k}$ of (156) are first computed as the unique solution of the linear system of six equations $F_j = 0, j = 0, 1, 2, 3, 4, 6$. Then the $J + 1 - 6$ remaining equations, $F_j = 0, j = 5, 7 : J$, which only depend on the fixed parameters (g_r, g_i, c), have the greatest common divisor (g.c.d.) $3g_i + 2g_r + 3c^2/50$, and this factor defines the first solution ((158) below). After division by this g.c.d., the system of three equations, $F_j = 0, j = 5, 7, 8$, provides two, and only two other solutions, see (159) and (160) below, with the respective constraints $(c = 0, g_i = 0)$ and $(g_r = 0, 50g_i - c^2 = 0)$, and all the remaining equations, $F_j = 0, j \geq 9$, are identically satisfied.

Therefore, with this lower bound, $m = 2$, one already recovers all the known first-order sub-equations. These are, with the series (154),

$$\left(M' + \frac{c}{5}M + \frac{c^3}{250}\right)^2 + 2\left(M + \frac{c^2}{50}\right)\left(M - \frac{c^2}{50} - \frac{2}{3}g_r\right)^2 = 0\,,$$

$$3g_i + 2g_r + \frac{3c^2}{50} = 0\,, \tag{158}$$

$${M'}^2 + 2(M - g_r)M^2 = 0\,, \quad c = 0\,, \quad g_i = 0\,, \tag{159}$$

$$\left(M' + \frac{c}{5}M\right)^2 + 2M^3 = 0\,, \quad g_r = 0\,, \quad g_i - \frac{c^2}{50} = 0\,. \tag{160}$$

Finally, for each of the three sub-equations, the fourth step finds a zero value for the genus and returns the general solution as a rational function of $e^{a(\xi-\xi_0)}$. Basic trigonometric identities then allow this to be converted to the second degree polynomials in $(k/2)\tanh k(\xi - \xi_0)/2$ listed in (80)–(82).

Similarly, with the other series (155), one obtains

$$\left(M' + \frac{c}{5}M - \frac{c^3}{500}\right)^2 - \left(M - \frac{c^2}{100}\right)\left(M + \frac{c^2}{100} - \frac{2}{3}g_r\right)^2 = 0\,,$$

$$3g_i - g_r + \frac{3c^2}{80} = 0\,, \tag{161}$$

$${M'}^2 - M\left(M - \frac{2}{3}g_r\right)^2 = 0\,, \quad c = 0\,, \quad 3g_i - g_r = 0\,, \tag{162}$$

$$\left(M' + \frac{c}{5}M\right)^2 - M^3 = 0\,, \quad g_r = 0\,, \quad g_i - \frac{c^2}{50} = 0\,. \tag{163}$$

With the correct two-family lower bound, $m = 4$, which corresponds to 18 unknowns, $a_{j,k}$, and at least 24 terms in the series, we have checked that there is no solution other than the above three. This situation is quite similar to the absence of a solution in the class (117), and it just reflects the difficulty of the CGL3 equation.

The case $m = 8$ (60 unknowns, $a_{j,k}$, and at least 66 terms in the series) is currently under investigation, but preliminary results seem to indicate the absence of any new solution, and we are now automating the computer algebra program, in order to handle much larger values of m.

7.3 Domain of Applicability of the Method

As we have seen, the sub-equation method contains the truncation methods and its cost is minimal, since the main step is a linear computation.

The two key assumptions behind this "sub-equation method" are:

1. a Laurent series should exist,
2. a first-order autonomous algebraic sub-equation should exist.

Its best applicability is therefore to nonintegrable N-th order autonomous nonlinear ODEs admitting a Laurent series which only depends on one movable constant, such as the CGL3 ODE (19) or the traveling wave reduction (22) of the Kuramoto-Sivashinsky equation [8, 41].

Two examples of inapplicability are

1. the Lorenz model, in which the Laurent series generically does not exist, and has to be replaced with a psi-series [36],
2. the autonomous ODE $u''' - 12uu' - 1 = 0$, which admits the first Painlevé transcendent as its general solution, a case in which no first-order sub-equation exists.

8 Conclusion

How do these two classes of methods (truncations and first-order sub-equations) really compare, independently of the amount of computation involved?

Let us first recall a preliminary, classical result.

The class (60) of polynomials of degree $-p$ in tanh obeys a first-order equation of degree $m = -p$. For instance, given the polynomial

$$u = \tanh^2 + 2a \tanh + b \,, \tag{164}$$

this amounts to eliminating tanh between the two algebraic equations

$$\begin{cases} \tanh^2 + 2a \tanh + b - u = 0, \\ 2(\tanh + a)(1 - \tanh^2) - u' = 0, \end{cases} \tag{165}$$

which results in[2]

[2] This formula, due to Sylvester, expresses the resultant of two polynomials of degrees m and n as a determinant of order $m + n$.

$$\begin{vmatrix} 0 & 0 & 1 & 2a & b-u \\ 0 & 1 & 2a & b-u & 0 \\ 1 & 2a & b-u & 0 & 0 \\ 0 & -2 & -2a & 2 & 2a-u' \\ -2 & -2a & 2 & 2a-u' & 0 \end{vmatrix} \tag{166}$$

$$= (u' - 4a(u - b + a^2))^2 - 4(u - b + 2a^2 - 1)^2(u - b + a^2) = 0\,, \tag{167}$$

an equation with degree $m = 2 = -p$, having genus zero.

Similarly, the class of polynomials of global degree $-p$ in $(\tanh, \mathrm{sech})$ or (σ, τ) obeys a first-order ODE with degree $m = -2p$.

Lastly, the class (131) of polynomials of $(\wp, \wp')$, of singularity degree at most equal to p, obeys a first-order ODE with degree $m = -p$.

Therefore, given a value of p (the singularity degree of the ODE) and a truncation considered in Sect. 6, there exists a value of m (either $-p$ or $-2p$) at which the result of the truncation can be found by the method of first-order sub-equations.

Conversely, given a value of m (the degree of a first-order sub-equation), the class of solutions of the method of first-order sub-equations consists of the rational functions (of $(\wp, \wp')$ or of $e^{k\xi}$ i.e. $(k/2)\tanh k\xi/2)$, a class richer than the polynomials.

This proves the *identity* of the two classes of methods, provided the truncations assume rational functions instead of polynomials.

However, from a practical point of view, considering the amount of computation involved, the increasing order of difficulty seems to be

1. Truncations of polynomials.
2. First-order sub-equations.
3. Truncations of rational functions.

How does this compare with the approach of Chow (see e.g. [6]) used to find solutions of PDEs in terms of elliptic functions? To be definite, let us consider a PDE in (x, t). The solutions which do depend on both x and t (i.e. which do not satisfy some ODE) are richer than those described here. As to the solutions of solitary wave type, $f(x - ct)$, the method of Chow belongs to the first class of methods, i.e. it may, or may not, find the most general elliptic solution which exists.

Acknowledgements

We gratefully acknowledge the Tournesol grant T2003.09.

9 Appendix. Classical Results on First-Order Autonomous Equations

The following results were mainly obtained by Briot and Bouquet, Fuchs and Poincaré, and were put in final form by Painlevé [29, pages 58–59].

Theorem. Given an algebraic first-order autonomous ODE (7), whose general solution is therefore (6), the following properties are equivalent.

1. Its general solution is single-valued.
2. Its general solution is an elliptic function, possibly degenerate.
3. The genus of the algebraic curve (7) is one or zero.
4. There is equivalence between knowledge of f and that of F.
5. There exist a positive integer m and $(m+1)^2$ complex constants $a_{j,k}$, with $a_{0,m} \neq 0$, such that the polynomial F, of two variables, necessarily has the form

$$F(u,u') \equiv \sum_{k=0}^{m} \sum_{j=0}^{2m-2k} a_{j,k} u^j {u'}^k = 0 \ , \ a_{0,m} \neq 0 \ . \tag{168}$$

6. If the genus is one, there exist two rational functions R_1, R_2, such that the general solution is

$$u = R_1(\wp) + \wp' R_2(\wp) \ , \tag{169}$$

in which $\wp = \wp(\xi - \xi_0, g_2, g_3)$ is the Weierstrass elliptic function characterized by (24).

If the genus is zero, there exists a (possibly zero) constant a and a rational function R, such that the general solution is

$$u = R(e^{a\xi}) \ , \tag{170}$$

with the degeneracy $u = R(\xi)$ in the case of a being zero.

References

1. M. J. Ablowitz and P. A. Clarkson, *Solitons, nonlinear evolution equations and inverse scattering* (Cambridge University Press, Cambridge, 1991).
2. G. P. Agrawal, *Nonlinear fiber optics*, 3rd edition (Academic press, Boston, 2001).
3. N. N. Akhmediev and V. V. Afanasjev, Novel arbitrary-amplitude soliton solutions of the cubic-quintic complex Ginzburg-Landau equation, Phys. Rev. Lett. **75**, 2320–2323 (1995).
4. N. Bekki and K. Nozaki, Formations of spatial patterns and holes in the generalized Ginzburg-Landau equation, Phys. Lett. A **110**, 133–135 (1985).
5. F. Cariello and M. Tabor, Painlevé expansions for nonintegrable evolution equations, Physica D **39**, 77–94 (1989).

6. Kwok W. Chow, A class of doubly periodic waves for nonlinear evolution equations, Wave Motion **35**, 71–90 (2002).
7. R. Conte, The Painlevé approach to nonlinear ordinary differential equations, *The Painlevé property, one century later*, 77–180, ed. R. Conte, CRM series in mathematical physics (Springer, New York, 1999). http://arXiv.org/abs/solv-int/9710020
8. R. Conte and M. Musette, Painlevé analysis and Bäcklund transformation in the Kuramoto-Sivashinsky equation, J. Phys. A **22**, 169–177 (1989).
9. R. Conte and M. Musette, Linearity inside nonlinearity: exact solutions to the complex Ginzburg-Landau equation, Physica D **69**, 1–17 (1993).
10. R. Conte and M. Musette, Analytic expressions of hydrothermal waves, Reports on mathematical physics **46**, 77–88 (2000). http://arXiv.org/abs/nlin.SI/0009022
11. R. Conte, A. P. Fordy and A. Pickering, A perturbative Painlevé approach to nonlinear differential equations, Physica D **69**, 33–58 (1993).
12. J.-D. Fournier, E. A. Spiegel and O. Thual, Meromorphic integrals of two non-integrable systems, *Nonlinear dynamics*, 366–373, ed. G. Turchetti (World Scientific, Singapore, 1989).
13. M. van Hecke, Building blocks of spatiotemporal intermittency, Phys. Rev. Lett. **80**, 1896–1899 (1998).
14. M. van Hecke, C. Storm and W. van Saarlos, Sources, sinks and wavenumber selection in coupled CGL equations and experimental implications for counter-propagating wave systems, Physica D **133**, 1–47 (1999). http://arXiv.org/abs/patt-sol/9902005.
15. Mark van Hoeij, package "algcurves", Maple V (1997). http://www.math.fsu.edu/~hoeij/algcurves.html.
16. N. A. Kudryashov, Exact soliton solutions of the generalized evolution equation of wave dynamics, Prikladnaia Matematika i Mekhanika **52**, 465–470 (1988) [English :Journal of applied mathematics and mechanics **52** (1988) 361–365]
17. N. A. Kudryashov, Exact solutions of a generalized equation of Ginzburg-Landau, Matematicheskoye modelirovanie **1**, 151–158 (1989).
18. N. A. Kudryashov, Exact solutions of the generalized Kuramoto-Sivashinsky equation, Phys. Lett. A **147**, 287–291 (1990).
19. Y. Kuramoto and T. Tsuzuki, Persistent propagation of concentration waves in dissipative media far from thermal equilibrium, Prog. Theor. Phys. **55**, 356–369 (1976).
20. J. Lega, Traveling hole solutions of the complex Ginzburg-Landau equation: a review, Physica D **152–153**, 269–287 (2001).
21. J. Lega, J. V. Moloney and A. C. Newell, Swift-Hohenberg equation for lasers, Phys. Rev. Lett. **73**, 2978–2981 (1994).
22. P. Manneville, *Dissipative structures and weak turbulence* (Academic Press, Boston, 1990). French adaptation: *Structures dissipatives, chaos et turbulence* (Aléa-Saclay, Gif-sur-Yvette, 1991).
23. P. Marcq, H. Chaté and R. Conte, Exact solutions of the one-dimensional quintic complex Ginzburg-Landau equation, Physica D **73**, 305–317 (1994). http://arXiv.org/abs/patt-sol/9310004
24. K. Maruno, A. Ankiewicz and N. N. Akhmediev, Exact soliton solutions of the one-dimensional complex Swift-Hohenberg equation, Physica D **176**, 44–66 (2003).

25. J. D. Moores, On the Ginzburg-Landau laser mode-locking model with fifth-order saturable absorber term, Optics Communications **96**, 65–70 (1993).
26. M. Musette, Painlevé analysis for nonlinear partial differential equations, *The Painlevé property, one century later*, 517–572, ed. R. Conte, CRM series in mathematical physics (Springer, New York, 1999).
27. M. Musette and R. Conte, Analytic solitary waves of nonintegrable equations, Physica D **181**, 70–79 (2003). http://arXiv.org/abs/nlin.PS/0302051
28. K. Nozaki and N. Bekki, Exact solutions of the generalized Ginzburg-Landau equation, J. Phys. Soc. Japan **53**, 1581–1582 (1984).
29. P. Painlevé, *Leçons sur la théorie analytique des équations différentielles* (Leçons de Stockholm, 1895) (Hermann, Paris, 1897). Reprinted, *Œuvres de Paul Painlevé*, vol. I (Éditions du CNRS, Paris, 1973).
30. N. R. Pereira and L. Stenflo, Nonlinear Schrödinger equation including growth and damping, Phys. Fluids **20**, 1733–1743 (1977).
31. A. Pickering, A new truncation in Painlevé analysis, J. Phys. A **26**, 4395–4405 (1993).
32. Y. Pomeau and P. Manneville, Stability and fluctuations of a spatially periodic flow, J. Physique Lett. **40**, L609–L612 (1979).
33. A. M. Samsonov, Nonlinear strain waves in elastic waveguides, *Nonlinear waves in solids*, 349–382 eds. A. Jeffrey and J. Engelbrecht (Springer-Verlag, Wien, 1994).
34. W. van Saarloos, Front propagation into unstable states, Physics reports **386**, 29–222 (2003).
35. W. van Saarloos and P. C. Hohenberg, Fronts, pulses, sources and sinks in generalized complex Ginzburg-Landau equations, Physica D **56**, 303–367 (1992). Erratum **69**, 209 (1993).
36. H. Segur, Solitons and the inverse scattering transform, *Topics in ocean physics*, 235–277, eds. A. R. Osborne and P. Malanotte Rizzoli (North-Holland publishing co., Amsterdam, 1982).
37. J. Swift and P. C. Hohenberg, Hydrodynamic fluctuations at the convective instability, Phys. Rev. A **15**, 319–328 (1977).
38. O. Thual and U. Frisch, Natural boundary in the Kuramoto model, *Combustion and nonlinear phenomena*, 327–336, eds. P. Clavin, B. Larrouturou and P. Pelcé (Éditions de physique, Les Ulis, 1986).
39. S. Toh, Statistical model with localized structures describing the spatio-temporal chaos of Kuramoto-Sivashinsky equation, J. Phys. Soc. Japan **56**, 949–962 (1987).
40. P. Winternitz, Symmetry reduction and exact solutions of nonlinear partial differential equations, *The Painlevé property, one century later*, 591–660, ed. R. Conte, CRM series in mathematical physics (Springer, New York, 1999).
41. Yee T.-l., R. Conte, and M. Musette, Sur la "solution analytique générale" d'une équation différentielle chaotique du troisième ordre, 195–212, *From combinatorics to dynamical systems*, eds. F. Fauvet and C. Mitschi, IRMA lectures in mathematics and theoretical physics **3** (de Gruyter, Berlin, 2003). http://arXiv.org/abs/nlin.PS/0302056 Journées de calcul formel, Strasbourg, IRMA, 21–22 mars 2002.

Stability Analysis of Pulses via the Evans Function: Dissipative Systems

T. Kapitula

Department of Mathematics and Statistics, University of New Mexico, Albuquerque, NM 87131, USA
kapitula@math.unm.edu

Abstract. Linear stability analysis of pulses is considered in this review chapter. The Evans function is an analytic tool whose zeros correspond to eigenvalues. Herein, the general manner of its construction shown. Furthermore, the construction is done explicitly for the linearization of the nonlinear Schrödinger equation about the 1-soliton solution. In an explicit calculation, it is shown how the Evans function can be used to track the non-zero eigenvalues arising from a dissipative perturbation of the nonlinear Schrödinger equation which arises in the context of pulse propagation in nonlinear optical fibers.

1 Introduction

One of the more exciting areas in applied mathematics is the study of the dynamics associated with the propagation of information. Phenomena of interest include the transmission of impulses in nerve fibers, the transmission of light down an optical fiber, and phase transitions in materials. The nature of the system dictates that the relevant and important effects occur along one axial direction. The models formulated in these areas exhibit many other effects, but it is these nonlinear waves that are the raison-d'être of the models. The demands on the mathematician for techniques to analyze these models may best be served by developing methods tailored to determining the local behavior of solutions near these structures. The most basic question along these lines is the stability of the waves relative to perturbations in the initial data. Only waves that are stable can be reasonably expected to be physically realizable. By the same token, the presence of any instability and understanding its source can be crucial if the goal is to control the wave to a stable configuration.

There have been some striking advances over recent years in the development of stability techniques for nonlinear waves such as fronts, pulses and wave-trains. A motivating force behind these stability methods is the desire for "rules-of-thumb" or "principles" that can be adopted by physical scientists. An example of such a rule would be the result that a travelling wave solution of a (scalar) reaction-diffusion equation is stable exactly when it is monotone (see [1]). In other words: fronts are stable, but pulses are not. This is a simple consequence of classical Sturm-Liouville theory applied to the

T. Kapitula: *Stability Analysis of Pulses via the Evans Function: Dissipative Systems*, Lect. Notes Phys. **661**, 407–428 (2005)
www.springerlink.com

eigenvalue equations of the equations linearized at the wave, and it yields a relationship between the structure of the wave itself and its stability properties. Armed with this information, one has a "rule-of-thumb" for how to discriminate between stable and unstable waves in this system. Another example is the Vakhitov-Kolokov criterion, which is applicable to systems such as the nonlinear Schrödinger equation. If we let $P(\omega)$ represent the energy of the underlying wave, then the wave is stable if $\mathrm{d}P/\mathrm{d}\omega > 0$, and unstable otherwise. The idea that such rules should exist in more general systems and for nonlinear waves with more complicated structure has driven much of the research in this area.

The key information for stability is contained in the linearization of the PDE about the wave. In many cases, location of the spectrum suffices to determine the stability, i.e., a spectrum in the left half-plane corresponds to stable directions and one in the right half-plane corresponds to unstable directions. In dissipative systems, this basic linear information is definitive. However, there are a number of interesting problems in which more subtle information about the linearized system is sought; these tend to occur in conservative or near-conservative systems for which there is considerable spectrum either on or touching the imaginary axis.

One tool in particular has come to stand out as central in stability investigations of nonlinear waves. The Evans function is an analytic function whose zeroes give the eigenvalues of the linearized operator, with the order of the zero and the multiplicity of the eigenvalue matching. The Evans function is a generalization of the transmission coefficient from quantum mechanics to systems of PDEs. It was first formulated by Evans for a specific class of systems – see [2, 3, 4, 5]. Evans was interested in the stability of nerve impulses and formulated a category of equations that he named "nerve impulse equations". This class of equations had a special property that made the formulation of this function straightforward. In his paper, he used the notation $D(\lambda)$ to recall the word "determinant", as it played the same role as the determinant of an eigenvalue matrix in finite-dimensional problems. [6] used Evans' idea to solve the problem for the stability of the travelling pulse (nerve impulse) of the FitzHugh-Nagumo system. Jones named it the "Evans function", and the notation $E(\lambda)$ is now in common usage. The first general definition of the Evans function was given by [7]. Although based on Evans' idea, these authors put it on a new conceptual plane in order to give a clear, general definition.

We will now focus on a representative problem which motivates much of the discussion in this review chapter. Pulse propagation in a standard single-mode optical fiber is modelled by an equation of the form

$$\mathrm{i}q_t + \frac{1}{2}q_{xx} + |q|^2 q = \epsilon R(x, t, q, q^*) \, . \tag{1}$$

Here q represents the slowly-varying envelope of the rapidly-varying wave, the term $|q|^2 q$ represents the nonlinear response of the fiber, and the perturbation term R incorporates additional (nonlinear) effects such as Raman scattering, phase amplification, spectral filtering, and impurities in the fiber. A pulse solution corresponds to a "bit" of information propagating down the optical fiber, and is realized as a homoclinic orbit for the underlying travelling wave ODE. Equation (1), in addition to being a physically realistic model, is amenable to an extensive and thorough analysis; furthermore, it incorporates many of the mathematical issues and difficulties present in more complex perturbed Hamiltonian systems.

The structure and stability of the pulse is well-known for the unperturbed problem, since equation (1) is an integrable system (see Sect. 4). The pulse can be described well by using four parameters: amplitude, wave speed, position, and phase. From this, it can be seen that the origin of the complex plane is an eigenvalue with geometric multiplicity (g.m.) two and algebraic multiplicity (a.m.) four. The fact that a.m. $\neq$ g.m. is a reflection of the fact that the system is Hamiltonian. Furthermore, the only other spectrum is essential, and it resides completely on the imaginary axis.

There are several fundamental stability issues that arise for the perturbed problem, assuming that some subset of the family of pulse solutions is chosen by the perturbation. The first is the fate of the spectrum of the origin. If the perturbation is dissipative but breaks neither of the symmetries, then the small bifurcating eigenvalues will typically be of $O(\epsilon)$. How does one systematically and generally capture the location of these small eigenvalues? This problem can be thought of as lying in the realm of classical bifurcation theory, and is addressed in Sect. 5.

A more subtle effect to be understood is the influence of the perturbation on the essential spectrum. The location of the essential spectrum after the perturbation is well-understood; however, what is less understood is determining the location of point eigenvalues in, or near, the essential spectrum which popped out after the perturbation. In the physics literature, these eigenvalues are often called internal modes, since, for perturbed Hamiltonian systems, they often reside on the imaginary axis. An understanding of this phenomenon is important, because any or all of these eigenvalues can lead to an instability. This creation of internal modes has been termed an "Edge Bifurcation" [8]. This name is due to the fact that, for equation (1), the bifurcation occurs only at the edge of the essential spectrum. The questions to be answered in this problem are: (a) from which points in the essential spectrum can an edge bifurcation occur? (b) how many eigenvalues will arise from a particular bifurcation? This problem can be thought of as lying outside classical bifurcation theory, primarily because it is not clear on the location of the bifurcation point in the spectral plane. This issue is briefly addressed in Sect. 4.1, and much more extensively in [9].

This outline of this chapter is as follows. In Sect. 2, we show how to construct the Evans function for a simple example. In Sect. 3, we give several equivalent constructions of the Evans function, and discuss its properties. In

Sect. 4, we use the Evans function to compute the spectrum for equation (1) in the unperturbed case. Finally, in Sect. 5, we compute the spectrum for equation (1) for a particular dissipative perturbation.

2 Basic Example

Consider the scalar reaction-diffusion equation

$$u_t = u_{xx} - u + 2u^3 \,, \tag{2}$$

where $(x,t) \in \mathbb{R} \times \mathbb{R}^+$. It is easy to check that equation (2) has a pulse solution which is given by $u(x) = U(x)$, where $U(x) = \operatorname{sech}(x)$. Linearizing equation (2) yields the linear eigenvalue problem

$$p'' - (1 - 6U^2(x))p = \lambda p \,, \quad ' = \frac{\mathrm{d}}{\mathrm{d}x} \,. \tag{3}$$

Since equation (2) is translationally invariant, $\lambda = 0$ is an eigenvalue, with associated eigenfunction $U'(x)$. Since $U(x)$ is a non-monotone solution, then, as an easy application of Sturm-Liouville theory, one can deduce that there is one positive eigenvalue, and hence that the wave is unstable. The purpose of this section is to show how the Evans function can be used to deduce the same conclusion. The idea is that, by examining a relatively simple problem, the reader will then have a "blueprint" for the theory and ideas that lie ahead.

Upon setting $\mathbf{Y} = (p,q)^{\mathrm{T}}$, write the eigenvalue problem equation (3) as the first-order system

$$\mathbf{Y}' = (M(\lambda) + R(x))\mathbf{Y} \,, \tag{4}$$

where

$$M(\lambda) = \begin{pmatrix} 0 & 1 \\ 1+\lambda & 0 \end{pmatrix} , \quad R(x) = \begin{pmatrix} 0 & 0 \\ -6U^2(x) & 0 \end{pmatrix} .$$

It is important to note that $\lim_{|x|\to+\infty} |R(x)| = 0$, and that the decay is exponentially fast. For the rest of this discussion, it will be assumed that $\operatorname{Re}\lambda > -1$. The eigenvalues of $M(\lambda)$ are given by

$$\mu^{\pm}(\lambda) = \pm\sqrt{1+\lambda} \,,$$

and the associated eigenvectors are

$$\eta^{\pm}(\lambda) = (1, \mu^{\pm}(\lambda))^T \,.$$

One can construct solutions $\mathbf{Y}^{\pm}(\lambda, x)$ to equation (4) which satisfy

$$\lim_{x\to\pm\infty} \mathbf{Y}^{\pm}(\lambda, x)\mathrm{e}^{-\mu^{\mp}(\lambda)x} = \eta^{\mp}(\lambda) \,;$$

note that the construction implies that $\lim_{x\to\pm\infty} |\mathbf{Y}^{\pm}(\lambda, x)| = 0$. The Evans function is given by

$$E(\lambda) = \det(\mathbf{Y}^{-}, \mathbf{Y}^{+})(\lambda, x) , \tag{5}$$

and, by Abel's formula, it is independent of x.

The importance of the manner in which the Evans function is constructed is seen from the following argument. Suppose that $E(\lambda_0) = 0$ for some λ_0 with Re $\lambda_0 > -1$. It is then clear that $\mathbf{Y}^{-}(\lambda_0, x) = \alpha\mathbf{Y}^{+}(\lambda_0, x)$ for some $\alpha \in \mathbb{C}$. Hence, there is a localized solution to equation (3) when $\lambda = \lambda_0$, so that λ_0 is an eigenvalue. Similarly, if λ_0 is an eigenvalue with Re $\lambda_0 > -1$, then it is not difficult to convince oneself that $E(\lambda_0) = 0$. The following proposition has then been almost proved.

Proposition 1. *Set $\Omega = \{\lambda \in \mathbb{C} : \text{Re } \lambda > -1\}$. The Evans function is analytic on Ω. Furthermore, $E(\lambda) = 0$ if and only if λ is an eigenvalue, and the order of the zero is equal to the algebraic multiplicity of the eigenvalue.*

It is of interest to relate the Evans function to the transmission coefficient associated with the Inverse Scattering Transform [10, 11]. For Re $\lambda > -1$ and $x \gg 1$, the solution $\mathbf{Y}^{-}(\lambda, x)$ has the asymptotics

$$\mathbf{Y}^{-}(\lambda, x) = a(\lambda)\eta^{+}(\lambda)e^{\mu^{+}(\lambda)x} + b(\lambda)\eta^{-}(\lambda)e^{\mu^{-}(\lambda)x} + O(x^{-1}) . \tag{6}$$

Here $a(\lambda)$ is the transmission coefficient, and $a(\lambda_0) = 0$ if and only if λ_0 is an eigenvalue. Using the definition of the Evans function given in equation (5) and letting $x \to +\infty$ yields

$$a(\lambda) = -\frac{E(\lambda)}{2\sqrt{1+\lambda}} ; \tag{7}$$

hence, for Re $\lambda > -1$, one has the result $E(\lambda) = 0$ if and only if $a(\lambda) = 0$.

We will now use the Evans function to show that the wave is unstable. For $\lambda \in \Omega$ with $|\lambda| \gg 1$, the system is essentially autonomous, i.e., the influence of the matrix $R(x)$ on the solutions to equation (4) becomes negligible (see [7] for the details). This is equivalent to the transmission coefficient being unity for large λ. As a consequence, it is easy to see, from equation (7), that for $\lambda \in \Omega$,

$$\lim_{|\lambda|\to+\infty} \frac{E(\lambda)}{\sqrt{1+\lambda}} = -2 . \tag{8}$$

Since $\lambda = 0$ is an eigenvalue, one has the result $E(0) = 0$. As a consequence of equation (8), it is seen that the Evans function is negative for large real positive λ; therefore, the wave will necessarily be unstable if $E'(0) > 0$. We now proceed to make that calculation.

By construction, one has $\mathbf{Y}^{-}(0, x) = \mathbf{Y}^{+}(0, x) = (U'(x), U''(x))^{\mathrm{T}}$. Taking the derivative with respect to λ yields

$$E'(0) = \det(\partial_\lambda(\mathbf{Y}^- - \mathbf{Y}^+)\,,\,\mathbf{Y}^+)(0,x)\,. \tag{9}$$

Now, upon observation of equation (4), it is easy to see that, at $\lambda = 0$,

$$(\partial_\lambda \mathbf{Y}^\pm)' = (M(0) + R(x))\partial_\lambda \mathbf{Y}^\pm + M'(0)\mathbf{Y}^\pm\,. \tag{10}$$

Note that

$$M'(0)\mathbf{Y}^\pm = \begin{pmatrix} 0 \\ U'(x) \end{pmatrix}.$$

Upon solving equation (10) via variation of parameters, one finds that

$$\partial_\lambda(\mathbf{Y}^- - \mathbf{Y}^+)(0,x) = \left(-\int_{-\infty}^{+\infty} (U'(x))^2\,\mathrm{d}x\right)\mathbf{u}_2(x) + C_1\mathbf{u}_1(x) \tag{11}$$

for some constant C_1. Here $\mathbf{u}_1(x) = \mathbf{Y}^-(0,x)$, and $\mathbf{u}_2(x)$ is another solution to equation (4) at $\lambda = 0$ such that $\det(\mathbf{u}_1,\,\mathbf{u}_2)(x) = 1$. Substitution of the result of equation (11) into the expression of equation (9) then yields

$$\begin{aligned} E'(0) &= \left(-\int_{-\infty}^{+\infty} (U'(x))^2\,\mathrm{d}x\right)\det(\mathbf{u}_2,\,\mathbf{u}_1)(x) \\ &= \int_{-\infty}^{+\infty} (U'(x))^2\,\mathrm{d}x\,. \end{aligned} \tag{12}$$

Thus, one has the existence of an odd number of real positive zeros. As previously mentioned, there is exactly one.

While we will not discuss the issue in any more detail here, this idea of proving the instability of a wave has turned out to be very fruitful. The interested reader should consult, for example, [12, 13, 14, 15, 16, 17, 18] and the references therein.

2.1 Alternative Definition

Recall the definition of the Evans function given in equation (5). There is an equivalent definition which was first used by [2, 3, 4, 5], and which was later used by [6] to determine the stability of the fast-travelling pulse of the Fitzhugh-Nagumo equation (also see [19, 20]). Consider the adjoint equation to equation (4) given by

$$\mathbf{Z}' = -(M(\lambda)^* + R(x))^{\mathrm{T}}\mathbf{Z}\,. \tag{13}$$

As in the construction of the solutions $\mathbf{Y}^\pm(\lambda, x)$ of equation (4), one can construct a solution $\mathbf{Z}^+(\lambda, x)$, analytic in λ for fixed x, of equation (13) which satisfies

$$\lim_{x\to+\infty} \mathbf{Z}^+(\lambda,x)\mathrm{e}^{-\mu^-(\lambda)^* x} = (\mu^-(\lambda)^*, 1)^{\mathrm{T}}\,.$$

The Evans function can be written as

$$E(\lambda) = \langle \mathbf{Y}^-(\lambda, x), \mathbf{Z}^+(\lambda, x) \rangle \,, \tag{14}$$

where $\langle \cdot, \cdot \rangle$ represents the standard inner product on $\mathbb{C}^2$. A generalization of this formulation will be given in the next section.

3 Construction of the Evans Function

Now that we have an idea as to how the Evans function is constructed for scalar reaction-diffusion equations, let us consider more general PDEs of the class

$$u_t = Bu_{xx} + f(u, u_x) \,,$$

where it is assumed that the initial value problem is well-posed. Upon setting $z = x - ct$, the PDE becomes

$$u_t = Bu_{zz} + cu_z + f(u, u_z) \,.$$

For the PDE at hand, it will be assumed that there is a travelling pulse $U(z)$ and a constant state U^* such that $|U(z) - U^*| \to 0$ exponentially fast as $|z| \to \infty$. If one desires, this restriction can be relaxed in order to consider travelling fronts which connect a constant state U_- to a state U_+; however, upon doing so, one only increases the notational complexity without increasing the generality of all that follows (see [7] for further details).

After linearizing the PDE about the travelling wave, the eigenvalue problem can be rewritten as the first-order system

$$\mathbf{Y}' = (M(\lambda) + R(z))\mathbf{Y} \,, \quad ' = \mathrm{d}/\mathrm{d}z \,, \tag{15}$$

where $\lambda \in \mathbb{C}$ is the eigenvalue parameter and the $n \times n$ matrix $|R(z)| \to 0$ exponentially fast as $|z| \to \infty$. If λ is not in the continuous spectrum, then the matrix $M(\lambda)$ has no purely imaginary eigenvalues. Denote the region of the complex plane for which this property is true by Ω, and assume that $\{\lambda : \mathrm{Re}\,\lambda > 0\} \subset \Omega$. This assumption implies that any temporal exponential instability will be due solely to the presence of the point spectrum. Assume, for $\lambda \in \Omega$, that $M(\lambda)$ has m eigenvalues with positive real part, say $\mu_i^+(\lambda)$ for $i = 1, \ldots, m$, and $n - m$ eigenvalues with negative real part, say $\mu_i^-(\lambda)$ for $i = 1, \ldots, n - m$. The eigenvectors associated with $\mu_i^\pm(\lambda)$ will be denoted by $\eta_i^\pm(\lambda)$.

We will now construct the Evans function using various formulations. The central idea in all cases will be that we wish to create an analytic function which vanishes precisely when there is a localized solution to the eigenvalue problem. This requires that we construct solutions of equation (15) which decay as $z \to \pm\infty$.

3.1 Construction with Simple Eigenvalues

Let us first assume that the eigenvalues $\mu_i^{\pm}(\lambda)$ are simple. The eigenvectors can then be chosen to be analytic ([21]). Then solutions $\mathbf{Y}_i^{\pm}(\lambda, z)$, analytic in λ for fixed z, of equation (15) can be constructed so that

$$\lim_{z \to -\infty} \mathbf{Y}_i^{-}(\lambda, z)\mathrm{e}^{-\mu_i^{+}(\lambda)z} = \eta_i^{+}(\lambda) , \quad i = 1, \ldots, m , \tag{16}$$

and

$$\lim_{z \to +\infty} \mathbf{Y}_i^{+}(\lambda, z)\mathrm{e}^{-\mu_i^{-}(\lambda)z} = \eta_i^{-}(\lambda) , \quad i = 1, \ldots, n-m \tag{17}$$

([7]). The Evans function, $E(\lambda)$, is given by the scaled Wronskian of these solutions, i.e.,

$$E(\lambda) = m(\lambda, z)\det(\mathbf{Y}_1^{-}, \cdots, \mathbf{Y}_m^{-}, \mathbf{Y}_1^{+}, \cdots, \mathbf{Y}_{n-m}^{+})(\lambda, z) , \tag{18}$$

where

$$m(\lambda, z) = \exp\left(-\int_0^x \mathrm{trace}(M(\lambda) + R(s))\,\mathrm{d}s\right) .$$

As a consequence of Abel's formula, the Evans function is independent of z. In the above context, the following theorem was proved by [7].

Theorem 1. *The Evans function is analytic on Ω. Furthermore, $E(\lambda) = 0$ if and only if λ is an eigenvalue, and the order of the zero is equal to the algebraic multiplicity of the eigenvalue.*

3.2 Construction Via Inner Products

Under a slight relaxation of the above criteria, i.e., if one assumes only that the eigenvalues $\mu_i^{-}(\lambda)$ are simple, then there is an equivalent formulation of the Evans function which was presented in [22]. The equivalence was shown in [23]; furthermore, this new formulation has been exploited in a series of problems [15, 16, 24] in which the eigenvalue problem has a Hamiltonian formulation . Consider the adjoint system associated with equation (15):

$$\mathbf{Z}' = -(M(\lambda)^* + R(z))^{\mathrm{T}}\mathbf{Z} . \tag{19}$$

The eigenvalues of $(-M(\lambda)^*)^{\mathrm{T}}$ are given by $-(\mu_i^{\pm}(\lambda))^*$; let the associated eigenvectors be given by $\zeta_i^{\pm}(\lambda)$. As above, one can construct solutions $\mathbf{Z}_i^{+}(\lambda, z)$ of equation (19) which satisfy

$$\lim_{z \to +\infty} \mathbf{Z}_i^{+}(\lambda, z)\mathrm{e}^{\mu_i^{+}(\lambda)^* z} = \zeta_i^{+}(\lambda) , \quad i = 1, \ldots, m . \tag{20}$$

The Evans matrix is the generalization of the formulation of equation (14), and is given by

$$\mathbb{D}(\lambda) = \begin{pmatrix} \langle \mathbf{Y}_1^-, \mathbf{Z}_1^+ \rangle & \cdots & \langle \mathbf{Y}_1^-, \mathbf{Z}_m^+ \rangle \\ \vdots & \ddots & \vdots \\ \langle \mathbf{Y}_m^-, \mathbf{Z}_1^+ \rangle & \cdots & \langle \mathbf{Y}_m^-, \mathbf{Z}_m^+ \rangle \end{pmatrix} (\lambda, z) , \quad \lambda \in \Omega , \tag{21}$$

where now $\langle \cdot, \cdot \rangle$ represents the inner product on $\mathbb{C}^n$.

Theorem 2. *Set* $D(\lambda) = \det(\mathbb{D}(\lambda))$. *There exists an analytic function* $C(\lambda) \neq 0$ *such that* $E(\lambda) = C(\lambda)D(\lambda)$.

Remark 1. Without loss of generality, it can henceforth be assumed that $C(\lambda) = 1$.

3.3 Construction via Exponential Dichotomies

Unfortunately, it turns out that, in many problems of interest, the assumption that the eigenvalues of $M(\lambda)$ are simple for $\lambda \in \Omega$ does not hold. As a consequence, in order to preserve the analyticity of the Evans function, solutions to equation (15) must be constructed in a different manner [25]. Denote the evolution associated with equation (15) by $\Phi(\lambda; z, y)$. As discussed in [25, 26] (also see [27, 28] and the references therein), one can construct projection operators, $P_s(\lambda)$ and $P_u(\lambda)$, analytic in $\lambda \in \Omega$, such that, for some $\kappa_s < 0 < \kappa_u$ and $K \geq 1$,

$$|\Phi(\lambda; z, 0)P_u(\lambda)| \leq K\mathrm{e}^{\kappa_u z}, \quad z \leq 0 ; \quad |\Phi(\lambda; z, 0)P_s(\lambda)| \leq K\mathrm{e}^{\kappa_s z}, \quad z \geq 0 .$$

Furthermore, these operators have the property that

$$\dim\mathrm{R}(P_u(\lambda)) = m , \quad \dim\mathrm{R}(P_s(\lambda)) = n - m ;$$

thus, they are maximal in the sense that they capture all of the initial data leading to exponentially-decaying solutions as $z \to \pm\infty$. Given a $\mathbf{Y}_0 \in \mathbb{C}^n$, set $\mathbf{Y}^{\pm}(\lambda, z)$ to be

$$\mathbf{Y}^-(\lambda, z) = \Phi(\lambda; z, 0)P_u(\lambda)\mathbf{Y}_0 , \quad \mathbf{Y}^+(\lambda, z) = \Phi(\lambda; z, 0)P_s(\lambda)\mathbf{Y}_0 .$$

Note that, if the eigenvalues of $M(\lambda)$ are simple, then

$$\mathbf{Y}^-(\lambda, z) \in \mathrm{Span}\{\mathbf{Y}_1^-(\lambda, z), \ldots, \mathbf{Y}_m^-(\lambda, z)\} ,$$

and

$$\mathbf{Y}^+(\lambda, z) \in \mathrm{Span}\{\mathbf{Y}_1^+(\lambda, z), \ldots, \mathbf{Y}_{n-m}^+(\lambda, z)\} .$$

It is clear that an initial condition will lead to a bounded solution if and only if $\mathbf{Y}^-(\lambda, 0) \cap \mathbf{Y}^+(\lambda, 0) \neq \{\mathbf{0}\}$.

Since the projections are analytic, then as a consequence of (see Chapter II.4.2 of [21]), one can choose analytic bases $\{b_1(\lambda), \ldots, b_m(\lambda)\}$ and $\{b_{m+1}(\lambda), \ldots, b_n(\lambda)\}$ of $\mathrm{R}(P_u(\lambda))$ and $\mathrm{R}(P_s(\lambda))$, respectively. If one defines the analytic matrix $B(\lambda) \in \mathbb{C}^{n\times n}$ via

$$B(\lambda) = (b_1(\lambda) \cdots b_m(\lambda)\, b_{m+1}(\lambda) \cdots b_n(\lambda)) ,$$

then, by construction, it is clear that for $\lambda \in \Omega$, a uniformly bounded solution to equation (15) will exist if and only if $\dim \mathrm{N}(B(\lambda)) \geq 1$. The Evans function can then be defined as:

$$E(\lambda) = \det(B(\lambda)) . \tag{22}$$

4 The Linearization of the Nonlinear Schrödinger Equation

Let us now apply the results of the previous section to a well-understood example. The nonlinear Schrödinger hierarchy is the class of nonlinear integrable Hamiltonian systems given by

$$\mathbf{u}_t = \mathcal{K}(\mathbf{u}) , \tag{23}$$

where $\mathbf{u} = (r, q) \in L^2(\mathbb{R}, \mathbb{C}^2)$,

$$\mathcal{K}(\mathbf{u}) = -2\sigma_3 \Omega(\mathcal{L}^A(\mathbf{u}))\mathbf{u} ,$$

σ_3 is the Pauli spin matrix

$$\sigma_3 = \begin{pmatrix} 1 & 0 \\ 0 & -1 \end{pmatrix} ,$$

$\mathcal{L}^A$ is the integro-differential operator

$$\mathcal{L}^A(\mathbf{u})\mathbf{v} = -\frac{\mathrm{i}}{2}\partial_x \mathbf{v} + \mathrm{i}\mathbf{u} \int_{-\infty}^{x} [q v_1 - r v_2] \,\mathrm{d}y ,$$

and $\Omega(\cdot) = \mathrm{i}P(\cdot)$, where $P(\cdot)$ is a polynomial with real-valued coefficients. The notation $\mathbf{v} = (v_1, v_2)$ is being used. If $P(k) = 1/2 + k^2$, then the evolution equation (23) is

$$\begin{aligned} q_t &= \mathrm{i}\left(\frac{1}{2} q_{xx} - q - q^2 r\right) \\ r_t &= -\mathrm{i}\left(\frac{1}{2} r_{xx} - r - q r^2\right) , \end{aligned} \tag{24}$$

which, upon setting $r = -q^*$, is the focusing nonlinear Schrödinger equation (NLS). The interested reader is referred to [10, 11] for further details.

Assume that $\mathbf{u}_0$ is a stationary 1-soliton solution of equation (23). The interest is in the spectrum of the linearization $\mathcal{K}'(\mathbf{u}_0)$ about $\mathbf{u}_0$ and, particularly, in eigenvalues λ for which

$$\mathcal{K}'(\mathbf{u}_0)\mathbf{u} = \lambda \mathbf{u}$$

has a non-zero solution $\mathbf{u}$ in $L^2(\mathbb{R}, \mathbb{C}^2)$. Note that the eigenvalue problem for the focusing NLS equation (24) reads

$$\begin{aligned} \lambda q = & \ \mathrm{i}\left(\frac{1}{2}q_{xx} - q - 2q_0 r_0 q - q_0^2 r\right) \\ \lambda r = & -\mathrm{i}\left(\frac{1}{2}r_{xx} - r - 2q_0 r_0 r - r_0^2 q\right) \end{aligned} \tag{25}$$

with $\mathbf{u}_0 = (q_0, r_0)$. This problem can be solved if we can construct, and compute, the Evans function associated with the operator $\mathcal{K}'(\mathbf{u}_0)$. The essential spectrum of $\mathcal{K}'(\mathbf{u}_0)$ is given by

$$\sigma_{\mathrm{e}}(\mathcal{K}'(\mathbf{u}_0)) := \{\lambda \in \mathbb{C};\ \lambda = \pm 2\Omega(k),\ k \in \mathbb{R}\} \subset \mathrm{i}\mathbb{R}\ .$$

Also, it was shown in [29] that $\lambda = 0$ is the only eigenvalue of $\mathcal{K}'(\mathbf{u}_0)$. These eigenvalues at $\lambda = 0$ are due to the invariances associated with equation (23). We will recover this result later in this section for the polynomial dispersion relation P that gives the nonlinear Schrödinger equation.

The key to calculating the Evans function is to exploit Inverse Scattering Theory, and this is possible since equation (23) is integrable. The underlying linear scattering problem associated with the nonlinear operator $\mathcal{K}(\mathbf{u})$ is the Zakharov-Shabat problem [11]

$$\mathbf{v}_x = \begin{pmatrix} -\mathrm{i}k & q_0(x) \\ -q_0^*(x) & \mathrm{i}k \end{pmatrix} \mathbf{v}$$

where $k \in \mathbb{C}$ is a complex parameter. The Jost functions are solutions of the Zakharov-Shabat eigenvalue problem that satisfy certain boundary conditions at $x = \pm\infty$. Appropriate quadratic combinations of the Jost functions define the adjoint squared eigenfunctions, which we denote[1] by $\Psi^{\mathrm{A}}(k, x)$, defined for $\mathrm{Im}k \geq 0$, and $\bar{\Psi}^{\mathrm{A}}(k, x)$, defined for $\mathrm{Im}k \leq 0$. The adjoint squared eigenfunctions are crucial ingredients when applying Soliton Perturbation Theory [30, 31]. As we shall see below, they can also be used to explicitly calculate the Evans function associated with $\mathcal{K}'(\mathbf{u}_0)$. For $k \in \mathbb{R}$, the adjoint squared eigenfunctions satisfy the identities

$$[\mathcal{L}^{\mathrm{A}}(\mathbf{u}) - k]\Psi^{\mathrm{A}}(k, x) = [\mathcal{L}^{\mathrm{A}}(\mathbf{u}) - k]\bar{\Psi}^{\mathrm{A}}(k, x) = 0\ .$$

Furthermore, they have the property that, for fixed x, $\Psi^{\mathrm{A}}(k, x)$ is analytic in k for $\mathrm{Im}k > 0$, while $\bar{\Psi}^{\mathrm{A}}(k, x)$ is analytic in k for $\mathrm{Im}k < 0$. In addition, they have the asymptotics

$$\lim_{x\to-\infty} \Psi^{\mathrm{A}}(k, x)\mathrm{e}^{2\mathrm{i}kx} = \begin{pmatrix} 0 \\ -1 \end{pmatrix}, \quad \lim_{x\to\infty} \Psi^{\mathrm{A}}(k, x)\mathrm{e}^{2\mathrm{i}kx} = a(k)^2 \begin{pmatrix} 0 \\ -1 \end{pmatrix} \tag{26}$$

[1] Notation: We denote by q^* the complex conjugate of a complex number q. Thus, $\bar{q}$ does *not* refer to the complex conjugate.

and

$$\lim_{x\to-\infty} \bar{\Psi}^{\mathrm{A}}(k,x)\mathrm{e}^{-2\mathrm{i}kx} = \begin{pmatrix} 1 \\ 0 \end{pmatrix}, \quad \lim_{x\to\infty} \bar{\Psi}^{\mathrm{A}}(k,x)\mathrm{e}^{-2\mathrm{i}kx} = \bar{a}(k)^2 \begin{pmatrix} 1 \\ 0 \end{pmatrix} \tag{27}$$

The functions $a(k)$ and $\bar{a}(k)$ are the transmission coefficients for the Zakharov-Shabat eigenvalue problem, and, for the 1-soliton of the NLS equation (24), are given by

$$a(k) = \frac{\sqrt{2}\,k - \mathrm{i}}{\sqrt{2}\,k + \mathrm{i}}, \quad \bar{a}(k) = \frac{\sqrt{2}\,k + \mathrm{i}}{\sqrt{2}\,k - \mathrm{i}} \tag{28}$$

[31, 32]. Furthermore, it is known (see [29] for references), that the adjoint squared eigenfunctions and the transmission coefficients can be extended analytically across the line $\mathrm{Im}k = 0$.

Lemma 1 ([29]). *The adjoint squared eigenfunctions satisfy*

$$\begin{aligned} &[\mathcal{K}'(\mathbf{u}_0) - 2\Omega(k)]\sigma_3\Psi^{\mathrm{A}}(k,x) = 0\,, && \mathrm{Im}k \geq 0 \\ &[\mathcal{K}'(\mathbf{u}_0) + 2\Omega(\bar{k})]\sigma_3\bar{\Psi}^{\mathrm{A}}(\bar{k},x) = 0\,, && \mathrm{Im}\bar{k} \leq 0\,. \end{aligned}$$

Remark 2. The interested reader should also consult [33, 34].

Thus, upon applying the matrix σ_3 to the squared eigenfunctions, one recovers eigenfunctions for the linearized problem. We are now in a position to calculate the Evans function. While it is possible to make this calculation for the full hierarchy (see [29] for the details), for the sake of clarity, only the specific dispersion relation

$$\Omega(\ell) = \mathrm{i}\left(\frac{1}{2} + \ell^2\right) \tag{29}$$

of the nonlinear Schrödinger equation will be considered. For this case, explicit expressions are known [31] for the adjoint squared eigenfunctions given in Lemma 1; however, as will be seen below, these expressions are not necessary in order to make a calculation.

Consider equation (29), and note that the associated continuous spectrum consists of the elements $\lambda = \pm\mathrm{i}(1 + 2\ell^2)$, where ℓ varies in $\mathbb{R}$, i.e., it is the imaginary axis minus the interval $(-\mathrm{i}, \mathrm{i})$. Since we wish to exploit Lemma 1, we choose for each $\lambda \notin \sigma_{\mathrm{e}}(\mathcal{K}'(\mathbf{u}_0))$ numbers k and $\bar{k}$ such that

$$\lambda = 2\Omega(k) = \mathrm{i}(1 + 2k^2)\,, \quad \lambda = -2\Omega(\bar{k}) = -\mathrm{i}(1 + 2\bar{k}^2)\,,$$

i.e.,

$$k(\lambda) = \frac{1}{\sqrt{2}}\mathrm{e}^{\mathrm{i}3\pi/4}\sqrt{\lambda - \mathrm{i}}\,, \quad \arg(\lambda - \mathrm{i}) \in (-3\pi/2, \pi/2]\,, \tag{30}$$

and

$$\bar{k}(\lambda) = \frac{1}{\sqrt{2}}\mathrm{e}^{-\mathrm{i}3\pi/4}\sqrt{\lambda + \mathrm{i}}\,, \quad \arg(\lambda - \mathrm{i}) \in (-\pi/2, 3\pi/2]\,. \tag{31}$$

The branch cuts have been chosen so that $\mathrm{Im}k > 0$ and $\mathrm{Im}\bar{k} < 0$, and so that k and $\bar{k}$ are analytic for $\lambda \notin \sigma_{\mathrm{e}}(\mathcal{K}'(\mathbf{u}_0))$. By Lemma 1, it is seen that $\sigma_3\Psi^{\mathrm{A}}(k,x)$ and $\sigma_3\bar{\Psi}^{\mathrm{A}}(\bar{k},x)$ are two solutions of the eigenvalue problem of equation (25). Both of these solutions decay exponentially fast to zero as $x \to -\infty$. Furthermore, as a consequence of equation (26) and equation (27), we know their asymptotics as $x \to \infty$. From now on, we regard k and $\bar{k}$ as functions of λ defined via equation (30) and equation (31).

Next, rewrite the eigenvalue problem of equation (25) as an ODE

$$\frac{\mathrm{d}\mathbf{Y}}{\mathrm{d}x} = [M(\lambda) + R(x)]\mathbf{Y} \ , \tag{32}$$

where $\mathbf{Y} = (q, r, q_x, r_x)^{\mathrm{T}} \in \mathbb{C}^4$, and

$$M(\lambda) = \begin{pmatrix} 0 & 0 & 1 & 0 \\ 0 & 0 & 0 & 1 \\ -2\mathrm{i}(\lambda+\mathrm{i}) & 0 & 0 & 0 \\ 0 & 2\mathrm{i}(\lambda-\mathrm{i}) & 0 & 0 \end{pmatrix}, \quad R(x) = 2\begin{pmatrix} 0 & 0 & 0 & 0 \\ 0 & 0 & 0 & 0 \\ 2q_0r_0 & q_0^2 & 0 & 0 \\ r_0^2 & 2q_0r_0 & 0 & 0 \end{pmatrix}.$$

Note that $|R(x)| \to 0$ as $|x| \to \infty$. The eigenvalues and eigenvectors of $M(\lambda)$ are analytic for $\lambda \notin \sigma_{\mathrm{e}}(\mathcal{K}'(\mathbf{u}_0))$. We can define two solutions[2]

$$\mathbf{Y}_1^{\mathrm{u}}(\lambda,x) = \begin{pmatrix} 1 \\ \partial_x \end{pmatrix} \sigma_3\Psi^{\mathrm{A}}(k(\lambda),x) \ , \qquad \mathbf{Y}_2^{\mathrm{u}}(\lambda,x) = \begin{pmatrix} 1 \\ \partial_x \end{pmatrix} \sigma_3\bar{\Psi}^{\mathrm{A}}(\bar{k}(\lambda),x)$$

that decay exponentially to zero as $x \to -\infty$. By analogy, there are two solutions to the adjoint equation associated with equation (32) which decay exponentially as $x \to +\infty$ for $\lambda \notin \sigma_{\mathrm{e}}(\mathcal{K}'(\mathbf{u}_0))$:

$$\begin{aligned} \lim_{x\to+\infty} \mathbf{Z}_1^{\mathrm{s}}(\lambda,x)\mathrm{e}^{2\mathrm{i}k(\lambda)^*x} &= (0, 2\mathrm{i}k(\lambda)^*, 0, 1)^{\mathrm{T}} \\ \lim_{x\to+\infty} \mathbf{Z}_2^{\mathrm{s}}(\lambda,x)\mathrm{e}^{-2\mathrm{i}\bar{k}(\lambda)^*x} &= (-2\mathrm{i}\bar{k}(\lambda)^*, 0, 1, 0)^{\mathrm{T}} \ . \end{aligned} \tag{33}$$

Following the formulation of Sect. 3.2, the Evans matrix is given by

$$\mathbb{D}(\lambda) = \begin{pmatrix} \langle \mathbf{Y}_1^{\mathrm{u}}, \mathbf{Z}_1^{\mathrm{s}} \rangle & \langle \mathbf{Y}_1^{\mathrm{u}}, \mathbf{Z}_2^{\mathrm{s}} \rangle \\ \langle \mathbf{Y}_2^{\mathrm{u}}, \mathbf{Z}_1^{\mathrm{s}} \rangle & \langle \mathbf{Y}_2^{\mathrm{u}}, \mathbf{Z}_2^{\mathrm{s}} \rangle \end{pmatrix} (\lambda,x) \ , \quad \lambda \notin \sigma_{\mathrm{e}}(\mathcal{K}'(\mathbf{u}_0)) \ , \tag{34}$$

and the Evans function is given by $E(\lambda) = \det(\mathbb{D}(\lambda))$. Now we evaluate the Evans function. As a consequence of equation (26) and equation (27), we have the asymptotics

$$\begin{aligned} \lim_{x\to+\infty} \mathbf{Y}_1^{\mathrm{u}}(\lambda,x)\mathrm{e}^{2\mathrm{i}k(\lambda)x} &= a(k(\lambda))^2\,(0,1,0,-2\mathrm{i}k(\lambda))^{\mathrm{T}} \\ \lim_{x\to+\infty} \mathbf{Y}_2^{\mathrm{u}}(\lambda,x)\mathrm{e}^{-2\mathrm{i}\bar{k}(\lambda)x} &= \bar{a}(\bar{k}(\lambda))^2\,(1,0,2\mathrm{i}\bar{k}(\lambda),0)^{\mathrm{T}} \ , \end{aligned} \tag{35}$$

[2] Note that the subscripts of $\mathbf{Y}_j^{\mathrm{u}}$ do *not* denote the components of the vectors.

where

$$a(k(\lambda)) = \frac{e^{i\pi/4}\sqrt{\lambda - i} - 1}{e^{i\pi/4}\sqrt{\lambda - i} + 1}, \quad \bar{a}(\bar{k}(\lambda)) = \frac{e^{-i\pi/4}\sqrt{\lambda + i} - 1}{e^{-i\pi/4}\sqrt{\lambda + i} + 1}.$$

Taking the limit $x \to \infty$ in equation (34) and using the asymptotics given in equation (33) and equation (35) yields

$$E(\lambda) = 8a(k(\lambda))^2 \bar{a}(\bar{k}(\lambda))^2 \sqrt{\lambda - i}\sqrt{\lambda + i}\,. \tag{36}$$

As a final remark, note that $a(k(0)) = \bar{a}(\bar{k}(0)) = 0$, and that these zeros are simple.

4.1 Edge Bifurcations

For the rest of the discussion in this section, assume that $\mathrm{Im}\lambda \geq 0$. An examination of the Evans function reveals that $\lambda = 0$ is a zero of order four, and that the only other zeros are at the branch points $\lambda = \pm i$. The eigenvalues at the origin are due to the symmetries associated with the NLS. Now consider $\lambda = i$. It is clear that $\sigma_3 \Psi^A(k(i), x)$ is uniformly bounded but non-decaying for all x, while $\sigma_3 \bar{\Psi}^A(\bar{k}(i), x)$ decays exponentially as $x \to -\infty$ and grows exponentially fast as $x \to +\infty$. Hence, $\lambda = i$ is not a true eigenvalue, as there exist no corresponding eigenfunctions which are localized in space.

What is the effect of this spurious zero? Set

$$\gamma^2 := \lambda - i\,. \tag{37}$$

This transformation defines a Riemann surface, and, with respect to the Evans function, the principal sheet is given by $\arg(\gamma) \in (-3\pi/4, \pi/4]$. Zeros of the Evans function on this sheet correspond to true eigenvalues, whereas zeros on the other sheet correspond to resonance poles. The Evans function on the Riemann surface is given by

$$E(\gamma) = 8a(k(\gamma))^2 \bar{a}(\bar{k}(\gamma))^2 \gamma \sqrt{\gamma^2 + 2i}\,.$$

The Evans function is analytic on the Riemann surface, and, in addition to the four zeros at $\gamma = e^{-i\pi/4}$ ($\lambda = 0$), it has a simple zero at $\gamma = 0$.

Now suppose that the NLS undergoes a smooth perturbation of $O(\epsilon)$. Assuming that the Evans function remains analytic on the Riemann surface, the zero at $\gamma = 0$ will generically move onto one of the two sheets, and will also be of $O(\epsilon)$. If the perturbed zero is on the principal sheet, then an eigenvalue has been created via an edge bifurcation, and, from equation (37), it is seen that it will be $O(\epsilon^2)$ from the branch point $\lambda = i$. Thus, upon perturbation, an eigenvalue can be created where once there was none. The spurious zero then leads to the potential creation of an eigenvalue upon a perturbation of the vector field. This is illustrated in Fig. 1.

While this is an interesting topic, it is beyond the scope of this chapter. The interested reader is referred to the review article [9] and the references therein. For a full treatment of the NLS hierarchy, one should consult [29].

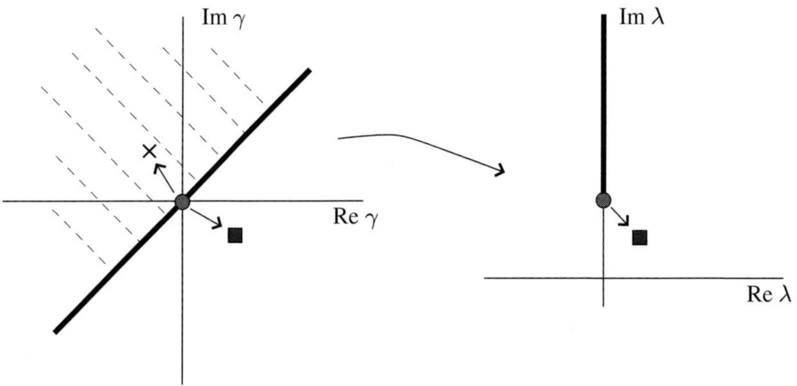

Fig. 1. The Riemann surface is given in the *left* panel, and the spectral plane is in the *right* panel. The physical sheet of the Riemann surface satisfies $\arg(\gamma) \in (-3\pi/4, \pi/4]$, and the boundary between the two sheets is given by $\mathrm{Re}\,\gamma = \mathrm{Im}\gamma$. The circle represents the zero of the Evans function for the unperturbed problem, and the squares and crosses represent the possible movement of the zero under perturbation. Note that an eigenvalue is created only if the zero on the Riemann surface moves onto the physical sheet

5 Dissipative Perturbations

In the previous sections, the spectrum was located via an Evans function calculation. The calculation was performed on a class of well-understood problems. In this section, we will start with the assumption that the unperturbed problem is well-understood, and then use the Evans function to understand the perturbed problem. In particular, we are interested in dissipative perturbations of Hamiltonian systems.

5.1 Theoretical Results

Let H be a Hilbert space with inner product $\langle\cdot,\cdot\rangle$, let J be an invertible skew-symmetric operator with bounded inverse, and consider the Hamiltonian system on H given by

$$\frac{\mathrm{d}v}{\mathrm{d}t} = JE'(v)$$

which respects a finite-dimensional abelian connected Lie group $\mathcal{G}$ with Lie algebra $\mathfrak{g}$ on H. Assume that $\dim(\mathfrak{g}) = n$. It will be assumed that the system is invariant under the action of a unitary representation T. We seek relative equilibria of the form

$$v(t) = T(\exp(\omega t))v_0\ ,$$

for appropriate $\omega \in \mathfrak{g}$. Therefore, change variables and consider

$$v(t) = T(\exp(\omega t))u(t)\ ,$$

so that u satisfies

$$\frac{\mathrm{d}u}{\mathrm{d}t} = JE_0'(u;\omega) , \tag{38}$$

where

$$E_0'(u;\omega) := E'(u) - J^{-1}T_\omega u .$$

Here T_ω is the skew-symmetric operator which is the generator of $\exp(\omega t)$. The new Hamiltonian depends therefore on ω. We assume that the steady-state equation

$$E_0'(u;\omega) = 0$$

has a smooth family $\Phi(\omega)$ of solutions, where ω varies in $\mathfrak{g}$.

Let the linear operator about the wave Φ be denoted by JE_0''. Since $\mathcal{G}$ is abelian, it is known that the operator JE_0'' will have a non-trivial kernel:

$$JE_0''(\Phi)T_{\omega_i}\Phi = 0 , \quad JE_0''(\Phi)\partial_{\omega_i}\Phi = T_{\omega_i}\Phi , \tag{39}$$

for $i = 1,\ldots,n$, with the set $\{T_{\omega_1}\Phi,\ldots,T_{\omega_n}\Phi\}$ being orthogonal. Here $\partial_\sigma := \partial/\partial\sigma$ for $\sigma \in \mathfrak{g}$. Furthermore, if the symmetric matrix $D_0 \in \mathbb{R}^{n\times n}$ given by

$$(D_0)_{ij} := \langle \partial_{\omega_j}\Phi, E_0''\partial_{\omega_i}\Phi\rangle \tag{40}$$

is invertible, then this set is a basis for the kernel. Thus, under the assumption that D_0 is non-singular, when considering the eigenvalue problem $JE_0''u = \lambda u$, one has the result that $\lambda = 0$ is an eigenvalue with geometric multiplicity n and algebraic multiplicity $2n$.

Now, consider the perturbed problem given by

$$\frac{\partial u}{\partial t} = JE_0'(u) + \epsilon E_1(u) , \tag{41}$$

where E_1 represents the dissipative perturbation. It will be assumed that, while the perturbation breaks the Hamiltonian structure, it does not remove any of the symmetries. When considering the persistence of the wave, a standard Lyapunov-Schmidt reduction reveals the following (see [35] for the case of Hamiltonian perturbations):

Lemma 2. *The wave $u = \Phi(\omega)+\epsilon\Phi_\epsilon+\mathrm{O}(\epsilon^2)$ will persist only if the condition*

$$\langle J^{-1}E_1(\Phi(\omega)), T_{\omega_j}\Phi\rangle = 0 , \quad j = 1,\ldots,n \tag{42}$$

holds for some $\omega \in \mathfrak{g}$. Furthermore,

$$\Phi_\epsilon = -(E_0''(\Phi(\omega))J^{-1}E_1(\Phi(\omega)) .$$

The above lemma yields a necessary, but not sufficient, condition for persistence. Since the perturbation does not destroy any of the symmetries associated with the original system, the eigenfunction $T_{\omega_j}\Phi$ will persist; in particular, one has the formal expansion $T_{\omega_j}\Phi + \epsilon T_{\omega_j}\Phi_\epsilon + \mathrm{O}(\epsilon^2)$. Now, write the linearization about the perturbed wave as $JE_0'' + \epsilon L_\epsilon + \mathrm{O}(\epsilon^2)$, and define the matrix $M \in \mathbb{R}^{n\times n}$ by

$$M_{ij} := \langle L_\epsilon \partial_{\omega_j} \Phi - T_{\omega_j} \Phi_\epsilon, J^{-1} T_{\omega_i} \Phi \rangle \ .$$

Lemma 3. *Suppose that M is non-singular. Then equation (42) is also sufficient.*

Since the Hamiltonian structure has been broken, n eigenvalues of $\mathrm{O}(\epsilon)$ will leave the origin. Another Lyapunov-Schmidt reduction, similar to that in [35, Theorem 4.4] (also see the more general case in [26, Sect. 4.3]), allows one to track the location of these eigenvalues.

Lemma 4. *The non-zero $\mathrm{O}(\epsilon)$ eigenvalues and associated eigenfunctions for the perturbed eigenvalue problem*

$$(JE_0'' + \epsilon L_\epsilon)u = \lambda u$$

are given by

$$\lambda = \epsilon\lambda_1 + \mathrm{O}(\epsilon^2) \ , \quad u = \sum_{j=1}^{n} v_j T_{\omega_j} \Phi + \mathrm{O}(\epsilon) \ ,$$

where λ_1 is the eigenvalue and $\mathbf{v}$ is the associated eigenvector for the generalized eigenvalue problem

$$(D_0\lambda_1 - M)\mathbf{v} = \mathbf{0} \ .$$

Remark 3. In [36], it was shown in specific examples that the perturbation expansion for the Evans function satisfies

$$E(\lambda, \epsilon) = \det(D_0\lambda - M\epsilon) + \mathrm{O}(\epsilon^2) \ .$$

In general, however, it is easier to find the eigenvalues of a matrix than it is to locate the zeros of its characteristic polynomial; hence the formulation in Lemma 4.

Now, it may also be possible for eigenvalues to pop out of the essential spectrum, creating internal modes via an edge bifurcation [8, 29, 37, 38] (also see Sect. 4.1). Since these eigenvalues will be of $\mathrm{O}(1)$, they will not be captured by the perturbation expansion given in Lemma 4. However, as will be seen in the example in the next section, this is not necessarily problematic.

5.2 Example: Nonlinear Schrödinger Equation

The theoretical results will now be applied to a particular dissipative perturbation of the nonlinear Schrödinger equation. Consider

$$\mathrm{i}\partial_t q + \frac{1}{2}\partial_x^2 q - \omega q + |q|^2 q = \mathrm{i}\epsilon R(q, q^*) , \tag{43}$$

where

$$R(q, q^*) := \frac{1}{2} d_1 \partial_x^2 q + d_2 q + d_3 |q|^2 q + d_4 |q|^4 q .$$

Here $d_j \in \mathbb{R}$ and $\omega \in \mathbb{R}^+$. Here $d_1 > 0$ describes spectral filtering, and a consequence of the sign is that the perturbed equation is dissipative. The parameter $d_2 (< 0)$ accounts for linear loss in the fiber, and a consequence of this sign is that any instability of a wave arises only from the point spectrum. The parameter d_3 accounts for nonlinear gain or loss, and d_4 represents a higher-order correction to the nonlinear gain or loss.

The unperturbed problem is well-understood (see Sect. 4). Solutions of equation (43) are invariant under the action

$$T(\xi, \theta) q = q(x + \xi) \mathrm{e}^{i\theta} ,$$

so that the unperturbed problem has two symmetries. The linearized problem associated with the unperturbed problem is as follows. After writing equation (43) in real and imaginary parts, it is seen that the linearization when $\epsilon = 0$ is JL_0, where

$$J = \begin{pmatrix} 0 & 1 \\ -1 & 0 \end{pmatrix} , \quad L_0 = \mathrm{diag}(L_r, L_i) ,$$

with

$$L_r := -\frac{1}{2}\partial_x^2 + \omega - 3Q_0(x)^2 , \quad L_i := -\frac{1}{2}\partial_x^2 + \omega - Q_0(x)^2 .$$

Here, the soliton is given by $Q_0(x) := \sqrt{2\omega}\,\mathrm{sech}(\sqrt{2\omega}\,x)$. Furthermore, one has the fact that the eigenfunctions given in equation (39) are

$$T_\xi Q_0 = \partial_x Q_0\, \mathbf{e}_1 , \quad \partial_\xi Q_0 = -x Q_0\, \mathbf{e}_2 ,$$

and

$$T_\phi Q_0 = Q_0\, \mathbf{e}_2 , \quad \partial_\phi Q_0 = \partial_\omega Q_0\, \mathbf{e}_1 .$$

Here $\mathbf{e}_j \in \mathbb{R}^2$ represents the jth unit vector.

Lemma 2 states a necessary condition for the persistence of the wave Q_0. A routine calculation shows that the condition reduces to

$$\langle R(Q_0, Q_0^*), Q_0 \rangle = 0 ,$$

i.e.,

$$e(\omega) := d_4 \omega^2 + \frac{5}{32}(4d_3 - d_1)\omega + \frac{15}{32} d_2 = 0 .$$

Define

$$d_4^* := \frac{5}{384} \frac{(4d_3 - d_1)^2}{d_2} .$$

Assuming that

$$4d_3 - d_1 > 0\,, \quad d_4^* < d_4 < 0\,,$$

$e(\omega) = 0$ has the solutions

$$\omega = \omega^{\pm} := -\frac{5}{64d_4}\left((4d_3 - d_1) \pm \sqrt{(4d_3 - d_1)^2 - \frac{384}{5}d_2 d_4}\right).$$

For $\omega = \omega^{\pm}$, the perturbed wave is given by

$$q = \begin{pmatrix} Q_0(x) \\ 0 \end{pmatrix} + \epsilon \begin{pmatrix} 0 \\ Q_0(x)\int_0^x \theta(s)\,\mathrm{d}s \end{pmatrix} + \mathrm{O}(\epsilon^2)\,, \tag{44}$$

where

$$\theta(s) = -\frac{1}{\sqrt{2\omega}}\tanh(\sqrt{2\omega}\,s)\left(d_1\omega + d_2 - \frac{8}{5}d_4\omega^2\mathrm{sech}^2(\sqrt{2\omega}\,s)\right).$$

Note that

$$\lim_{d_4 \to d_4^*} \omega^{\pm} = -6\frac{d_2}{4d_3 - d_1}\,.$$

Thus, when considering the existence problem, there is a saddle-node bifurcation at $d_4 = d_4^*$. One expects that, at most, only one of these perturbed waves will be stable. Also, note that the condition $d_4 \neq 0$ is crucial for the bifurcation to occur. The calculations given below will show that if $d_4 = 0$, in which case

$$\omega = \lim_{d_4 \to 0} \omega^- = -3\frac{d_2}{4d_3 - d_1}\,,$$

then this wave will be unstable.

Now let us consider the stability of the perturbed wave. The matrix D_0, given in Lemma 4, is given by

$$\begin{aligned} D_0 &= \mathrm{diag}(\langle \Phi_x, -x\Phi\rangle, -\partial_\omega\langle \Phi, \Phi\rangle) \\ &= 2\mathrm{diag}((2\omega)^{1/2}, -(2\omega)^{-1/2})\,, \end{aligned}$$

The first-order correction of the linearized operator is given by

$$\begin{aligned} L_\epsilon = (\frac{1}{2}d_1\partial_x^2 + d_2)\mathrm{id} + d_3 Q_0^2\begin{pmatrix} 3 & 0 \\ 0 & 1 \end{pmatrix} + d_4 Q_0^4\begin{pmatrix} 5 & 0 \\ 0 & 1 \end{pmatrix} \\ +2Q_0^2\int_0^x \theta(s)\,\mathrm{d}s\begin{pmatrix} -1 & 0 \\ 0 & 1 \end{pmatrix}. \end{aligned}$$

Upon using the perturbation expansion in equation (44), the matrix M, used in the stability calculation presented in Lemma 4, is then given by

$$M(\omega^{\pm}) = \mathrm{diag}(C_1 d_1, \mp C_2\sqrt{(4d_3 - d_1)^2 - 384d_2d_4/5})\,,$$

where C_i is a positive constant for $i = 1, 2$. Thus, upon using the result of Lemma 4, the location of the $\mathrm{O}(\epsilon)$ eigenvalues is given by

$$\begin{aligned} \lambda_{\mathrm{rot}} &= -C_{\mathrm{rot}} d_1 \epsilon + \mathrm{O}(\epsilon^2) \\ \lambda_{\mathrm{amp}} &= C_{\mathrm{amp}} \begin{cases} -\sqrt{(4d_3 - d_1)^2 - 384 d_2 d_4/5}\, \epsilon + \mathrm{O}(\epsilon^2)\,, & \omega = \omega^+ \\ \sqrt{(4d_3 - d_1)^2 - 384 d_2 d_4/5}\, \epsilon + \mathrm{O}(\epsilon^2)\,, & \omega = \omega^-\,. \end{cases} \end{aligned} \tag{45}$$

In the above, the constants are again positive. The eigenvalue λ_{rot} is always negative, whereas the sign of the eigenvalue λ_{amp} depends upon the amplitude of the wave. It should be noted that the adiabatic approach yields the same conclusion as above [39, 40]. The bifurcation diagram is given in Fig. 2.

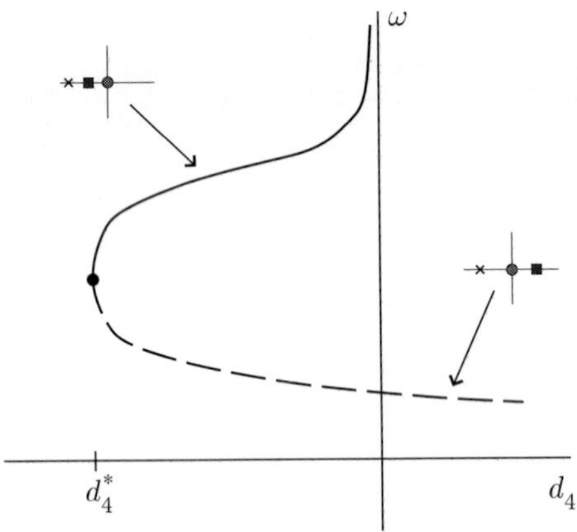

Fig. 2. The set $e(\omega) = 0$ for $d_1 > 0$, $d_2 < 0$, and $d_3 > d_1/4$ fixed. The insets give the location of the $\mathrm{O}(\epsilon)$ eigenvalues for $\omega = \omega^+$ (*solid line*) and $\omega = \omega^-$ (*dashed line*). The cross represents λ_{rot} and the square represents λ_{amp}

Unfortunately, the result of equation (45) is not enough to conclude that the perturbed wave with $\omega = \omega^+$ is stable, as all of the possible instability mechanisms have not yet been captured. As discussed in Sect. 4.1, it is possible for a single eigenvalue to leave the edge of the continuous spectrum upon perturbation. The location of this eigenvalue must now be determined.

The radiation modes themselves are readily computed. The upper branch is given by the dispersion relation

$$\lambda = \mathrm{i}\left(\omega + \frac{1}{2}k^2\right) + \epsilon\left(d_2 - \frac{1}{2}d_1\right)k^2\,, \quad k \in \mathbb{R}\,.$$

Since $d_2 < 0$ and $d_1 > 0$ by assumption, the continuous spectrum moves into the left-half of the complex plane and does not act to destabilize the wave.

Now, as in Sect. 4.1, one must track the zero of the Evans function on the appropriate Riemann surface to determine if any eigenvalue arises from an edge bifurcation. As is seen in [8], the appropriate Riemann surface is defined by

$$\gamma^2 := (1 - \mathrm{i}\epsilon d_1)\lambda - \mathrm{i}\omega - \epsilon(d_1\omega + d_2) + \mathrm{i}\epsilon^2 d_1 d_2 \,. \tag{46}$$

Since the zero γ_{edge} on the Riemann surface will satisfy $\gamma_{\mathrm{edge}} = \mathrm{O}(\epsilon)$, by solving the above for λ, it is seen that if an eigenvalue pops out of the continuous spectrum due to an edge bifurcation, then it is given by

$$\lambda_{\mathrm{edge}} = \mathrm{i}\omega + \epsilon d_2 + \mathrm{O}(\epsilon^2) \,.$$

Thus, if $d_2 = \mathrm{O}(1)$, then it is necessarily true that λ_{edge} is contained in the left-half of the complex plane, and hence does not contribute to an instability.

However, now suppose that $d_2 = \mathrm{O}(\epsilon)$. It is then possible to have $\mathrm{Re}\,\lambda_{\mathrm{edge}} > 0$, which implies that it may be possible for the adiabatically stable wave with $\omega = \omega^+$ to destabilize via a Hopf bifurcation. In order to track this eigenvalue, one needs to explicitly compute the Evans function on the Riemann surface given in equation (46), and then track the zero as it moves under the perturbation. This tedious calculation is carried out in [41]. It is determined therein that the zero moves onto the second sheet of the Riemann surface, and hence an eigenvalue is not created via an edge bifurcation. The perturbed wave with $\omega = \omega^+$ is then spectrally stable, and since the system is dissipative, this then implies that it is stable.

This work was partially supported by the National Science Foundation under grant DMS-0304982, and by the Army Research Office under grant ARO 45428-PH-HSI.

References

1. P. Fife and J. McLeod, Arch. Rat. Mech. Anal., **65**, 335, (1977).
2. J. Evans, Indiana U. Math. J., **21**, 877, (1972).
3. J. Evans, Indiana U. Math. J., **22**, 75, (1972).
4. J. Evans, Indiana U. Math. J., **22**, 577, (1972).
5. J. Evans, Indiana U. Math. J., **24**, 1169, (1975).
6. C.K.R.T. Jones, Trans. AMS, **286**, 431, (1984).
7. J. Alexander, R. Gardner and C.K.R.T. Jones, J. Reine Angew. Math., **410**, 167, 1990
8. T. Kapitula and B. Sandstede, Physica D, **124**, 58, (1998).
9. T. Kapitula and B. Sandstede, (to appear in Disc. Cont. Dyn. Sys.), (2003).
10. M. Ablowitz and P. Clarkson, *Solitons, Nonlinear Evolution Equations, and Inverse Scattering*, (London Math. Soc. Lecture Note Series, **149**, Cambridge U. Press, 1991)
11. M. Ablowitz, D. Kaup, A. Newell and H. Segur, Stud. Appl. Math., **53**, 249, (1974)
12. J. Alexander and C.K.R.T. Jones, J. Reine Angew. Math., **446**, 49, (1994).
13. J. Alexander and C.K.R.T. Jones, Z. Angew. Math. Phys., **44**, 189, (1993).
14. A. Bose and C.K.R.T. Jones, Indiana U. Math. J., **44**, 189, (1995).

15. T. Bridges and G. Derks, Proc. Royal Soc. London A, **455**, 2427, (1999).
16. T. Bridges and G. Derks, Arch. Rat. Mech. Anal., **156** 1, (2001).
17. R. Gardner and K. Zumbrun, Comm. Pure Appl. Math., **51**, 797, (1998).
18. T. Kapitula, Physica D, **116**, 95, (1998).
19. R. Pego and M. Weinstein, Phil. Trans. R. Soc. Lond. A, **340**, 47, (1992).
20. R. Pego and M. Weinstein, Evans" function, and Melnikov's integral, and solitary wave instabilities, In: *Differential Equations with Applications to Mathematical Physics*, (Academic Press, Boston, 1993), pp. 273–286,
21. T. Kato, *Perturbation Theory for Linear Operators*, (Springer-Verlag, Berlin, 1980).
22. J. Swinton, Phys. Lett. A, **163**, 57, (1992).
23. T. Bridges and G. Derks, Phys. Lett. A, **251**, 363, (1999).
24. T. Bridges, Phys. Rev. Lett., **84**, 2614, (2000).
25. B. Sandstede, *Handbook of Dynamical Systems*, (Elsevier Science, North-Holland, 2002), Vol. 2, Chapter 18, pp. 983–1055.
26. T. Kapitula, J.N. Kutz and B. Sandstede, Indiana U. Math. J., **53**, 1095, (2004).
27. W.A. Coppel, Dichotomies in stability theory, Lecture Notes in Mathematics 629, (Springer-Verlag, Berlin, 1978).
28. D. Peterhof, B. Sandstede and A. Scheel, J. Diff. Eq., **140**, 266, (1997).
29. T. Kapitula and B. Sandstede, SIAM J. Math. Anal., **33**, 1117, (2002).
30. D. Kaup, SIAM J. Appl. Math., **31**, 121, (1976).
31. D. Kaup, Phys. Rev. A, **42**, 5689, (1990).
32. D. Kaup, J. Math. Anal. Appl., **54**, 849, (1976).
33. J. Yang, J. Math. Phys., **41**, 6614, (2000).
34. J. Yang, Phys. Lett. A, **279**, 341, (2001).
35. T. Kapitula, P. Kevrekidis and B. Sandstede, Physica D, **195**, 263, (2004).
36. T. Kapitula, SIAM J. Math. Anal., **30**, 273, (1999).
37. Y. Kivshar, D. Pelinovsky, T. Cretegny and M. Peyrard, Phys. Rev. Lett., **80**, 5032, (1998).
38. D. Pelinovsky, Y. Kivshar and V. Afanasjev, Physica D, **116**, 121, (1998)
39. W. Van Saarloos and P. Hohenberg, Physica D, **560**, 303, (1992).
40. Y. Kodama, M. Romagnoli and S. Wabnitz, Elect. Lett., **28**,1981, (1992).
41. T. Kapitula and B. Sandstede, J. Opt. Soc. Am. B, **15**, 2757, (1998).

Bifurcations and Strongly Amplitude-Modulated Pulses of the Complex Ginzburg-Landau Equation

S.R. Choudhury

Department of Mathematics, MAP207 University of Central Florida, Orlando, FL 32816-1364, USA
choudhur@longwood.cs.ucf.edu

Abstract. In this chapter, we consider a theoretical framework for analyzing the strongly-amplitude modulated numerical pulse solutions recently observed in the complex Ginzburg-Landau Equation, which is a canonical model for dissipative, weakly-nonlinear systems. As such, the chapter also reviews background concepts of relevance to coherent structures in general dissipative systems (i.e. in regimes where such structures are stable and dominate the dynamics). This framework allows a comprehensive analysis of various bifurcations leading to transitions from one type of coherent structure to another as the system parameters are varied. It will also form a basis for future theoretical analysis of the great diversity of numerically-observed solutions, even including the spatially-coherent structures with temporally quasi-periodic or chaotic envelopes observed in recent simulations.

1 Introduction

Numerous attempts have been made to extend the well-developed concept of soliton interactions in integrable, conservative systems [1, 2] to more realistic active or dissipative media which are governed by non-integrable model equations. The reason is that the complicated spatio-temporal dynamics of such coherent structure solutions are governed by simple systems of ordinary differential equations, or low-dimensional dynamical systems, rather by the original complex nonlinear partial differential equation model.

There are situations [1, 2, 3, 4] where this approach is appropriate, particularly where the dynamics of various active or dissipative systems is primarily governed by localized coherent structures such as pulses (solitary waves) and kinks (fronts or shocks). Since such structures correspond to spatial modulations, they are also known as spatially-localized "patterns". The speeds and locations of the coherent structures may vary in a complex manner as they interact, but their spatial coherence is preserved in such situations. It is tempting to apply this approach to any system which admits pulse and/or kink solutions, but caution is necessary. Coherent structures may be transitory [5, 6] when they are unstable to small disturbances in their neighborhood. Also, only some of them may actually be selected, due to such stability considerations.

S.R. Choudhury: *Bifurcations and Strongly Amplitude-Modulated Pulses of the Complex Ginzburg-Landau Equation*, Lect. Notes Phys. **661**, 429–443 (2005)
www.springerlink.com

This area has become a large and diverse one over the past decade. In this introductory chapter, addressed to beginners in the field, we shall not attempt a comprehensive treatment. Rather, we shall survey only some important salient features of the field by focusing on some recently-obtained classes of complicated amplitude-modulated pulse solutions of the complex Ginzburg-Landau equation (abbreviated as the CGLE, and also known as the Newell-Whitehead equation in Fluid Dynamics). This equation is a canonical model for weakly-nonlinear, dissipative systems [7, 8, 9, 10], and, for this reason, arises in a variety of settings, including Nonlinear Optics, Fluid Dynamics, Chemical Physics, Mathematical Biology, Condensed Matter Physics, and Statistical Mechanics [7, 8, 9, 10]. It also displays many of the generic features of the dynamics of coherent structure patterns in such dissipative models, thus making it a suitable choice for our survey purposes.

In keeping with the intended level and readership, we shall omit many of the more technical details, although enough references are provided so that the motivated reader has access to them. The aim is to provide a broad-brush, but non-comprehensive and introductory, review. The remainder of this chapter is organized as follows. In Sect. 2, we briefly review some well-known properties of the coherent structure solutions of this equation. Sect. 3 summarizes some basic categories of numerically-determined pulse solutions, with complicated dynamical behaviors. It also considers a simple dynamical systems approach to recovering these pulse solutions using a coherent structure formulation. In Sect. 4, we review the sequence of bifurcations in the CGLE model in an attempt to understand and delineate the domain of existence of the complex pulse-type solutions. Following this, we conclude with some comments about other approaches that have been brought to bear on this problem, as well as additional treatments that may be worth attempting.

2 Background Properties of Coherent Structures of the CGLE Equation

We shall begin with the canonical governing equation for dissipative systems, i.e. the Ginzburg-Landau equation with both cubic and quintic nonlinear terms (i.e. the so-called "cubic-quintic CGLE"):

$$\partial_t A = \epsilon A + (1 + \mathrm{i}c_1)\partial_x^2 A + (1 + \mathrm{i}c_3)|A|^2 A - (1 - \mathrm{i}c_5)|A|^4 A \,. \tag{1}$$

In this section, we shall briefly review some standard background results on the dynamics of coherent structure solutions of this equation, closely following the treatment in [3]. These will be needed in subsequent sections. The reader is referred to [3] for further details.

The primary physical question we wish to address is: suppose we start with the spatially-uniform state $A = 0$ at $t = 0$, and introduce a small localized perturbation or disturbance. What will the system solution evolve to at long times? The most likely outcomes are:

i. the system decays back to $A = 0$.
ii. a localized pulse, or train of pulses, is formed. If so, is their speed zero, constant, periodic, or chaotic?
iii. a stable, finite amplitude state grows by the creation and propagation of a front which invades the $A = 0$ state. In this case, we need to determine the wave-vector, speed, and frequency of the finite amplitude front.
iv. the system becomes chaotic everywhere.

Note that we shall concentrate on the coherent structure states and not focus on the chaotic or turbulent states admitted by (1). One physical reason is that we shall attempt to understand very interesting classes of numerically-observed coherent structure solutions. Another is that it is the coherent structure solutions which are of potential interest as information carriers in areas such as nonlinear optical transmission (in a manner analogous to solitons for integrable systems).

We shall first consider uniformly-translating coherent structure solutions of the form

$$\begin{aligned} A(x,t) &= \mathrm{e}^{-\mathrm{i}\omega t}\hat{A}(x - vt) \\ \hat{A}(\xi) &= a(\xi)\mathrm{e}^{\mathrm{i}\phi(\xi)} \\ \xi &\equiv x - vt \end{aligned} \tag{2}$$

In particular, note that the structures in (2) may have their amplitudes vary in an arbitrary manner with respect to the travelling-wave variable, or pseudo-time, ξ. Also, the travelling wave speed, v, could be generalized to be a function of ξ for more general coherent structure solutions. The quantities v and ω are eigenvalues and are determined by the initial conditions for large or "spatially extended" systems, as well as boundary conditions for spatially moderately-extended systems.

We shall define pulses or solitary waves as structures which have the same values at both ends of the ξ axis, and fronts as ones which have different values at each end. In particular, further differentiations are sometimes made between simple fronts (connecting a zero value at one end of the ξ axis to a non-zero value at the other) and domain walls or nonlinear fronts (connecting different non-zero values at each end). We shall not consider this in any detail in this introductory treatment. Nor shall we discuss the distinction between sources and sinks based on the signs of the group velocity at the two ends.

Defining the variables

$$q(\xi) \equiv \partial_\xi \theta \tag{3a}$$

$$\kappa(\xi) = \partial_\xi a/a \tag{3b}$$

insertion of (2) in (1) yields the ordinary differential equations

$$\partial_\xi a = \kappa a \tag{4a}$$
$$\partial_\xi q = G(a, q, \kappa) \tag{4b}$$
$$\partial_\xi \kappa = H(a, q, \kappa) \tag{4c}$$

with

$$G = -\tilde{b}_1(\omega + vq) + \tilde{c}_1(\epsilon + v\kappa) - 2\kappa q - (\tilde{b}_1 c_3 + \tilde{c}_1 b_3)a^2 - (\tilde{b}_1 c_5 + \tilde{c}_1 b_5)a^4 \tag{5a}$$
$$H = -\tilde{c}_1(\omega + vq) - \tilde{b}_1(\epsilon + v\kappa) + q^2 - \kappa^2 + (\tilde{b}_1 b_3 - \tilde{c}_1 c_3)a^2 + (\tilde{b}_1 b_5 - \tilde{c}_1 c_5)a^4 \tag{5b}$$
$$\tilde{b}_1 = b_1(b_1^2 + c_1^2)^{-1} \tag{5c}$$
$$\tilde{c}_1 = c_1(b_1^2 + c_1^2)^{-1}. \tag{5d}$$

Following the terminology in [3], we may classify the fixed points of (4) as:

i. the linear fixed points L

$$a_L = 0\,, \qquad \kappa_L \neq 0 \tag{6}$$

and ii. the nonlinear fixed point(s) N

$$a_N \neq 0\,, \qquad \kappa_N = 0\,. \tag{7}$$

In general, there are two linear fixed points, while the number of nonlinear fixed points may vary from zero to four, [3] depending on the parameters. We shall denote the L fixed points with subscripts + or −, depending on the sign of κ and also enumerate them with numerical subscripts. Similarly, we shall enumerate the nonlinear or N fixed points with subscripts. A standard linear stability analysis of the linear fixed points (see [11] for instance) yields the following results. In the following, the signs of the eigenvalues of the Jacobian matrix at each fixed point are given in parentheses (eigenvalues with positive real parts correspond, as usual, to unstable directions):

$$\epsilon < 0 : L_{-1}(++-), L_{+2}(+--) \quad \text{for all } v \tag{8}$$
$$\epsilon > 0 : L_{-1}(++-), L_{+2}(+--), \quad \text{for } 0 < v < v_{cL} \tag{9}$$
$$L_{-1}(++-), L_{-2}(---), \quad \text{with } |\kappa_{L1}| > |\kappa_{L2}| \quad \text{for } v > v_{cL}\,. \tag{10}$$

with

$$v_{CL} = |c_1\epsilon - \omega b_1|/(b_1\epsilon)^{1/2}\,. \tag{11}$$

Similarly, for the N fixed points, we have:

$$\begin{aligned} v < v_{CN1} &: N_1(+--) \quad \text{and} \quad N_2(++-) \\ v_{CN1} < v < v_{CN2} &: N_1(+--) \quad \text{and} \quad N_2(---) \\ v > v_{CN2} &: N_1(+--) \quad \text{and} \quad N_2(+--) \end{aligned} \tag{12}$$

where the precise expressions for v_{CN1} and v_{CN2} will not be necessary for our subsequent discussion.

We shall now define three basic classes of coherent structures, *viz:*

a. pulses, connecting an L_+ to an L_- fixed point
b. fronts, going from an N to an L_-, or from an L_+ to an N fixed point
c. domain walls, connecting two N fixed points

Necessary conditions for the existence of such orbits are obtained by requiring that the orbit be orthogonal to any unstable direction at an incoming fixed point. For instance, for a pulse orbit coming into an $L_-(++-)$ fixed point, the eigenvalues v and ω must typically be adjusted to make it orthogonal to both unstable eigendirections. We may obtain an upper bound on the number of coherent structures by using the above idea. Suppose we consider an orbit flowing from an N to an L fixed point. If the N fixed point has n_N unstable eigenvectors, then there are $n_N - 1$ free parameters on the n_N dimensional unstable subspace. Together with the free eigenvalue parameters v and ω, this yields $n_N + 1$ free parameters. If the L fixed point tha the orbit flows into has n_L unstable directions, the requirement that the orbit be orthogonal to those directions now leaves a total of

$$n = (n_N + 1) - n_L \tag{13}$$

as the possible number of free parameters. In the absence of any additional symmetries on the system, this serves as an upper limit to the muliplicity of such trajectories or coherent structures. Thus, we may have an n-parameter family (if $n > 0$), a discrete set of such structures ($n = 0$), or no structure at all ($n < 0$).

Going through such counting arguments yields the following upper bounds [3]:

a. a discrete set of N to L_- fronts for all ϵ, and a two-parameter family of such fronts for $\epsilon > 0$.
b. a discrete set of pulses, including one with zero speed.
c. families of sources and sinks, which we do not consider here

The above counting arguments have recently been generalized to coupled Ginzburg-Landau systems [12], including the existence of counter-propagating coherent structure solutions.

In addition to the above, [3] also considers limited analytic solutions, perturbative approaches, and numerical solutions to consider the coherent structures of (1). An additional idea developed in [3] is the concept of "linear marginal stability" and its use in formulating selection criteria for the final selected coherent structure state to which localized initial conditions evolve. The reader is referred to the original paper for further details.

Having laid out some necessary background for the coherent structure solutions, we shall now begin considering some intriguing pulse coherent structures of (1), recently obtained via numerical simulations.

3 Numerical Pulse Solutions

Very interesting classes of pulse solutions of (1) have recently been obtained numerically. Note that there are additional numerical solutions, such as spatially chaotic or turbulent states, which are not considered here. As stated earlier, our primary focus is on attempting to analyze the stable coherent structure solutions which may be considered as the basic building blocks for the dynamics of the system. They may also be regarded as possible information carriers at some future date, in the same way as solitons in the context of integrable systems. We shall primarily discuss the solutions found in the context of Nonlinear Optics [13, 14]. However, they have also been observed in other numerical simulations, as well as in experiments.

In particular, Akhmediev et al. [13, 14] consider the cubic-quintic CGLE in the optical form (with space and time swapped):

$$\mathrm{i}\psi_z + \frac{D}{2}\psi_{tt} + |\psi|^2\psi + \nu|\psi|^4\psi = \mathrm{i}\delta\psi + \mathrm{i}\epsilon|\psi|^2\psi + \mathrm{i}\beta\psi_{tt} + \mathrm{i}\mu|\psi|^4\psi \,. \tag{14}$$

Note that the parameters in this equation are different from those in (1), but the connections between the two are clear from a direct comparison. Note, too, that two parameters in (1) have been scaled to unity, as may always be done (see [3] for details).

Akhmediev et al. observe plain pulses or solitary waves with constant amplitudes. However, and more interestingly, they observe spatially-localized coherent structures with pulsating amplitudes ("pulsating solitons", Fig. 1), and some of these double and then quadruple in period as the parameter ϵ is varied (Fig. 2). In addition, they observe interesting multi-peaked coherent structure solutions, such as the "creeping soliton" (Fig. 3) and the "slug"

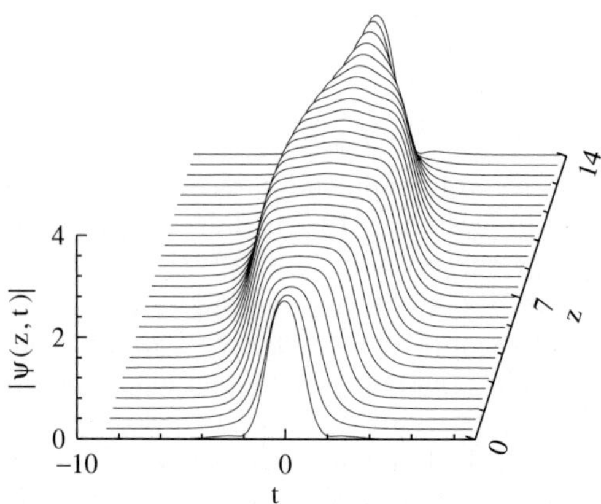

Fig. 1. Pulsating solitary wave of (14) for $D = 1, \epsilon = 0.66, \delta = -0.1, \beta = 0.08, \mu = -0.1$, and $\nu = -0.1$

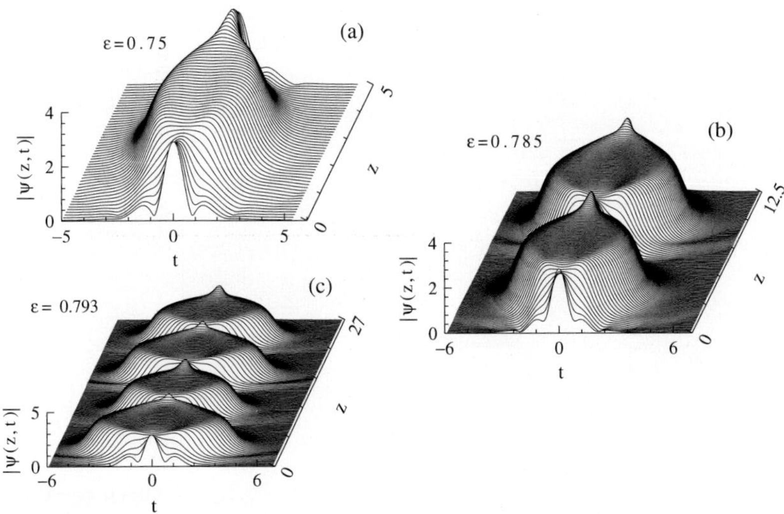

Fig. 2. Period doubling of pulsating solitary waves of (14) with $D = 1, \delta = -0.1, \beta = 0.08, \mu = -0.1$, and $\nu = -0.07$ as ϵ is varied. The ϵ values are (**a**) $\epsilon = 0.75$, (**b**) $\epsilon = 0.785$, and (**c**) $\epsilon = 0.793$

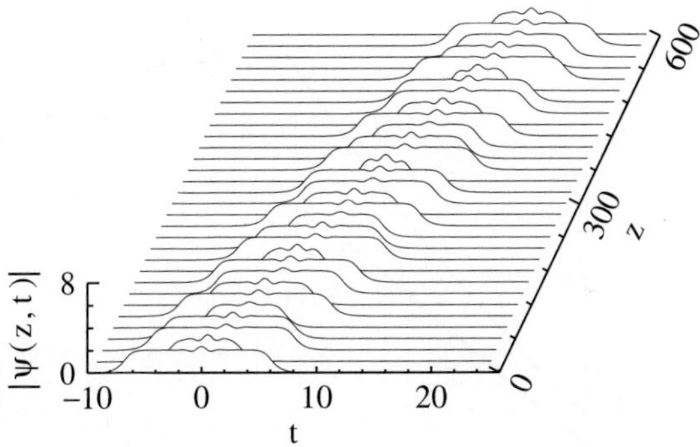

Fig. 3. Creeping solitary wave of (14) for $D = 1, \epsilon = 1.3, \delta = -0.1, \beta = 0.101, \mu = -0.3$, and $\nu = -0.101$

(Fig. 4). Note that the amplitudes vary in a complex manner in the last two figures, but the structure is still spatially-localized in a coherent fashion.

In addition, Akhmediev et al. have carried out a systematic analysis of the parameter space to classify the domains where various kinds of coherent structures are observed. Two typical examples are shown in Figs. 5 and 6 in the (ν, ϵ) and (β, ϵ) planes respectively. Note the dark areas in each figure corresponding to the period doubling and period quadrupling referred to earlier. Although not stated, in these regions it is clear that one has Feigenbaum's infinite sequence of period doublings, eventually leading to chaotically-varying

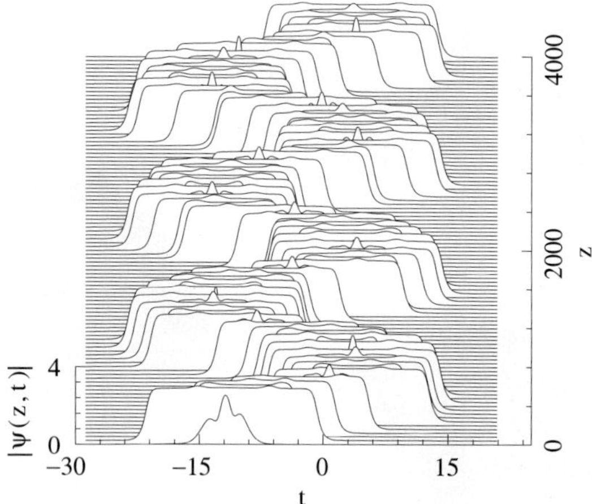

Fig. 4. "Slug" solitary wave of (14) for $D = 1, \delta = -0.1, \beta = 0.08, \mu = -0.11, \nu = -0.08$, and $\epsilon = 0.835$

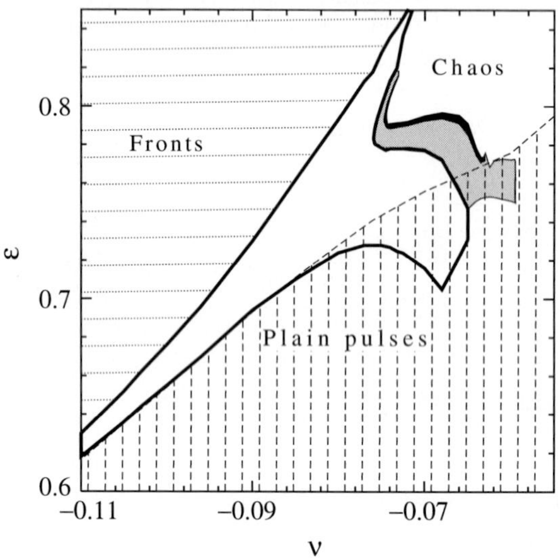

Fig. 5. Plain pulses, pulsating solitary waves (*enclosed in heavy lines*), period-doubled solutions (*gray area*), and chaotic solutions in the (ν, ϵ) plane for $D = 1, \delta = -0.1, \beta = 0.08$, and $\mu = -0.1$

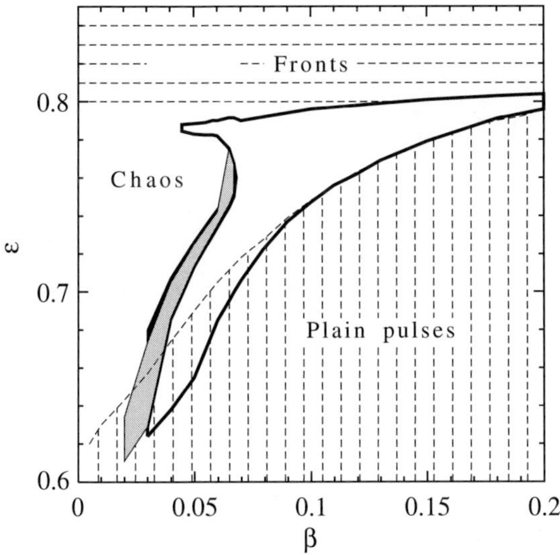

Fig. 6. The same as Fig. 5 in the (β, ϵ) plane for $D = 1, \delta = -0.1, \mu = -0.1$, and $\nu = -0.08$

envelopes. More will be said about this subsequently, in this and the following section.

One may attempt to directly analyze such structures within the coherent structure formalism outlined in the previous section. Note that this formalism is valid, since the numerics reveal that we are indeed dealing with long-lived, stable coherent entities which are principal organizing centers for the dynamics. Also note that, in order to treat the case of the "slug" in Fig. 4, one would need to generalize the ansatz (2) and the resulting low-dimensional system of ODEs (4) to allow the translation speed v of the coherent structure to vary with ξ.

The details of this analysis are too long and involved for this chapter, and will be published elsewhere. However, we may illustrate the idea concretely and simply by considering the following ODE (obtained by a travelling-wave or coherent structure reduction of a PDE using notation analogous to (2)):

$$\frac{dA}{d\xi} = -\mu A + \mathrm{i}\Omega A + \mathrm{i}FA^* + \mathrm{i}\beta|A|^2 A + N|A|^2 A + iF\Gamma(A^3 + 3|A|^2 A^*) - \mathrm{i}h|A|^4 A \,. \qquad (15)$$

This equation may be split into real and imaginary parts and solved numerically. For each set of parameters, the numerical solutions for the amplitude A may then be categorized as limit cycles (periodic), period-doubled limit cycles, quasi-periodic, or chaotic, using standard numerical diagnostics [15, 16]. A typical plot, summarizing the nature of the solutions, is given in Fig. 7 in the (Ω, F) plane and for specific values of the other parameters.

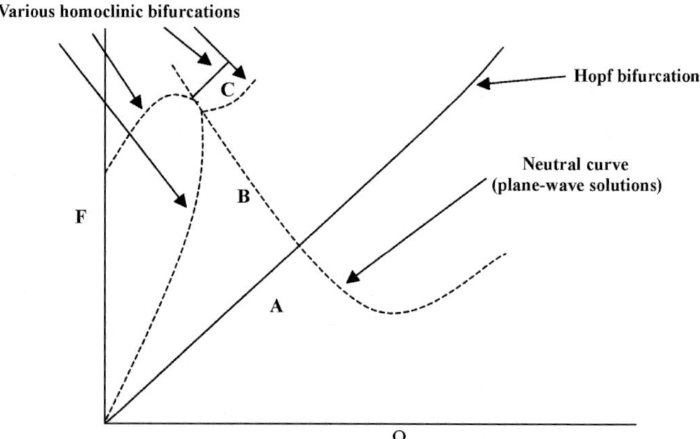

Fig. 7. Regions of plain pulses (A), periodically pulsating solitary waves (B), and period-doubled pulsating solitary waves (C) of (13) in the (Ω, F) plane for $\mu = 1, \beta = 0.4179, N = 0.1739, h = -0.0973$, and $\Gamma = 0.0278$

Note the regions A of regular or plain pulses (stable fixed points), region B with periodically-pulsating solutions (stable limit cycle behavior of the amplitude or envelope A, corresponding to periodically-pulsating solitons of the PDE), and region C containing period-doubled limit cycles (corresponding to period-doubled pulsating solutions). Note the similarity of many features with those in Figs. 5 and 6. Clearly, an analogous analysis of the numerical solutions of Akhmediev et al., summarized in Figs. 5 and 6, may be carried out in a similar manner using the coherent structure formulation in Sect. 2 (and allowing the travelling wave speed v to vary with ξ).

Having considered numerically-obtained families of complex pulse coherent structure solutions of (14), let us next turn to a variety of other approaches to this problem.

4 Bifurcations in the CGLE and Various Theoretical Approaches

4a. Bifurcations in the CGLE

In this section, we shall briefly summarize some of the important bifurcations relevant to the numerical solutions of the previous section. We shall also briefly consider some theoretical approaches which have been brought to bear on this problem. In keeping with the intended level of the article, we shall again omit many of the more technical details. The interested reader is advised to consult the primary references, as well as the general comprehensive technical review of the cubic CGLE in [8] for further details and additional solution behaviour. Once again, our primary focus here is on pulse coherent structures.

Considering a simple stationary pulsating solution,

$$A(x,t) = \mathrm{e}^{\mathrm{i}c_3 t}\,, \tag{16}$$

one may show that it undergoes a so-called modulational, or Benjamin-Feir, instability for

$$c_1 c_3 \geq 1\,. \tag{17}$$

A more relevant instability is, however, that of plane wave solutions. The simplest relevant solutions of the coherent structure ODEs (3) correspond to $(a, q, \kappa) = ((1-r^2)^{1/2}, r, 0)$ relating to plane-wave solutions (of wavenumber r)

$$A = \sqrt{1-r^2}\,\mathrm{e}^{\mathrm{i}(rx-\omega_r t)} \tag{18}$$

with angular frequency

$$\omega_r = -c_3 + q^2(c_1 + c_3)\,. \tag{19}$$

A linear stability analysis of these solutions may be performed by considering the perturbed solution

$$\tilde{A}_r = \left(\sqrt{1-r^2} + \delta a\right)\mathrm{e}^{\mathrm{i}(rx-\omega_r t)} \tag{20a}$$

$$\delta a \;\alpha\; e^{\sigma t}\,\mathrm{e}^{\mathrm{i}kx}. \tag{20b}$$

Using these, and linearizing in the small perturbation δa, it is straightforward to show that

$$\sigma(k) = -k^2 - 2\mathrm{i}rc_1k - (1-r^2) \pm \{(1+c_3^2)(1-r^2)^2 - [c_1k^2 - 2\mathrm{i}rk - c_3(1-r^2)]^2\}^{1/2}\,. \tag{21}$$

For long-wavelength perturbations (or wavenumber $k \to 0$), the “growth rate” σ becomes positive (and the perturbation δa begins to grow) for

$$r = r_E = \sqrt{(1-c_1c_3)/[2(1+c_3^2) + 1 - c_1c_3]}\,. \tag{22}$$

Note that, unlike the Benjamin-Feir instability, where all wavelengths become unstable, only a certain band of long wavelengths are unstable here. This is the so-called Eckhaus instability, and it immediately leads to plane wave solutions with time-dependent amplitudes. Clearly, this instability is relevant to the pulses with modulated amplitudes which were considered in the previous section.

In fact, it is straightforward to show that the Eckhaus instability leads to the onset of quasi-periodic solutions. Suppose that $\sigma = 0$ and the perturbations are neutrally stable on the arbitrarily-shaped neutral curve [7, 8, 9, 10] shown in Fig. 8. As the parameter (say c_3) is changed and one crosses the neutral curve into the unstable regime, two wavenumbers k_1 and k_2 simultaneously become unstable. In general, their ratio is incommensurate or irrational, i.e.

$$\frac{k_1}{k} \neq \frac{m}{n} , \tag{23}$$

where m and n are integers. Thus, via (18), the perturbed unstable solution contains two modes with wavenumbers of incommensurate spatial periods (wavelengths) and is thus a two-mode quasi-periodic solution. Subsequent secondary bifurcations may occur and may be analyzed similarly.

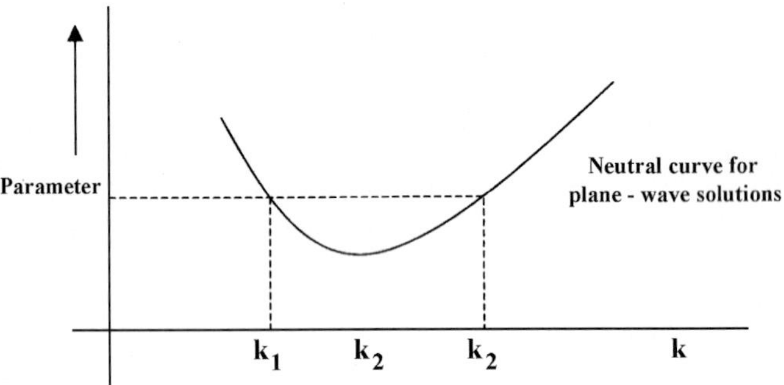

Fig. 8. A typical neutral curve in the (wavenumber, parameter) space. Note that the region above the curve is assumed to be unstable here, although one may reverse this and the curvature

Provided the growth rate, σ, of the perturbations remains small, one may perform a perturbative weakly nonlinear analysis [7, 8, 9, 10] of the Eckhaus instability. Such an analysis [17] reveals that weakly nonlinear effects may sometimes stabilize or saturate the emerging quasi-periodic solutions. These solutions thus co-exist with the original linearly-unstable (but nonlinearly-stabilized) plane wave (16). For obvious reasons, such quasi-periodic stable solutions are referred to as modulated amplitude waves (MAWs).

As already noted, the Eckhaus instability thus leads into a regime of modulated-amplitude plane-wave solutions that is reminiscent of the modulated-amplitude pulse coherent structures which we discussed in Sect. 3. However, in order to treat general coherent structures with arbitrarily-varying amplitudes well into the strongly nonlinear regime of this instability, we must abandon the starting plane-wave solutions in (16) and return to the general coherent structure equations (2)–(5). One may now start from initial conditions corresponding to the initial plane wave, i.e. $(a, q, \kappa) = ((1-r^2)^{1/2}, r, 0)$, and consider the evolution of the solutions as governed by (4)/(5). This has been done in [18] using the branch-following bifurcation software AUTO. The details are quite technical and are thus omitted here. However, we shall summarize the results, since they will give us an alternative perspective on the pulse solutions of Sect. 3, as well as clearly delineate their region of existence in parameter space. Background material on the various bifurcations that we shall encounter may be found in [11, 16].

Starting from initial conditions corresponding to the plane wave, i.e., $(a, q, \kappa) = ((1 - r^2)^{1/2}, r, 0)$, and varying the parameter c_3, one first has the Eckhaus instability (a Hopf bifurcation) where the smallest wavenumber $k = 2\pi/P$, with P being the wavelength or spatial period, is destabilized. Two branches now occur (see [18]) and the software adjusts the eigenvalues ω and v automatically to follow them. As discussed above, this Eckaus/Hopf bifurcation leads to the onset of MAWs or localized pulses with modulated envelope amplitudes, i.e. the "pulsating" solitons in the numerical results of Sect. 3. These MAWs emerge at this bifurcation with zero travelling wave speed v if the value of the average phase gradient parameter (over a spatial period or wavlength P)

$$\nu \equiv \frac{1}{P} \int_0^P dx \; \phi_x \tag{24}$$

is zero, and with non-zero v values otherwise. In the former case, they undergo a subsequent drift-pitchfork bifurcation leading to non-zero v values.

Following this, as c_3 is increased further, they undergo repeated spatial period-doubling bifurcations (as seen in the numerical results in Fig. 2). At each bifurcation, the dynamics is driven to the shorter period MAWs since the emergent double-period MAWs have linearly- unstable eigenvalues. Also, the MAWs begin to acquire a multi-peaked or multi-humped structure due to these period-doublings. The "creeping soliton" of Fig. 3 is such a multi-peaked MAW with $v > 0$. Similarly, the "slug" in Fig. 4 is one with v being periodic; as mentioned earlier, it may be captured by generalizing the coherent structure formulation of (2)–(5) to allow variable v.

For large or "spatially extended" systems, the system may undergo an infinite sequence of spatial period doublings (as seen in the purely numerical simulations in Figs. 5 and 6) leading to a spatially disordered or random ensemble of MAWs. This is similar to the chaotic regime near the top right-corner of Fig. 5 and the bottom left-corner of Fig. 6, except that the actual chaotic regime depends on which of the several parameters are being varied, and which are kept fixed.

As c_3 is increased further, the upper and lower branches meet and terminate in a saddle-node bifurcation (where two fixed points/branches annihilate). This marks the end of the regime of existence of MAWs, or the amplitude-modulated pulses of Sect. 3.

If other parameters are varied, as was done in the numerical simulations of Sect. 3, the actual details of the bifurcation sequence differ somewhat. However, the general picture outlined above remains valid and serves to delineate the regime of existence of the MAWs, or spatially-localized pulses with complex amplitude modulations, of Sect. 3.

4b. Various Theoretical Approaches

Various other theoretical approaches have been brought to bear on this problem. We shall mention them only very briefly, since they are quite specialized. The interested reader may refer to [8] for a general review. One may examine the stability of the MAWs using Floquet/Bloch theory [19]. Some series expansions of MAWs have been derived [20], but they do not generalize easily. Linear stability analysis of MAWs near the Eckhaus threshold [21], and singular perturbation analyses of their interactions [22], have been performed. There are also normal form analyses of the various bifurcations we have considered above, as well as some spectral analysis describing how the spectrum of the linearized equation around a coherent structure controls its stability, as well as the overall dynamics. Many of these approaches are still being applied to coherent structures in various important dissipative and active pattern-forming nonlinear PDEs.

5 Summary

In this article, we have summarized some topical and important numerical and theoretical results on complex pulse solutions of the cubic and cubic-quintic complex Ginzburg-Landau equation. As mentioned earlier, this is a canonical equation for weakly nonlinear, dissipative systems. For this reason, the results have relevance to general dissipative systems, particularly in regimes where their dynamics is dominated by stable coherent structures. As this volume attests, the area of Dissipative Systems is a large and extremely active one, and likely to remain so for some time to come as we continue to understand more about the dynamics of such systems. It is to be hoped that a better understanding of coherent structures will contribute fairly substantially to that endeavour.

Acknowledgement

The author would like to sincerely thank Nail Akhmediev for generously sharing his understanding and his numerical results. His invitation to write this chapter is also gratefully acknowledged.

References

1. P. G. Drazin and R. S. Johnson, *Solitons: an Introduction*, (Cambridge, Cambridge, 1989).
2. J. D. Murray, *Mathematical Biology*, (Springer-Verlag, Berlin, 1989).
3. W. van Saarloos and P. C. Hohenberg, Physica, **D 56**, 303 (1992).

4. N. J. Balmforth, Ann. Rev. Fluid Mechanics **27**, 335 (1995).
5. C. Jayaprakash, F. Hayot and R. Pandit, Phys. Rev. Lett., **71**, 12 (1993).
6. C. V. Conrado and T. Bohr, Phys. Rev. Lett., **72**, 3522, (1994).
7. R. K. Dodd, J. C. Eilbeck, J. D. Gibbon and H. C. Morris, *Solitons and Nonlinear Wave Equations*, (Academic, London, 1982).
8. I. S. Aranson and L. Kramer, Rev. Mod. Phys., **74**, 99, (2002).
9. M. C. Cross and P. C. Hohenberg, Rev. Mod. Phys., **65**, 851, (1993).
10. L. Debnath and S. Roy Choudhury (Eds), *Nonlinear Instability Analysis*, (Computational Mechanics Publishers, Southampton, 1997).
11. S. Strogatz, *Nonlinear Dynamics and Chaos*, (Addison-Wesley, Reading, 1994).
12. N. Akhmediev, J. Soto-Crespo, and G. Town, Phys. Rev. E **63**, 1 (2001).
13. R. J. Deissler and H.R. Brand, Phys. Rev. Lett. **72**, 478 (1994).
14. S. Roy Choudhury, Chaos, Solitons and Fractals **2**, 393 (1992).
15. A. H. Nayfeh and B. Balachandran, *Applied Nonlinear Dynamics*, (Wiley, New York, 1995).
16. B. Janiaud, A. Pumir, D. Bensimon V. Croquette, H. Richter, and L. Kramer, Physica **D 55**, 269 (1992).
17. L. Brusch, A. Torcini, and M. Bar, Physica **D 174**, 152 (2003).
18. M. Buttiker and H. Thomas, Phs. Rev. **A24**, 2635 (1981).
19. A. V. Porubov and M.G. Velarde, J. Math. Phys. **40**, 884 (1999).
20. D. E. Bar and A. A. Nepomnyashchy, Physica **D 86**, 586 (1995); H. H. Chang, E. A. Demekhin and E. Kalaidin, SIAM J. Appl. Math. **58**, 1246 (1998).
21. C. Elphick, E. Meron and E.A. Spiegel, SIAM J. Appl. Math. **50**, 490 (1990).
22. M. Or-Guil, I. G. Kevrekidis and M. Bar, Physica **D135**, 154 (2000).

Index

Lecture Notes in Physics

For information about Vols. 1–614
please contact your bookseller or Springer
LNP Online archive: springerlink.com

Vol.615: J. Büchner, C.T. Dum, M. Scholer (Eds.), Space Plasma Simulation.

Vol.616: J. Trampetic, J. Wess (Eds.), Particle Physics in the New Millenium.

Vol.617: L. Fernández-Jambrina, L. M. González-Romero (Eds.), Current Trends in Relativistic Astrophysics, Theoretical, Numerical, Observational

Vol.618: M.D. Esposti, S. Graffi (Eds.), The Mathematical Aspects of Quantum Maps

Vol.619: H.M. Antia, A. Bhatnagar, P. Ulmschneider (Eds.), Lectures on Solar Physics

Vol.620: C. Fiolhais, F. Nogueira, M. Marques (Eds.), A Primer in Density Functional Theory

Vol.621: G. Rangarajan, M. Ding (Eds.), Processes with Long-Range Correlations

Vol.622: F. Benatti, R. Floreanini (Eds.), Irreversible Quantum Dynamics

Vol.623: M. Falcke, D. Malchow (Eds.), Understanding Calcium Dynamics, Experiments and Theory

Vol.624: T. Pöschel (Ed.), Granular Gas Dynamics

Vol.625: R. Pastor-Satorras, M. Rubi, A. Diaz-Guilera (Eds.), Statistical Mechanics of Complex Networks

Vol.626: G. Contopoulos, N. Voglis (Eds.), Galaxies and Chaos

Vol.627: S.G. Karshenboim, V.B. Smirnov (Eds.), Precision Physics of Simple Atomic Systems

Vol.628: R. Narayanan, D. Schwabe (Eds.), Interfacial Fluid Dynamics and Transport Processes

Vol.629: U.-G. Meißner, W. Plessas (Eds.), Lectures on Flavor Physics

Vol.630: T. Brandes, S. Kettemann (Eds.), Anderson Localization and Its Ramifications

Vol.631: D. J. W. Giulini, C. Kiefer, C. Lämmerzahl (Eds.), Quantum Gravity, From Theory to Experimental Search

Vol.632: A. M. Greco (Ed.), Direct and Inverse Methods in Nonlinear Evolution Equations

Vol.633: H.-T. Elze (Ed.), Decoherence and Entropy in Complex Systems, Based on Selected Lectures from DICE 2002

Vol.634: R. Haberlandt, D. Michel, A. Pöppl, R. Stannarius (Eds.), Molecules in Interaction with Surfaces and Interfaces

Vol.635: D. Alloin, W. Gieren (Eds.), Stellar Candles for the Extragalactic Distance Scale

Vol.636: R. Livi, A. Vulpiani (Eds.), The Kolmogorov Legacy in Physics, A Century of Turbulence and Complexity

Vol.637: I. Müller, P. Strehlow, Rubber and Rubber Balloons, Paradigms of Thermodynamics

Vol.638: Y. Kosmann-Schwarzbach, B. Grammaticos, K.M. Tamizhmani (Eds.), Integrability of Nonlinear Systems

Vol.639: G. Ripka, Dual Superconductor Models of Color Confinement

Vol.640: M. Karttunen, I. Vattulainen, A. Lukkarinen (Eds.), Novel Methods in Soft Matter Simulations

Vol.641: A. Lalazissis, P. Ring, D. Vretenar (Eds.), Extended Density Functionals in Nuclear Structure Physics

Vol.642: W. Hergert, A. Ernst, M. Däne (Eds.), Computational Materials Science

Vol.643: F. Strocchi, Symmetry Breaking

Vol.644: B. Grammaticos, Y. Kosmann-Schwarzbach, T. Tamizhmani (Eds.) Discrete Integrable Systems

Vol.645: U. Schollwöck, J. Richter, D.J.J. Farnell, R.F. Bishop (Eds.), Quantum Magnetism

Vol.646: N. Bretón, J. L. Cervantes-Cota, M. Salgado (Eds.), The Early Universe and Observational Cosmology

Vol.647: D. Blaschke, M. A. Ivanov, T. Mannel (Eds.), Heavy Quark Physics

Vol.648: S. G. Karshenboim, E. Peik (Eds.), Astrophysics, Clocks and Fundamental Constants

Vol.649: M. Paris, J. Rehacek (Eds.), Quantum State Estimation

Vol.650: E. Ben-Naim, H. Frauenfelder, Z. Toroczkai (Eds.), Complex Networks

Vol.651: J.S. Al-Khalili, E. Roeckl (Eds.), The Euroschool Lectures of Physics with Exotic Beams, Vol.I

Vol.652: J. Arias, M. Lozano (Eds.), Exotic Nuclear Physics

Vol.653: E. Papantonoupoulos (Ed.), The Physics of the Early Universe

Vol.654: G. Cassinelli, A. Levrero, E. de Vito, P. J. Lahti (Eds.), Theory and Appplication to the Galileo Group

Vol.655: M. Shillor, M. Sofonea, J.J. Telega, Models and Analysis of Quasistatic Contact

Vol.656: K. Scherer, H. Fichtner, B. Heber, U. Mall (Eds.), Space Weather

Vol.657: J. Gemmer, M. Michel, G. Mahler (Eds.), Quantum Thermodynamics

Vol.658: K. Busch, A. Powell, C. Röthig, G. Schön, J. Weissmüller (Eds.), Functional Nanostructures

Vol.659: E. Bick, F.D. Steffen (Eds.), Topology and Geometry in Physics

Vol.660: A.N. Gorban, I.V. Karlin, Invariant Manifolds for Physical and Chemical Kinetics

Vol.661: N. Akhmediev, A. Ankiewicz (Eds.) Dissipative Solitons

Printing: Strauss GmbH, Mörlenbach
Binding: Schäffer, Grünstadt